P9-ECN-336

Foundations of Digital Signal Processing and Data Analysis

JAMES A. CADZOW

Arizona State University

Foundations of Digital Signal Processing and Data Analysis

Macmillan Publishing Company

New York

Collier Macmillan Publishers

London

I wish to dedicate this book to my wife Alice who has participated in my teaching and research career in a very supportive manner.

Printed in the United States of America.

Macmillan Publishing Company
866 Third Avenue, New York, New York 10022

Collier Macmillan Canada, Inc.

Library of Congress Cataloging in Publication Data

Cadzow, James A.
Foundations of digital signal processing and data analysis.

Includes bibliographies and index.
1. Signal processing—Digital techniques.
2. Data analysis. I. Title.
TK5102.5.C23 1987 621.38′043 86-8796

Printing: 1 2 3 4 5 6 7 8 Year: 7 8 9 0 1 2 3 4 5

ISBN 0-02-318010-2

PREFACE

Digital signal processing has evolved over the past two decades into an important contemporary discipline that encompasses an ever-growing number of interest areas such as frequency-discrimination filtering, spectral analysis and estimation, signal modeling, optimum mean squared error filtering, and system identification. Underlying each of these distinctive areas is a set of fundamental principles that constitute the discipline of digital signal processing. The primary purpose of this book is that of providing a thorough treatment of these fundamental concepts and illustrating their use in typical applications. The ability to use contemporary signal-processing techniques effectively is realized only after one has mastered these fundamental notions.

This book is intended for use in a first course on digital signal processing. It is written in a fashion so as to be useful in either an undergraduate course or in a first-year graduate course. The emphasis taken is that of stressing fundamental issues and not of giving an encyclopedic treatment of the many signal-processing algorithms available. The reasons for taking this approach are pedagogically based. It is felt that the most expedient and beneficial means for becoming proficient in signal processing (or in any discipline, for that matter) is to have an in-depth knowledge of the underlying basics. With such knowledge, one may readily read the literature directed toward specific signal-processing techniques and also develop novel signal-processing algorithms that may be required in a particular application. Thus we shall herein stress fundamental principles. A companion textbook emphasizes the algebraic approach to signal processing that is currently being used in many contemporary signal-processing algorithms.

There are a number of unique features in this book. For instance, the treatment of frequency-discrimination filtering is achieved in an efficient manner whereby the low-pass filter serves as the basic building block that may be used to construct other forms of filters, such as high-pass and bandpass filters. Using this viewpoint, we then concentrate our effort toward developing effective nonrecursive and recursive linear low-pass filter synthesis methods. Of more fundamental importance, however, a thorough treatment of the basic concepts used in characterizing random signals is provided. This is in recognition of the fact that in virtually all practical applications, the data to be processed are random in nature. The treatment of random signals herein contained is typically not available in a single textbook.

The primary goal of this book is to provide a self-contained treatment of the fundamental principles of signal processing in which the underlying signals may be deterministic or nondeterministic. It is hoped that this goal has been reasonably achieved. To supplement the learning process, a list of selected answers to the many problems that appear at the end of each chapter and a set of computer program problems illustrating the theoretical concepts are included.

The first six chapters of the book are directed toward fundamental deterministic concepts that are used in signal-processing applications. Particular emphasis is directed toward the transform domain approach to characterizing signals and linear signal operators (i.e., algorithms or systems). The z-transform and the Fourier transform are the principal tools for this purpose; they are examined in Chapters 2 through 4. The truncated Fourier transform plays a prominent role in this development, and it is treated in Chapter 5. This is in recognition of the fact that in all practical signal-processing applications only a finite amount of data is ever available for processing. The ramifications of this practical consideration are thoroughly described.

The notion of frequency-discrimination filtering plays a central role in many signal-processing applications. With this importance in mind, fundamental issues related to the analysis and synthesis of low-pass, high-pass, bandpass, and band-reject filters are developed in Chapters 5 and 6. Emphasis is again placed on the underlying principles of frequency-discrimination filters. In particular, it is shown that the low-pass filter forms the basic building block of frequency-discrimination filters. Specifically, one may synthesize high-pass, bandpass, and band-reject filters by an appropriate interconnection of low-pass filters. With this in mind, a number of procedures for synthesizing linear nonrecursive and recursive filters are presented in Chapter 6.

In any meaningful signal-processing application, the data to be analyzed are nondeterministic in nature. A useful characterization of such data and the development of signal-processing algorithms to process such data must then take into account the basic nature of this randomness. Probability theory is the mathematical tool for this characterization, and the concept of correlation is central to this treatment. It is desirable that the reader have a previous exposure to an elementary treatment of probability theory. Nonetheless, the basic concepts of probability theory needed for our purpose are given in Chapters 7, 8, and 9. These concepts are then used to develop a probabilistic treatment of random signals and the effects of linear operations on random signals in Chapters 10 and 11. Particular emphasis is given to the class of wide-sense stationary signals. This theory is then used to develop signal-processing algorithms that optimally process noise-contaminated data. The general notion of minimum mean-squared error filtering (i.e., Wiener filtering) and the matched filter are highlighted in this development.

I wish to take this opportunity to thank the many people who have graciously helped in making this book possible. Mr. Dong-Chang Shiue has been particularly helpful in both proofreading and in preparing the solution manual. Dr. Otis Solomon has also made a number of valuable suggestions in reviewing the manuscript. Similarly, Professor Yungong Sun has generated a number of performance plots that illustrate the capabilities of various signal-processing algorithms. Professor Guy Sohie has made a number of useful comments relative to the book's topic content and also prepared the computer program problems. In addition, I wish to thank the many students who have made useful suggestions and provided invaluable help in proofreading. The research support provided by ONR, AFOSR, and NOSC is also sincerely acknowledged. This support provided me with the opportunity to concentrate my energies on signal processing and hopefully to make some contributions to this area. Last but by no means least, I wish to thank Mrs. Linda Arneson, Mrs. Georgeann Becker, and Mrs. Chrina Darrington, who typed the original notes and the many revisions that followed in a cheerful and professional manner.

J.A.C.

CONTENTS

CHAPTER 1

Signals and Systems

1.1 INTRODUCTION

When analyzing the behavior of dynamic phenomena, fundamental concepts from system theory are invariably employed. Typically, a system's approach entails the introduction of a mathematical model that adequately represents the physical laws governing the phenomenon under study and is compatible with empirically observed measurements related to that phenomenon. For example, this viewpoint is taken when modeling the relationship between the voltage across and the current through such common analog circuit elements as resistors, capacitors, and inductors. In system theory, the two concepts of *signal* and *signal operator* play central roles. Signals are functions used to describe the time-varying nature of variables that in some sense convey important information concerning the phenomenon under study. On the other hand, signal operators are mechanisms used to change (or transform) signals into other signals. They are employed to depict subsystem operations that when taken as a whole form the system representation of the phenomenon being modeled.

For purposes of illustration, let us now consider two typical situations in which system models are commonly encountered. In a stereo system, the audio waveform appearing at the input terminal of an audio amplifier is an electrical signal that conveys information concerning the time-varying nature of a vocal or a musical passage. In turn, the stereo amplifier is a signal operator, since its primary function is to increase the audio waveform's power level as well as to alter its frequency content. The output waveform of the audio amplifier (also a signal) is therefore an altered version of the input waveform. This output signal will, in turn, serve as an input signal to the speaker subsystem (a signal operator), which will produce the ultimate sound signal. In a similar fashion, the sequence of daily deposits and withdrawals made in a bank savings account can be also interpreted as a signal, since it conveys time-varying information relative to the savings account system. The computer and attendant software used to calculate the savings account balance on a daily basis would constitute a signal operator. This computer system uses the deposit–withdrawal daily information as well as the prevailing interest rate to compute the savings account balance daily. With these examples illustrating the general concepts of signal and signal operator, the following formal definition is offered.

Definition 1.1. A *signal* (or function) is used to represent the time-varying behavior of a variable that, in part, describes a phenomenon. A *signal operator* represents a well-defined mechanism for changing one signal into another signal.

A system can be viewed as an interconnection of signal operators in which signals flow from one signal operator into another in a well-defined configuration.

In signal theory terminology, the signal that is being changed (or operated upon) by a signal operator is often referred to as the *input signal* or the *excitation*. Similarly, the signal that is produced by a signal operation is commonly known as the *output signal* or the *response*. It will be convenient to view this process in terms of the block diagram shown in Figure 1.1, in which the excitation signal "flows into" the box representing the signal operator while the response signal "flows out." This *cause-and-effect* interpretation of signal operation is of fundamental importance. Whenever possible, we shall represent the excitation and response signals by the symbols $x(t)$ and $y(t)$, respectively. The independent variable t is here being explicitly used to emphasize the fact that the signal typically varies as a function of a variable t, which will be called *time*. We shall now examine some of the more fundamental issues concerning signals and signal operators.

1.2 SIGNALS AND TIME SERIES

In the audio system and savings account examples just described, the associated signals are of a fundamentally different nature. The audio waveform is observed to have an amplitude that changes in a continuous manner as time evolves. As such, this signal type is usually referred to as being a *continuous-time signal*. On the other hand, the daily balance of a savings account sequence represents information that changes only at specific instants of time. Signals of this nature

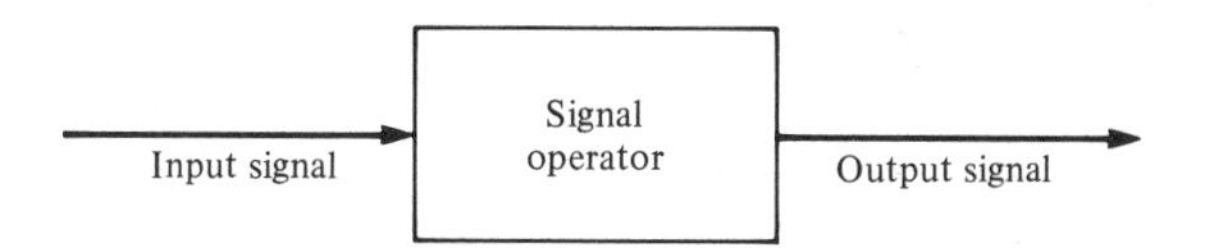

FIGURE 1.1 Concept of signal operation.

do not evolve in a continuous-time manner and therefore are called *discrete-time signals*. The distinction between continuous- and discrete-time signals is depicted in Figure 1.2. It should also be noted that although we have described signals as having amplitudes that vary as a function of time, many examples can be given in which the independent variable represents a physical quantity other than time. As examples, the measure of a steel beam's deflection along its length has distance as the independent variable, while temperature serves this role in an experiment concerned with the recording of gas pressure inside a closed container as temperature is varied. Despite such variants, we shall adhere to the convention of referring to the independent variable as "time," fully recognizing that there will be situations in which a different descriptor would be more appropriate.

Definition 1.2. The signal $x(t)$ is said to be a *continuous-time signal* if the independent time variable t takes on all values in an interval(s). When the time variable takes on only a discrete set of values t_n for an appropriate set of integers n, the signal $x(t_n)$ is said to be a *discrete-time signal.*

Analog-to-Digital Conversion

Although our primary interest will be directed toward discrete-time signals, it is to be noted that these signals are often generated through the process of

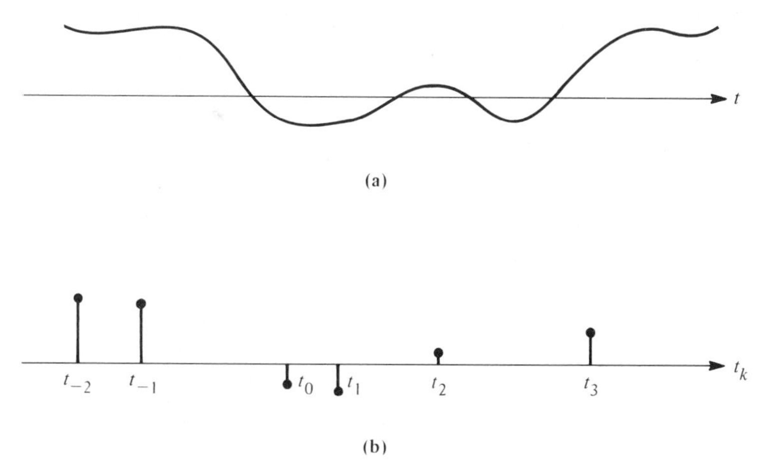

FIGURE 1.2 Signal types: (a) continuous and (b) discrete-time signals.

sampling a continuous-time signal. Specifically, in applications where a digital computer is to be employed for the processing of continuous-time signals, it will be necessary to convert such signals into a format that is compatible with digital computation. In effect, this entails changing a continuous-time signal into a number sequence which retains the essential features of that signal. This number sequence can then be manipulated by numerical algorithms that are implementable on a digital computer.

Changing a continuous-time signal into a discrete-time signal is commonly referred to as *sampling* or *analog-to-digital conversion*. The process of sampling may be visualized as shown in Figure 1.3, whereby the normally open switch is instantaneously closed at the *sample instants* t_n to produce the associated discrete-time signal $x(t_n)$. If this number sequence is to adequately represent the continuous-time signal, it is intuitively clear that the sampling instants t_n must be selected prudently. The rate of sampling should in some sense be made proportionate to the rapidity with which the continuous-time signal varies as a function of time. The sampling rate selection will be studied further in Section 1.14. In that study, the sampling instants are spaced uniformly so that $t_n = nT$, where T denotes the fixed time between adjacent time samples. *Uniform sampling* is used extensively in real-world applications. It is to be noted that the discrete-time signal depicted in Figure 1.2b may be interpreted as being a *nonuniformly* sampled version of the continuous-time signal shown in Figure 1.2a.

With the comments above in mind, a discrete-time signal will then be used to describe either of the following:

1. An experimental outcome of a phenomenon that is inherently discrete time in nature.
2. A number sequence that arises through the process of sampling a continuous-time signal.

Independent of how the discrete-time signal arises, we shall hereafter be concerned primarily with the characterization of such signals. This study will include the analysis of numerical algorithms that are used for the signal processing (or manipulation) of discrete-time signals. In recognition of the growing role that statistics is playing in the signal-processing field, investigators are increasingly using the statistician's term *time series* in place of the more traditional engineering expression *discrete-time signal*. These equivalent terminologies are used interchangeably throughout this book.

Time-Series Notation

From the remarks above, a time series (or discrete-time signal) may be interpreted as being a sequence of numbers that is ordered by a discrete-time variable. We shall denote a time series in the following manner:

$$\begin{aligned}\mathbf{x} &= \{x(t_n)\} \\ &= \{\ldots, x(t_{-2}), x(t_{-1}), x(t_0), x(t_1), \ldots\}\end{aligned} \tag{1.1}$$

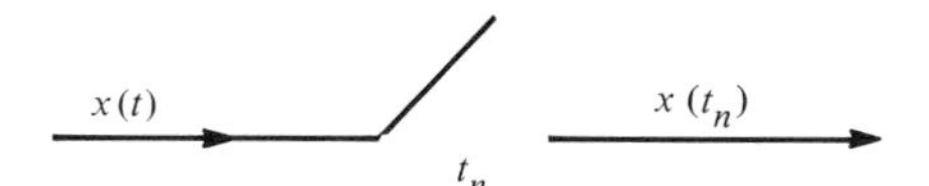

FIGURE 1.3 Sampling process.

where the equivalent symbols **x** and $\{x(t_n)\}$ are compact representations for the generally infinite-length time series. We can interpret the specific element $x(t_n)$ as the value of the time series (or discrete-time signal) **x** evaluated at the time instant t_n. This representation of a number sequence is notationally somewhat cumbersome due to the explicit appearance of the sampling instants t_n. To alleviate this notational burden, we adopt the following shorthand representation:

$$\begin{aligned} \mathbf{x} &= \{x(n)\} \\ &= \{\ldots, x(-2), x(-1), x(0), x(1), \ldots\} \end{aligned} \tag{1.2}$$

in which the integer variable n has been substituted for the discrete-time instant t_n. When using this more compact notation, the reader is cautioned to remember that there is always an underlying set of discrete-time instants and that the variable n is to be interpreted as being t_n. With this serving as background, henceforth we use the convenient time series representation (1.2). Unless indicated otherwise, the sampling instants will be hereafter taken to be uniform so that $t_n = nT$.

From the discussion so far, it is apparent that a time series will be used primarily to describe the time-varying nature of a variable that in part characterizes an experimental outcome. The nature of the experiment and the descriptive variable can take on distinctly different forms depending on the specific application at hand. For example, the experiment might be of an economic, scientific, or numerical algorithmic nature. Whatever the case, however, the time series used to describe the experimental outcome will have the same structural attribute of producing a sequence of ordered numerical-valued measurements. This concept is depicted in Figure 1.4. A list of several situations that are susceptible to a time series description follows.

1. Monthly automobile sales in the United States.
2. Hourly traffic flow at a given highway intersection.
3. Quarterly gross national product (GNP) figures for the United States.
4. Range bin measurements obtained from radar returns.
5. Sampled measurements of a seismological linear array.
6. Equally spaced samples of an electrocardiogram recording.

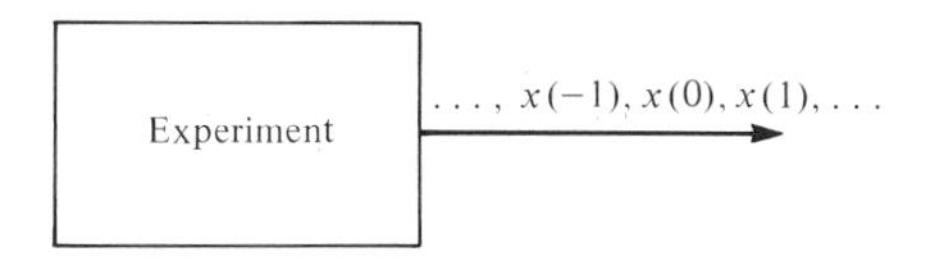

FIGURE 1.4 Experimental outcome described by a time series.

The first three examples describe phenomena whose outcomes are inherently discrete-time in nature. The remaining examples describe experiments in which a continuous-time signal is sampled to generate a related time series.

EXAMPLE 1.1

Let the experiment under study consist of a company manufacturing and selling a certain product (e.g., automobiles). Two time series that provide a partial characterization of this experiment's outcome have the following description:

$x_1(n)$ = number of product items manufactured on the nth day of the year

$x_2(n)$ = number of product items sold on the nth day of the year

1.3 DETERMINISTIC AND NONDETERMINISTIC TIME SERIES

When examining the intrinsic nature of a specific time series, it will generally be possible to classify it as being either deterministic or nondeterministic (random) in behavior. A *deterministic time series* can be thought of as arising from an experiment whose outcome is predetermined before the experiment is conducted. Although such a situation might seem contrived, this is certainly not the case. As an example, the time series describing the monthly home mortgage payment schedule is completely predetermined once a bank home loan is initiated. Similarly, the sampled trajectory of a natural orbiting body (e.g., our moon) is for all practical purposes known for all future times. It must be acknowledged, however, that most experiments encountered in the real world will not be exactly predictable in this fashion. In the more normal case, one is unable to predict with exactness the experimental outcome that is ultimately to be measured. Time series that fall into this category are said to be *random* or *nondeterministic*. For instance, the daily closing price of IBM stock on the New York Stock Exchange is not perfectly predictable from knowledge of its previous closing-price history. One of the more interesting applications of signal processing is that of using such known information (i.e., previous closing-price data) to make predictions of future outcomes. With these thoughts in mind, the following formal definition is offered.

Definition 1.3. The time series or discrete-time signal $\{x(n)\}$ is said to be

(a) *Deterministic* if its element values $x(n)$ are completely specified before the experiment that gives rise to it has been conducted. Such time series will often be described by a convenient analytical formula (e.g., unit-step and sinusoidal time series).

(b) *Nondeterministic* (or random) if its element values $x(n)$ are known only after the experiment that gives rise to it has been conducted. Such time series are generally described using probabilistic methods.

As might be anticipated, an analysis of the two fundamentally distinct classes of deterministic and nondeterministic time series will entail different approaches. Although a deterministic time series may appear to be of little practical value, this is certainly not the case. Deterministic time series will play a vital role in the analysis and synthesis of important forms of digital signal processing algorithms. This will be made apparent in Chapters 5 and 6, where we discuss the synthesis of frequency-discrimination digital filters using purely deterministic tools. On the other hand, the primary tool used in the study of the more practical nondeterministic time series will be probability theory. How probability theory is to be used in describing and processing random time series is the subject of Chapters 7 to 14.

1.4 STANDARD DETERMINISTIC TIME SERIES

With the comments above in mind, some of the more commonly encountered deterministic time series (signals) will now be introduced. Invariably, these standard time series will be characterized by simple expressions describing their functional dependency on time. Moreover, these prototype deterministic time series will often serve as appropriate *test* excitation time series used to characterize the dynamical behavior of linear operators.

Unit-Impulse Time Series

In studies related to linear operators, the *unit-impulse* (also known as the Kronecker delta and unit-sample) *time series* plays a central role. This time series (or signal) is formally defined by

$$\delta(n) = \begin{cases} 1 & n = 0 \\ 0 & n \neq 0 \end{cases} \tag{1.3}$$

and is seen to be composed of all zeros except for a 1 that appears at time $n = 0$. The unit-impulse time series is depicted in Figure 1.5a. This time series is used so extensively in linear operator investigations that the special notation, δ, has been reserved for its representation. The unit-impulse function $\{\delta(n)\}$ is seen to map any integer argument into either the value 0 or 1. When the unit-impulse function's argument is 0, it takes on the value 1. Otherwise, it takes on the value 0. With this interpretation, the related shifted time series $\{\delta(n - m)\}$, in which m is a fixed integer, is seen to be 0 everywhere except at time $n = m$, where it takes on the value 1. As such, $\{\delta(n - m)\}$ is said to be an m right-shifted version of $\{\delta(n)\}$. When interpreting time series by formulas such as that given in expression (1.3), inequalities of the form $n \neq 0$ are to be read as "being equal to all integers satisfying the inequality." Thus the noninteger number $\frac{1}{2}$ does not belong to the set $n \neq 0$ even though $\frac{1}{2} \neq 0$.

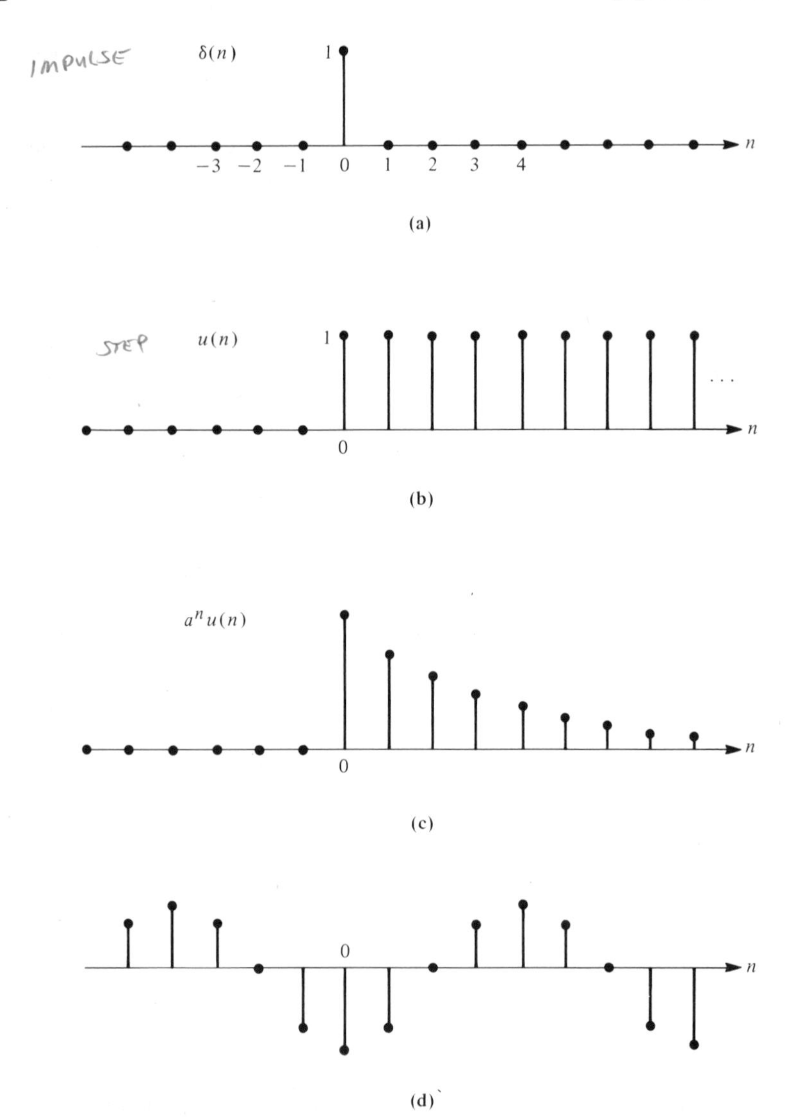

FIGURE 1.5 Standard deterministic signals: (a) unit-impulse; (b) unit-step; (c) one-sided exponential; (d) two-sided real sinusoidal.

Unit-Step Time Series

It will often happen that in a given application, a linear operator will be subjected to an excitation that appears suddenly and is of constant level thereafter. For example, this describes the situation where a person opens a savings account plan in which regular deposits of a fixed amount are to be made. To describe such frequently occurring happenings, the *unit-step time series (signal)* is employed, where

$$u(n) = \begin{cases} 1 & \text{for } n \geq 0 \\ 0 & \text{for } n < 0 \end{cases} \tag{1.4}$$

A visual depiction of this time series is given in Figure 1.5b. The unit step is composed of an infinite sequence of contiguous zeros followed by an infinite sequence of contiguous ones, with the transition in signal level occurring at time $n = 0$. The special symbol u is here used to designate the unit-step function. This function maps integer arguments into either 0 or 1, depending on whether the argument is a negative or a nonnegative integer, respectively. It is a simple matter to show that the unit-step and unit-impulse time series are interrelated according to

$$\delta(n) = u(n) - u(n-1) \quad \text{and} \quad u(n) = \sum_{k=0}^{\infty} \delta(n-k)$$

It is to be noted that the prototype unit-impulse and unit-step deterministic time series each has a transition in amplitude level that occurs at time $n = 0$. Although any other transition instant could have been chosen (e.g., $n = 1$), the time origin is particularly convenient for this purpose.

Exponential Time Series

The *exponential time series (signal)* is of particular value in describing more complex time series in the form of a Fourier series. A general two-sided exponential time series will be specified by the simple analytical expression

$$x(n) = b(a)^n \tag{1.5}$$

where a and b are fixed scalars that may be complex valued. The time-varying nature of such time series is seen to be dependent on the scalar a. This time series is said to be two-sided because the exponential behavior (i.e., a^n) is seen to hold for both negative and positive times.

In characterizing an important class of linear operators, it will be found that the one-sided counterpart of the two-sided exponential time series above will arise in a natural manner. The following two time series are said to be of a one-sided exponential behavior:

$$x_c(n) = b(a)^n u(n) \tag{1.6}$$

$$x_a(n) = b(a)^n u(-n-1) \tag{1.7}$$

The first one-sided exponential time series (1.6) is seen to first become nonzero at time $n = 0$, while the second (1.7) becomes identically zero for values of time $n \geq 0$. A one-sided exponential time series would appear as shown in Figure 1.5c for the case where a is a real positive number of size less than 1.

Sinusoidal Time Series

A special class of exponential-type time series is useful in describing the frequency-discrimination characteristics of linear operators. This class of time series (signals) is described by the *complex-valued sinusoid*

$$x(n) = be^{j(\omega n + \theta)} \tag{1.8}$$

and the associated *real-valued sinusoid*

$$x(n) = b \cos(\omega n + \theta) \tag{1.9}$$

in which the real parameters b, ω, and θ identify the sinusoid's amplitude, radian frequency, and phase angle, respectively. The expressions above constitute the two-sided sinusoidal representations. Their corresponding one-sided representations are obtained simply by multiplying the right sides of these expressions by either $u(n)$ or $u(-n - 1)$. Due to the behavior of a sinusoidal time series, it is readily shown that the two radian frequencies ω and $\omega + 2\pi$ will produce the same sequence of values in expression (1.8) or (1.9) for any choice ω. This being the case, our interest will be concerned primarily with the "principal" range of digital frequencies that lie in the interval $0 \leq \omega < 2\pi$. The time-varying nature of the two-sided real sinusoidal time series is shown in Figure 1.5d.

1.5 ADDITION AND SCALAR MULTIPLICATION OPERATIONS

In signal-processing analysis, the two operations of time series addition and the multiplication of a time series by a scalar are of fundamental importance. This being the case, we now briefly study these two basic operations. The time series $\{z(n)\}$ that arises upon adding the two time series $\{x(n)\}$ and $\{y(n)\}$ is formally defined by

$$\begin{aligned} \{z(n)\} &= \{x(n)\} + \{y(n)\} \\ &= \{x(n) + y(n)\} \end{aligned} \tag{1.10}$$

Thus the general nth element of the "sum" time series is simply equal to the sum of the nth elements of the two time series being added. In a similar fashion, the time series $\{w(n)\}$ that arises upon multiplying the time series $\{x(n)\}$ by the scalar α is defined by

$$\begin{aligned} \{w(n)\} &= \alpha\{x(n)\} \\ &= \{\alpha x(n)\} \end{aligned} \tag{1.11}$$

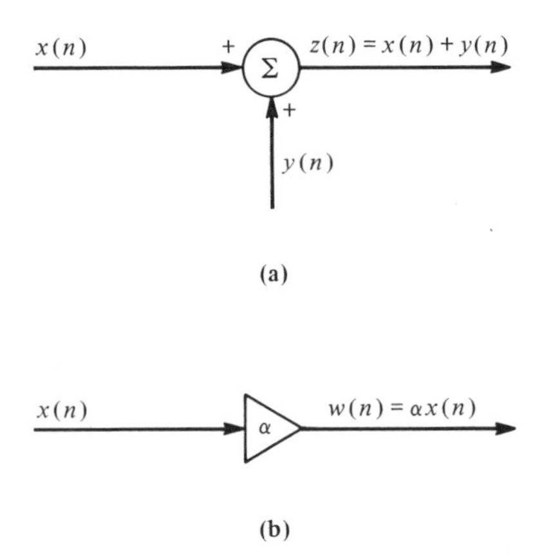

FIGURE 1.6 Fundamental operations: (a) signal summation; (b) scalar multiplication.

indicating that the element values are related by $w(n) = \alpha x(n)$ for all n. These two fundamental time series operations are intuitively appealing in that they make use of the standard algebraic rules for the addition and multiplication of numbers. As we shall subsequently show, these two operations form the building blocks for an important class of signal-processing algorithms. We shall use the summation node and the multiplier unit configurations shown in Figure 1.6 to represent these two signal operations.

EXAMPLE 1.2

Evaluate the time series that is specified by

$$\{y(n)\} = 2\{x_1(n)\} - 3\{x_2(n)\} + 5\{x_3(n)\}$$

in which the individual time series are specified by $x_1(n) = u(n)$, $x_2(n) = u(n-1)$, and $x_3(n) = u(n-2)$, where $u(n)$ denotes the unit-step sequence defined in expression (1.4). In accordance with the fundamental operations (1.10) and (1.11), it follows that the required time series has its element values given by

$$\begin{aligned} y(n) &= 2x_1(n) - 3x_2(n) + 5x_3(n) \\ &= 2u(n) - 3u(n-1) + 5u(n-2) \end{aligned}$$

or equivalently,

$$y(n) = \begin{cases} 0 & n < 0 \\ 2 & n = 0 \\ -1 & n = 1 \\ 4 & n \geq 2 \end{cases}$$

Using the unit-step and unit-impulse time series definitions, this expression may be equivalently represented by the formula

$$y(n) = 4u(n) - 2\delta(n) - 5\delta(n - 1)$$ ■

Although we shall not pursue this point here, it is readily established that the set of all time series in conjunction with the two operations (1.10) and (1.11) satisfy the axioms of a vector space (see Problem 1.8). It is then possible to employ the considerable theory that has been developed for characterizing general vector spaces to yield a penetrating study of time series and linear operations (transformations) on time series. We mention this possibility to emphasize the potential value of general linear vector space concepts in system theory applications such as those represented by signal processing.

1.6 MAGNITUDE, ENERGY, AND POWER

In analytically based studies, the ability to measure the *size* of a time series (signal) will be of paramount importance. Unfortunately, the concept of time series size is rather ambiguous and is open to many interpretations. For the purpose of this book, however, we shall be concerned primarily with three distinct measures. In the first measure, the size of the time series $\{x(n)\}$ is set equal to its largest magnitude element, that is,

$$\mathcal{M}_x = \max_{-\infty<n<\infty} |x(n)| \tag{1.12}$$

where the symbol $\mathcal{M}_x$ with appended subscript x is used to denote this particular measure.† A time series will be said to be *bounded* whenever this magnitude measure is finite. When studying the stability characteristics of linear time-invariant operators, this time series size measure will play a prominent role.

Motivated by the concepts of signal energy and signal power that are basic to continuous-time signal studies, we now introduce analogous time series size measures. In particular, the *energy* contained in the time series $\{x(n)\}$ will be defined by

$$\mathscr{E}_x = \sum_{n=-\infty}^{\infty} |x(n)|^2 \tag{1.13}$$

†For certain time series, there will not exist an element with largest magnitude [e.g., $x(n) = [1 - (10.5)^n]u(n)$]. For such situations, $\mathcal{M}_x$ is set equal to the smallest real number M such that $M \geq |x(n)|$ for all n. It is mathematically proper to use the smallest upper-bound operation "sup" in place of "max."

Similarly, the *power* associated with this time series will be specified by the quantity

$$\mathcal{P}_x = \lim_{N\to\infty} \frac{1}{2N+1} \sum_{n=-N}^{N} |x(n)|^2 \tag{1.14}$$

It is readily established that if a time series has finite energy, it will have zero power and that a time series with nonzero power will have infinite energy. Moreover, the time series $\{x_1(n)\}$ will be said to have more power than the time series $\{x_2(n)\}$ if $\mathcal{P}_{x_1} > \mathcal{P}_{x_2}$. Similar statements hold for the relative magnitudes (1.12) and energies (1.13) of time series.

In subsequent developments, we will frequently be called upon to provide convenient closed-form representations for *geometric summations* of the form $1 + \alpha + \alpha^2 + \cdots + \alpha^n$. Such an ability will enable us to make useful theoretical interpretations of the intrinsic nature of time series and linear operators. With this in mind, the following theorem is offered.

Theorem 1.1. The closed-form representation for the *finite geometric series*

$$\sum_{k=0}^{n} \alpha^k = \frac{1-\alpha^{n+1}}{1-\alpha} \tag{1.15}$$

holds for all values of the generally complex-valued scalar α, where n is a nonnegative integer. Moreover, the closed-form representation of the *infinite geometric series,*

$$\sum_{k=0}^{\infty} \alpha^k = \frac{1}{1-\alpha} \qquad \text{for } |\alpha| < 1 \tag{1.16}$$

holds as long as $|\alpha| < 1$.

To prove the finite geometric-summation closed-form representation, let us set $S_n = 1 + \alpha + \alpha^2 + \cdots + \alpha^n$. It is then readily shown that

$$(1-\alpha)S_n = 1 - \alpha^{n+1}$$

from which identity (1.15) follows directly. Upon taking the limit of this identity as n approaches plus infinity, result (1.16) is obtained. The importance of these simple identities to signal processing cannot be overemphasized and they should therefore be committed to memory. To gain a facility for their use, let us consider the following example.

EXAMPLE 1.3

Compute the time series size measures $\mathcal{M}_x$, $\mathcal{E}_x$, and $\mathcal{P}_x$ associated with the one-sided exponential

$$x(n) = a^n u(n)$$

Using the appropriate definitions, we find that for $|a| < 1$,

$$\mathcal{M}_x = 1$$

$$\mathcal{E}_x = \sum_{n=-\infty}^{\infty} |a^n u(n)|^2 = \sum_{n=0}^{\infty} |a|^{2n} u^2(n)$$

From identity (1.16), this summation simplifies to

$$\mathcal{E}_x = \sum_{n=0}^{\infty} (|a|^2)^n = \frac{1}{1 - |a|^2}$$

Finally, it is seen that

$$\mathcal{P}_x = 0$$

On the other hand, for $|a| > 1$, each of these measures is found to be plus infinity. ■

1.7 CAUSAL AND ANTICAUSAL TIME SERIES

When examining the various attributes of time series (signals) and linear operators, the concept of causality invariably appears in some form. This being the case, we shall now briefly examine this salient concept.

Definition 1.4. The time series $\{x(n)\}$ is said to be *causal* if its elements are identically zero for negative time, that is,

$$x(n) = 0 \qquad \text{for } n = -1, -2, -3, \ldots \tag{1.17}$$

and to be *anticausal* if its elements are identically zero for nonnegative time, that is,

$$x(n) = 0 \qquad \text{for } n = 0, 1, 2, \ldots \tag{1.18}$$

Both causal or anticausal time series are said to be *one-sided*, since in each case there exists an infinite set of contiguous time integers for which the signal is identically zero. It often happens, however, that a time series may be neither purely causal nor purely anticausal in nature. In such situations, the mixed time series may always be uniquely decomposed into a sum composed of its causal and anticausal time components as given by

$$\begin{aligned} x(n) &= x(n)u(n) + x(n)u(-n-1) \\ &= x_c(n) + x_a(n) \end{aligned} \tag{1.19}$$

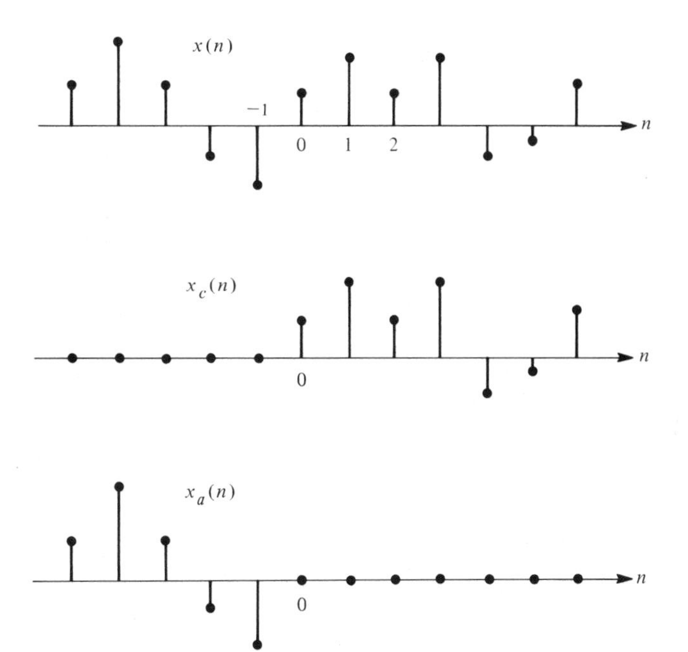

FIGURE 1.7 Causal and anticausal components of a time series.

The subscripts c and a have been here appended to denote the causal and anticausal time series components, respectively. The ability to decompose a time series into its causal and anticausal components will be of particular value when generating the z-transform associated with a two-sided signal. This capability will be required in Chapters 10 and 11 when we study the power spectral density of a random signal. Figure 1.7 depicts the decomposition of a general time series into its causal and anticausal components.

1.8 TIME SERIES OPERATORS

A *time series operator* provides a systematic procedure whereby one time series (called the *excitation*) is converted into another time series (called the *response*) through a well-defined rule. This operator can be implemented by means of a computer program or hardwired digital circuitry. Whatever the case, this process may be described mathematically by

$$\mathbf{y} = T\mathbf{x}$$

or equivalently,

$$\{y(n)\} = T\{x(n)\} \tag{1.20}$$

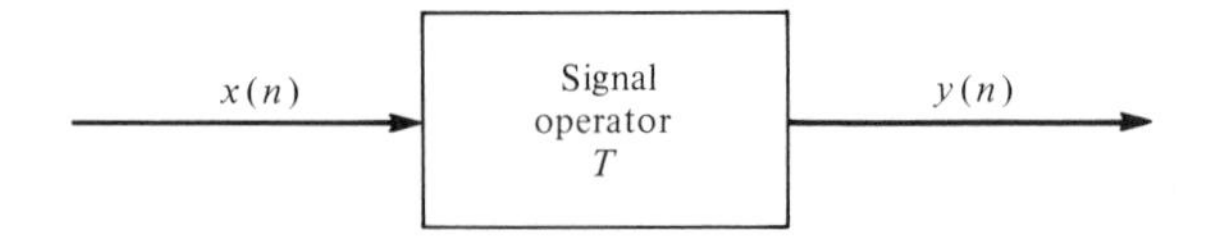

FIGURE 1.8 Signal operation.

where T represents the time series operator that when applied to the excitatio $\{x(n)\}$ produces the response $\{y(n)\}$. The operator rule T is said to be well define if to each possible excitation, an associated unique response is produced. A depiction of this cause-and-effect relationship is shown in Figure 1.8.

EXAMPLE 1.4

It is possible to postulate a simple model for a bank savings account operat using elementary reasoning. Specifically, let the excitation time series $\{x(n$ represent the sum of the deposits minus the sum of the withdrawals made i the nth period (i.e., day, month, quarter) for a given account. Furthermor let the balance of funds in this account at the completion of the nth period b denoted by the response $\{y(n)\}$. For a commonly used savings account pla these excitation and response time series elements are interrelated accordin to the first-order difference equation.

$$y(n) = x(n) + \left(1 + \frac{r}{N}\right)y(n - 1)$$

where the parameter N denotes the number of interest-compounding period in a year (e.g., $N = 4$ corresponds to compounding quarterly) and r corr sponds to the prevailing interest rate being compounded N times per year f the savings account. Clearly, given any deposit–withdrawal history $\{x(n)$ this recursive relationship may be used to systematically compute the saving account balance.

EXAMPLE 1.5

In numerical analysis, various procedures for iteratively computing the squa root of a positive number exist. One such algorithm that is widely used ma be represented as a time series operation. In particular, consider the weighte unit step given by

$$x(n) = au(n) \qquad \text{for } n \geq 0$$

in which the positive quantity a represents the number whose square root i being sought. If this time series is taken to be the excitation for the tim series operator governed by

$$y(n) = \frac{1}{2}\left[y(n - 1) + \frac{x(n)}{y(n - 1)}\right] \qquad \text{for } n = 0, 1, 2, \ldots$$

it can be shown that under general conditions, the corresponding response $\{y(n)\}$ will approach $\sqrt{a}$ for large positive time [i.e., $y(n)$ converges to $\sqrt{a}$ as n becomes large positive]. When this algorithm is started at $n = 0$, it is found that $y(0) = \frac{1}{2}[y(-1) + a/y(-1)]$. The "initial condition" $y(-1)$ required to compute $y(0)$ will be taken to be an initial guess of $\sqrt{a}$ [e.g., $y(-1) = 1$ will suffice]. Once $y(0)$ has been computed, the remaining response elements $y(1)$, $y(2)$, $y(3)$, . . . may be computed sequentially using the signal operator rule above. It will be found that successive values of $y(n)$ get closer to the required square root of a. ■

Although the operator rules used in the examples above are of a simple nature, they depict fairly well the dynamical process of a time series operation. Specifically, elements of the response time series are computed by means of a mathematical rule (or algorithm) that depends explicitly on excitation elements and possibly on previously computed response elements. This mathematical rule must always be formulated so that no ambiguity exists in the computational scheme.

1.9 LINEAR TIME-INVARIANT OPERATORS

The class of well-defined time series operators is rather large and encompasses the simplest to the most complex of numerical algorithms. Within this large class, however, the important subclass of *linear time-invariant operators* has found particular favor in the signal-processing community. A disposition toward this time series operator class arises primarily from the observations that (1) such operators perform remarkably well in many relevant applications, and (2) they are susceptible to a thorough analysis, which generally yields useful dynamical characterization insight. With these practical and theoretical reasons serving as motivation, the formal description of a linear time-invariant signal operator is now given.

Definition 1.5. A time series operator is said to be *linear time invariant* if the relationship between its excitation and response elements is expressible in the *linear convolution summation* format

$$y(n) = \sum_{k=-\infty}^{\infty} h(k)x(n - k) \tag{1.21a}$$

The *weighting sequence* elements $h(k)$ that compose this convolution summation completely identify the dynamics of the linear operator. Moreover, upon making the change of summation variables $m = n - k$, this convolution summation relationship may be represented equivalently as

$$y(n) = \sum_{m=-\infty}^{\infty} h(n - m)x(m) \tag{1.21b}$$

These two equivalent convolution summation operations are often represented by the compact notation

$$\{y(n)\} = \{h(n)\} * \{x(n)\}$$

in which the asterisk represents the convolution summation operation.

Upon examination of either convolution summation representation (1.21), it is seen that a linear time-invariant operator will be one for which the response elements are linear combinations of excitation elements. If the time instant n is thought of as being *present time,* we may decompose the response element $y(n)$ into a term exclusively involving present and past excitations elements [i.e., $x(n), x(n - 1), x(n - 2), \ldots$] and a term exclusively involving future excitation elements [i.e., $x(n + 1), x(n + 2), \ldots$]. This particular decomposition will be useful in studying general linear time-invariant operations and is given by

$$y(n) = \underbrace{\sum_{k=0}^{\infty} h(k)x(n - k)}_{\text{past and present excitations}} + \underbrace{\sum_{k=-\infty}^{-1} h(k)x(n - k)}_{\text{future excitations}}$$

In real-time applications, access to *future* excitations is not possible. This necessitates that any associated linear time-invariant operation be constrained to use only the first decomposition term, which exclusively involves present and past excitation elements. This is equivalent to requiring that the operator be such that its associated weighting sequence be constrained so that $h(k) = 0$ for $k < 0$. Such operators are said to be *causal* in nature, due to the requirement that the operator's *weighting sequence* $\{h(n)\}$ must be a causal time series. We shall now formalize this concept.

Definition 1.6. A *causal linear time-invariant operator* is characterized by the linear convolution summation

$$y(n) = \sum_{k=0}^{\infty} h(k)x(n - k) = \sum_{k=-\infty}^{n} h(n - k)x(k) \tag{1.22}$$

Similarly, an *anticausal linear operator* is specified by

$$y(n) = \sum_{k=-\infty}^{-1} h(k)x(n - k) = \sum_{k=n+1}^{\infty} h(n - k)x(k) \tag{1.23}$$

Properties of Linear Time-Invariant Operators

The general linear time-invariant operator (1.21) possesses the two important properties of *linearity* and *time invariance*. To describe these properties math ematically, let the response $\{y(n)\}$ of such an operator to the excitation $\{x(n)\}$ be denoted operationally by

$$\{y(n)\} \fallingdotseq L\{x(n)\} \tag{1.24}$$

in which the symbol L represents any general linear time-invariant operation that can be put into form (1.21). By using this operator notation representation, it is readily established that the following two properties hold:

Linearity: $$L(a_1\{x_1(n)\} + a_2\{x_2(n)\}) = a_1L\{x_1(n)\} + a_2L\{x_2(n)\} \tag{1.25}$$

Time invariance: $$\{y(n - m)\} = L\{x(n - m)\} \tag{1.26}$$

for any and all selections of the scalars a_1 and a_2, excitations $\{x_1(n)\}$, $\{x_2(n)\}$, and $\{x(n)\}$, and integer values of m. The ability to make a rather thorough analysis of linear time-invariant operators is a direct consequence of these two properties. As a matter of fact, a linear time-invariant operator is formally defined to be an operator that satisfies these two properties. It happens that any such operator may always be put into the linear convolution summation format (1.21).

The linearity property indicates that the response associated with any linear combination of excitations is equal to the same linear combination of the responses due to the individual excitations. Thus a linear operation is seen to preserve linear combinations. On the other hand, the time-invariance property indicates that a linear time-invariant signal operator's response to any excitation will be the same independent of when that excitation was applied except for a trivial time shift.

EXAMPLE 1.6

Following are two examples of linear time-invariant operators:

$$y(n) = -2x(n + 1) + 7.5x(n) - 3x(n - 2)$$

$$y(n) = x(n) - 0.5y(n - 1)$$

The second expression is recursive in nature, since $y(n)$ is explicitly dependent on the most recently computed response element $y(n - 1)$, and it is therefore not in the convolution summation form (1.21) that is characteristic of linear operators. It is shown in Example 1.9, however, that this second operator expression may equivalently be expressed as

$$y(n) = \sum_{k=0}^{\infty} (-0.5)^k x(n - k)$$

and is therefore a linear time-invariant operator. ■

As suggested by the terminology used to this point, there exist linear time series operators whose dynamics vary with time.† Although the class of linear time-varying linear operators is important in its own right, we shall concentrate our effort on the study of time-invariant operators. With this in mind, the concise terminology *linear operator* will henceforth be used in place of the more cumbersome phrase *linear time-invariant operator*. In those instances where a time-varying operator is being examined, the term *time-varying* will be explicitly appended.

1.10 UNIT-IMPULSE AND UNIT-STEP RESPONSES

The dynamical characteristics of a linear operator may be determined from knowledge of its responses to either the unit-impulse or unit-step excitations. To demonstrate why this is so, let us consider the general linear operator (time-invariant)

$$y(n) = \sum_{k=-\infty}^{\infty} h(k)x(n-k)$$

Upon setting the excitation signal equal to the unit-impulse [i.e., $x(n) = \delta(n)$] and using the unit-impulse function definition (1.3), the corresponding response is found to be

$$\begin{aligned} y_\delta(n) &= \sum_{k=-\infty}^{\infty} h(k)\delta(n-k) \\ &= h(n) \end{aligned} \tag{1.27}$$

where the subscript δ has been appended to y so as to explicitly designate that it is the unit-impulse response. Thus the response of any linear operator (time-invariant) to the unit-impulse excitation is seen to be equal to the operator's weighting sequence. This result provides a particularly attractive means of determining the weighting sequence which governs a linear operator that is not in a convolution summation format (e.g., a linear recursive operator). This operational relationship is depicted in Figure 1.9.

† A linear time-varying operator is characterized by a convolution summation of the form

$$y(n) = \sum_{k=-\infty}^{\infty} h(n, k)x(k)$$

where the *weighting sequence* $h(n, k)$ is seen to be dependent on the observation time n, and, the excitation application time k. A time-invariant operator is seen to have a weighting sequence that depends only on the difference of these two times [i.e., $h(n, k) = h(n-k)$].

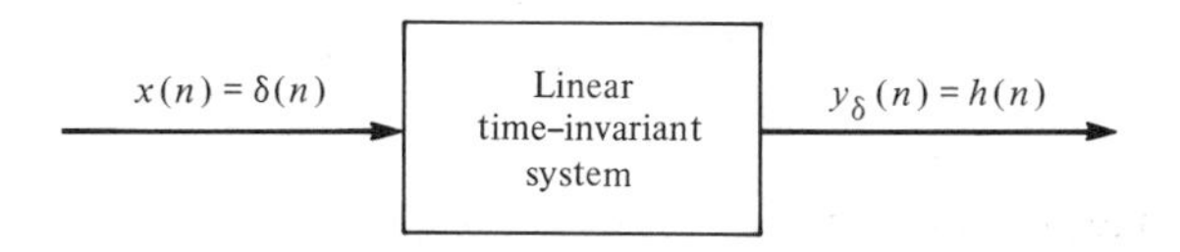

FIGURE 1.9 Weighting sequence characterization of a linear time-invariant system.

In applications where a constant level of excitation is relevant, the unit-step response is of direct interest. Upon setting the excitation equal to the unit step, the linear operator's is found to be

$$y_u(n) = \sum_{k=-\infty}^{\infty} h(k)u(n - k) = \sum_{k=-\infty}^{n} h(k) \tag{1.28}$$

where the subscript u has been appended to y so as to explicitly designate that it is the unit-step response. Again this particular response is seen to convey information relative to the linear operator's weighting sequence. In particular, it is seen that the weighting sequence may be extracted from the unit-step response by applying the *first difference operator* to that response, that is,

$$y_u(n) - y_u(n - 1) = y_\delta(n) = h(n)$$

To illustrate the utility of the unit-impulse and unit-step responses, let us now consider a practical application.

EXAMPLE 1.7

In Example 1.4 the recursive model

$$y(n) = x(n) + \left(1 + \frac{r}{N}\right)y(n - 1)$$

was postulated for a savings account system. This signal operator is readily shown to be linear and time-invariant [i.e., it satisfies relationships (1.25) and (1.26)] and therefore has an equivalent convolution summation representation (1.21). To determine this operator's weighting sequence, we may conceptually apply the unit-impulse excitation and then compute the corresponding response, which by relationship (1.27) equals $\{h(n)\}$. It is a simple matter to show that the response of this signal operator to the unit-impulse sequence is

$$y_\delta(n) = h(n) = \left(1 + \frac{r}{N}\right)^n u(n)$$

The initial condition $y(-1)$ was set to zero in arriving at this result, since the excitation prior to $n = 0$ was zero [i.e., $\delta(n) = 0$ for $n < 0$], thereby also causing the response to be zero [i.e., $y(n) = 0$ for $n < 0$]. It is apparent from this expression that the weighting sequence becomes unbounded in size as n approaches infinity, since r/N is positive.

Let us now examine the case in which a person opens a savings account plan and systematically deposits d dollars each conversion period and makes no withdrawals. The excitation therefore corresponds to the scaled step signal $x(n) = du(n)$. According to expression (1.28), the savings account value is then given by

$$y_u(n) = \sum_{k=-\infty}^{n} d\left(1 + \frac{r}{N}\right)^k u(k)$$

$$= d\,\frac{(1 + r/N)^{n+1} - 1}{r/N}\,u(n)$$

where the finite geometric series identity (1.15) has been used. ■

1.11 STABLE LINEAR OPERATORS

A time series operator is said to be stable in the *bounded excitation-bounded response* sense if its response to any bounded excitation is itself bounded. It will be recalled that the time series $\{x(n)\}$ is said to be bounded if there exists a finite scalar M such that

$$|x(n)| \leq M \qquad \text{for all } n \tag{1.29}$$

That is, the magnitude of the signal's elements never gets larger than some prescribed level. We now examine the question of what conditions on the general linear operator expression

$$y(n) = \sum_{k=-\infty}^{\infty} h(k)x(n - k) \tag{1.30}$$

must be imposed to ensure its stability. A little thought should convince oneself that this stability characterization must be linked to the behavior of the operator's weighting sequence $\{h(n)\}$. This linkage will now be established.

If we use the following standard magnitude inequality and equality

$$|a + b| \leq |a| + |b| \qquad |ab| = |a| \cdot |b|$$

which hold for all numbers a and b, it is seen that the general response element (1.30) is bounded from above by

$$\begin{aligned}|y(n)| &\leq \sum_{k=-\infty}^{\infty} |h(k)| \cdot |x(n-k)| \\ &\leq M \sum_{k=-\infty}^{\infty} |h(k)|\end{aligned}$$

where use of the assumed excitation bounding (1.29) has been incorporated. Since this upper bound holds for all values of n, we have established the necessary condition for linear operator stability, as expressed by the following theorem.

Theorem 1.2. A linear operator characterized by the weighting sequence $\{h(n)\}$ is *bounded excitation-bounded response* stable if and only if

$$\sum_{n=-\infty}^{\infty} |h(n)| < \infty \tag{1.31}$$

Thus a linear operator is stable if its weighting sequence elements $h(n)$ decay to zero sufficiently rapidly as n approaches plus and minus infinity so that condition (1.31) is satisfied.

1.12 FREQUENCY RESPONSE

The manner in which a linear operator responds to a sinusoidal excitation is of particular interest in signal-processing applications. Linear operators are often used for the purposes of effecting a prescribed frequency-discrimination behavior as exemplified by the widespread use of low-pass, bandpass, and high-pass digital filters. With these applications in mind, we now examine the response of the general stable linear operator

$$y(n) = \sum_{k=-\infty}^{\infty} h(k)x(n-k)$$

to the complex sinusoidal excitation

$$x(n) = ae^{j(\omega n + \theta)}$$

Inserting this excitation into the linear operator's convolution summation representation, we have

$$y(n) = \sum_{k=-\infty}^{\infty} h(k) a e^{j(\omega n - \omega k + \theta)}$$
$$= a e^{j(\omega n + \theta)} \sum_{k=-\infty}^{\infty} h(k) e^{-j\omega k} \tag{1.32}$$

The term multiplying the summation in this expression is recognized as being the excitation signal. Thus the linear operator's response is seen to be equal to the product of the excitation signal and the entity

$$H(e^{j\omega}) = \sum_{k=-\infty}^{\infty} h(k) e^{-j\omega k}$$
$$= |H(e^{j\omega})| e^{j\phi(\omega)} \tag{1.33}$$

The entity $H(e^{j\omega})$ will be referred to as the *frequency response* associated with the linear operator and is seen to be completely dependent on the operator's weighting sequence $\{h(n)\}$ and the sinusoidal excitation's radian frequency ω. We have represented the frequency response in its polar form in expression (1.33), where $|H(e^{j\omega})|$ and $\phi(\omega)$ denote the *magnitude* and *phase* functions, respectively, of the generally complex-valued frequency response function $H(e^{j\omega})$.

Upon inserting the *frequency response* above into the response expression (1.32), it is seen that

$$y(n) = |H(e^{j\omega})| a e^{j(\omega n + \theta + \phi(\omega))}$$

Thus the linear operator is seen to impart an amplitude change of $|H(e^{j\omega})|$ and a phase shift of $\phi(\omega)$ on the sinusoidal excitation. This being the case, the possibility arises of prudently selecting the linear operator's weighting sequence so as to achieve a predetermined frequency-discrimination characteristic. As a matter of fact, it will be shown in Chapters 5 and 6 that linear operators may effectively be used to implement such operations as low-pass, bandpass, high-pass, and notch filtering. To depict such behavior visually, plots of the frequency response's magnitude, $|H(e^{j\omega})|$, and phase angle, $\phi(\omega)$, may be displayed as functions of radian frequency ω. Values of ω for which the magnitude function $|H(\omega)|$ is relatively large (near zero) are said to constitute the passband (stopband) of the linear operator. We now summarize these results.

Theorem 1.3. The *frequency response* associated with a linear operator with weighting sequence $\{h(n)\}$ is given by

$$H(e^{j\omega}) = \sum_{n=-\infty}^{\infty} h(n) e^{-j\omega n}$$
$$= |H(e^{j\omega})| e^{j\phi(\omega)} \tag{1.34}$$

Moreover, this linear operator has the following sinusoidal excitation–response pair relationships:

$$\text{(a) If } x(n) = ae^{j(\omega n+\theta)}, \text{ then } y(n) = a|H(e^{j\omega})|e^{j(\omega n+\theta+\phi(\omega))} \tag{1.35}$$

$$\begin{aligned}\text{(b) If } x(n) &= a \sin(\omega n + \theta), \text{ then} \\ y(n) &= a|H(e^{j\omega})| \sin[\omega n + \theta + \phi(\omega)]\end{aligned} \tag{1.36}$$

where in the real sinusoidal excitation–response pair (1.36) it is tacitly assumed that the linear operator's weighting sequence is real.

One may readily establish the validity of expression (1.36) by first representing the given excitation in its Euler identity form

$$a \sin(\omega n + \theta) = a\frac{e^{j(\omega n+\theta)} - e^{-j(\omega n+\theta)}}{2j}$$

This expression is next substituted into the linear operator's convolution summation representation and then straightforward manipulations are employed to achieve (1.36). This proof is left to the reader as an exercise (see the problem section).

EXAMPLE 1.8

Consider a linear operator whose weighting sequence $\{h(n)\}$ is specified by

$$h(n) = \begin{cases} \dfrac{1}{N} & 0 \le n \le N - 1 \\ 0 & \text{otherwise} \end{cases}$$

where N is a fixed positive integer. The frequency response corresponding to this linear operator is then given by

$$\begin{aligned} H(e^{j\omega}) &= \sum_{n=0}^{N-1} \frac{1}{N} e^{-j\omega n} \\ &= \frac{1}{N}\sum_{n=0}^{N-1} (e^{-j\omega})^n = \frac{1}{N}\frac{1 - e^{-j\omega N}}{1 - e^{-j\omega}} \end{aligned}$$

where use of the finite geometric series identity (1.15) with $\alpha = e^{-j\omega}$ has been made. We may put this function into the more useful form

$$\begin{aligned} H(e^{j\omega}) &= \frac{1}{N}\frac{e^{-j\omega N/2}}{e^{-j\omega/2}}\frac{e^{j\omega N/2} - e^{-j\omega N/2}}{e^{j\omega/2} - e^{-j\omega/2}} \\ &= \frac{1}{N} e^{j\omega(N-1)/2}\frac{\sin(\omega N/2)}{\sin(\omega/2)} \end{aligned}$$

so that

$$|H(e^{j\omega})| = \frac{1}{N}\left|\frac{\sin(\omega N/2)}{\sin(\omega/2)}\right|$$

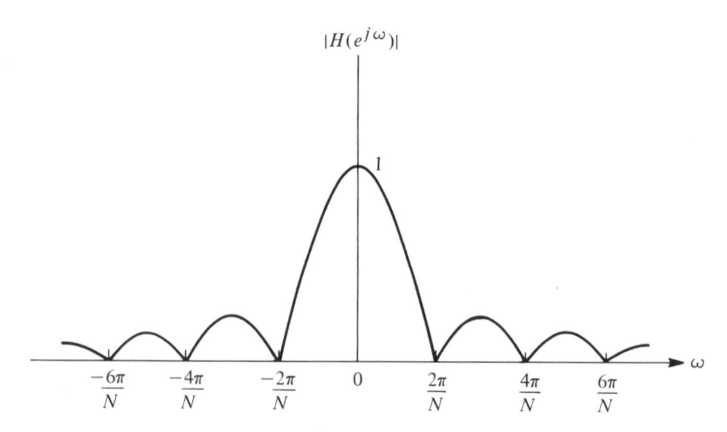

FIGURE 1.10 Sketch of frequency response magnitude.

A sketch of this periodic magnitude function in the fundamental period $-\pi \leq \omega \leq \pi$ is depicted in Figure 1.10. For relatively large values of N, this magnitude behavior suggests that low-frequency sinusoidal signals (i.e., ω less than $2\pi/N$ in radian value) will readily be transmitted by the linear operator, whereas higher-frequency sinusoids will not. The linear operator with the prescribed weighting sequence is therefore said to be a low-pass filter. ■

1.13 NONRECURSIVE AND RECURSIVE LINEAR OPERATORS

The two most widely employed signal-processing algorithms are the linear *nonrecursive* and *recursive* operators. A linear noncursive operator is governed by a linear convolution summation that contains only a *finite* number of nonzero weighting sequence elements. On the other hand, although a recursive operator has an infinite-length weighting sequence, it is governed by a finite parameter algorithm which incorporates feedback response terms. A brief presentation of these two important operator forms will now be made. A more thorough development of their properties will be made in subsequent chapters.

Nonrecursive Linear Operator

A signal operator is said to be *linear nonrecursive* if it is expressible as

$$y(n) = \sum_{k=s}^{q} h(k)x(n - k) \tag{1.37}$$

where q and s are fixed integers with $q \geq s$. The response element of this linear nonrecursive operator is seen to be a linear combination of $q + 1 - s$ excitation elements. Given any excitation, the corresponding response arising from this nonrecursive relationship is well defined and unique. It will subsequently be shown that proper selection of the coefficients $h(s)$, $h(s + 1)$, . . . , $h(q)$ will enable this operator to implement various types of useful signal-processing operations. In particular, the ability to effect frequency discrimination is made evident upon examining the frequency response

$$H(e^{j\omega}) = \sum_{k=s}^{q} h(k)e^{-j\omega k} \tag{1.38}$$

which is associated with this nonrecursive linear operator.

As suggested earlier, in real-time signal-processing applications, we have available excitation elements only up to the present time n for computing the response element $y(n)$. In such commonly occurring situations, we are then restricted to the use of a causal signal operation. The most general *causal linear nonrecursive* operator is of the form

$$y(n) = h(0)x(n) + h(1)x(n - 1) + \cdots + h(q)x(n - q) \tag{1.39}$$

in which q is a fixed positive integer. This causal linear nonrecursive operator is also referred to as being a *transversal filter*, a *finite impulse response filter*, or a *moving-average operator* of order q.

Recursive Linear Operator

An operator that is governed by a difference equation of the form

$$\sum_{k=p_1}^{p_1} a_k y(n - k) = \sum_{k=q_1}^{q_2} b_k x(n - k) \tag{1.40}$$

is said to be a *linear recursive operator*. Given any excitation, its corresponding response must be such that the right- and left-side summations constituting this recursive expression are equal for all values of time n. Unlike its nonrecursive counterpart, however, this linear recursive operation is not unambiguously defined. A different response to a given excitation will arise if we interpret recursive expression (1.40) as being causal, anticausal, or mixed causal in nature. This point will be clarified in what follows. Whichever interpretation is rendered, however, linear recursive operators are extensively employed in a variety of signal-processing applications. To provide one reason why this is so, it will be shown in Chapter 2 that the frequency response associated with linear recursive operator (1.40) is given by the ratio of polynomials

$$H(e^{j\omega}) = \frac{\sum_{k=q_1}^{q_2} b_k e^{-j\omega k}}{\sum_{k=p_1}^{p_2} a_k e^{-j\omega k}} \tag{1.41}$$

provided that the recursive operator is stable. By prudently selecting the operator coefficients a_k and b_k, this frequency response may be made to take on various desired frequency-discrimination behaviors (e.g., low-pass filtering).

Causal Interpretation

In most practical applications, the signal-processing algorithm is required to be causal in nature. The most general *causal linear recursive operator* will be governed by a relationship of the form

$$y(n) = \sum_{k=0}^{q} b_k x(n - k) - \sum_{k=1}^{p} a_k y(n - k) \tag{1.42}$$

where q and p are fixed nonnegative and positive integers, respectively. In this causal operation, the "present" response element $y(n)$ is computed by evaluating the foregoing two summations, which involve the present and the most recent past q excitation elements and the most recently computed past p response elements. The response elements are then systematically generated by incrementing the time variable n by $+1$ at each computation stage. This will result in a sequential generation of the response elements $y(n)$ as n evolves in the standard positive time direction. Under this causal interpretation, if an excitation first becomes nonzero at n_0 [i.e., $x(n) = 0$ for $n < n_0$], then the response elements prior to n_0 must all be nonzero [i.e., $y(n) = 0$ for $n < n_0$]. The justification for setting the elements $y(n) = 0$ for $n < n_0$ arises from the observation that since the excitation was identically zero up to n_0, the response of the causal linear operator (1.42) must of necessity also be zero over this interval. Thus the resultant response will be specified by

$$y(n) = \begin{cases} 0 & n < n_0 \\ b_0 x(n_0) & n = n_0 \\ b_0 x(n_0 + 1) + b_1 x(n_0) - a_1 y(n_0) & n = n_0 + 1 \\ \quad \text{and so on} & \end{cases}$$

Anticausal Interpretation

In situations where the time series to be processed is stored on tape or some other mechanism, it is possible to use anticausal or mixed causal–anticausal signal-processing algorithms. Based on this observation, we now briefly examine the operational characteristics of the general *anticausal linear recursive operator*

$$y(n) = \sum_{k=0}^{q} b_k x(n + k) - \sum_{k=1}^{p} a_k y(n + k) \tag{1.43}$$

where the parameters p and q are fixed positive and nonnegative integers, respectively. In this anticausal operation, the present response element $y(n)$ is computed by evaluating the foregoing two summations, which depend on the present and the most immediate future q excitation elements and the most immediate future p response elements. The term *future* is used here in the standard sense that times larger than the present time n are considered future times. Since the excitation time series being operated on is stored, however, we have access to all future, past, and present excitation elements, and the normal inability to use future data is therefore not applicable.

The response elements (1.43) are systematically computed by incrementing the time variable n by -1 at each iteration. This will result in a sequential generation of $y(n)$ as n evolves in the nonstandard negative time direction. Let us now consider the situation in which the excitation is identically zero for all times greater than n_0 [i.e., $x(n) = 0$ for $n > n_0$]. In this anticausal case, owing to the lack of an excitation, the response elements will also be identically zero for $n > n_0$. This being the case, the corresponding response will then be generated according to

$$y(n) = \begin{cases} 0 & n > n_0 \\ b_0 x(n_0) & n = n_0 \\ b_0 x(n_0 - 1) + b_1 x(n_0) - a_1 y(n_0) & n = n_0 - 1 \\ \quad \text{and so on} & \end{cases}$$

Causal and anticausal operators are seen to have a mirror-image behavior in the sense that they give rise to response elements that evolve in positive and negative time directions, respectively.

Mixed Causal–Anticausal Recursive Operators

From a signal-processing viewpoint, significant performance improvements can be accrued when using a *mixed causal–anticausal linear recursive operator*. Of course, operators of this nature will be useful only in situations where the time series (or signal) to be processed is stored on tape or some other medium. A mixed causal–anticausal linear recursive operation will be characterized by the expression

$$y(n) = y_c(n) + y_a(n) \tag{1.44}$$

in which the causal response term $y_c(n)$ is generated by a causal linear recursive operator of form (1.42), while the anticausal response term $y_a(n)$ is computed using an anticausal linear recursive operator of form (1.43). To generate the mixed causal–anticausal recursive response, we simply compute the respective causal and anticausal responses separately and then add them. This mixed operation is depicted in Figure 1.11.

EXAMPLE 1.9

Let us now consider the simple linear recursive operator as governed by

$$y(n) + 0.5y(n - 1) = x(n)$$

This expression identifies the relationship that must be satisfied by the oper-

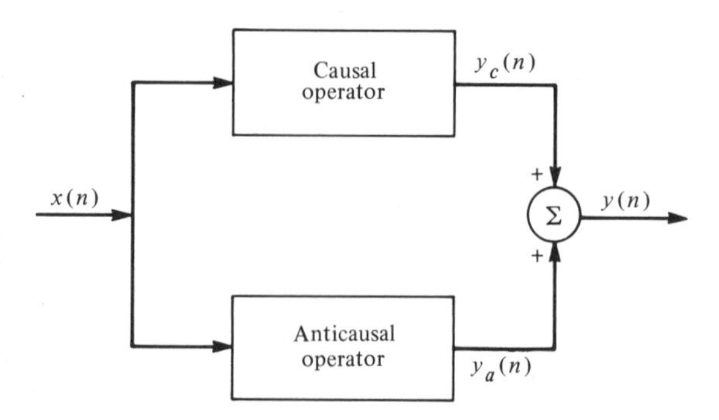

FIGURE 1.11 Implementation of a mixed causal–anticausal signal operation.

ator's excitation and response elements for all times n. Although this may appear to be an unambiguous relationship, we shall show that this is not the case. Specifically, the causal interpretation of this recursive operator as governed by

$$y(n) = x(n) - 0.5y(n - 1)$$

will have a distinctly different dynamical behavior than its associated noncausal interpretation

$$y(n - 1) = 2x(n) - 2y(n)$$

To prove this point, let us now find the convolution summation representation for each of these related linear recursive operators. In accordance with the remarks of Section 1.10, the required weighting sequence governing the convolution summation is obtained by computing the linear operator's response to the unit-impulse excitation $x(n) = \delta(n)$. If this excitation is inserted into the causal interpretation, we have

$$y_\delta(n) = \delta(n) - 0.5y_\delta(n - 1)$$

In this causal interpretation, the response elements are computed as n is incrementally increased. For negative values of n, the excitation is identically zero, implying that $y_\delta(n) = 0$ for $n < 0$. At $n = 0$ it is seen that $y_\delta(0) = \delta(0) = 1$, where the fact that $y_\delta(-1) = 0$ has been incorporated. Continuing in this recursive updating manner, we find that the resultant unit-impulse response is given by

$$y_\delta(n) = h_c(n) = (-0.5)^n u(n)$$

Thus the causal recursive operator interpretation is seen to yield a stable causal weighting sequence.

If the unit-impulse excitation is now inserted into the recursive operator's anticausal interpretation, it is found that

$$y_\delta(n - 1) = 2\delta(n) - 2y_\delta(n)$$

We compute this response as n is incrementally decreased. It is seen that since the unit-impulse excitation is identically zero for $n > 0$, the corresponding anticausal response will also be zero for $n \geq 0$. At $n = 0$ we have $y_\delta(-1) = 2\delta(0) = 2$, where use has been made of the fact that $y_\delta(0) = 0$. Continuing in this manner as n is incremented in the negative time direction, we find that the resultant unit-impulse response is given by

$$y_\delta(n) = h_a(n) = -(-2)^{-n}u(-n - 1)$$

The weighting sequence associated with the noncausal interpretation of the given linear recursive operator is seen to be an unstable anticausal time series. This particular excitation–response behavior illustrates the general point that causal and anticausal interpretations of a given linear recursive operator are distinct signal operations. ■

Although linear nonrecursive and recursive operators may be used to implement the same signal-processing operations (e.g., low-pass filtering), they typically achieve this in a distinctively different manner. It will often be necessary to use *many more* parameters for a given signal-processing operation when using a nonrecursive operation compared to its recursive counterpart. This disadvantage may be offset by the observations that (1) a nonrecursive linear operator is guaranteed always to be stable, whereas a recursive linear operator can be unstable if its a_k parameters are improperly selected, and (2) for a given signal-processing application, it is often much easier to synthesize the required nonrecursive linear operator.

1.14 SAMPLING OF BANDLIMITED CONTINUOUS-TIME SIGNALS

As indicated in Section 1.2, a time series will often arise through the process of sampling a continuous-time signal. The critical question of how to select the sampling instants so as not to lose information through the sampling process will now be addressed. *Fourier analysis* will serve as the primary tool for this study. In particular, let $x(t)$ denote the continuous-time signal that is to be sampled. The *Fourier transform* of this signal is then defined by

$$X(\omega) = \int_{-\infty}^{\infty} x(t)e^{-j\omega t}\, dt \tag{1.45}$$

and the inverse *Fourier transform* is specified by

$$x(t) = \frac{1}{2\pi}\int_{-\infty}^{\infty} X(\omega)e^{j\omega t}\,d\omega \tag{1.46}$$

Using these two integral transforms, we may examine a continuous-time signal in either the natural *time-domain* or the auxiliary *frequency-domain*. In many situations there is a distinct advantage to be gained by taking the frequency-domain approach.

It is well known that the only case in which information is not lost through the process of sampling occurs when the continuous-time signal is *bandlimited*. To show why this is true, let it be assumed that the signal $x(t)$ is *bandlimited*, which implies that its Fourier transform (1.45) is identically zero for frequencies larger than some prescribed value ω_c, that is,

$$X(\omega) = 0 \qquad \text{for } |\omega| > \omega_c \tag{1.47}$$

Thus the Fourier transform $X(\omega)$ has nonzero behavior only in the frequency interval $[-\omega_c, \omega_c]$, as depicted in Figure 1.12a. Let us now consider a *periodic extension* of this nonzero behavior as denoted by $\tilde{X}(\omega)$ and shown in Figure

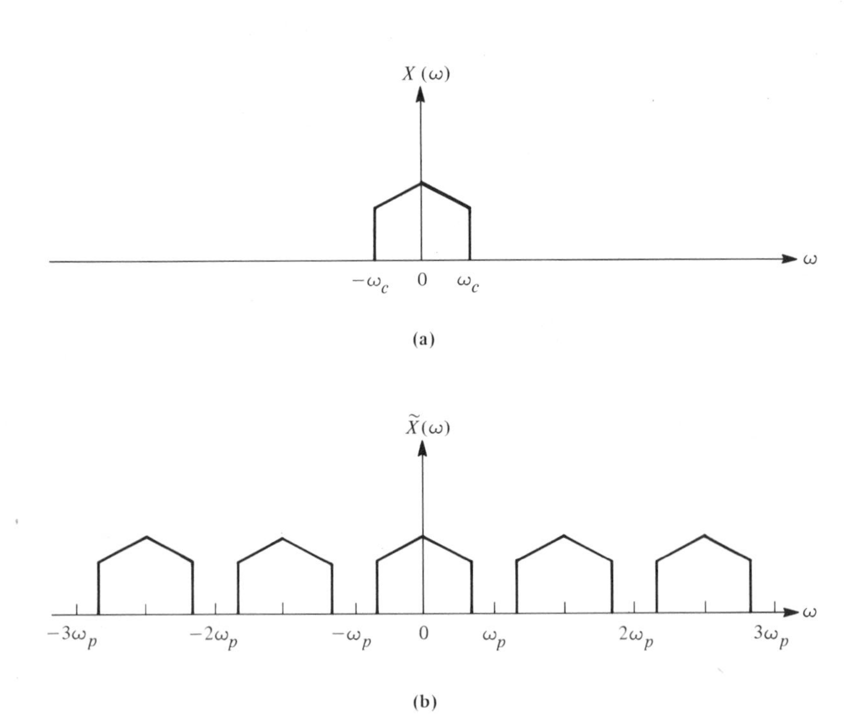

FIGURE 1.12 Uniform samplings: (a) bandlimited signal $X(\omega)$; (b) periodic extension of $X(\omega)$.

1.12b. The only restriction on the parameter ω_p is that it be larger than ω_c so that the fundamental period $[-\omega_p, \omega_p]$ will contain the entire nonzero portion of $X(\omega)$. Since $\tilde{X}(\omega)$ is a periodic function of ω with fundamental period of $2\omega_p$, it will have the Fourier series representation

$$\begin{aligned} \tilde{X}(\omega) &= \sum_{n=-\infty}^{\infty} x_n e^{-j2\pi n\omega/2\omega_p} \\ &= \sum_{n=-\infty}^{\infty} x_n e^{-j\omega nT} \end{aligned} \tag{1.48}$$

where the associated parameter $T = \pi/\omega_p$ has the units of time and will subsequently be identified as the sampling interval. The Fourier coefficients in this representation are given by the standard integral expression

$$\begin{aligned} x_n &= \frac{1}{2\omega_p} \int_{-\omega_p}^{\omega_p} X(\omega) e^{j\omega nT}\, d\omega \\ &= Tx(nT) \end{aligned}$$

Expression (1.46) and the fact that $X(\omega)$ is bandlimited have been employed in arriving at this relationship. Thus the Fourier series coefficients used to represent $\tilde{X}(\omega)$ are seen to equal the sampled continuous-time signal elements $x(nT)$ multiplied by the sampling interval T.

Upon examination of Figure 1.12 it is clear that the original Fourier transform function $X(\omega)$ and its periodic extension $\tilde{X}(\omega)$ will be identical on $[-\omega_p, \omega_p]$. This being the case, it follows that the bandlimited Fourier transform function can be represented as

$$X(\omega) = \begin{cases} T \displaystyle\sum_{n=-\infty}^{\infty} x(nT) e^{-j\omega nT} & |\omega| \leq \omega_p \\ 0 & \text{otherwise} \end{cases} \tag{1.49}$$

Thus the bandlimited Fourier transform function is seen to be *completely specified* by the set of uniformly spaced samples of the associated continuous-time signal $x(t)$, that is,

$$\{x(nT)\} = \{\ldots, x(-2T), x(-T), x(0), x(T), \ldots\}$$

Moreover, the original continuous-time signal may also be reconstructed from this set of sampled values. This is readily effected by substituting expression (1.49) into the inverse Fourier transform relationship (1.46) to obtain

$$x(t) = \frac{1}{2\pi} \int_{-\omega_p}^{\omega_p} \left[T \sum_{n=-\infty}^{\infty} x(nT) e^{-j\omega nT} \right] e^{j\omega t}\, d\omega$$

After interchanging the order of summation and integration, this expression is found to simplify to

$$x(t) = \sum_{n=-\infty}^{\infty} x(nT) \frac{\sin[(t - nT)\pi/T]}{(t - nT)\pi/T} \tag{1.50}$$

Relationships (1.49) and (1.50) indicate that in cases where the continuous-time signal $x(t)$ is bandlimited, its associated Fourier transform $X(\omega)$ may be reconstructed perfectly only from knowledge of its sampled values $\{x(nT)\}$. The only requirement in this reconstruction is that the sampling interval T be selected so that $\omega_p > \omega_c$ or, in other words,

$$T = \frac{\pi}{\omega_p} < \frac{\pi}{\omega_c} = \frac{1}{2f_c} \tag{1.51}$$

That is, if the sampling interval is chosen to be smaller than the reciprocal of the *Nyquist rate* $2f_c$, no information will be lost through the process of sampling. In other words, the continuous-time signal must be sampled at a rate which at least exceeds twice that of its highest frequency content. If a larger sampling interval is chosen (i.e., $T \geq \frac{1}{2}f_c$), however, *frequency aliasing* will result and it will be impossible to recover the original continuous-time signal from its uniformly sampled version. Frequency aliasing arises when there exists an overlap of the nonzero parts of $X(\omega)$ and its shifted versions that form the periodic extension shown in Figure 1.12b. It should be noted that an analogous sampling theorem can be developed for the nonuniform sampling of bandlimited signals (Freeman, 1965).

1.15 SUMMARY

The concepts of *signals* and *signal operators* that are fundamental to system theory have been introduced. A signal is defined as a time-varying function that describes important aspects of a phenomenon under examination. On the other hand, a signal operator represents a mechanism (or algorithm) that transforms one signal into another signal according to a well-defined rule. A system is then composed of a set of signal operators connected together in some configuration in which signals flow from and into the various signal operators.

SUGGESTED READINGS

CADZOW, J. A., *Discrete-Time Systems*. Englewood Cliffs, N.J.: Prentice-Hall, Inc., 1973.

FREEMAN, H., *Discrete-Time Systems*. New York: John Wiley & Sons, Inc., 1965.

GABEL, R. A., AND R. A. ROBERTS, *Signals and Linear Systems*. New York: John Wiley & Sons, Inc., 1980.

HAMMING, R. W., *Digital Filters*. Englewood Cliffs, N.J.: Prentice-Hall, Inc., 1977.

KAILATH, T., *Linear Systems*. Englewood Cliffs, N.J.: Prentice-Hall, Inc., 1980.

OPPENHEIM, A. V., AND A. S. WILLSKY, WITH I. T. YOUNG, *Signals and Systems*. Englewood Cliffs, N.J.: Prentice-Hall, Inc., 1983.

PAPOULIS, A., *Signal Analysis*. New York: McGraw-Hill Book Company, 1977.

ZEIMER, R. E., W. H. Tranter, and D. R. Fannis, *Signals and Systems*. New York: Macmillan Publishing Company, 1983.

PROBLEMS

1.1 Classify the following signals as being either deterministic or nondeterministic in nature.

(a) The number of automobiles sold by General Motors on a monthly basis for the next 10 years.
(b) The number of days in a year for the years A.D. 1999 to 3005.
(c) The number of infants born in the United States in a given year for the years A.D. 1999 to 3005.
(d) The daily high temperature in New York City for the year A.D. 2213.
(e) The number of times the earth rotates about the sun for the years A.D. 1999 to 3005.

1.2 Determine the number sequence that arises when the following continuous-time signals are sampled at the time instants $t_n = nT$ seconds, where T is a constant and n is an integer variable.

(a) $x(t) = 3e^{-2t} - 5$

(b) $x(t) = \begin{cases} 3 + t & t \geq 0 \\ 1 - t^2 & \text{for } t < 0 \end{cases}$

(c) $x(t) = \sin(62\pi t)$

1.3 Repeat Problem 1.2 for the case in which the sampling instants are specified by

$$t_n = n^3 T \qquad -\infty < n < \infty$$

1.4 Sketch the time series governed by the following expressions.

(a) $x_1(n) = \delta(n + 6)$
(b) $x_2(n) = u(n - 3)$
(c) $x_3(n) = (0.9)^n u(n - 1)$
(d) $x_4(n) = 2\delta(n + 3) - 5\delta(n - 1)$
(e) $x_5(n) = nu(n)$
(f) $x_6(n) = x_5(n) - x_5(n - 3)$
(g) $x_7(n) = u(-n) - u(n - 5)$

1.5 Show that for the time series characterized by $x(n) = b(a)^n u(n)$, the following identity holds:

$$x(n) = b\delta(n) + ax(n - 1)$$

in which $x(-1) = 0$.

1.6 Evaluate the elements of the time series specified by

$$\{y(n)\} = \{x_1(n)\} - 2\{x_2(n)\} + \{x_3(n)\}$$

for the case in which:
(a) $x_1(n) = x_2(n + 1) = x_3(n + 2) = u(n)$
(b) $x_1(n) = u(n),\ x_2(n) = \delta(n),\ x_3(n) = u(-n)$

1.7 What time series must be added to the time series given by $x(n) = (-1)^n$ so that the sum time series is the unit step? Make a sketch of this time series.

1.8 A vector space X is a set of elements (called vectors) and two operations that satisfy a set of axioms. The first operation is vector addition, which associates with any two vectors $\mathbf{x}$ and $\mathbf{y}$ contained in X a vector denoted by $\mathbf{x} + \mathbf{y}$, which is also contained in X. The second operation is scalar multiplication, which for any vector $\mathbf{x}$ contained in X and scalar α associates a vector denoted by $\alpha\mathbf{x}$ which is also contained in X. If the set of elements X and these two operations satisfy the following axioms:

1. $\mathbf{x} + \mathbf{y} = \mathbf{y} + \mathbf{x}$
2. $(\mathbf{x} + \mathbf{y}) + \mathbf{z} = \mathbf{x} + (\mathbf{y} + \mathbf{z})$
3. there exists a null vector $\boldsymbol{\theta}$ in X such that $\mathbf{x} + \boldsymbol{\theta} = \mathbf{x}$ for all $\mathbf{x}$ in X
4. to every vector x in X there exists a unique vector $-x$ such that $x + (-x) = \theta$
5. $\alpha(\mathbf{x} + \mathbf{y}) = \alpha\mathbf{x} + \alpha\mathbf{y}$
6. $(\alpha + \beta)\mathbf{x} = \alpha\mathbf{x} + \beta\mathbf{x}$
7. $(\alpha\beta)\mathbf{x} = \alpha(\beta\mathbf{x})$
8. $0\mathbf{x} = \boldsymbol{\theta}$
9. $1\mathbf{x} = \mathbf{x}$

then the set X is said to be a vector space. Show that the set composed of all the time series in conjunction with the two operations (1.10) and (1.11) satisfy the axioms above and that therefore the set of time series is a vector space.

1.9 Determine the time series measures $\mathcal{M}_x$, $\mathcal{E}_x$, and $\mathcal{P}_x$ for the following signals.
(a) $x_1(n) = 3(\frac{1}{2})^n u(n)$
(b) $x_2(n) = 3(\frac{1}{2})^{|n|}$
(c) $x_3(n) = nu(n)$
(d) $x_4(n) = n$
(e) $x_5(n) = [1 - (\frac{1}{2})^n]u(n)$
(f) $x_6(n) = 2\cos(2\pi n/3)$

1.10 Determine the time series measures $\mathcal{M}_x$, $\mathcal{E}_x$, and $\mathcal{P}_x$ for the following signals.
(a) $x_1(n) = -5(\frac{1}{3})^n u(n) + 2(\frac{1}{4})^n u(n)$
(b) $x_2(n) = 2(\frac{1}{5})^n u(n) - \frac{1}{7}u(n - 2)$
(c) $x_3(n) = 5(\frac{1}{2})^n u(n) - 3(4)^n u(-n)$

1.11 Prove that the closed-form identity

$$\sum_{k=0}^{n} k(\alpha)^k = \frac{\alpha}{(1 - \alpha)^2}(1 - \alpha^n - n\alpha^n + n\alpha^{n+1})$$

holds for all values of the scalar α where n is a positive integer. *Hint:* Use the fact that

$$a \frac{d}{da}(1 + a + a^2 + \cdots + a^n) = a + 2a^2 + \cdots + na^n$$

From the identity above, conclude that

$$\sum_{k=0}^{\infty} k(\alpha)^k = \frac{\alpha}{(1 - \alpha)^2}$$

holds for all scalars α such that $|\alpha| < 1$.

1.12 For each of the time series considered in Problem 1.4, indicate whether they are causal, anticausal, or mixed. Determine their causal and anticausal components.

1.13 Repeat Problem 1.12 for the time series considered in Problem 1.10.

1.14 Using the square-root algorithm postulated in Example 1.5, compute the first five iterations for $\sqrt{2}$ in which $y(-1) = 1$.

1.15 Prove that the operators characterized by
(a) $y(n) = x(n) - 2x(n - 1) + x(n - 2)$
(b) $y(n) = x(n) + 0.5y(n - 1)$
possess the linearity and time-invariance property as defined by expressions (1.25) and (1.26), respectively.

1.16 Prove that the linear operator characterized by the convolution summation (1.21) satisfies the linearity and time-invariance properties (1.25) and (1.26), respectively.

1.17 Give a closed-form representation for the response of the following operators to the unit-impulse excitation $x(n) = \delta(n)$.
(a) $y(n) = x(n) - 2x(n - 1) + x(n - 2)$
(b) $y(n) = x(n) - x(n - 5)$
(c) $y(n) = x(n) + 0.5y(n - 1)$
Give a justification for the assumption that $y(n) = 0$ for $n < 0$ in part (c).

1.18 Repeat Problem 1.17 for the unit-step excitation $x(n) = u(n)$.

1.19 Consider the savings account system treated in Example 1.4. Give a closed-form representation for the system response in the case where systematic deposits of d dollars are made each compounding period [i.e., $x(n) = du(n)$ and the initial conditions $y(n) = 0$ hold for $n < 0$].

1.20 Demonstrate that the two linear operators

$$(1)\ y(n) = x(n) + 0.5y(n - 1)$$

$$(2)\ y(n) = \sum_{k=0}^{n} (\tfrac{1}{2})^k x(n - k)$$

will provide the same response to the following excitations: **(a)** the unit-impulse **(b)** the unit-step. In system (1), assume that the initial conditions $y(n) = 0$ for $n < 0$ hold.

1.21 Provide closed-form expressions for the response of the following linear operators to the (1) unit-impulse and (2) unit-step excitations.
(a) $y(n) = x(n) + x(n + 1) - 2x(n - 1)$
(b) $y(n) = x(n) + x(n - 1) + \cdots + x(n - 5)$
(c) $y(n) = x(n) - x(n - 1) + 0.25y(n - 1)$
(d) $y(n) = x(n) + 3x(n - 1) + 0.25y(n - 1)$
In parts (c) and (d), assume that the initial conditions $y(n) = 0$ for $n < 0$ hold. Give a justification for the validity of this assumption.

1.22 Determine the unit-impulse response in closed form for the related linear operators
(a) $y(n) = x(n) - 0.25y(n - 1)$
(b) $y(n - 1) = 4x(n) - 4y(n)$
Assume that the following initial conditions hold: (a) $y(n) = 0$ for $n < 0$, and (b) $y(n) = 0$ for $n > 0$. How are these two operators related?

1.23 Give a closed-form representation for the frequency response associated with the linear operator

$$y(n) = \sum_{k=0}^{N-1} (-1)^k x(n - k)$$

Determine and give a sketch of the associated magnitude function $|H(e^{j\omega})|$. From this sketch, classify this filter as to type (i.e., low pass, high pass, etc.).

1.24 For a linear time-invariant operator with a real-valued weighting sequence, prove that expression (1.36) characterizes the real sinusoidal excitation–response pair.

1.25 Determine the response of the linear operator governed by

$$y(n) - \tfrac{1}{4}y(n - 1) = x(n)$$

to the excitation $x(n) = 3\delta(n) - 2\delta(n - 1)$ when the operator is taken to be **(a)** causal; **(b)** anticausal.

.26 Repeat Problem 1.25 for the linear operator specified by

$$y(n) - \tfrac{1}{3}y(n - 1) = x(n) + \tfrac{2}{3}x(n - 1)$$

.27 Determine the weighting sequence for the linear operator governed by

$$y(n) - \tfrac{1}{2}y(n - 1) = x(n) + 2x(n - 1) + x(n - 2)$$

when the operator is taken to be **(a)** causal; **(b)** anticausal. Which of the interpretations yields a stable operator?

1.28 Repeat Problem 1.27 for the system governed by

$$y(n) - 2y(n - 1) = x(n) + 2x(n - 1) + x(n - 2)$$

1.29 Give a derivation of expression (1.50) starting with the Fourier series representation (1.49) for $X(\omega)$.

CHAPTER 2

z-Transform

2.1 INTRODUCTION

The characterization of discrete-time signals (or time series) and the study and implementation of linear operators are most naturally carried out in the time domain. As we shall see in this and the remaining chapters, however, there are often significant benefits to be accrued by using a transform-domain approach. This invariably entails transforming (mapping) a time series into a function of a complex variable by means of a well-defined rule. The most widely used of these methods is the *z-transform*. In this chapter we examine the fundamental concepts involved when utilizing the z-transform. This is followed by a study of the properties possessed by the z-transform. The most relevant signal-processing property is that which replaces the cumbersome time-domain convolution summation operation by a simple product of z-transforms. Finally, the concept of the inverse z-transform is developed to give us the ability to recover the time series that generates a given z-transform.

2.2 z-TRANSFORM

When using the z-transform and the Fourier transform to a time series, it is tacitly assumed that the time interval between its elements $x(n)$ and $x(n + 1)$

is the same for all integers n. The z-transform associated with the time series $\{x(n)\}$ is formally defined by

$$X(z) = \sum_{n=-\infty}^{\infty} x(n)z^{-n} \quad (2.1)$$

where z is a complex-valued variable. Here we have adopted the standard convention of denoting the z-transform by the uppercase letter (i.e., X) corresponding to the time series lowercase-letter designation (i.e., x). On occasion, however, it is convenient to depict the z-transform by the equivalent operator symbol $Z\{x(n)\}$. Whatever the case, the z-transform is seen to be a mechanism whereby a time series (or discrete-time signal) is mapped into a function of the complex variable z. The value of this function at the point z is seen to be totally dependent on the time series elements $x(n)$. Hence the notation $X(z)$ for this functional relationship is indeed natural. The more mathematically oriented reader will recognize that the z-transform is the *Laurent series expansion* of the function $X(z)$ about the origin $z = 0$ in which the time series elements $x(n)$ are the Laurent series expansion coefficients.

It is often beneficial to provide another interpretation to the z-transform. That is, the entity z^{-n} may alternatively be viewed as a time marker which identifies the location of the time series element $x(n)$ that it multiplies. With this in mind, the tag z^{-4} locates the element $x(4)$, z^7 identifies the element $x(-7)$, and so on. This interpretation is of particular value when studying the response of linear operators to excitations via the z-transform.

EXAMPLE 2.1

Determine the z-transform of the simple time series

$$x(n) = 4\delta(n+1) - 2\delta(n) + \pi\delta(n-5)$$

which is seen to be identically zero for all values of n except -1, 0, and 5. In accordance with definition (2.1), the corresponding z-transform is

$$\begin{aligned} X(z) &= 4z - 2z^0 + \pi z^{-5} \\ &= 4z - 2 + \pi z^{-5} \end{aligned}$$

Two observations are worth mentioning at this point. First, this z-transform is seen to be bounded [i.e., $|X(z)| < \infty$] for all finite values of z except at $z = 0$, where the term z^{-5} becomes unbounded. Second, we are able to reconstruct the time series $\{x(n)\}$ uniquely from knowledge of its corresponding z-transform $X(z)$ by employing the aforementioned time-marker concept. For instance, the transform term πz^{-5} is seen to correspond to the time series element $x(5) = \pi$, and vice versa. ■

Although the concepts illustrated in Example 2.1 are enlightening, they are also quite deceptive because of the simplicity of the time series. When the time series is of infinite length, however, more delicate issues must be addressed and understood. In the sections that follow, we shall explore some of these importan finer points. In this analysis, extensive use of the following *geometric serie identities* is made.

Lemma 2.1. The closed-form expression for the *finite geometric series*

$$\sum_{k=0}^{n} \alpha^k = \frac{1 - \alpha^{n+1}}{1 - \alpha} \tag{2.2}$$

holds for any complex value of the parameter α. Moreover, it follows that

$$\sum_{k=0}^{\infty} \alpha^k = \frac{1}{1 - \alpha} \qquad \text{for } |\alpha| < 1 \tag{2.3}$$

We introduced these geometric series identities in Chapter 1, but they are re peated here for convenience as well as to emphasize its utility in z and Fourie transform theory.

EXAMPLE 2.2

To illustrate a typical application of the geometric series identities, let us fin the z-transform of the causal and anticausal exponential time series:

$$x_1(n) = a^n u(n)$$

$$x_2(n) = -a^n u(-n - 1)$$

where a is a generally complex-valued constant. From the z-transforms defi nition (2.1), we have

$$X_1(z) = \sum_{n=-\infty}^{\infty} a^n u(n) z^{-n} = \sum_{n=0}^{\infty} (az^{-1})^n$$

Using identity (2.3) with $\alpha = az^{-1}$, we find that

$$X_1(z) = \frac{1}{1 - az^{-1}} = \frac{z}{z - a} \qquad \text{for } |z| > |a|$$

where use of the readily established fact that $|az^{-1}| < 1$ and $|z| > |a|$ are equivalent inequalities has been made. The set of z satisfying $|z| > |a|$ is seen to consist of all complex-valued z that lie outside a circle of radius $|a|$ centered at the origin of the z-plane.

In a similar fashion, it is seen that

$$X_2(z) = \sum_{n=-\infty}^{\infty} -a^n u(-n - 1) z^{-n} = -\sum_{n=-\infty}^{-1} (a^{-1}z)^{-n}$$

We now use the standard trick of adding and subtracting the term $(a^{-1}z)^0 = 1$ so as to put this summation into form (2.3), that is,

$$X_2(z) = 1 - \sum_{n=0}^{\infty} (a^{-1}z)^n = 1 - \frac{1}{1 - a^{-1}z}$$

$$= \frac{z}{z - a} \qquad \text{for } |z| < |a|$$

where use of the equivalency $|a^{-1}z| < 1$ and $|z| < |a|$ has been made. The inequality $|z| < |a|$ is satisfied by all complex values of z that lie inside a circle of radius $|a|$ centered at the origin of the z-plane.

From a superficial comparison of the two z-transforms $X_1(z)$ and $X_2(z)$, one might incorrectly conclude that they are identical. Although they each have the same functional form of $z/(z - a)$, they are defined in different regions

TABLE 2.1 Commonly Used *z*-Transform Pairs

$x(n)$	$X(z)$	Region of Convergence
1. $\delta(n)$	1	All z
2. $u(n)$	$\dfrac{z}{z - 1}$	$\|z\| > 1$
3. $a^n u(n)$	$\dfrac{z}{z - a}$	$\|z\| > \|a\|$
4. $-a^n u(-n - 1)$	$\dfrac{z}{z - a}$	$\|z\| < \|a\|$
5. $na^n u(n)$	$\dfrac{az}{(z - a)^2}$	$\|z\| > \|a\|$
6. $-na^n u(-n)$	$\dfrac{az}{(z - a)^2}$	$\|z\| < \|a\|$
7. $\dfrac{(n + k - 1)!}{n!(k - 1)} a^n u(n)$	$\dfrac{z^k}{(z - a)^k}$	$\|z\| > \|a\|$
8. $a^{\|n\|}$	$\dfrac{a^2 - 1}{a} \dfrac{z}{(z - a)(z - 1/a)}$	$\|a\| < \|z\| < \|a\|^{-1}$
9. $\sin(n\omega)u(n)$	$\dfrac{z \sin \omega}{z^2 - 2z \cos \omega + 1}$	$\|z\| > 1$
10. $\cos(n\omega)u(n)$	$\dfrac{z(z - \cos \omega)}{z^2 - 2z \cos \omega + 1}$	$\|z > 1$
11. $a^n \sin(n\omega)u(n)$	$\dfrac{az \sin \omega}{z^2 - 2az \cos \omega + a^2}$	$\|z\| > \|a\|$
12. $a^n \cos(n\omega)u(n)$	$\dfrac{z(z - a \cos \omega)}{z^2 - 2az \cos \omega + a^2}$	$\|z\| > \|a\|$

of the complex z-plane. $X_1(z)$ is defined outside the circle of radius $|a|$ centered at the origin, and $X_2(z)$ is defined inside that circle. They each represent *different* z-transforms. This is reassuring, since the time series that gave rise to each are themselves not equal.

The process of generating a z-transform by means of evaluating the infinite summation (2.1) is conceptually straightforward. This evaluation can be tedious however, and is prone to human error. As such, it is common to appeal to a z-transform pair table such as given in Table 2.1 for many widely encountered time series. A required z-transform may often be obtained by direct lookup from this table without going through the cumbersome summation evaluation process.

2.3 REGION OF CONVERGENCE

For infinite-length time series, the value of the transform $X(z)$ at the point z is seen to be composed of an infinite summation of the terms $x(n)z^{-n}$. Since the usefulness of the z-transform will be confined to those values of z for which this summation is bounded, we shall now investigate the set of z for which $|X(z)|$ is finite in value. This set of z is formally defined by

$$R = \left\{ z: \left| \sum_{n=-\infty}^{\infty} x(n)z^{-n} \right| < \infty \right\} \tag{2.4}$$

and is referred to as the *region of convergence* for the z-transform.†

To determine this region of convergence, it is prudent to decompose the z-transform (2.1) into its *causal* and *anticausal* generated components, that is,

$$\begin{aligned} X(z) &= \sum_{n=0}^{\infty} x(n)z^{-n} + \sum_{n=-\infty}^{-1} x(n)z^{-n} \\ &= X_c(z) + X_a(z) \end{aligned} \tag{2.5}$$

where we have attached the subscripts c and a to denote the causal and anticausal components, respectively. From this decomposition it is apparent that the causal component $X_c(z)$ tends to be bounded for large magnitude values of z because of the negative exponent $-n$ to which z is raised. In a similar fashion, the anticausal component $X_a(z)$ tends to be bounded for small magnitude values of z. By using standard calculus arguments, it can be shown that the regions of absolute convergence for the causal and anticausal components must be of the form

†Expression (2.4) is to be formally read as "the set of complex values z for which the magnitude of the transform $X(z)$ is finite."

$$|X_c(z)| < \infty \qquad \text{for } |z| > r_c$$

and $$\tag{2.6}$$

$$|X_a(z)| < \infty \qquad \text{for } |z| < r_a$$

for appropriate choices of the nonnegative scalars r_c and r_a. The z-transform $X_c(z)$ of the purely causal time series component $\{x(n)u(n)\}$ is then seen to converge outside a circle of radius r_c centered at the origin of the complex z-plane as depicted in Figure 2.1a. In a similar fashion, the z-transform $X_a(z)$ of the purely anticausal time series component $\{x(n)u(-n - 1)\}$ converges inside a circle of radius r_a centered at the origin as shown in Figure 2.1b. Our ultimate objective, however, is to determine the region of convergence for the complete z-transform. From expression (2.5) it is apparent that this region is given by the set of z lying in the intersection of the regions of convergence for

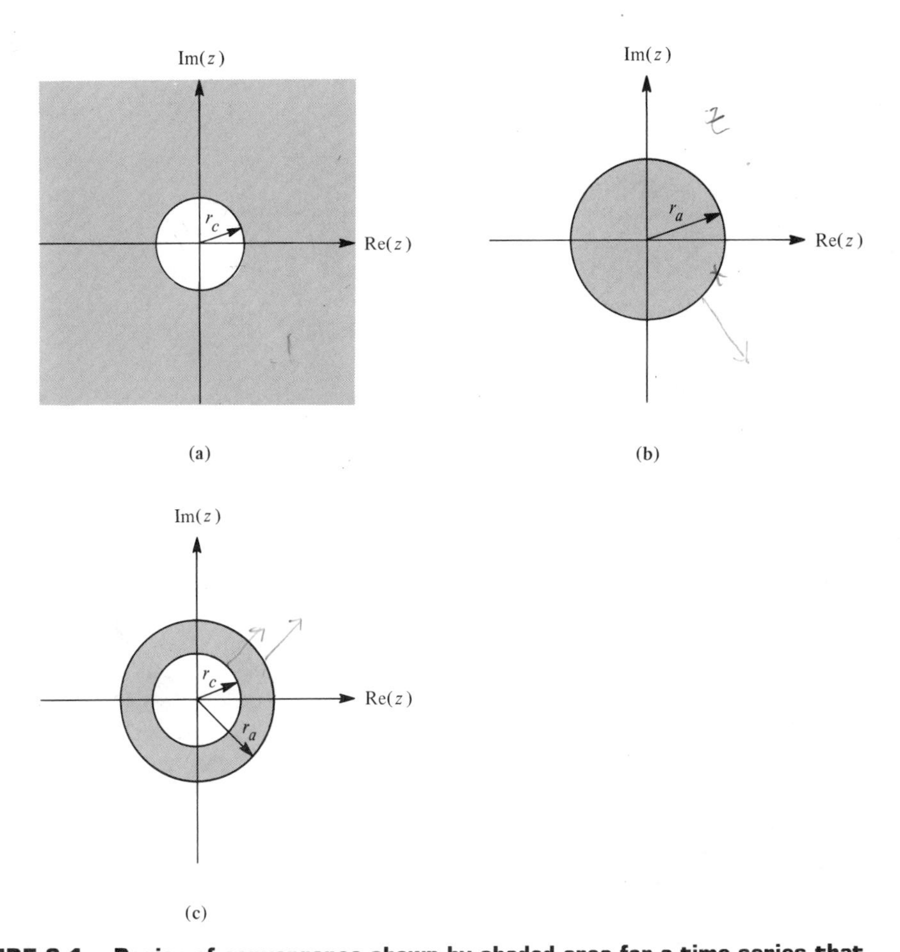

FIGURE 2.1. Region of convergence shown by shaded area for a time series that (a) is causal, (b) is anticausal, and (c) contains both a causal and an anticausal component.

the causal and anticausal z-transforms, that is,

$$R = \{z: r_c < |z| < r_a\} \tag{2.7}$$

For this region to be nonempty, it is clear that we must have $r_a > r_c$ if the time series $\{x(n)\}$ is to have a z-transform. This region of convergence is seen to be composed of an annulus centered at the origin as shown in Figure 2.1c. The inner circle's radius r_c can be as small as zero, which occurs only when the time series is strictly anticausal. Similarly, the outer circle's radius r_a can be as large as positive infinity, which implies that the time series is strictly causal in nature. If the time series being considered is strictly causal or anticausal, its associated region of convergence is as shown in Figure 2.1a or b, respectively.

EXAMPLE 2.3

Determine the z-transform of the time series

$$x(n) = (\tfrac{1}{2})^n u(n) + 3(4)^n u(-n-1)$$

and its associated region of convergence. The z-transform is formally given by

$$X(z) = \sum_{n=0}^{\infty} (\tfrac{1}{2})^n z^{-n} + \sum_{n=-\infty}^{-1} 3(4)^n z^{-n}$$

Upon combining summand terms with equal exponents, we have the causal and anticausal components.

$$X_c(z) = \sum_{n=0}^{\infty} (\tfrac{1}{2}z^{-1})^n = \frac{1}{1 - \frac{1}{2}z^{-1}} \qquad \text{for } |z| > \tfrac{1}{2}$$

and

$$X_a(z) = \sum_{n=-\infty}^{-1} 3(4z^{-1})^n = -3 + 3\sum_{n=0}^{\infty} (\tfrac{1}{4}z)^n$$

$$= -3 + 3\frac{1}{1 - \frac{1}{4}z} \qquad \text{for } |z| < 4$$

where the standard technique of adding and subtracting the $n = 0$ term in $X_a(z)$ has been applied. In arriving at these closed-form expressions for $X_c(z)$ and $X_a(z)$, we have used identity (2.3) and the inequality equivalencies

$$|\tfrac{1}{2}z^{-1}| < 1 \Leftrightarrow |z| > \tfrac{1}{2} \qquad \text{and} \qquad |\tfrac{1}{4}z| < 1 \Leftrightarrow |z| < 4$$

The required z-transform and its associated annulus region of convergence are then given by

$$X(z) = \frac{1}{1 - \frac{1}{2}z^{-1}} - 3 + 3\frac{1}{1 - \frac{1}{4}z}$$

$$= -\frac{2z(z + \frac{5}{4})}{(z - \frac{1}{2})(z - 4)} \qquad \text{for } \tfrac{1}{2} < |z| < 4$$

■

2.4 RATIONAL *z*-TRANSFORM

For an important class of time series, the associated z-transforms are expressible as a ratio of polynomials in the variable z^{-1}, that is,

$$X(z) = \frac{b_0 + b_1 z^{-1} + \cdots + b_q z^{-q}}{1 + a_1 z^{-1} + \cdots + a_m z^{-m}} \tag{2.8}$$

This transform can be converted into an equivalent ratio of polynomials in z simply by multiplying the numerator and denominator by z^m if $m \geq q$ or z^q if $q > m$. Any transform that has this ratio of polynomials structure will be referred to as a *rational z-transform*. Some of the more commonly encountered time series that possess rational z-transforms were listed in Table 2.1. Using a fundamental theorem from algebra, we may always factor each of these polynomials into their first-order product forms to obtain the equivalent representation

$$X(z) = \frac{b_0(1 - z_1 z^{-1})(1 - z_2 z^{-1}) \cdots (1 - z_q z^{-1})}{(1 - p_1 z^{-1})(1 - p_2 z^{-1}) \cdots (1 - p_m z^{-1})} \tag{2.9}$$

The parameters z_k and p_k that arise from this factorization are referred to as the *zeros* and the *poles* of the z-transform $X(z)$, respectively. It is to be noted that when the function $X(z)$ is evaluated at a z corresponding to one of its zeros (i.e., at $z = z_k$), the function itself is zero. Hence the terminology zero for the parameter z_k is indeed natural. On the other hand, the function $X(z)$ becomes unbounded when it is evaluated at a z corresponding to one of its poles (i.e., at $z = p_k$). It is beneficial to display this zero–pole information by placing tiny circles at the zero locations z_k and tiny crosses at the pole locations p_k. As shown subsequently, these locations convey valuable information concerning the intrinsic underlying nature of the time series.

EXAMPLE 2.4

In Example 2.3 it was shown that the z-transform associated with the time series

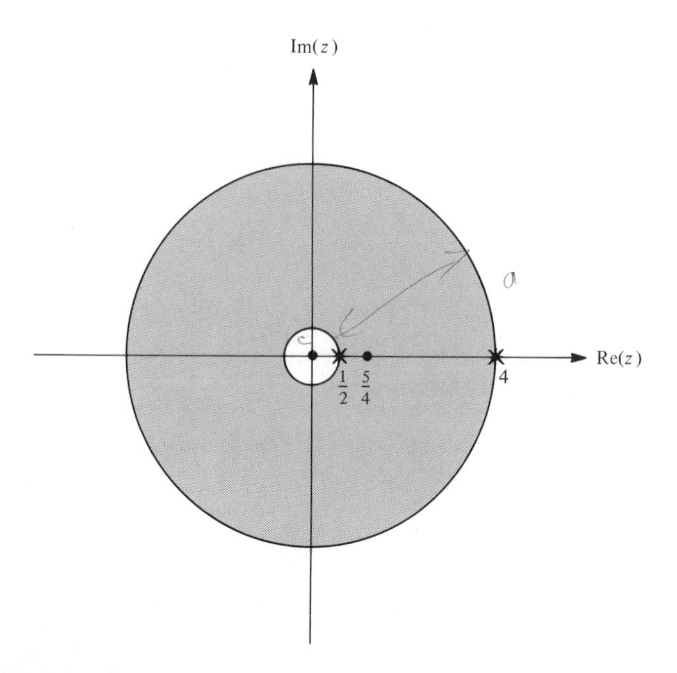

FIGURE 2.2. Region of convergence and zero–pole location for the *z*-transform of Example 2.4.

$$x(n) = (\tfrac{1}{2})^n u(n) + 3(4)^n u(-n-1)$$

was given by

$$X(z) = -\frac{2z(z + \frac{5}{4})}{(z - \frac{1}{2})(z - 4)} \qquad \text{for } \tfrac{1}{2} < |z| < 4$$

A plot of these rational z-transforms zeros (i.e., $z_1 = 0$, $z_2 = -\frac{5}{4}$) and poles (i.e., $p_1 = \frac{1}{2}$, $p_2 = 4$) is shown in Figure 2.2 together with the annulus shaped region of convergence.

It is readily established that a rational z-transform always has at least one pole located on any *boundary* separating its regions of convergence and divergence. In addition, there is never a pole located *inside* the region of convergence. These concepts were illustrated in Example 2.4, where the poles $p_1 = \frac{1}{2}$ and $p_2 = 4$ were found to lie on the inner and outer circles defining the annulus region of convergence. The following theorem characterizes the pole location of a rational z-transform.

Theorem 2.1. Let the rational z-transform $X(z)$ be decomposed into its rational causal and anticausal components

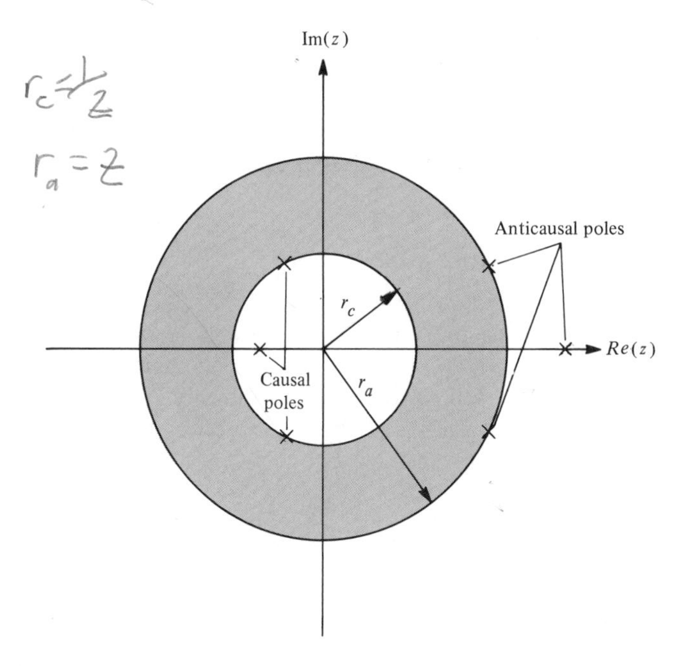

FIGURE 2.3. Causal and anticausal pole locations for a rational z-transform.

$$X(z) = X_c(z) + X_a(z) \qquad \text{for } r_c < |z| < r_a$$

It then follows that the poles of $X_c(z)$ all have magnitudes less than or equal to r_c, with at least one of these poles having a magnitude of r_c. Similarly, the poles of $X_a(z)$ all have magnitudes greater than or equal to r_a, with at least one of these poles having a magnitude of r_a.

A visual depiction of this theorem's results is given in Figure 2.3 relative to the annulus region of convergence, where all the *causal poles* are seen to lie on or inside the inner circle, while the *anticausal poles* all lie on or outside the outer circle.

2.5 PROPERTIES OF THE z-TRANSFORM

The primary reason for the rather extensive use of the z-transform in linear system and signal-processing theory arises from the many utilitarian properties that it possesses. Primary among these are the linearity, shifting, and convolution summation properties. In this section we develop these three properties in recognition of their importance to our future needs. This development is followed in turn by Table 2.2, which lists other relevant properties that are widely employed in theoretical studies and practical applications.

Linearity Property

In many signal-processing applications, a time series (or discrete-time signal) is expressible as a linear combination of other time series. For instance, let us consider the case in which the time series being examined is specified as a linear combination of two other time series, that is,

$$x(n) = aw(n) + by(n) \tag{2.10}$$

in which a and b are scalar multipliers. Let it now be desired to obtain the z-transform of the time series $\{x(n)\}$ in which this linear-combination representation is to be utilized. The desired z-transform is formally given by expression (2.1), which after substituting in the equivalent expression (2.10) for $x(n)$ becomes

$$X(z) = \sum_{n=-\infty}^{\infty} [aw(n) + by(n)]z^{-n}$$

A straightforward manipulation of this summation is seen to result in

$$X(z) = a \sum_{n=-\infty}^{\infty} w(n)z^{-n} + b \sum_{n=-\infty}^{\infty} y(n)z^{-n} \tag{2.11}$$

The terms multiplying the scalars a and b are recognized as being the z-transforms of the component time series $\{w(n)\}$ and $\{y(n)\}$, respectively. We may therefore express $X(z)$ as

$$X(z) = aW(z) + bY(z) \tag{2.12}$$

It has therefore been shown that if a time series is expressible as a linear combination of two (or finitely more) constituent time series, the z-transform of that time series is equal to the same linear combination of the constituent time series z-transforms. This property is commonly referred to as *linearity*.

When establishing the *linearity property* above, we did not address the important issue of the region of convergence. From expression (2.12) it is apparent that $X(z)$ is bounded for values of z in which $W(z)$ and $Y(z)$ are bounded simultaneously. Thus the region of convergence of $X(z)$ is at least as large as the intersection of the regions of convergence for the z-transforms $Y(z)$ and $W(z)$. It must be noted, however, that it is possible to construct examples in which the region of convergence for $X(z)$ is larger than this overlapping region. As an example, for rational z-transforms, this can occur when a pole of either $W(z)$ or $Y(z)$ that lies on a region of convergence is exactly canceled by a zero of the linear combination $aW(z) + bY(z)$.

Shifting Property

When analyzing the properties of a linear signal processing algorithm (or filter), it is often necessary to determine the z-transforms of shifted versions of the time series. Specifically, let us examine the time series characterized by

$$y(n) = x(n - m) \tag{2.13}$$

in which m is a fixed integer. The time series $\{y(n)\}$ is seen to be an m-delayed (or right-shifted) version of $\{x(n)\}$. The z-transform of this shifted time series is obtained in the following straightforward fashion:

$$Y(z) = \sum_{n=-\infty}^{\infty} [x(n - m)]z^{-n}$$

where the term in brackets is recognized as being $y(n)$. When we make the change of summation variables $k = n - m$, this transform expression becomes

$$\begin{aligned} Z\{x(n - m)\} &= \sum_{k=-\infty}^{\infty} x(k)z^{-(k+m)} \\ &= z^{-m} \sum_{k=-\infty}^{\infty} x(k)z^{-k} \\ &= z^{-m}X(z) \end{aligned} \tag{2.14}$$

where the exponential rule $z^{-(k+m)} = z^{-k}z^{-m}$ and the fact that z^{-m} is a multiplicative constant relative to the summation index k have been used in going from the first line to the second line. The shifting property conveyed in expression (2.14) simply states that the z-transform of an m-right-shifted time series is equal to the z-transform of the unshifted time series multiplied by z^{-m}. The regions of convergence of $X(z)$ and $Z\{x(n - m)\}$ are readily shown to be the same.

Convolution Summation Property

From a linear system theory viewpoint, the *convolution summation property* represents the most valuable of the z-transform properties. To appreciate why this is so, it will be recalled from Chapter 1 that the response of a (time-invariant) linear operator to the excitation $\{x(n)\}$ is governed by the convolution summation

$$y(n) = \sum_{k=-\infty}^{\infty} h(k)x(n - k) \tag{2.15}$$

in which $\{h(k)\}$ denotes the linear operator's weighting sequence. Let us now determine the z-transform associated with this response time series. Formally this transform is given by

$$Y(z) = \sum_{n=-\infty}^{\infty} \left[\sum_{k=-\infty}^{\infty} h(k)x(n-k) \right] z^{-n}$$

in which the term in brackets is recognized as being $y(n)$. When we interchange the order of summations and make the change of summation variables $p = n - k$, where k is taken as fixed, this double summation becomes

$$Y(z) = \sum_{k=-\infty}^{\infty} h(k) \sum_{p=-\infty}^{\infty} x(p)z^{-(p+k)}$$

$$= \sum_{k=-\infty}^{\infty} h(k)z^{-k} \sum_{p=-\infty}^{\infty} x(p)z^{-p}$$

The first and second summations in this relationship are recognized as being the z-transforms of the time series $\{h(n)\}$ and $\{x(n)\}$, respectively. We have therefore shown that the z-transform of the time series that arises from linearly convolving $\{h(n)\}$ with $\{x(n)\}$ is simply equal to the product of their respective z-transforms, that is,

$$Y(z) = H(z)X(z) \tag{2.16}$$

The z-transform of the signal operator's weighting sequence $\{h(n)\}$ is denoted by $H(z)$ and is commonly referred to as the operator's *transfer function*, where

$$H(z) = \sum_{n=-\infty}^{\infty} h(n)z^{-n}$$

Thus a linear operator's response has a z-transform that is equal to the product of the operator's transfer function and the z-transform of the excitation. A visual depiction of this salient relationship is shown in Figure 2.4.

This is a rather remarkable result in the sense that the relatively cumbersome time domain convolution summation operation (2.15) is replaced by the much

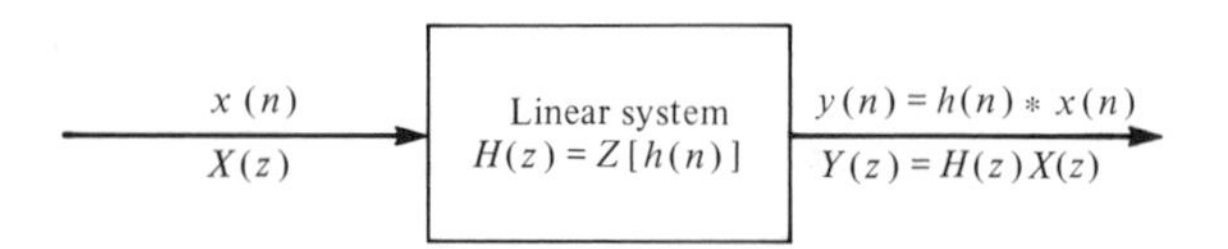

FIGURE 2.4. Transfer function relationship for linear operators.

simpler z-domain multiplication operation (2.16). It is precisely because of this feature that the analysis of linear signal operations are often best approached from the z-domain rather than the time domain. The region of convergence for the z-transform $Y(z)$ is again seen to be at least as large as the intersection of the regions of convergence associated with the z-transforms $H(z)$ and $X(z)$. As in the linearity property, however, it can happen that when a pole lying on a boundary of the region of convergence is canceled by a zero, this region of convergence can be enlarged.

EXAMPLE 2.5

Consider the causal linear operator governed by

$$y(n) = x(n) + \tfrac{1}{2}y(n-1)$$

Determine the z-transform of its response to the unit-step excitation

$$x(n) = u(n)$$

We now use the convolution summation property (2.16) to find $Y(z)$. This first necessitates determination of this operator's weighting sequence. In accordance with Section 1.10, this weighting sequence is equivalent to the operator's unit-impulse response which is readily found to be

$$h(n) = (\tfrac{1}{2})^n u(n)$$

Using the time series $\{x_1(n)\}$ treated in Example 2.2 with $a = \frac{1}{2}$ and 1, respectively, we find that the operator's transfer function is specified by

$$H(z) = \frac{z}{z - \frac{1}{2}} \qquad \text{for } |z| > \tfrac{1}{2}$$

while the excitation's z-transform is

$$X(z) = \frac{z}{z - 1} \qquad \text{for } |z| > 1$$

Thus the response's z-transform as given by expression (2.16) becomes

$$Y(z) = \frac{z^2}{(z - \frac{1}{2})(z - 1)} \qquad \text{for } |z| > 1$$

where the region of convergence in this case is given by the intersection of the regions of convergence for $H(z)$ and $X(z)$. ■

TABLE 2.2 Properties of the z-Transform

Property	Time Series	z-Transform	Region of Convergence
	$x(n)$ $y(n)$	$X(z)$ $Y(z)$	$r_{cx} < \|z\| < r_{ax}$ $r_{cy} < \|z\| < r_{ay}$
Linearity	$ax(n) + by(n)$	$aX(z) + bY(z)$	At least max $(r_{cx}, r_{cy}) < \|z\| < \min (r_{ax}, r_{ay})$
Time shift	$x(n - m)$	$z^{-m}X(z)$	$r_{cx} < \|z\| < r_{ax}$
Convolution	$\sum_{k=-\infty}^{\infty} x(k)y(n - k)$	$X(z)Y(z)$	At least max $(r_{cx}, r_{cy}) < \|z\| < \min (r_{ax}, r_{ay})$
Exponential multiplication	$a^n x(n)$	$X(a^{-1}z)$	$\|a\|r_{cx} < \|z\| < \|a\|r_{ax}$
Time multiplication	$nx(n)$	$-z \dfrac{dX(z)}{dz}$	$r_{cx} < \|z\| < r_{ax}$
Product	$x(n)y(n)$	$\dfrac{1}{2\pi j} \oint_C X(w)Y\left(\dfrac{z}{w}\right) w^{-1}\, dw$	$r_{cx}r_{cy} < \|z\| < r_{ax}r_{ay}$
Correlation	$\sum_{k=-\infty}^{\infty} x(k)y(n + k)$	$X(z^{-1})Y(z)$	$\max (r_{ax}^{-1}, r_{cy}) < \|z\| < \min (r_{cx}^{-1}, r_{ay})$
Time transpose	$x(-n)$	$X(z^{-1})$	$r_a^{-1} < \|z\| < r_c^{-1}$

Other Properties

As suggested earlier, the z-transform possesses a number of other properties which are of value in various applications. A list of some of the more important of these properties is given in Table 2.2.

2.6 CASCADING OF LINEAR OPERATORS

One of the important features related to time-invariant linear operators (or systems) is their use as subsystems in a more complex configuration. As an example, let us consider the *cascaded* connection shown in Figure 2.5a. These two linear operators (or systems) are said to be cascaded, since the response of the first system serves as the excitation to the second system. We shall henceforth

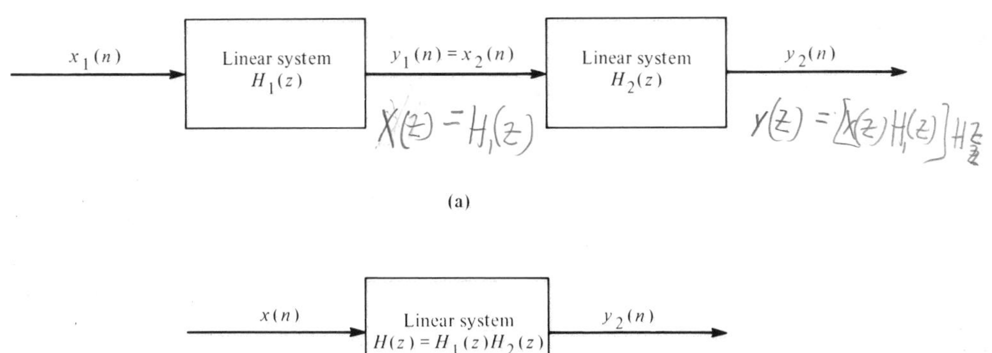

FIGURE 2.5. (a) Cascaded linear systems; (b) equivalent linear system.

use the equivalent term *operators* and *systems* interchangeably. It is readily established that this cascaded system is itself linear. Moreover, the *transform function* of this cascaded system is simply equal to the product of the transfer functions associated with the constituent linear systems; that is,

$$H(z) = H_1(z)H_2(z) \tag{2.17}$$

This equivalency is depicted in Figure 2.5b. In terms of excitation–response behavior, we are unable to differentiate between the cascaded configuration and its equivalent single system as represented by transfer function (2.17).

To establish the validity of transfer function relationship (2.17), we now make use of identity (2.16), which holds for each of the linear systems constituting the cascaded configuration; that is,

$$Y_1(z) = H_1(z)X_1(z)$$

$$Y_2(z) = H_2(z)X_2(z) = H_2(z)Y_1(z)$$

In the second expression, use has been made of the cascading condition that $x_2(n) = y_1(n)$. Upon substituting the first expression into the second, we obtain

$$Y_2(z) = H_2(z)H_1(z)X_1(z)$$

This relationship indicates that the z-transforms of the cascaded configuration's excitation and response are related in a simple multiplicative form. It follows directly from transfer function relationship (2.16) that the transfer function of the cascaded configuration must be given by identity (2.17).

2.7 INVERSE z-TRANSFORM

The beauty of the z-transform resides in its utility for obtaining results straightforwardly in the z-domain that would be much more cumbersome to come by if a strictly time-domain approach were taken. The convolution summation property provides a particularly important illustration of this principle. Since our ultimate objective is almost always to obtain a result in the time domain, however, it is apparent that the z-transform approach is useful only if a mechanism is available for recovering the time series that generates a given z-transform. This process is commonly referred to as taking an *inverse z-transform* because it reverses the operation effected by the z-transform

$$X(z) = \sum_{n=-\infty}^{\infty} x(n)z^{-n} \qquad \text{for } z \in R \tag{2.18}$$

where R denotes the z-transform's region of convergence. In this section we present the *integral inversion* and *partial-fraction expansion* methods for performing the inverse z-transform operation.

Integral Inversion Method

In theoretical studies, the integral inversion method is used extensively. Its development makes use of the Cauchy integral theorem from complex variables, which states that the contour integral

$$\frac{1}{2\pi j} \oint_C z^{n-1}\, dz = \begin{cases} 1 & \text{for } n = 0 \\ 0 & \text{for } n \neq 0 \end{cases}$$

holds for any simple counterclockwise closed path C that encloses the origin of the z-plane. To obtain the integral inversion equation, we simply multiply each side of the z-transform expression (2.18) by z^{k-1} and then integrate that product over any counterclockwise closed path C that encloses the origin and lies totally within the region of convergence R of $X(z)$. After interchanging the order of integration and summation and applying the Cauchy integral theorem, it is found that

$$x(n) = \frac{1}{2\pi j} \oint_C X(z) z^{n-1}\, dz \tag{2.19}$$

where the integer variable k has been replaced by n. This integral expression provides a mechanism for recovering the time series $\{x(n)\}$ that is associated with the given z-transform $X(z)$. When $X(z)$ is rational in nature, the evaluation of this inversion integral may be obtained using complex-variable residue theory. We shall not pursue this approach, since the partial-fraction expansion method to be now examined is of much more practical value.

Partial-Fraction Expansion Method

The partial-fraction expansion method is the most widely employed procedure for inverting a rational z-transform. It is assumed here that the z-transform to be inverted is in the rational form

$$X(z) = \frac{b_0 z^m + b_1 z^{m-1} + \cdots + b_m}{z^m + a_1 z^{m-1} + \cdots + a_m} \qquad r_c < |z| < r_a \tag{2.20}$$

which tacitly requires that the order of the numerator polynomial not be larger than that of the denominator polynomial. If this condition is not satisfied, the required form may be effected by dividing the numerator polynomial by the denominator polynomial until the remainder polynomial's order is less than or equal to m (e.g., see Example 2.6). The next step in the partial-fraction expansion method entails factoring the denominator polynomial to give the factored form

$$X(z) = \frac{b_0 z^m + b_1 z^{m-1} + \cdots + b_m}{(z - p_1)(z - p_2) \cdots (z - p_m)} \qquad \text{for } r_c < |z| < r_a \tag{2.21}$$

We next decompose this relatively complex structured z-transform into a sum of simpler z-transforms whose inverse z-transforms are known. The nature of this decomposition is dependent on whether the poles p_k obtained in the factorization above are simple (i.e., all distinct) or are multiple. With this in mind, we treat these two possibilities separately.

Distinct Poles

When the poles of the rational z-transform (2.21) are distinct, it is always possible to express this z-transform equivalently as the sum of m first-order elementary terms and a constant, that is,

$$X(z) = \frac{b_m}{a_m} + \alpha_1 \frac{z}{z - p_1} + \alpha_2 \frac{z}{z - p_2} + \cdots + \alpha_m \frac{z}{z - p_m} \qquad \text{for } r_c < |z| < r_a \tag{2.22}$$

This decomposition is commonly referred to as a *partial-fraction expansion*. The value of the general α_k coefficient appearing in this expansion is readily obtained by multiplying each side of this expression by $(z - p_k)/z$ and then setting $z = p_k$. Each term on the right side then goes to zero except the kth term, thereby indicating that

$$\alpha_k = \frac{z - p_k}{z} X(z) \bigg|_{z=p_k}$$

Upon substituting expression (2.21) into the right-hand side of this equation and canceling the common $(z - p_k)$ term appearing in the numerator and denominator, the appropriate value of α_k is specified by

$$\alpha_k = \frac{b_0(p_k)^m + b_1(p_k)^{n-1} + \cdots + b_m}{p_k \prod_{\substack{n=1 \\ \neq k}}^{m} (p_k - p_n)} \qquad \text{for } 1 \leq k \leq m \tag{2.23}$$

With the rational function $X(z)$ decomposed into the elementary first-order terms through the partial-fraction expansion (2.22), we are in a position to generate the required inverse z-transform. This makes use of the following previously established z-transform pairs:

$$1 \leftrightarrow \delta(n)$$

$$\frac{z}{z - p_k} \leftrightarrow \begin{cases} (p_k)^n u(n) & \text{for a causal time-series interpretation} \\ -(p_k)^n u(-n - 1) & \text{for an anticausal time-series interpretation} \end{cases}$$

In regard to the second z-transform pair, the question arises as to whether a causal or an anticausal interpretation is to be made on the individual elementary terms $z/(z - p_k)$. In accordance with the remarks made in Section 2.4, it is clear that the appropriate interpretation is given by

$$\begin{aligned} &\text{causal if} && |p_k| \leq r_c \\ &\text{anticausal if} && |p_k| \geq r_a \end{aligned} \tag{2.24}$$

Thus, depending on the magnitude of the pole p_k, we associate either a causal or an anticausal time series to the expansion component $z/(z - p_k)$. If we use the linearity property of the z-transform, it follows that the time series associated with the rational function $X(z)$ with partial-fraction expansion (2.22) is given by

$$x(n) = \frac{b_m}{a_m}\delta(n) + \sum_{|p_k| \leq r_c} \alpha_k (p_k)^n u(n) - \sum_{|p_k| \geq r_a} \alpha_k (p_k)^n u(-n - 1) \tag{2.25}$$

where it is noted that there are a total of m terms in the two summations. The first summation is composed of strictly causal terms, and the second summation contains strictly anticausal terms.

In summary, the partial-fraction expansion method as applied to rational z-transforms with distinct poles consists of five steps:

1. Put the z-transform into *standard form,* whereby the order of the numerator polynomial does not exceed the order of the denominator polynomial.
2. Factor the denominator polynomial.
3. Evaluate the α_k coefficients appearing in the partial-fraction expansion using expression (2.23).
4. Separate the expansion terms into causal and anticausal terms in accordance with the root magnitude criterion (2.24).
5. The required inverse z-transform is then given by expression (2.25).

To illustrate this inversion procedure, let us consider the following example.

EXAMPLE 2.6

Determine the time series whose z-transforms are given by

$$\text{(a)}\ X_1(z) = -\frac{2z^2 + 2.5z}{z^2 - 4.5z + 2} \qquad \text{for } \tfrac{1}{2} < |z| < 4$$

$$\text{(b)}\ X_2(z) = \frac{2z^4 - 5z^3}{z^2 - 1} \qquad \text{for } |z| > 1$$

(a) In this example the annulus region of convergence is specified by $r_c = \frac{1}{2}$ and $r_a = 4$ and $X_1(z)$ is in standard form. We begin the inversion process by factoring the denominator polynomial to obtain

$$X_1(z) = -\frac{2z^2 + 2.5z}{(z - \frac{1}{2})(z - 4)}$$

Since the poles are distinct, a partial-fraction expansion of form (2.22) yields

$$X_1(z) = \frac{z}{z - \frac{1}{2}} - 3\frac{z}{z - 4}$$

where it is noted that $b_m = b_2 = 0$ in this case. It is now noted that the first term's pole $p_1 = \frac{1}{2}$ is causal in nature since $|p_1| \le r_c = \frac{1}{2}$, while the second term's pole $p_2 = 4$ is anticausal, since $|p_2| \ge r_a = 4$. Thus, according to expression (2.25), the resultant inverse z-transform is given by

$$x_1(n) = (\tfrac{1}{2})^n u(n) + 3(4)^n u(-n - 1)$$

which is consistent with the results of Example 2.3.

(b) In this case the numerator polynomial's order exceeds that of the denominator, thereby necessitating long division to put $X(z)$ into the standard form, that is,

$$X_2(z) = \frac{2z^4 - 5z^3}{z^2 - 1} = 2z^2 - 5z + \frac{2z^2 - 5z}{(z - 1)(z + 1)}$$

A partial-fraction expansion of the rightmost term then gives

$$X_2(z) = 2z^2 - 5z - \frac{3}{2}\frac{z}{z - 1} + \frac{7}{2}\frac{z}{z + 1}$$

The time series corresponding to each of these elementary terms then indicate that the desired inverse z-transform is given by

$$x_2(n) = 2\delta(n + 2) - 5\delta(n + 1) - \tfrac{3}{2}u(n) + \tfrac{7}{2}(-1)^n u(n)$$

in which a causal interpretation to the two rightmost terms has been made because the region of convergence is specified by $|z| > 1$. ■

Multiple Poles

In the case where the rational transform $X(z)$ has multiple poles, a simple modification to the distinct-pole procedure is required. Specifically, suppose that this factorization produces the result

$$X(z) = \frac{b_0 z^m + b_1 z^{m-1} + \cdots + b_m}{(z - p_1)^{m_1}(z - p_2)^{m_2} \cdots (z - p_q)^{m_q}} \quad \text{for } r_c < |z| < r_a \tag{2.26}$$

where the kth pole p_k has multiplicity m_k in which $m_1 + m_2 + \cdots + m_q = m$. In the multiple-pole case, at least one of the integers m_k is greater than 1. For such situations, the required partial-fraction expansion is of the form

$$\begin{aligned} X(z) = \frac{b_m}{a_m} &+ \alpha_1 \frac{z}{z - p_1} + \alpha_2 \frac{z^2}{(z - p_1)^2} + \cdots + \alpha_{m_1} \frac{z^{m_1}}{(z - p_1)^{m_1}} \\ &+ \beta_1 \frac{z}{z - p_2} + \beta_2 \frac{z^2}{(z - p_2)^2} + \cdots + \beta_{m_2} \frac{z^{m_2}}{(z - p_2)^{m_2}} \\ &+ \cdots + \xi_1 \frac{z}{z - p_q} + \xi_2 \frac{z^2}{(z - p_q)^2} + \cdots + \xi_{m_q} \frac{z^{m_q}}{(z - p_q)^{m_q}} \end{aligned} \tag{2.27}$$

where the $\alpha_k, \beta_k, \ldots, \xi_k$ coefficients may be evaluated by using any one of several techniques. It is important to note and appreciate the fact that the expansion contain $m + 1$ elementary terms and associated coefficients where m corresponds to the order of $X(z)$'s denominator polynomial. The basic steps of this partial-fraction expansion method are outlined in Table 2.3.

TABLE 2.3 Steps of Partial-Fraction Expansion Method

1. Put the z-transform into *standard form*, whereby the numerator's polynomial order is less than or equal to that of the denominator's order.
2. Factor the denominator polynomial of $X(z)$ to determine its roots, some of which may be multiple.
3. Make a partial-fraction expansion of $X(z)$ of form (2.22) in the distinct-pole case or of form (2.27) for the multiple-pole case.
4. Use the z-transform pair Table 2.4 to determine the associated component inverse z-transforms. A causal selection is to be made for all $|p_k| \leq r_c$, while an anticausal selection is taken for all $|p_k| \geq r_a$.

EXAMPLE 2.7

Using the partial-fraction expansion method, evaluate the inverse z-transform of

$$X(z) = \frac{-2z^3 + \frac{38}{3}z^2 + 6z - 18}{z^3 - \frac{17}{3}z^2 + 7z + 3} \qquad \text{for } \frac{1}{3} < |z| < 3$$

where the annulus region of convergence is specified by $r_c = \frac{1}{3}$ and $r_a = 3$. Factoring the denominator polynomial yields

$$X(z) = \frac{-2z^3 + \frac{38}{3}z^2 + 6z - 18}{(z + \frac{1}{3})(z - 3)^2}$$

Since a pole of multiplicity 2 exists at $z = 3$, we must use the partial-fraction expansion form (2.27), which yields

$$X(z) = -6 + 5\frac{z}{z + \frac{1}{3}} - 3\frac{z}{z - 3} + 2\frac{z^2}{(z - 3)^2}$$

TABLE 2.4 z-Transform Elementary Pairs Used in the Partial-Fraction Expansion Method

Elementary Terms	Causal Selection	Anticausal Selection
$\frac{z}{z-a}$	$a^n u(n)$	$-a^n u(-n-1)$
$\frac{z^2}{(z-a)^2}$	$(n+1)a^n u(n)$	$-(n+1)a^n u(-n-2)$
$\frac{z^3}{(z-a)^3}$	$\frac{(n+1)(n+2)a^n}{2!}u(n)$	$-\frac{(n+1)(n+2)}{2!}a^n u(-n-3)$
$\frac{z^k}{(z-a)^k}$	$\frac{(n+k-1)!}{n!(k-1)!}a^k u(n)$	$-\frac{(-1)^k(-n-1)!}{(-n-k)!(k-1)!}a^n u(-n-k)$

Finally, the elementary term with pole $p_1 = -\frac{1}{3}$ is given a causal interpretation, since $|p_1| \leq r_c = \frac{1}{3}$ and the elementary terms with poles at $p_2 = 3$ are given anticausal interpretations, since $|p_2| \geq r_a = 3$. If we use Table 2.4, it then follows that the inverse z-transform is specified by

$$\begin{aligned} x(n) &= -6\delta(n) + 5(-\tfrac{1}{3})^n u(n) + 3(3)^n u(-n - 1) \\ &\qquad - 2(n + 1)(3)^n u(-n - 1) \\ &= -6\delta(n) + 5(-\tfrac{1}{3})^n u(n) + [1 - 2n](3)^n u(-n - 1) \end{aligned}$$ ■

EXAMPLE 2.8

Some of the entries of Table 2.4 appear rather complicated and the reader has probably wondered how one goes about obtaining such z-transform pairs. Typically, it is necessary to employ specialized techniques that are directed toward the particular z-transform being examined. To illustrate this point, let us now consider the kth order exponential entry four of Table 2.4, which shall be designated by

$$X_k(z) = \frac{z^k}{(z - a)^k} \qquad \text{for } k \geq 0$$

where $\{x_k(n)\} = Z^1\{X_k(z)\}$. From this relationship, the following z-domain recursive relationship between successive ordered exponentials is obtained

$$X_k(z) = \frac{z}{z - a} X_{k-1}(z) \qquad \text{for } k \geq 1$$

After multiplying each side of this expression by $(z - a)$ and then taking an inverse z-transform, the following time-domain recursive expression is obtained

$$x_k(n + 1) - ax_k(n) = x_{k-1}(n) \qquad \text{for } k \geq 1$$

We now use this time domain recursion to obtain the desired closed form expression for the signal $\{x_k(n)\}$ when a causal interpretation of $X_k(z)$ is made. A similar approach can be taken for an anticausal interpretation. Specifically, we begin by setting $k = 1$ and using the trivial identity $x_0(n) = a(n)$. Under these conditions, an evaluation of this recursive equation as n is incremented through the integer values 0, 1, 2, 3, . . . is found ato yield the first-order sequence $x_1(n) = a^n u(n)$. In a similar fashion, setting $k = 2$ is found to result in $x_2(n) = (n + 1)a^n u(n)$ and then letting $k = 3$ yields $x_3(n) = 0.5(n + 1)$ $(n = 2)a^n u(n)$. Continuing in this manner for increasing values of k, we are led to hypothesize the causal time domain formula given in column 2 of entry four. To verify the correctness of this conjecture, we simply substitute this hypothesized solution into the above time-domain recursion and find that it satisfies this relationship. ■

2.8

TRANSFER FUNCTION OF A LINEAR RECURSIVE OPERATOR

In signal-processing applications, linear recursive operators are used extensively. With this in mind, it is quite natural to inquire into the possible usefulness of z-transform theory for characterizing and understanding this class of relevant linear operators. In this section it is found that a straightforward procedure exists for writing down directly the transfer function associated with the recursive operator.

$$\sum_{k=0}^{p} a_k y(n-k) = \sum_{k=0}^{q} b_k x(n-k) \tag{2.28}$$

To achieve this transfer function characterization, we first multiply each side of equality (2.28) by z^{-n} and then sum these equalities over the range $-\infty < n < \infty$ to give

$$\sum_{n=-\infty}^{\infty} \left[\sum_{k=0}^{p} a_k y(n-k) \right] z^{-n} = \sum_{n=-\infty}^{\infty} \left[\sum_{k=0}^{q} b_k x(n-k) \right] z^{-n}$$

Upon interchanging the order of summation operations relative to the indices k and n, we have

$$\sum_{k=0}^{p} a_k \left[\sum_{n=-\infty}^{\infty} y(n-k) z^{-n} \right] = \sum_{k=0}^{q} b_k \left[\sum_{n=-\infty}^{\infty} x(n-k) z^{-n} \right]$$

Next, we use the shifting property (2.14) and then factor out the common terms $Y(z)$ and $X(z)$ to conclude that

$$\left[\sum_{k=0}^{p} a_k z^{-k} \right] Y(z) = \left[\sum_{k=0}^{q} b_k z^{-k} \right] X(z)$$

Finally, dividing each side of this expression by the polynomial multiplying $Y(z)$, we obtain the response–excitation relationship

$$Y(z) = H(z)X(z) \tag{2.29}$$

where the *transfer function* $H(z)$ is specified by

$$H(z) = \frac{b_0 + b_1 z^{-1} + \cdots + b_q z^{-q}}{a_0 + a_1 z^{-1} + \cdots + a_p z^{-p}} \tag{2.30}$$

From expression (2.29) it is seen that the linear recursive operator's response and excitation z-transforms are interrelated through a simple multiplicative relationship. The *rational transfer function* $H(z)$ that characterizes this multiplication relationship is seen to depend exclusively on the parameters governing the linear recursive operator (2.28). Given the parameters b_k and a_k associated with the linear recursive operator, it is then a trivial matter to construct the associated transfer function (2.30).

Upon comparing expression (2.29) with the convolution summation property (2.16), it is apparent that the linear recursive operator's transfer function must be equal to the z-transform of that operator's associated weighting sequence, that is,

$$Z\{h(n)\} = \frac{b_0 + b_1 z^{-1} + \cdots + b_q z^{-q}}{a_0 + a_1 z^{-1} + \cdots + a_p z^{-p}} \tag{2.31}$$

In Section 2.7, procedures for obtaining the time series that corresponds to a given z-transform were developed. With this ability we may use relationship (2.31) to obtain a convenient *closed-form* representation for linear recursive operator's (2.28) weighting sequence $\{h(n)\}$. This capability is generally not readily achievable by using the unit-impulse response procedure discussed earlier. Of equal importance, the transfer function relationship (2.29) may be used to compute in closed form a linear recursive operator's response to any rational-type excitation.

EXAMPLE 2.9

Determine the transfer function and weighting sequence associated with the linear recursive operator

$$y(n) - \tfrac{1}{2}y(n-1) = x(n)$$

Upon using expression (2.30), the required transfer function is given directly by

$$H(z) = \frac{1}{1 - \frac{1}{2}z^{-1}} = \frac{z}{z - \frac{1}{2}}$$

This is the same result as that obtained in Example 2.5 using a much more cumbersome approach. If a causal interpretation is given to this z-transform, the resulting stable weighting sequence of the corresponding causal system is

$$h(n) = (\tfrac{1}{2})^n u(n)$$

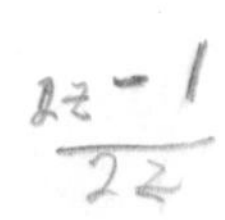

On the other hand, an anticausal interpretation would yield the unstable weighting sequence of the associated anticausal system

$$h(n) = -(\tfrac{1}{2})^n u(-n - 1)$$ ■

Stability of a Linear Recursive Operator

The stability characteristics of a recursive linear operator are directly obtainable upon factoring the denominator polynomial of the transfer function (2.30). In particular, let this factorization be given by

$$A(z) = \sum_{k=0}^{p} a_k z^{-k} = a_0 \prod_{k=1}^{p} (1 - p_k z^{-1})$$

For the same reasoning as that used in Section 2.7, the following stability theorem results.

Theorem 2.2. The linear recursive operator (2.28) will be stable if and only if its transfer function (2.30) as decomposed by $H(z) = H_c(z) + H_a(z)$ has all causal poles of magnitude less than 1 and all anticausal poles of magnitude greater than 1.

When a recursive operator is unstable, its response to bounded excitation invariably results in an unbounded response. The following example illustrates this behavior.

EXAMPLE 2.10

For the linear operator characterized by

$$y(n) - \tfrac{1}{2}y(n - 1) = x(n)$$

determine its response to the unit-step excitation $x(n) = u(n)$ when this operator is taken to be (a) causal; (b) anticausal.

(a) For a causal operator interpretation, the prevailing recursive expression is to be read $y(n) = x(n) + \frac{1}{2}y(n - 1)$. The z-transform of the resulting unit-step response is then given by expression (2.29), that is,

$$Y(z) = \frac{z}{z - \frac{1}{2}} \frac{z}{z - 1}$$

$$= \frac{z}{z - 1} - \frac{z}{z - \frac{1}{2}} \qquad \text{for } |z| > 1$$

The inverse z-transform is then

$$y(n) = u(n) - (\tfrac{1}{2})^n u(n)$$

(b) When the operator is taken to be anticausal, the prevailing recursive expression is to be read $y(n - 1) = -2x(n) + 2y(n)$. The z-transform of the resulting unit-step response is then

$$Y(z) = \frac{z}{z - \frac{1}{2}} \frac{z}{z - 1} \qquad \text{for } 1 < |z| < \tfrac{1}{2}$$

However, the region of convergence is seen to be empty, indicating that the response's z-transform does not exist. This simply reflects the fact that the unit-step response of the unstable anticausal linear operator is unbounded for all values of time n. ■

EXAMPLE 2.11

Consider the stable mixed causal–anticausal operator whose transfer function is specified by

$$H(z) = \frac{-z^2 + 5z}{(z - \frac{1}{2})(z + 4)}$$

Write the recursive relationships that describe this operator. Upon making a partial-fraction expansion of this transfer function, we find

$$H(z) = \frac{z}{z - \frac{1}{2}} - 2\frac{z}{z + 4}$$

Since the operator is required to be stable, Theorem 2.2 indicates that the first term of this expansion must be causal in nature, while the second term is anticausal. Separating the operator response into its causal and anticausal components yields

$$\begin{aligned} Y(z) &= Y_c(z) + Y_a(z) = H(z)X(z) \\ &= [H_c(z) + H_a(z)]X(z) \end{aligned}$$

We therefore have

$$Y_c(z) = \frac{z}{z - \frac{1}{2}} X(z) \qquad \text{and} \qquad Y_a(z) = -\frac{2z}{z + 4} X(z)$$

Taking the inverse z-transforms of these expressions and writing them in their causal and anticausal forms gives

$$y_c(n) = x(n) + \tfrac{1}{2}y_c(n - 1)$$

and

$$y_a(n + 1) = -2x(n + 1) - 4y_a(n)$$

Given a specific excitation, we then compute the causal and anticausal responses separately using these recursive expressions. The total operator's response is then formed according to

$$y(n) = y_c(n) + y_a(n)$$ ■

2.9 EXPONENTIAL SIGNAL ANNIHILATION

One of the more fundamental concepts in signal processing is that of *signal annihilation*. In such diverse applications as spectral estimation, modeling, and, speech processing, the underlining requirement is that of identifying a nontrivial linear system that annihilates a given set of data. To introduce the concept of annihilation, let us consider the following sum of complex exponential signals:

$$x(n) = \sum_{k=1}^{m} \alpha_k (p_k)^n u(n) \tag{2.32}$$

in which the p_k elements may be complex valued. It is to be noted that signals which have real-valued sinusoidal components can be so represented.

It is now desired to synthesize a linear nonrecursive operator which, when excited by the exponential type signal (2.32), produces a response that is zero for all but a finite transient period. A linear operator that achieves this objective is said to have *annihilated* the excitation signal. With this in mind, let us now identify the weighting coefficients of the nonrecursive linear operator

$$y(n) = \sum_{k=0}^{q} h(k)x(n - k) \tag{2.33}$$

so that it annihilates the given exponential signal. To characterize the requirements on the $h(n)$ elements, it is beneficial to use the corresponding transfer function relationship

$$Y(z) = H(z)X(z) \tag{2.34}$$

in which the system transfer function may be put into the factored form

$$H(z) = \sum_{n=0}^{q} h(n)z^{-n}$$

$$= h(0) \prod_{k=1}^{q} (1 - z_k z^{-1}) \tag{2.35}$$

To compute the response using relationship (2.34), we must first generate the sum of exponential signal's (2.32) z-transform. Using the third entry of the z-transform table (Table 2.1) and the linearity of the z-transform operator, we obtain

$$X(z) = \sum_{k=1}^{m} \alpha_k \frac{1}{1 - p_k z^{-1}}$$

$$= \frac{\beta_0 + \beta_1 z^{-1} + \cdots + \beta_{m-1} z^{-m+1}}{(1 - p_1 z^{-1})(1 - p_2 z^{-1}) \cdots (1 - p_m z^{-1})} \tag{2.36}$$

where the β_k coefficients are a by-product of putting $X(z)$ into a common denominator format. Substitution of these z-transforms into response relationship (2.34) results in the revealing expression

$$Y(z) = h(0) \prod_{k=1}^{q} (1 - z_k z^{-1}) \frac{\beta_0 + \beta_1 z^{-1} + \cdots + \beta_{m-1} z^{-m+1}}{\prod_{i=1}^{m} (1 - p_i z^{-1})} \tag{2.37}$$

To obtain the response's time-domain behavior, we simply take the inverse z-transform of this expression. Using the partial-fraction expansion inversion method, in general, will lead to terms of the form $(p_i)^n u(n)$ in the response. For signal annihilation, however, these terms cannot be present, for otherwise the response signal never goes to zero. Fortunately, examination of expression (2.37) reveals a straightforward means for annihilating these exponential terms: namely, zeros of the system transfer function should be chosen so as to cancel the poles of the excitation's z-transform. The simplest such transfer function is specified by

$$H_{\min}(z) = \prod_{k=1}^{m} (1 - p_k z^{-1}) \tag{2.38}$$

whereby $h_{\min}(0) = 1$ and the system order q has been set equal to m. For this choice of $H(z)$, the corresponding response's z-transform corresponds to the numerator of expression (2.37) after the pole–zero cancellation has been carried out on $H_{\min}(z)X(z)$. In the time domain, the resultant finite-length response is then

$$y(n) = \begin{cases} \beta_n & 0 \le n \le m - 1 \\ 0 & \text{otherwise} \end{cases}$$

Thus, after the transient response phase $0 \le n \le m - 1$, the response is identically zero and the required annihilation has been achieved.

Although system transfer function (2.38) characterizes the simplest annihilating linear operator, it is clear that any all-zero transfer function which has m of its zeros located at p_k for $1 \le k \le m$ will have the same annihilating effect. The transfer function of such a general annihilating operator must therefore be of the form $H_{min}(z)G(z)$, where $G(z)$ is an arbitrary finite polynomial in the variable z^{-1}. When applying the annihilating concept to signal-processing problems, there are good reasons for using an annihilating operator $H_{min}(z)G(z)$ that is not of minimal order.

EXAMPLE 2.12

Determine the minimal-order annihilating operator associated with the general damped sinusoid signal

$$x(n) = \alpha(r)^n \cos(\omega_0 n + \theta)u(n)$$

where α, r, ω_0, and θ are real-valued parameters. To synthesize this operator, it is first necessary to represent this signal as a sum of exponentials. The Euler identity achieves this, whereby

$$x(n) = \frac{\alpha}{2}[e^{j\theta}(re^{j\omega_0})^n + e^{-j\theta}(re^{-j\omega_0})^n]u(n)$$

from which the corresponding z-transform poles are seen to be $p_1 = re^{j\omega_0}$ and $p_2 = re^{-j\omega_0}$. In accordance with expression (2.38), the minimal-order annihilating operator has the transfer function

$$\begin{aligned} H(z) &= (1 - re^{j\omega_0}z^{-1})(1 - re^{-j\omega_0}z^{-1}) \\ &= 1 - (2r\cos\omega_0)z^{-1} + r^2z^{-2} \end{aligned}$$

and is therefore governed by the nonrecursive operation

$$y(n) = x(n) - (2r\cos\omega_0)x(n-1) + r^2x(n-2)$$

We may straightforwardly show that the response of this nonrecursive operator to the given damped sinusoid is the length 2 annihilated sequence

$$y(n) = [\alpha\cos(\theta)]\delta(n) - [\alpha r\cos(\omega_0 - \theta)]\delta(n-1)$$ ■

2.10 SUMMARY

We have examined some of the more fundamental issues involved in using the z-transform method of analysis. Of particular note was the concept of the transform's existence. This led in a natural manner to the notion of regions of convergence in the complex plane. In general, if a time series has a z-transform, the corresponding region of convergence will always be an annulus centered at the origin. Another fundamental development here concerned the inverse z-transform, which was shown to involve a contour integration in the complex z-plane. For rational z-transforms, however, the inverse z-transform is better carried out by using the partial-fraction expansion method.

SUGGESTED READINGS

AHMED, N., AND T. NATARAJAN, *Discrete-Time Signals and Systems*. Reston, Va Reston Publishing Co. Inc., 1983.

BOWEN, B. A., AND W. R. BROWN, *VLSI Systems Design for Digital Signal Processing* Vol. 1. Englewood Cliffs, N.J.: Prentice-Hall, Inc., 1982.

CHEN, C. T., *One-Dimensional Digital Signal Processing*. New York: Marcel Dekker Inc., 1979.

GABEL, R. A., AND R. A. ROBERTS, *Signals and Linear Systems,* 2nd ed. New York John Wiley & Sons, Inc., 1980.

HAMMING, R. W., *Digital Filters,* Englewood Cliffs, N.J.: Prentice-Hall, Inc., 1977.

JACKSON, L. B., *Digital Filters and Signal Processing,* Hingham, Mass.: Kluwer Academic Publishers, 1985.

OPPENHEIM, A. V., AND R. W. SCHAFER, *Digital Signal Processing*. Englewood Cliffs N.J.: Prentice-Hall, Inc., 1975.

PAPOULIS, A., *Signal Analysis*. New York: McGraw-Hill Book Company, 1977.

PELED, A. P., AND B. LIU, *Digital Signal Processing*. New York: John Wiley & Sons Inc., 1976.

STANLEY, W. D., G. R. DOUGHERTY, AND R. DOUGHERTY, *Digital Signal Processing* 2nd ed. Reston, Va.: Reston Publishing Co., Inc., 1984.

STEARNS, S. D., *Digital Signal Analysis*. Rochelle Park, N.J.: Hayden Book Company Inc., 1975.

TRETTER, S. A., *Discrete-Time Signal Processing*. New York: John Wiley & Sons, Inc. 1976.

PROBLEMS

2.1 Determine the z-transforms and the corresponding regions of convergence for the following time series.

(a) $(\frac{1}{3})^n u(n-1)$ **(b)** $(2)^n u(-n)$
(c) $na^n u(n)$ **(d)** $-na^n u(-n-1)$
(e) $\cos(\omega_0 n)u(n)$ **(f)** $\delta(n-5)$
(g) $u(n) - u(n-9)$ **(h)** $e^{j\omega_0 n}u(n)$
(i) $e^{j\omega_0 n}u(-n-1)$ **(j)** $a^n\delta(n-6)$

2.2 The following two *window* sequences of length N are commonly used in the discrete Fourier transform. Determine their z-transforms, sketch their associated regions of convergence, and locate their zeros and poles.

(a) Rectangular window: $w_r(n) = u(n) - u(n - N)$

(b) Bartlett window: $w_b(n) = \begin{cases} n + 1 & 0 \le n \le [(N - 1)/2] \\ N - n & [(N - 1)/2] < n \le N - 1 \\ 0 & \text{otherwise} \end{cases}$

where $[(N - 1)/2]$ denotes the integer part of $(N - 1)/2$.

2.3 Determine the z-transforms, sketch the regions of convergence, and locate the zeros and poles for the following time series.

(a) $2\left(\dfrac{1 + j}{4}\right)^n u(n) - 5(2 - 3j)^n u(-n - 1) + 5\delta(n)$

(b) $(4)^n u(n) - 3(-3)^n u(-n - 1)$

(c) $-7(0.7)^n u(n) + 6(7)^n u(-n - 1)$

2.4 Show that the z-transform of the time series $x(n) = x_1(n) + x_2(n)$, where

$$x_1(n) = 2u(n) \qquad \text{and} \qquad x_2(n) = -2u(n - N)$$

will have a region of convergence that is greater than the intersection of the regions of convergence for $X_1(z)$ and $X_2(z)$. Why is this so?

2.5 Show that the region of convergence for the sequence $\{y(n)\}$ generated by the convolution summation

$$y(n) = \sum_{k=-\infty}^{\infty} h(k)x(n - k)$$

in which

$$h(n) = 2\delta(n) - (\tfrac{1}{4})^n u(n) \qquad \text{and} \qquad x(n) = (\tfrac{1}{2})^n u(n)$$

is greater than the region of convergence for the intersection of the regions of convergence for $H(z)$ and $X(z)$. Why is this so? To understand these results fully, evaluate $y(n)$ directly using the convolution summation.

2.6 Let us now generalize the results of Example 2.5. In particular, let $\{h(n)\}$ and $\{x(n)\}$ be two causal time series with rational transforms given by

$$H(z) = \alpha \frac{\prod_{k=1}^{q_1} (1 - z_k z^{-1})}{\prod_{k=1}^{p_1} (1 - p_k z^{-1})} \qquad X(z) = \beta \frac{\prod_{k=1}^{q_2} (1 - \tilde{z}_k z^{-1})}{\prod_{k=1}^{p_2} (1 - \tilde{p}_k z^{-1})}$$

In what way may the convolved z-transform $Y(z) = H(z)X(z)$ have a region of convergence larger than the intersection of the regions of convergence for $H(z)$ and $Y(z)$?

2.7 Prove that the z-transform possesses the following properties, in which $X(z) = Z\{x(n)\}$ and $Y(z) = Z\{y(n)\}$.

(a) $Z\{a^n x(n)\} = X(a^{-1}z)$

(b) $Z\{nx(n)\} = -z\dfrac{dX(z)}{dz}$

(c) $Z\{x(n)y(n)\} = \dfrac{1}{2\pi j}\oint X(w)Y\left(\dfrac{z}{w}\right)w^{-1}\,dw$

(d) $Z\left\{\sum_{k=-\infty}^{\infty} x(k)y(n+k)\right\} = X(z^{-1})Y(z)$

(e) $Z\{x(-n)\} = X\left(\dfrac{1}{z}\right)$

2.8 Let $x(n)$ be a symmetric time series of length N as governed by

$$x(n) = \begin{cases} x(N-n-1) & \text{for } 0 \le n \le N-1 \\ 0 & \text{otherwise} \end{cases}$$

Show that the zeros of $X(z)$ occur in reciprocal pairs [i.e., if $X(z_0) = 0$, then $X(z_0^{-1}) = 0$]. Furthermore, if the time series is real, the roots will occur in complex-conjugate reciprocal sets.

2.9 Let the time series $\{x(n)\}$ have a rational z-transform $X(z)$ with finite zeros at z_k for $1 \le k \le q$ and poles at p_k for $1 \le k \le p$. Where are the zeros and poles of the related time series $\{(-1)^n x(n)\}$ located?

2.10 If $\{x(n)\}$ is a causal time series whose associated z-transform is given by $X(z)$, prove the initial value theorem, which states that $x(0) = \lim_{z\to\infty} X(z)$.

2.11 If the time series $\{y(n)\}$ has its elements specified by

$$y(n) = \sum_{k=-\infty}^{n} x(k)$$

where $\{x(n)\}$ is any associated time series, show that

$$Y(z) = \frac{1}{1 - z^{-1}} X(z)$$

Hint: Use the fact that $y(n) - y(n-1) = x(n)$.

2.12 Using the Cauchy integral theorem (specified in Section 2.7), give a detailed proof of the inverse z-transform contour integral relationship (2.19).

2.13 Determine the time series that corresponds to the following z-transforms when the underlying z-transform is taken to be (1) causal; (2) anticausal.

(a) $\dfrac{1 + z}{z^2 + \frac{5}{2}z - \frac{3}{2}}$

(b) $\dfrac{2 + 4z}{z^2 - \frac{3}{2}z + 1}$

(c) $\dfrac{1 - 4z^{-1}}{z^3 - \frac{7}{3}z^2 + \frac{5}{3}z - \frac{1}{3}}$

2.14 Determine the time series that correspond to the following z-transforms.

(a) $\dfrac{z^2}{(z - 1)(z + 1)}$ for $|z| < 1$

(b) $\dfrac{z^2}{(z - 1)(z + 1)}$ for $|z| > 1$

(c) $\dfrac{z(z - 3)}{(z - 1)(z - 4)}$ for $1 < |z| < 4$

(d) $\dfrac{z + 1}{(z + 1)^2(z + \frac{1}{2})}$ for $\frac{1}{2} < |z| < 1$

2.15 Determine the inverse z-transform associated with the following z-transforms.

(a) $\dfrac{z^4 - 2z^2 - 1}{z^2 - z + \frac{1}{4}}$ $|z| > \frac{1}{2}$

(b) $\dfrac{z^4 - 2z^2 - 1}{z^2 - z + \frac{1}{4}}$ $|z| < \frac{1}{2}$

(c) $\dfrac{5z^3 - 3z}{z^2 - z + 1}$ $|z| > 1$

2.16 Determine the transfer function of the following linear operators and their associated weighting sequences.

(a) $y(n) - y(n - 1) + \frac{1}{4}y(n - 2) = x(n + 2) - 2x(n) - x(n - 2)$ in which an essentially *causal*-type operation is intended.

(b) The same operator as in part (a), in which an essentially anticausal-type operation is intended.

(c) $y(n) - y(n - 1) + y(n - 2) = 5x(n + 1) - 3x(n - 1)$ where a basically causal-type operation is intended.

2.17 Determine the unit-step and unit-ramp [i.e., $nu(n)$] responses of the causal linear operators

(a) $y(n) + \frac{5}{2}y(n - 1) - \frac{3}{2}y(n - 2) = x(n - 1) + x(n - 2)$

(b) $y(n) - \frac{3}{2}y(n - 1) + y(n - 2) = 4x(n - 1) + 2x(n - 2)$

using the z-transform approach.

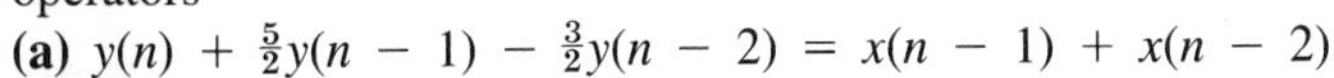

2.18 Given the transfer function

$$H(z) = \frac{z^3 - 2z + 4}{z^3 + 2z^2 + 5z - 3}$$

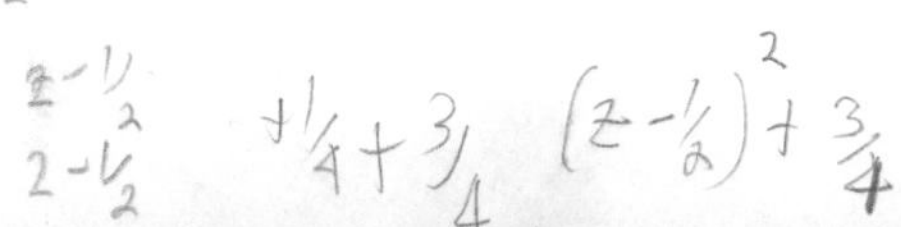

write the linear recursive operator expression associated with **(a)** a causal and **(b)** an anticausal interpretation.

2.19 Determine the minimal-order annihilating operator associated with the exponential-type sequences
(a) $x(n) = [3(\frac{1}{4})^n - 2(-\frac{1}{2})^n]u(n)$
(b) $x(n) = [1 - 2\sin(\pi n/4)]u(n)$
Compute the associated response of the annihilating operator.

2.20 Show that the operator

$$y(n) = x(n) - \tfrac{3}{4}x(n-1) + \tfrac{1}{8}x(n-2)$$

annihilates the excitation $x(n) = (\frac{1}{4})^n u(n)$. Explain why this is so.

2.21 Carry out the details outlined in Example 2.8 to show the validity of entry four of Table 2.4 for the causal interpretation of $z^k/(z - a)^k$.

2.22 Repeat Problem 2.21 for an anticausal interpretation of $z^k/(z - a)^k$.

CHAPTER 3

Fourier Transform

3.1 INTRODUCTION

The intrinsic nature of a time series is often best displayed in the *frequency* (or *spectral*) domain rather than in the more natural time domain. An invaluable tool for this purpose is the *Fourier transform,* a formal procedure for representing a time series as a continuum of sinusoidal components. In applying this approach, a time series $\{x(n)\}$ is mapped into a function $X(e^{j\omega})$ of the real-valued radian frequency variable ω. The behavior of the Fourier transform $X(e^{j\omega})$ as a function of ω often leads to significantly simpler explanations of phenomenon than would be otherwise possible in a time-domain setting. Without doubt, the Fourier transform plays a dominant role in contemporary signal processing.

In this chapter we first explore some of the more important theoretical features of the Fourier transform as related to signal processing. We next direct our attention to more practical considerations. Of particular note is a treatment of the important application in which only a finite set of time series observations is available for effecting a frequency-domain characterization. This situation describes virtually all practical data-processing applications. The *truncated Fourier transform* is used to achieve the required spectral characterization, in which data outside the observation window are assumed to be identically zero. To evaluate the truncated Fourier transform numerically, the concept of the *discrete Fourier transform* (DFT) and the associated *fast Fourier transform* (FFT) algorithm are introduced and examined in the next chapter.

3.2 FOURIER TRANSFORM

The discrete-time Fourier transform is related in a simple fashion to its more general z-transform counterpart. In particular, the discrete-time *Fourier transform* of the time series $\{x(n)\}$ is formally defined by

$$X(e^{j\omega}) = \sum_{n=-\infty}^{\infty} x(n)e^{-j\omega n} \tag{3.1}$$

where $j = \sqrt{-1}$ and ω is the real-valued *frequency variable* measured in radians. It is tacitly assumed that the time spacing between the elements $x(n)$ and $x(n+1)$ is the same for all integers n. The Fourier transform will be designated alternatively by the operator notation $\mathcal{F}\{x(n)\}$. This transformation provides a systematic procedure for mapping the time series $\{x(n)\}$ into a function of the frequency variable ω. In what follows it is shown that the behavior of this Fourier transform provides insightful interpretations of the effective frequency content of a time series. In turn, this interpretation provides a meaningful basis for introducing the concept of frequency discrimination filters, as is done in Chapters 5 and 6. We now describe some salient properties of the Fourier transform.

Periodicity

From relationship (3.1) it is apparent that the Fourier transform is in general a complex-valued function of ω that is periodic with period 2π. That is, the Fourier transform of any time series satisfies the periodic relationship

$$X(e^{j(\omega+2\pi)}) = X(e^{j\omega}) \tag{3.2}$$

for all values of the frequency variable. This property is a consequence of the appearance of the terms $e^{j\omega n}$ (complex sinusoids) in the summand of the Fourier transform expression. Since each of these terms satisfies a relationship of the form (3.2), the periodicity of $X(e^{j\omega})$ directly follows. Due to this periodicity the frequency interval

$$-\pi < \omega \leq \pi \tag{3.3}$$

is often referred to as the set of *primary digital frequencies*.

Magnitude and Phase Functions

Since the Fourier transform is a complex-valued function of ω, it is often beneficial to represent $X(e^{j\omega})$ in polar form. This representation is expressible as

$$X(e^{j\omega}) = |X(e^{j\omega})|e^{j\phi(\omega)} \tag{3.4}$$

where $|X(e^{j\omega})|$ and $\phi(\omega)$ are real-valued functions of ω referred to as the Fourier transform's *magnitude* and *phase functions,* respectively. A desirable feature of this polar representation arises from the capability thus derived of depicting the behavior of the Fourier transform by plots of the magnitude and phase functions versus ω. A typical plot of this pair of functions would appear as shown in Figure 3.1. It is to be noted that each of these functions is also periodic, so that one need only plot their behavior over the fundamental set of digital frequencies $-\pi < \omega \leq \pi$. A time series under Fourier analysis is said to be rich (or weak)

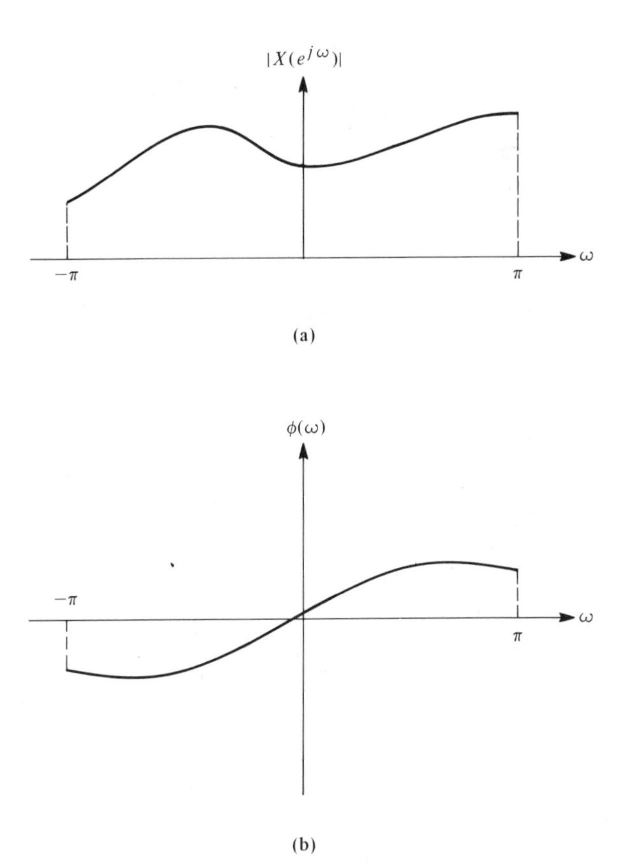

FIGURE 3.1. Sketch of Fourier transform's magnitude and phase functions in the primary digital frequency range $-\pi < \omega \leq \pi$.

in spectral content for those frequencies in which its Fourier transform's magnitude $|X(e^{j\omega})|$ takes on relatively large (small) values.

Real-Valued Time Series

When the time series under analysis is real valued, the magnitude and phase functions of the associated Fourier transform possess revealing symmetric properties. These properties are made evident upon inserting the Euler identity for $e^{-j\omega n}$ into expression (3.1) to obtain

$$\begin{aligned} X(e^{j\omega}) &= \sum_{n=-\infty}^{\infty} x(n)\cos(\omega n) + j\sum_{n=-\infty}^{\infty} -x(n)\sin(\omega n) \\ &= X_R(e^{j\omega}) + jX_I(e^{j\omega}) \end{aligned}$$

The components $X_R(e^{j\omega})$ and $X_I(e^{j\omega})$ of this Fourier transform are real-valued functions of ω due to the assumed real valuedness of the terms $x(n)$.† Moreover, the *real component* $X_R(e^{j\omega})$ is an even function of ω due to the even behavior of $\cos(\omega n)$. Similarly, the *imaginary component* $X_I(e^{j\omega})$ is an odd function of ω because of the summand terms $\sin(\omega n)$. From these properties it therefore follows that the magnitude function

$$|X(e^{j\omega})| = \sqrt{X_R^2(e^{j\omega}) + X_I^2(e^{j\omega})} = |X(e^{-j\omega})| \tag{3.5}$$

is an even function of ω, while the phase function

$$\phi(\omega) = \tan^{-1}\frac{X_I(e^{j\omega})}{X_R(e^{j\omega})} = -\phi(-\omega) \tag{3.6}$$

is an odd function of ω. This implies that the behavior of the Fourier transform of a real-valued time series may be described completely by plots of $|X(e^{j\omega})|$ and $\phi(\omega)$ versus ω in the frequency interval $0 \le \omega \le \pi$.

EXAMPLE 3.1

Determine the Fourier transform of the causal exponential time series

$$x(n) = a^n u(n)$$

where a is a fixed scalar. After substituting this time series into the Fourier transform expression (3.1), we have

†If the time series $\{x(n)\}$ is complex valued, the associated functions $X_R(e^{j\omega})$ and $X_I(e^{j\omega})$ will generally be complex-valued functions of ω.

$$X(e^{j\omega}) = \sum_{n=-\infty}^{\infty} a^n u(n) e^{-j\omega n} = \sum_{n=0}^{\infty} (ae^{-j\omega})^n$$

This summation is seen to be a geometric series and from identity (2.3). It is therefore concluded that for $|ae^{-j\omega}| = |a| < 1$, this summation simplifies to the convenient closed form

$$X(e^{j\omega}) = \frac{1}{1 - ae^{-j\omega}} \qquad \text{for } |a| < 1$$

On the other hand, the given time series does not have a Fourier transform if $|a| > 1$. It is readily shown that if a is real with $-1 < a < 1$, the magnitude and phase functions associated with this Fourier transform are given by

$$|X(e^{j\omega})| = \frac{1}{\sqrt{1 + a^2 - 2a\cos\omega}}$$

$$\phi(\omega) = \tan^{-1}\left(\frac{a\sin\omega}{1 - a\cos\omega}\right)$$

From these expressions it is seen that $|X(e^{j\omega})|$ and $\phi(\omega)$ are even and odd functions of ω, respectively. If the parameter a is complex, however, this symmetrical behavior does not follow. ■

3.3 RELATIONSHIP TO THE z-TRANSFORM

It has probably occurred to the reader that the Fourier and z-transforms share remarkably similar structures. As alluded to earlier, the Fourier transform is a special case of the z-transform. This is made apparent by evaluating the z-transform relationship (2.1) at the value $z = e^{j\omega}$, that is,

$$X(e^{j\omega}) = X(z)\big|_{z=e^{j\omega}} = \sum_{n=-\infty}^{\infty} x(n)e^{-j\omega n} \tag{3.7}$$

Thus the Fourier transform is seen to correspond to an evaluation of the z-transform on the circle of radius 1 centered at the origin of the complex z-plane (i.e., the unit circle). This follows because the unit circle is completely described by the relationship $z = e^{j\omega}$ as the frequency variable spans the primary digital frequency range $-\pi < \omega \leq \pi$.

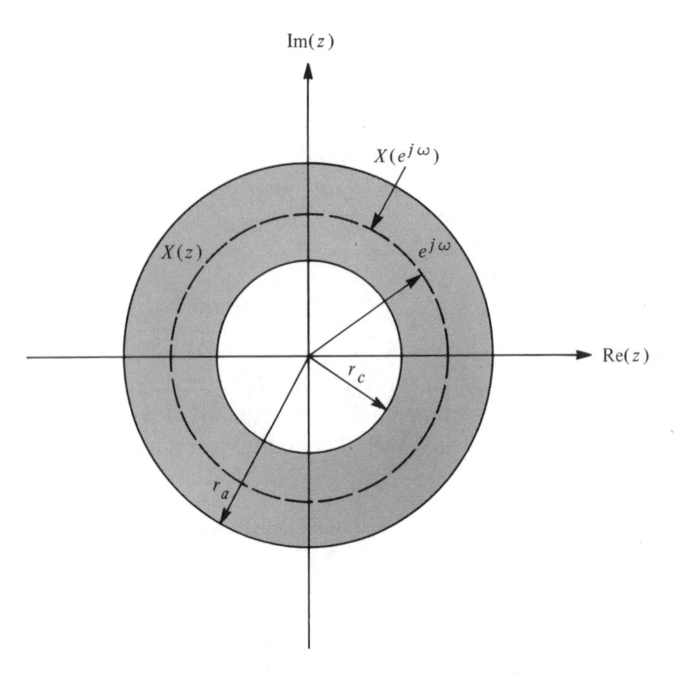

FIGURE 3.2. z-Transform's region of convergence is shown shaded; the Fourier transform is evaluated on the unit circle.

The foregoing interpretation of the Fourier transform is depicted in Figure 3.2. A number of fundamental by-products arise from this transformation linkage. From a practical viewpoint, the most important consequence is the ability thus derived to readily generate Fourier transforms of relevant time series through the use of available z-transform tables. That is, if a time series $\{x(n)\}$ has a z-transform whose region of convergence includes the unit circle, then the associated Fourier transform is obtained simply by replacing z by $e^{j\omega}$ everywhere it appears in $X(z)$. To illustrate this procedure, let us consider the exponential time series $\{n\, a^n u(n)\}$ whose associated z-transform is entry 5 of Table 2.1. The corresponding Fourier transform is then specified by

$$\mathcal{F}\{na^n u(n)\} = \left.\frac{az}{(z - a)^2}\right|_{z=e^{j\omega}}$$

$$= \frac{ae^{j\omega}}{(e^{j\omega} - a)^2} \qquad |a| < 1$$

where the script letter $\mathcal{F}$ has been used as a shorthand notation for the Fourier transform operation as defined by expression (3.1). It has been tacitly assumed that $|a| < 1$, which ensures that the region of convergence for $X(z)$ (i.e., $|z| > |a|$) includes the unit circle as required.

TABLE 3.1 Commonly Used Fourier Transform Pairs

	Time Series	Fourier Transform, $-\pi < \omega \leq \pi$
1.	$\delta(n)$	1
2.	$u(n)$	$\dfrac{1}{1 - e^{-j\omega}} + \pi\delta(\omega)$
3.	1	$2\pi\delta(\omega)$
4.	$a^n u(n), \quad \lvert a \rvert < 1$	$\dfrac{1}{1 - ae^{-j\omega}}$
5.	$\dfrac{(n + k - 1)!}{n!(k - 1)!} a^n u(n), \quad \lvert a \rvert < 1$	$\dfrac{1}{(1 - ae^{-j\omega})^k}$
6.	$\cos(\omega_1 n), \quad \lvert \omega_1 \rvert < \pi$	$\pi[\delta(\omega - \omega_1) + \delta(\omega + \omega_1)]$
7.	$\sin(\omega_1 n), \quad \lvert \omega_1 \rvert < \pi$	$j\pi[\delta(\omega + \omega_1) - \delta(\omega - \omega_1)]$
8.	$e^{j\omega_1 n}, \quad \lvert \omega_1 \rvert < \pi$	$2\pi\delta(\omega - \omega_1)$
9.	$\dfrac{\sin(\omega_1 n)}{\pi n}, \quad 0 < \omega_1 < \pi$	$u(\omega + \omega_1) - u(\omega - \omega_1)$
10.	$u(n + N) - u(n - N)$	$\dfrac{\sin[\omega(N + 0.5)]}{\sin(\omega/2)}$

Using the foregoing philosophy, we can directly generate a table of Fourier transform pairs from an associated z-transform pair table. This approach has the obvious attractiveness of avoiding the unnecessary duplicative effort of evaluating Fourier transforms through direct use of relationship (3.1). Moreover, it reinforces in the user's mind the close relationship that exists between these two important transformation tools. Using this philosophy and the z-transform pair table (Table 2.1), we can directly construct the Fourier transform pair table (Table 3.1).

3.4 PROPERTIES OF THE FOURIER TRANSFORM

Since the Fourier transform is a special case of the more general z-transform, it directly follows that all the properties possessed by the latter transform are also shared by the former. With this in mind, it is unnecessary to reestablish Fourier transform properties that have already been proved for the z-transform. Using this philosophy, we find that many of the Fourier transform properties listed in Table 3.2 follow immediately from those listed in Table 2.2. It is tacitly assumed here that the time series that appear each have z-transforms whose regions of convergence include the unit circle.

TABLE 3.2 Properties of the Fourier Transform

Property	Signal	Fourier Transform
Linearity	$ax(n) + by(n)$	$aX(e^{j\omega}) + bY(e^{j\omega})$
Time shift	$x(n - m)$	$e^{-j\omega m}X(e^{j\omega})$
Convolution	$\sum_{k=-\infty}^{\infty} x(k)y(n - k)$	$X(e^{j\omega})Y(e^{j\omega})$
Exponential multiplication	$a^n x(n), \quad r_{cx} < \lvert a\rvert^{-1} < r_{ax}$	$X(a^{-1}e^{j\omega})$
Frequency shift	$x(n)e^{j\omega_1 n}$	$X(e^{j(\omega-\omega_1)})$
Time multiplication	$nx(n)$	$-z \left.\frac{dX(z)}{dz}\right\rvert_{z=e^{j\omega}}$
Product	$x(n)y(n)$	$\frac{1}{2\pi}\int_{-\pi}^{\pi} X(e^{j\theta})Y(e^{j\theta}Y(e^{j(\omega-\theta)})\, d\theta$
Correlation	$\sum_{k=-\infty}^{\infty} x(k)y(n + k)$	$X(e^{-j\omega})Y(e^{j\omega})$
Time transpose	$x(-n)$	$X(e^{-j\omega})$

To illustrate a typical procedure for proving these properties directly, let us consider the case in which the time series $\{x(n)\}$ and $\{y(n)\}$ are related by

$$y(n) = x(n)e^{j\omega_1 n}$$

It should be noted that a relationship of this form is fundamental to many communication applications as well as to frequency-discrimination filter synthesis. The Fourier transform of $\{y(n)\}$ is formally given by

$$Y(e^{j\omega}) = \sum_{n=-\infty}^{\infty} x(n)e^{j\omega_1 n}e^{-j\omega n}$$

$$= \sum_{n=-\infty}^{\infty} x(n)e^{-j(\omega-\omega_1)n}$$

Upon examination of this expression, the right side is recognized as being of the Fourier transform form in which the frequency variable is replaced by $\omega - \omega_1$. We have therefore proved the following useful *frequency-shifting* property:

$$\mathcal{F}\{x(n)e^{j\omega_1 n}\} = X[e^{j(\omega-\omega_1)}]$$

which is entry 5 of Table 3.2. Thus multiplication in the time domain by $e^{j\omega_1 n}$ is seen to produce a frequency shift of ω_1 radians in the frequency domain. This feature is put to use in Chapter 5, where a prototype low-pass filter is used to generate bandpass filters through the simple mechanism of a time-domain multiplication.

3.5 INVERSE FOURIER TRANSFORM

It is possible, using elementary manipulations, to recover the time series that generates a given Fourier transform function. Specifically, upon multiplying each side of the Fourier transform expression (3.1) by the term $e^{j\omega k}$ and then integrating this product for $-\pi < \omega \leq \pi$, it is eventually found that

$$x(n) = \frac{1}{2\pi} \int_{-\pi}^{\pi} X(e^{j\omega}) e^{j\omega n} d\omega \tag{3.8}$$

This integral expression constitutes the *inverse Fourier transform,* which in conjunction with relationship (3.1) enables us to convert from a time-domain representation to a frequency-domain representation and back again. As a matter of fact, these Fourier transform pair relationships have another useful interpretation. The Fourier transform (3.1) can also be thought of as being a Fourier series representation of the periodic function $X(e^{j\omega})$, in which the time series elements $x(n)$ are the Fourier coefficients. These coefficients are then obtained through the inverse Fourier integral (3.8).

EXAMPLE 3.2

To illustrate the ability of the Fourier transform to identify signals that are rich in spectral content, let us examine the pure real sinusoidal time series

$$x(n) = A \cos (\omega_0 n + \theta)$$

where A, ω_0, and θ are real parameters that specify the sinusoid's amplitude, frequency, and phase, respectively. This time series does not have a Fourier transform in the formal sense that the infinite summation (3.1) converges. It directly follows, however, that the frequency function composed of *Dirac delta functions*†

$$X(e^{j\omega}) = A\pi[e^{j\theta}\delta(\omega - \omega_0) + e^{-j\theta}\delta(\omega + \omega_0)] \qquad \text{for } -\pi < \omega \leq \pi$$

†The continuous-time *Dirac delta function* $\delta(x)$ is used extensively in system theory. It satisfies the integral sampling property

$$\int_{-\pi}^{\pi} f(x)\delta(x - x_0)\, dx = f(x_0)$$

if $f(x)$ is continuous at x_0. The Dirac delta function $\delta(x - x_0)$ may be visualized as a function with unit area that is everywhere zero except at $x = x_0$, where it is unbounded.

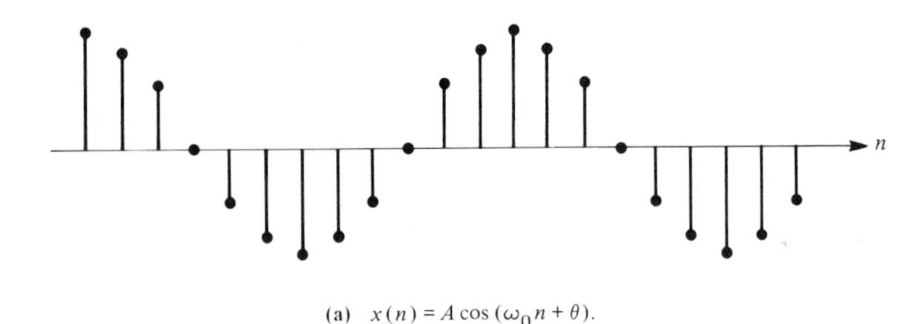

(a) $x(n) = A\cos(\omega_0 n + \theta)$.

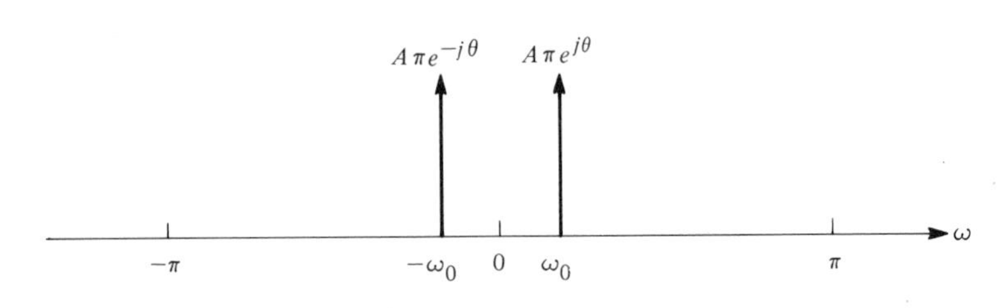

(b) $X(e^{j\omega}) = A\pi\,[e^{j\theta}\delta(\omega - \omega_0) + e^{-j\theta}\delta(\omega + \omega_0)]$ for $-\pi < \omega \leq \pi$.

FIGURE 3.3. Fourier transform pairs: (a) pure sinusoidal time series: (b) associ ated Fourier transform consisting of Dirac delta impulses.

when substituted into the inverse Fourier transform integral (3.8) will reproduce the foregoing pure sinusoidal signal. The function $X(e^{j\omega})$ is therefore the Fourier transform of this pure sinusoidal signal and can be visualized as being zero everywhere in $(-\pi, \pi]$ except at the two discrete frequencies $\pm\omega_0$, where its amplitude is unbounded. A sketch of this time series and its associated Fourier transform is depicted in Figure 3.3, where it is apparent that the frequency (or spectral) plot is of a much simpler nature than the time-domain plot. For such *narrow-band* time series, there is much to be gained by making a frequency-domain analysis through the Fourier transform. ■

3.6 FREQUENCY RESPONSE OF LINEAR OPERATORS

Although the Fourier transform provides an appealing mechanism for measurin the spectral content of a time series, it is also useful in characterizing the fre quency-discrimination capabilities of linear time-invariant operators. This is mad apparent through the convolution summation property (2.16), which after sub stituting $z = e^{j\omega}$ indicates that

$$Y(e^{j\omega}) = H(e^{j\omega})X(e^{j\omega}) \tag{3.9}$$

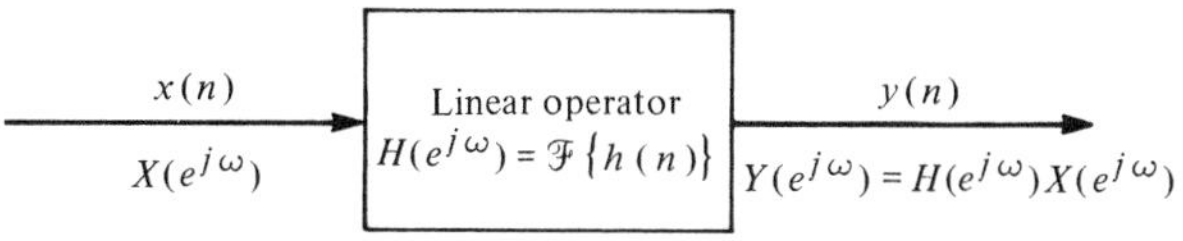

$x(n)$ →	Linear operator	$y(n)$ →
$X(e^{j\omega})$	$H(e^{j\omega}) = \mathcal{F}\{h(n)\}$	$Y(e^{j\omega}) = H(e^{j\omega})X(e^{j\omega})$

FIGURE 3.4. Linear operator's excitation-response behavior in the frequency domain.

provided that the regions of convergence for both $H(z)$ and $X(z)$ contain the unit circle. That is, the Fourier transform of a linear operator's response is equal to the product of the excitation and the operator's unit-impulse response Fourier transforms. This dynamical process is depicted in Figure 3.4. The entity $H(e^{j\omega})$ is referred to as the linear operator's *frequency response* and is specified formally by

$$H(e^{j\omega}) = \sum_{n=-\infty}^{\infty} h(n)e^{-j\omega n} \tag{3.10}$$

By prudently selecting the behavior of the linear operator's frequency response, we may therefore accentuate or deemphasize selected spectral components of the excitation time series. Deterministic procedures for designing such prototype digital filter operators will be developed in Chapters 5 and 6. In particular, prototype ideal low-pass, high-pass, and bandpass digital filters will have the frequency response depicted in Figure 3.5. Any spectral component of the excitation that lies within the ideal filter's passband where $|H(e^{j\omega})| = 1$ (stopband where $H(e^{j\omega}) = 0$) is perfectly transmitted (rejected) by the filter.

EXAMPLE 3.3

The classical ideal low-pass digital filter shown in Figure 3.5a has a frequency response that is given by

$$H_{lp}(e^{j\omega}) = \begin{cases} 1 & |\omega| \leq \omega_c \\ 0 & \omega_c < |\omega| < \pi \end{cases}$$

where ω_c is the filter's *cutoff frequency*. The unit-impulse response associated with this ideal filter is obtained by appealing to the inverse Fourier transform expression (3.8), that is,

$$h_{lp}(n) = \frac{1}{2\pi}\int_{-\omega_c}^{\omega_c} e^{j\omega n}\, d\omega = \frac{\sin(\omega_c n)}{\pi n}$$

This unit-impulse response is seen to be neither strictly causal or anticausal. In most applications it is required to have a strictly causal filter operation. Methods for designing causal filters that approximate the ideal low-pass behavior above will be examined in Chapters 5 and 6. ■

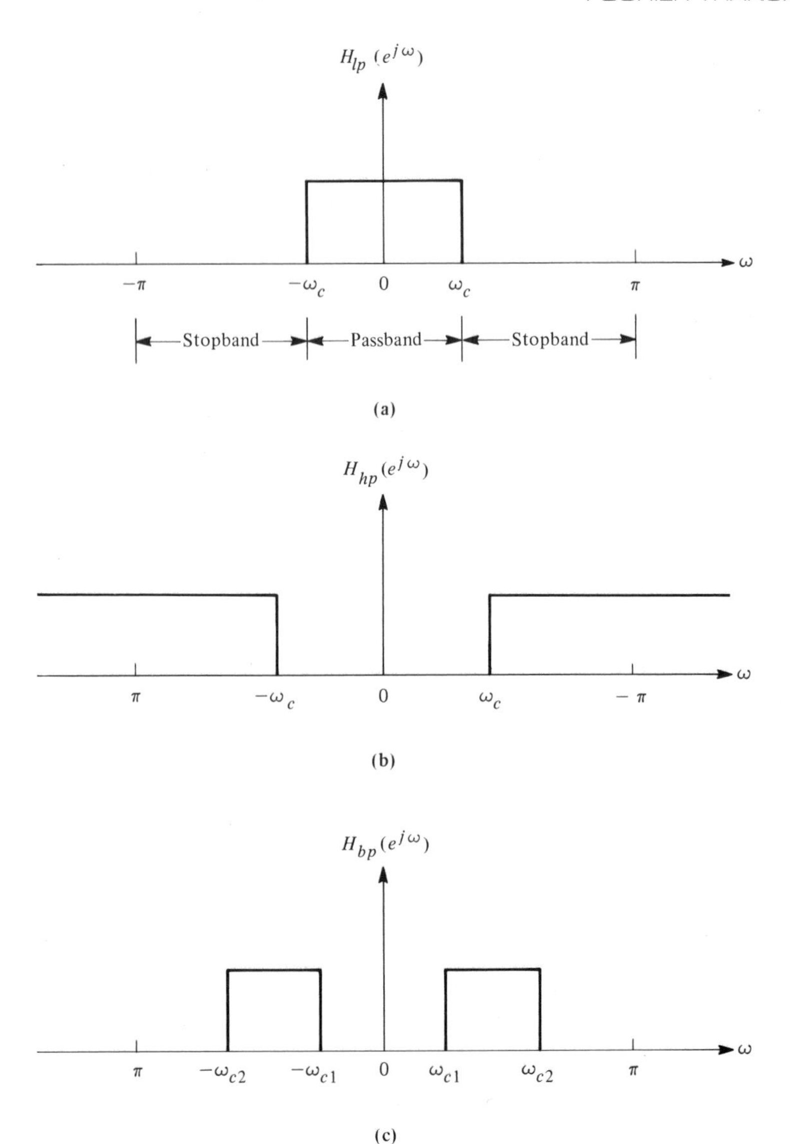

FIGURE 3.5. Ideal prototype digital filters: (a) low-pass; (b) high-pass; (c) band-pass.

Typical Filter Application

To illustrate these filtering concepts, let us consider a time series $\{x(n)\}$ that is composed of a signal plus noise component. Furthermore let the Fourier transform of $X(e^{j\omega})$ be as depicted in Figure 3.6. Moreover, suppose that the useful information (i.e., the signal) in the time series is confined to the frequency interval $-\omega_c \leq \omega \leq \omega_c$, while the higher-frequency components constitute

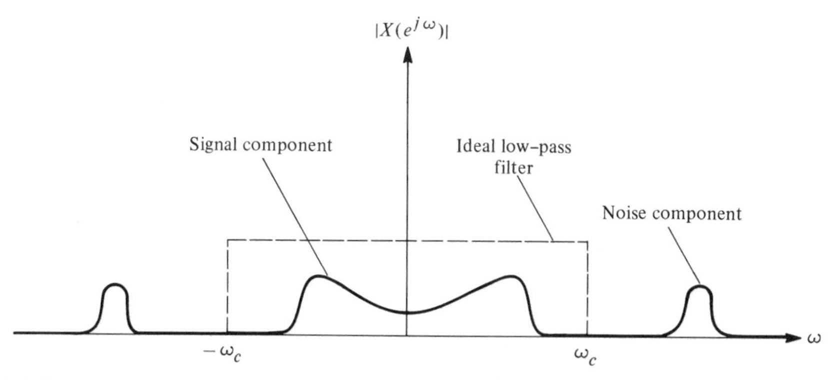

FIGURE 3.6. Application of ideal low-pass filter for frequency-discrimination filtering.

unuseful information (i.e., the noise) and lies outside this interval. Upon passing $\{x(n)\}$ through a low-pass filter with cutoff frequency ω_c, the useful signal component is transmitted without distortion, while the noise component is filtered out (i.e., eliminated). This picture captures the very spirit of frequency-discrimination filtering and is applicable in those situations where the signal and noise components have Fourier transforms that are essentially nonzero over distinct frequency intervals. We shall explore further in Chapters 5 and 6 the fundamental idea of frequency-discrimination filtering.

3.7 FREQUENCY RESPONSE DETERMINATION: GEOMETRICAL PROCEDURE

It has been shown in Chapter 2 that when a stable system with transfer function $H(z)$ is driven by the sinusoidal input

$$x(n) = \sin(n\omega + \theta_0) \qquad \text{for } n = 0, 1, 2, \ldots$$

the resultant steady-state response is given by

$$y(n) = |H(e^{j\omega})| \sin[n\omega + \theta + \phi(\omega)]$$

where $\phi(\omega) = \text{angle}\,\{H(e^{j\omega})\}$. Hence the steady-state sinusoidal response is readily obtained by evaluating the system's transfer function at $z = e^{j\omega}$, where ω is the radian frequency of the input sinusoid. The entity $H(e^{j\omega})$ was referred to in the preceding section as the *frequency response* associated with the linear operator. As we now show, a linear system's frequency response behavior as a function of ω may be determined in a straightforward manner by employing geometrical reasoning. One of the primary advantages accrued in using this geometrical approach is that of establishing a relationship between the location of the transfer function's zeros and poles and its magnitude and phase behavior. This will readily enable us to design a system whose magnitude factor approximates any desired frequency discrimination behavior.

Frequency Response: First-Order System

We shall first demonstrate this procedure by using a simple first-order system and then expand the method to more general systems. In particular, let us consider the causal first-order system

$$y(n) = b_0 x(n) + b_1 x(n - 1) + p_1 y(n - 1)$$

where the coefficients b_0, b_1, and p_1 are real, with p_1 having a magnitude of less than 1 for stability purposes. The transfer function of this system is seen to be

$$H(z) = b_0 \frac{z - z_1}{z - p_1}$$

in which $z_1 = -b_1/b_0$. The frequency response of this system is obtained by substituting $e^{j\omega}$ for z, giving

$$H(e^{j\omega}) = b_0 \frac{e^{j\omega} - z_1}{e^{j\omega} - p_1}$$
$$= b_0 \frac{\alpha_1(\omega)}{\beta_1(\omega)}$$

in which the complex-valued functions of ω

$$\alpha_1(\omega) = e^{j\omega} - z_1 \qquad \text{and} \qquad \beta_1(\omega) = e^{j\omega} - p_1$$

have been introduced for notational simplification. If we use standard complex number properties, the magnitude and angle function corresponding to this frequency response $H(e^{j\omega})$ are then given by

$$|H(e^{j\omega})| = |b_0| \cdot \frac{|\alpha_1(\omega)|}{|\beta_1(\omega)|} \tag{3.11a}$$

and

$$\text{angle}\,\{H(e^{j\omega})\} = \theta_1(\omega) - \phi_1(\omega) + \text{angle}\,\{b_0\} \tag{3.11b}$$

where $\theta_1(\omega)$ and $\phi_1(\omega)$ are the angles associated with the polar representations of the complex quantities $\alpha_1(\omega)$ and $\beta_1(\omega)$, respectively. The entity angle $\{b_0\}$ is zero if b_0 is positive and is π if b_0 is negative.

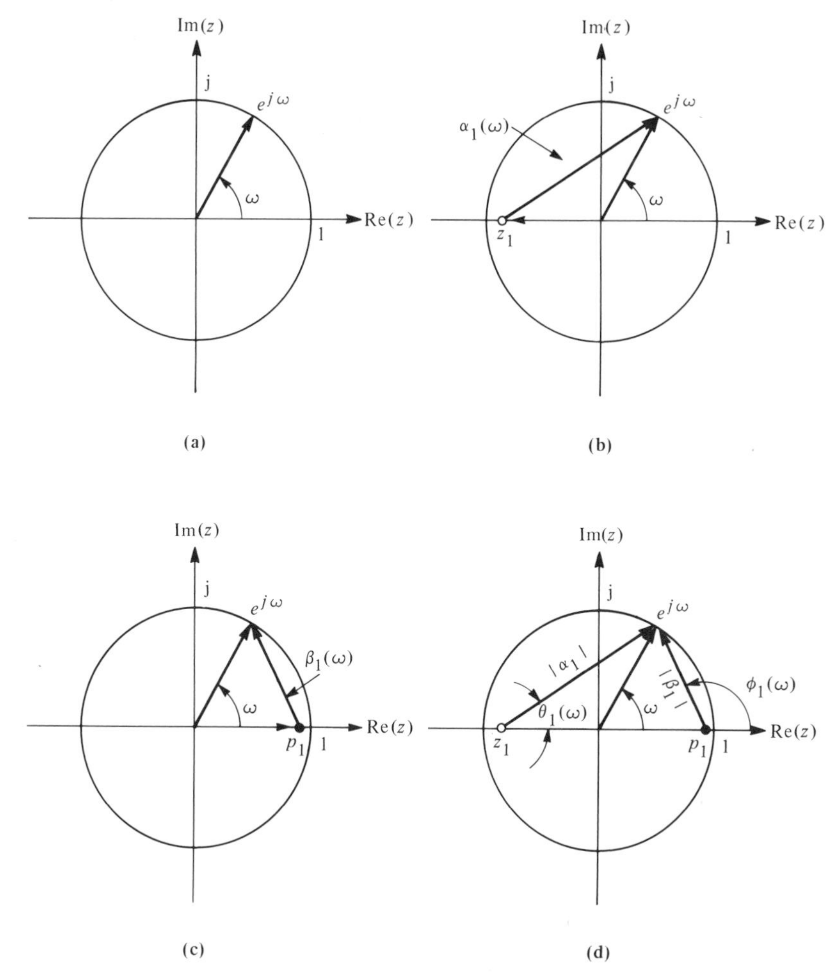

FIGURE 3.7. Geometric construction of the frequency response: (a) vector $e^{j\omega}$ on the unit circle; (b) zero-generated vector $\alpha_1(\omega)$; (c) pole-generated vector $\beta_1(\omega)$; (d) relevant lengths and angles.

Since the complex quantities $\alpha_1(\omega)$ and $\beta_1(\omega)$ each involve the complex number $e^{j\omega}$, it is expedient to give a vector interpretation to this number. That is, the quantity $e^{j\omega}$ may be represented as a vector of unit length and angle ω that extends from the origin of the complex z-plane to the unit circle, as shown in Figure 3.7a. In determining the quantity $\alpha_1(\omega)$, we must first form the difference $e^{j\omega} - z_1$. The real number z_1 may be geometrically interpreted as the vector extending from the origin to the point $(z_1, 0)$ in the complex z-plane. From the relationship $\alpha_1(\omega) = e^{j\omega} - z_1$, it follows that $\alpha_1(\omega)$ is constructed by joining the point $(z_1, 0)$ to the point $e^{j\omega}$. Expressed in a more suggestive manner, $\alpha_1(\omega)$ is represented by the vector extending from the zero of the transfer function (i.e., at $z = z_1$) to the point $e^{j\omega}$ as depicted in Figure 3.7b. In a similar fashion, the vector $\beta_1(\omega)$ is constructed by connecting the point $(p_1, 0)$ to the point $e^{j\omega}$ as shown in Figure 3.7c.

To determine the magnitude and phase function values at ω, we simply read off the lengths and angles of the vectors $\alpha_1(\omega)$ and $\beta_1(\omega)$ and then use relationships (3.11). All this information is contained in the consolidated geometric interpretation shown in Figure 3.7d. By varying the frequency variable ω over the range $-\pi < \omega \leq \pi$, we are able to geometrically determine and plot the behavior of the gain and phase functions versus ω.

Frequency Response: General System

More generally, we are interested in determining the magnitude and phase functions associated with an nth-order system whose transfer function is expressible in the factored form

$$H(z) = \frac{b(z - z_1)(z - z_2) \cdots (z - z_m)}{(z - p_1)(z - p_2) \cdots (z - p_n)} \tag{3.12}$$

where the zero and pole entities z_i and p_i may be complex numbers. Evaluating this function at $z = e^{j\omega}$ yields the associated frequency response

$$\begin{aligned} H(e^{j\omega}) &= \frac{b(e^{j\omega} - z_1)(e^{j\omega} - z_2) \cdots (e^{j\omega} - z_m)}{(e^{j\omega} - p_1)(e^{j\omega} - p_2) \cdots (e^{j\omega} - p_n)} \\ &= \frac{b\alpha_1(\omega)\alpha_2(\omega) \cdots \alpha_m(\omega)}{\beta_1(\omega)\beta_2(\omega) \cdots \beta_n(\omega)} \end{aligned} \tag{3.13}$$

in which the complex functions $\alpha_i(\omega)$ and $\beta_i(\omega)$ are given by

$$\alpha_i(\omega) = e^{j\omega} - z_i \qquad \text{for } i = 1, 2, \ldots, m \tag{3.14}$$

$$\beta_i(\omega) = e^{j\omega} - p_i \qquad \text{for } i = 1, 2, \ldots, n \tag{3.15}$$

A vector interpretation of the general vector term $\alpha_i(\omega)$ is obtained by connecting the tip of vector z_i [a zero of $H(z)$] to the tip of vector $e^{j\omega}$ as shown in Figure 3.8. The vector $\alpha_i(\omega)$ has length $|\alpha_i|$ and angle θ_i. In a similar manner, the vector $\beta_i(\omega)$ is generated by connecting the pole of $H(z)$ located at p_i to the tip of vector $e^{j\omega}$. This vector has length $|\beta_i|$ and angle ϕ_i. From this geometrical procedure, we determine the polar representations of $\alpha_i(\omega)$ and $\beta_i(\omega)$ as

$$\alpha_i(\omega) = |\alpha_i(\omega)|e^{j\theta_i(\omega)} \qquad \text{and} \qquad \beta_i(\omega) = |\beta_i(\omega)|e^{j\phi_i(\omega)}$$

From expression (3.13) it follows that the *system magnitude (gain) function* is

$$|H(e^{j\omega})| = \frac{|b| \cdot |\alpha_1(\omega)| \cdot |\alpha_2(\omega)| \cdots |\alpha_m(\omega)|}{|\beta_1(\omega)| \cdot |\beta_2(\omega)| \cdots |\beta_n(\omega)|} \tag{3.16}$$

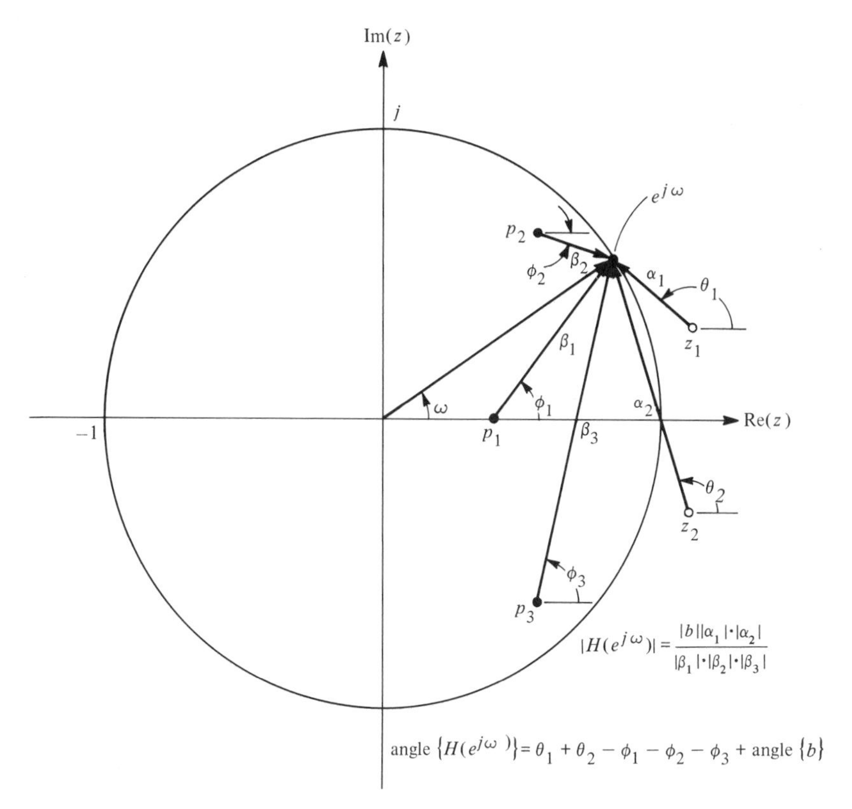

FIGURE 3.8. Geometrical construction of the frequency response for a linear system with three poles and two zeros.

whereas the *system phase function* is

$$\begin{aligned}\text{angle }\{H(e^{j\omega})\} &= \theta_1(\omega) + \theta_2(\omega) + \cdots + \theta_m(\omega) \\ &\quad - \phi_1(\omega) - \phi_2(\omega) - \cdots - \phi_n(\omega) + \text{angle }\{b\}\end{aligned} \tag{3.17}$$

where angle $\{b\}$ is 0 when b is positive and is π for b negative.

Important Observations

The correlation between the frequency behavior of the magnitude function and the location of the transfer function's zeros and poles has obvious implications. The zeros and poles of the transfer function closest to the unit circle have the most profound effect on the frequency behavior of the magnitude function. In particular, a zero located close to $z = e^{j\omega_0}$ has the effect of making the magnitude function small for ω in a close neighborhood of ω_0. Similarly, a pole located close to $z = e^{j\omega_0}$ has the effect of making the magnitude function large for

TABLE 3.3 Geometrical Procedure for Obtaining the Magnitude Function

1. Determine the point $e^{j\omega}$ on the unit circle with angle ω.
2. Connect the lines from all the zeros and poles of the transfer function to the point $e^{j\omega}$.
3. Measure the corresponding distances and angles of these lines.
4. Calculate the magnitude and phase functions using expressions (3.16) and (3.17), respectively.

ω in a close neighborhood of ω_0. Based on these observations, it is possible to design a digital filter whose magnitude function may be made to approximate any desired frequency behavior by appropriately locating the zeros and poles of its transfer function. This is an extremely powerful synthesis tool that has great utility in present-day filter design. In conclusion, we may obtain the steady-state sinusoidal frequency response at radian frequency ω by following the procedure outlined in Table 3.3. The results of this procedure may conveniently be displayed by plots of the magnitude and phase functions versus ω for $-\pi < \omega \leq \pi$.

3.8 SEQUENCE PRODUCT

In various applications of signal processing, a sequence is often formed through the operation of multiplying two other sequences, that is,

$$y(n) = w(n)x(n) \tag{3.18}$$

We shall now examine how the Fourier transforms of the three sequences forming this product are related. This will have important implications in designing frequency-discrimination filters and in characterizing the concept of data windowing.

The Fourier transform of the product sequence $\{y(n)\}$ generated in the foregoing fashion is formally given by

$$Y(e^{j\omega}) = \sum_{n=-\infty}^{\infty} w(n)x(n)e^{-j\omega n}$$

in which $w(n)x(n)$ has been substituted for $y(n)$ in the Fourier transformation definition. Using the inverse Fourier transform expression (3.8) for $x(n)$ in this relationship, we have

$$Y(e^{j\omega}) = \sum_{n=-\infty}^{\infty} w(n) \left[\frac{1}{2\pi} \int_{-\pi}^{\pi} X(e^{jv})e^{jvn}\, dv \right] e^{-j\omega n}$$

Under the assumption that the operations of integration and summation may be interchanged, this expression becomes

$$Y(e^{j\omega}) = \frac{1}{2\pi}\int_{-\pi}^{\pi} X(e^{jv}) \left[\sum_{n=-\infty}^{\infty} w(n)e^{-j(\omega - v)n}\right] dv$$

The term in brackets is recognized as being equal to the Fourier transform of $w(n)$ evaluated at the frequency argument $\omega - v$. We have therefore established the following important Fourier transform pair relationship:
If

$$y(n) = w(n)x(n) \tag{3.19a}$$

then

$$\begin{aligned} Y(e^{j\omega}) &= \frac{1}{2\pi}\int_{-\pi}^{\pi} X(e^{jv})W(e^{j(\omega - v)})\, dv \\ &= X(e^{j\omega}) * W(e^{j\omega}) \end{aligned} \tag{3.19b}$$

Namely, the Fourier transform of the product-generated sequence $\{w(n)x(n)\}$ is equal to the convolution of the Fourier transforms of $\{w(n)\}$ and $\{x(n)\}$. Thus multiplication in the time domain is equivalent to convolution in the frequency domain.

EXAMPLE 3.4

In the windowing of the data sequence $\{x(n)\}$, the second sequence $\{w(n)\}$ in the product (3.18) is often selected to be the *symmetrical* rectangular window

$$w(n) = \begin{cases} 1 & -\frac{N}{2} \le n \le \frac{N}{2} \\ 0 & \text{otherwise} \end{cases}$$

This reflects the fact that we are typically able to observe and use only a finite amount of data in any practical application. The Fourier transform of the *rectangular window* sequence $w(n)$ is readily found to be given by

$$\begin{aligned} W(e^{j\omega}) &= \sum_{n=-N/2}^{N/2} e^{-j\omega n} \\ &= \frac{\sin\,[\omega(N + 1)/2]}{\sin\,(\omega/2)} \end{aligned} \tag{3.20}$$

Thus the Fourier transform of the rectangular windowed data $y(n) = w(n)x(n)$, according to expression (3.19b), is given by

$$Y(e^{j\omega}) = \frac{1}{2\pi} \int_{-\pi}^{\pi} X(v) \frac{\sin [(\omega - v)(N + 1)/2]}{\sin [(\omega - v)/2]} dv$$

We shall examine this important relationship extensively in future chapters. ■

Parseval's Relation

The Fourier transform of the product sequence $\{w(n)x(n)\}$ has been shown to be given by

$$\sum_{n=-\infty}^{\infty} w(n)x(n)e^{-jvn} = \frac{1}{2\pi} \int_{-\pi}^{\pi} X(e^{j\omega})W(e^{j(v-\omega)})\, d\omega$$

This equality holds for all values of the frequency variable v. In particular, upon setting $v = 0$, the useful *Parseval's relation* arises; namely,

$$\sum_{n=-\infty}^{\infty} w(n)x(n) = \frac{1}{2\pi} \int_{-\pi}^{\pi} X(e^{j\omega})W(e^{-j\omega})\, d\omega \tag{3.21}$$

Parseval's relation provides a convenient means for evaluating the left-side infinite summation when the corresponding Fourier transforms $W(e^{j\omega})$ and $X(e^{j\omega})$ are such as to render the integral on the right tractable. Its greatest utility is found for the case in which the sequence $w(n) = \bar{x}(n)$, where the overbar denotes the operation of complex conjugation. In this special situation, the Parseval relation becomes

$$\sum_{n=-\infty}^{\infty} |x(n)|^2 = \frac{1}{2\pi} \int_{-\pi}^{\pi} |X(e^{j\omega})|^2\, d\omega \tag{3.22}$$

The energy in the sequence $\{x(n)\}$ as measured by the left-side summation is therefore related to the right-side spectral integral. It is recalled that when $\{x(n)\}$ is a real sequence, the function $\overline{X}(e^{j\omega})$ that appears in the right-side integrand equals $X(e^{-j\omega})$, thereby causing the integrand to be given by $X(e^{j\omega})X(e^{-j\omega})$. Thus, for $\{x(n)\}$ real, we have

$$\sum_{n=-\infty}^{\infty} |x(n)|^2 = \frac{1}{2\pi} \int_{-\pi}^{\pi} X(e^{j\omega})X(e^{-j\omega})\, d\omega \tag{3.23}$$

The time-domain energy in time series $\{x(n)\}$ as measured by the left-side summation is seen to equal the integral of the squared Fourier transform's magnitude function $|X(e^{j\omega})|^2$ over $-\pi < \omega \leq \pi$. We can thus interpret $|X(e^{j\omega})|^2$ as providing a distribution of time-series energy as a function of frequency. This is a most profound and useful observation. The time series $\{x(n)\}$ is therefore said to be rich (poor) in spectral content for ω in which $|X(e^{j\omega})|$ is relatively large (small). In addition to this spectral interpretation, Parseval relationships (3.21) and (3.22) are useful in establishing summation identities and in providing an analysis tool in filter synthesis problems. The following example provides a demonstration of this point.

EXAMPLE 3.5

Consider the Fourier transform given by

$$X(e^{j\omega}) = \begin{cases} 1 & |\omega| \leq \omega_1 \\ 0 & \omega_1 < |\omega| \leq \pi \end{cases}$$

which corresponds to an ideal low-pass filter function. The corresponding inverse Fourier transform of this function is

$$x(n) = \frac{1}{2\pi} \int_{-\omega_1}^{\omega_1} e^{j\omega n} \, d\omega = \frac{\sin (\omega_1 n)}{\pi n}$$

In accordance with Parseval relationship (3.22), the following closed-form summation expression has been established:

$$\sum_{n=-\infty}^{\infty} \left[\frac{\sin (\omega_1 n)}{\pi n} \right]^2 = \frac{1}{2\pi} \int_{-\omega_1}^{\omega_1} 1 \, d\omega = \frac{\omega_1}{\pi}$$

■

3.9 TRUNCATED FOURIER TRANSFORM: FINITE TIME SERIES OBSERVATIONS

The Fourier transform method for spectrally characterizing a time series is of fundamental theoretical importance. To incorporate this spectral approach in relevant situations, however, modifications necessitated by practical considerations must be introduced. To help appreciate this point, let us consider the typical case in which only a finite set of time series data is available for extracting the spectral information desired. Specifically, let it be assumed that we have the N time series elements

$$x(0), x(1), \ldots, x(N-1) \tag{3.24}$$

to effect the spectral identification. There is nothing restrictive in using the specific *observation interval* $0 \leq n \leq N - 1$, since data of length N defined over any other interval can be treated similarly. Since the Fourier transform (3.1) requires the entire set of time series elements $x(n)$ for $-\infty < n < \infty$, the critical question naturally arises as how do we account for the unobserved time series elements that lie outside the measurement interval $0 \leq n \leq N - 1$. It is extremely important for the user of Fourier transform methods to understand fully the implications resulting when accounting for this *missing data*.

Truncated Fourier Transform

The almost universally employed procedure for filling in the missing data is to assume that they are identically zero [i.e., $x(n) = 0$ for $n < 0$ and $n \geq N$]. Under this typically *unrealistic* assumption on the data, the associated *truncated Fourier transform* is formally defined by

$$X_N(e^{j\omega}) = \sum_{n=0}^{N-1} x(n)e^{-j\omega n} \tag{3.25}$$

The subscript N has been attached to recognize explicitly the distinction between this *truncated Fourier transform,* which is calculable from the given time series observations (3.24), and the untruncated Fourier transform (3.1), which is dependent on the entire time series including elements outside the observation interval $0 \leq n < N$. It is a simple matter to establish a revealing relationship between these two transforms. This entails substituting the inverse Fourier integral expression (3.8) for $x(n)$ into the equation above, that is,

$$X_N(e^{j\omega}) = \sum_{n=0}^{N-1} \left[\frac{1}{2\pi} \int_{-\pi}^{\pi} X(e^{jv})e^{jvn}\, dv \right] e^{-j\omega n}$$

Let us now interchange the order of the summation and integration operations. After a few elementary manipulations, the desired result

$$\begin{aligned} X_N(e^{j\omega}) &= \frac{1}{2\pi} \int_{-\pi}^{\pi} X(e^{jv})W_r(e^{j(\omega - v)})\, dv \\ &= \frac{1}{2\pi} X(e^{j\omega}) * W_r(e^{j\omega}) \end{aligned} \tag{3.26}$$

is obtained. The transform function $W_r(e^{j\omega})$ that appears in the integrand is referred to as the *rectangular window transform*, since it is equal to the Fourier transform of the rectangular window sequence $w_r(n) = u(n) - u(n - N)$, that is,

$$W_r(e^{j\omega}) = \sum_{n=0}^{N-1} e^{-j\omega n} = e^{-j\omega(N-1)/2} \frac{\sin(\omega N/2)}{\sin(\omega/2)} \tag{3.27}$$

These results are in agreement with those found in Section 3.8 when it is recognized that the truncated data are specified by $x(n)w_r(n)$.

3.10

TRUNCATED FOURIER TRANSFORM: SPECTRAL DISTORTION

The effect of using only a finite amount of data to estimate the underlying Fourier transform is seen to manifest itself in the form of the convolution integral (3.26). If the truncated Fourier transform $X_N(e^{j\omega})$ is to equal $X(e^{j\omega})$ exactly, as is desired, it is apparent from this convolution integral expression that the window's Fourier transform $W_r(e^{j\omega})$ should be equal to the Dirac delta function $2\pi\delta(\omega)$ in the interval $(-\pi, \pi)$. To measure the degree to which this is achieved, a plot of the amplitude-variable component of the window as specified by

$$\text{amp}\,[W_r(e^{j\omega})] = \frac{\sin(\omega N/2)}{\sin(\omega/2)} \tag{3.28}$$

is given in Figure 3.9. The window function is seen to be composed of a number of pulselike waveforms centered at the frequencies 0 and $k\pi/N$ for $k = 3, 5, 7, \ldots$. These pulselike waveforms are generally referred to as *lobes*, with the largest lobe centered at $\omega = 0$ being called the *main lobe*, while the other lobes, called *side lobes*, are smaller in magnitude. From this plot and the properties associated with the Fourier transform, it is apparent that:

1. The window's main lobe has a width of $4\pi/N$ and an amplitude of N and the side lobes each have a width of $2\pi/N$.
2. The window's main and side lobes have widths (amplitudes) that decrease (increase) to zero (infinity) as N increases.
3. The area under the window transform (3.28) over the interval $-\pi \le \omega \le \pi$ equals the constant 2π for all positive integer values of N.

From these characteristics it can be inferred that the window's transform behaves like the desired Dirac delta function $2\pi\delta(\omega)$ as N approaches infinity, thereby

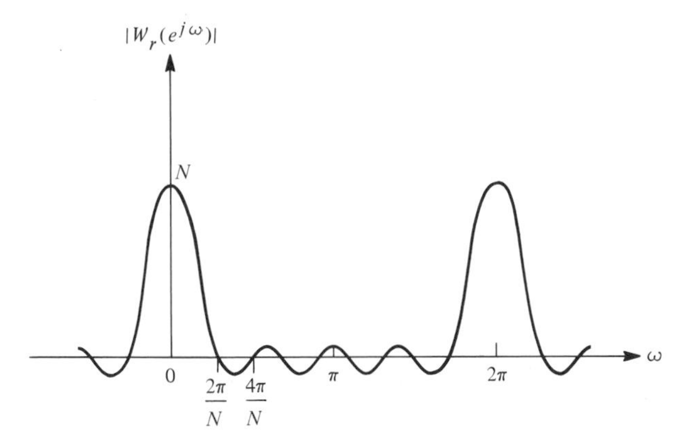

FIGURE 3.9. Rectangular window transform magnitude plot for $N = 8$.

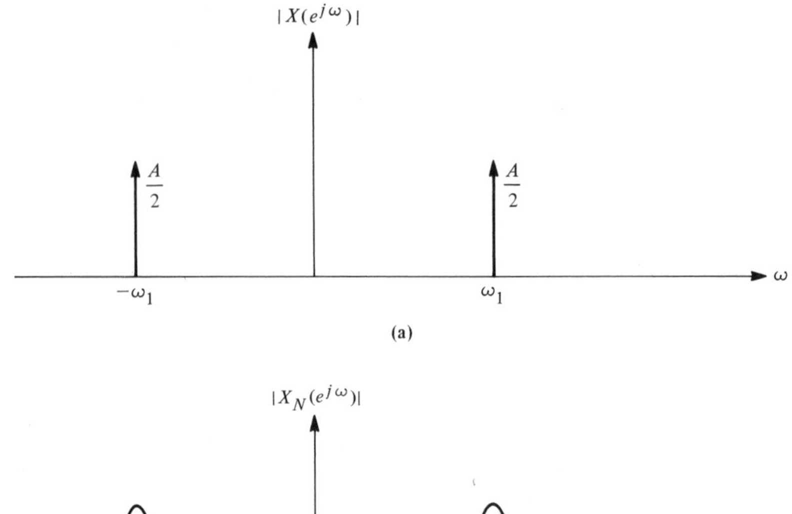

FIGURE 3.10. Magnitude of the Fourier transform associated with: (a) untruncated time series $x(n) = A \cos(\omega_1 n)$; (b) truncated time series $X_t(n) = A \cos(\omega_1 n)[u(n) - u(n - N)]$.

causing $X_N(e^{j\omega}) \approx X(e^{j\omega})$. Unfortunately, in most data-processing applications, the data length N is relatively small, indicating that $X_N(e^{j\omega})$ is a distorted version of $X(e^{j\omega})$. We shall now discuss two primary consequences of this distortion.

Leakage

To determine the effects of using only a finite amount of data in the Fourier transform, let us first consider the case in which the underlying signal is the pure sinusoid

$$x(n) = A \cos(\omega_1 n) \tag{3.29a}$$

The corresponding Fourier transform of this time series has been shown previously to be equal to

$$X(e^{j\omega}) = \pi A[\delta(\omega - \omega_1) + \delta(\omega + \omega_1)] \qquad \text{for } \pi < \omega \leq \pi \tag{3.29b}$$

which is composed of two Dirac impulse functions, as shown in Figure 3.10a. In accordance with relationship (3.26), the spectrum of the associated truncated Fourier transform (3.25) is obtained by convolving $X(e^{j\omega})$ with $W_r(e^{j\omega})$. This convolution produces the function

$$\begin{aligned} X_N(e^{j\omega}) &= A\, e^{-j(\omega-\omega_1)(N-1)/2} \frac{\sin[(\omega - \omega_1)N/2]}{\sin[(\omega - \omega_1)/2]} \\ &\quad + A\, e^{j(\omega+\omega_1)(N-1)/2} \frac{\sin[(\omega + \omega_1)N/2]}{\sin[(\omega + \omega_1)/2]} \end{aligned} \tag{3.30}$$

which is seen to be composed of two frequency-shifted versions of the window transform (3.27) centered at ω_1 and $-\omega_1$, as shown in Figure 3.10b. Thus, by using only a finite segment of the sinusoidal signal (i.e., windowing), the corresponding spectrum is seen to be a smeared version of the original impulse-type spectrum. This effect is commonly referred to as spectral *leakage* or *smearing,* whereby a pure sinusoid is made to appear as a continuum of spectral components in a region about the sinusoid's frequencies, ω_1 and $-\omega_1$. This leakage is a direct consequence of using only a finite segment of the underlying infinite-length signal in the Fourier transform generation.

Spectral Smoothing and the Ripple Effect

The utilization of only a finite amount of data also has a distorting effect on signals whose spectrum have sharp discontinuities. To illustrate this point, let us consider the shifted by $N/2$ infinite-length time series

$$x(n) = \frac{\sin(\omega_1[n - N/2])}{\pi(n - N/2)} \tag{3.31a}$$

which has the corresponding Fourier transform

$$X(e^{j\omega}) = \begin{cases} e^{-j\omega N/2} & |\omega| \le \omega_1 \\ 0 & \omega_1 < |\omega| \le \pi \end{cases} \tag{3.31b}$$

A plot of this transform's magnitude is shown in Figure 3.11a by the dashed-line plot, where discontinuities at $\pm\omega_1$ are evident. When the truncated Fourier transform (3.25) associated with time series (3.31a) is computed, the results shown superimposed in Figure 3.11a by the solid-line plot are obtained. Although this truncated Fourier transform has roughly the same features as those of the untruncated Fourier transform, it is apparent that the effect of using only a finite amount of time-series data has caused:

1. A smoothing of the discontinuities at $\pm\omega_1$
2. A rippling effect.

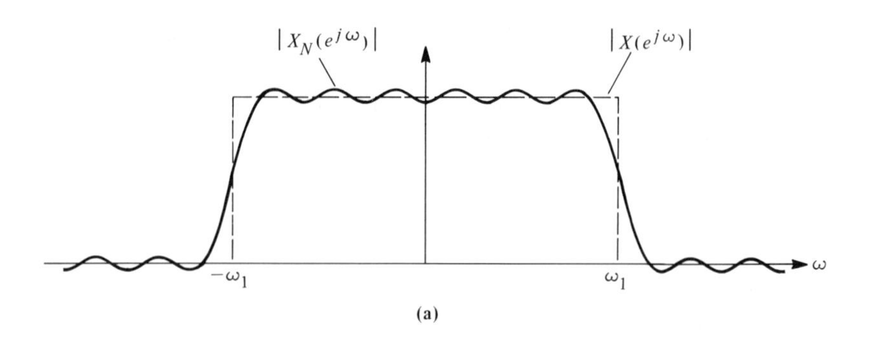
$|X_N(e^{j\omega})|$
$|X(e^{j\omega})|$
$-\omega_1$
ω_1
ω

(a)

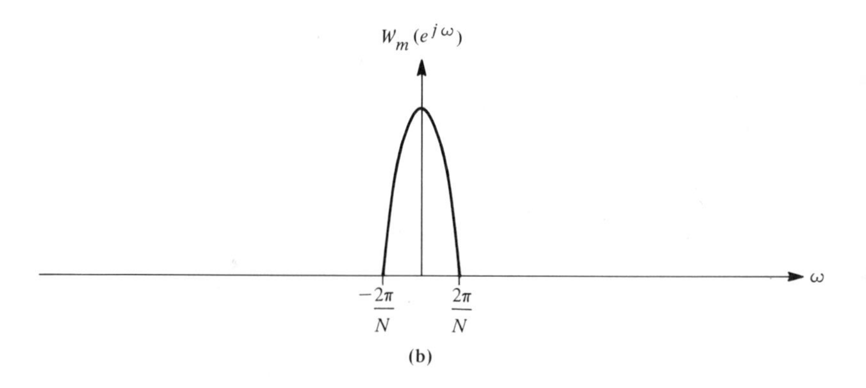
$W_m(e^{j\omega})$
$-\frac{2\pi}{N}$
$\frac{2\pi}{N}$
ω

(b)

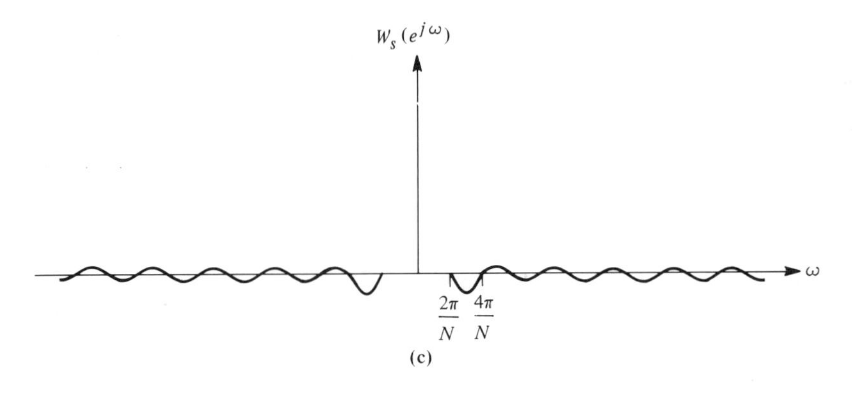
$W_s(e^{j\omega})$
$\frac{2\pi}{N}$
$\frac{4\pi}{N}$
ω

(c)

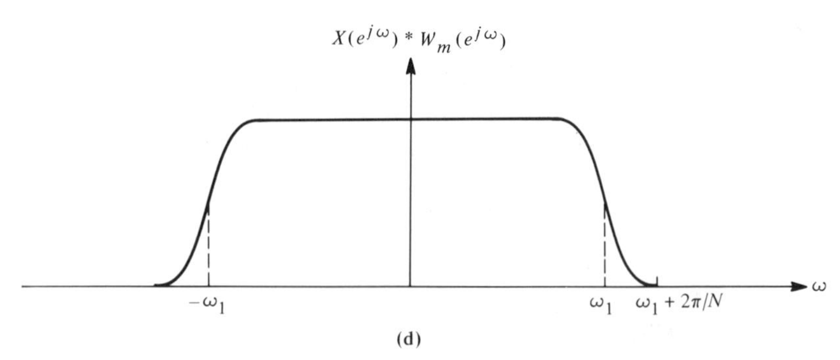
$X(e^{j\omega}) * W_m(e^{j\omega})$
$-\omega_1$
ω_1
$\omega_1 + 2\pi/N$
ω

(d)

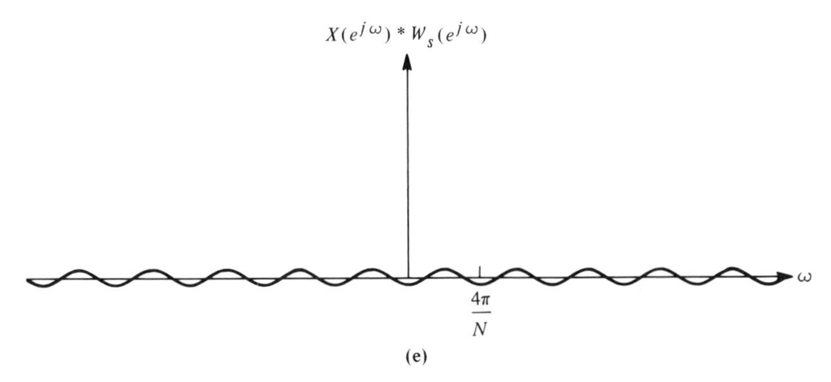

FIGURE 3.11. Ideal low-pass filter: (a) $X(e^{j\omega})$ and $X_N(e^{j\omega})$ plots; (b) window's main lobe; (c) window's side lobes; (d) $X(e^{j\omega})$ convolved with $W_m(e^{j\omega})$; (e) $X(e^{j\omega})$ convolved with $W_s(e^{j\omega})$.

To describe the sources of these distortions, it is beneficial to decompose the window transform as

$$W_r(e^{j\omega}) = W_m(e^{j\omega}) + W_s(e^{j\omega}) \tag{3.32}$$

where $W_m(e^{j\omega})$ and $W_s(e^{j\omega})$ denote the main-lobe and side-lobe components as shown in Figure 3.11b and c, respectively. With this decomposition, the truncated Fourier transform convolution integral relationship (3.26) is given by

$$2\pi X_N(e^{j\omega}) = W_m(e^{j\omega}) * X(e^{j\omega}) + W_s(e^{j\omega}) * X(e^{j\omega}) \tag{3.33}$$

As is now demonstrated, the first term on the right side of this expression gives rise to the smoothing distortion, while the second term produces the rippling distortion.

Since the main lobe of the window's transform is dominant in terms of window energy, it follows that the convolution component $W_m(e^{j\omega}) * X(e^{j\omega})$ constitutes the primary effect arising from using only a finite amount of data in the Fourier transform. This is made evident in Figure 3.11d, where this convolution component is depicted. In particular, from Figure 3.11a and it is clear that when evaluating $X(e^{j\omega}) * W_m(e^{j\omega})$ using expression (3.26) for frequency shifts $0 \leq |\omega| \leq \omega_1 - 2\pi/N$, the shifted main lobe $W_m(e^{j(\omega-\nu)})$ is wholly contained within the region where $X(e^{j\nu})$ is equal to one.† For this range of ω, the convolution integral $X(e^{j\omega}) * W_m(e^{j\omega})$ is a constant equal to the area under the main lobe. For frequency shifts in the range $\omega_1 - 2\pi/N < |\omega| < \omega_1 + 2\pi/N$, however, the shifted main lobe $W_m(e^{j(\omega-\nu)})$ only partially overlaps that segment of $X(e^{j\nu})$ which is equal to one. The convolution integral $X(e^{j\omega}) * W_m(e^{j\omega})$ for this range of frequencies is therefore equal to only a fraction of the main-lobe area. For

†It is here being implicitly assumed that $\omega_1 \geq 2\pi/N$. For smaller values of ω_1, similar arguments can be made.

instance, only half the area is generated at the frequency shifts $\pm\omega_1$. Finally this convolution integral equals zero for frequencies $\omega_1 + 2\pi/N \leq |\omega| \leq \pi$ since the shifted main lobe $W_m(e^{j(\omega - \nu)})$ is wholly contained in the region where $X(e^{j\nu})$ is equal to zero. The convolution integral component $X(e^{j\omega}) * W_m(e^{j\omega})$ thus generated is depicted in Figure 3.11d. This component is seen to be a modified version of the original rectangular function (3.31b) in which the discontinuities at $\pm\omega_1$ have been changed into gradual transitions of width $4\pi/N$ (the main lobe's width).

When the side-lobe component $W_s(e^{j\omega})$ is convolved with $X(e^{j\omega})$, the resulting function appears as shown in Figure 3.11e. This function is seen to contain a rippled effect throughout 'the frequency range $-\pi < \omega \leq \pi$, in which the ripples oscillate with a fundamental frequency of $4\pi/N$. This effect is explained by observing that the area under $X(e^{j\nu})W_s(e^{j(\omega - \nu)})$ may be positive or negative depending on the choice of ω and that a positive (or negative) area will result over shifts of length $2\pi/N$.

The total convolution integral is simply equal to the sum of the components $X(e^{j\omega}) * W_m(e^{j\omega})$ and $X(e^{j\omega}) * W_s(e^{j\omega})$ as depicted by the solid-line plot in Figure 3.11a. From these developments it is apparent that using only a finite amount of data has given arise to the following effects on the truncated Fourier transform:

1. A smearing or leakage of narrow-band components (e.g., pure sinusoids).
2. A smoothing of the underlying Fourier transform whereby discontinuities became gradual transitions of width equal to the window's main-lobe width.
3. A relatively low amplitude ripple effect of fundamental frequency $4\pi/N$ throughout the entire frequency range.

3.11 RESOLUTION OF TWO EQUAL-AMPLITUDE SINUSOIDS

The Fourier transform is used extensively in signal-processing applications for detecting the presence of two or more closely spaced narrow-band processes (e.g., sinusoids). The attendant spectral smoothing caused by using only a finite data length makes this identification difficult if not impossible when the narrow-band processes are sufficiently close in frequency. To illustrate this point, let us consider a time series composed of two *equal-amplitude* sinusoids,

$$x(n) = A \cos(\omega_1 n) + A \cos[(\omega_1 + \Delta)n]$$

which are separated in frequency by Δ radians. The Fourier transform of this infinite-length time series is readily shown to be

$$X(e^{j\omega}) = \pi A [\delta(\omega - \omega_1) + \delta(\omega + \omega_1) + \delta(\omega - \omega_1 - \Delta) + \delta(\omega + \omega_1 + \Delta)]$$

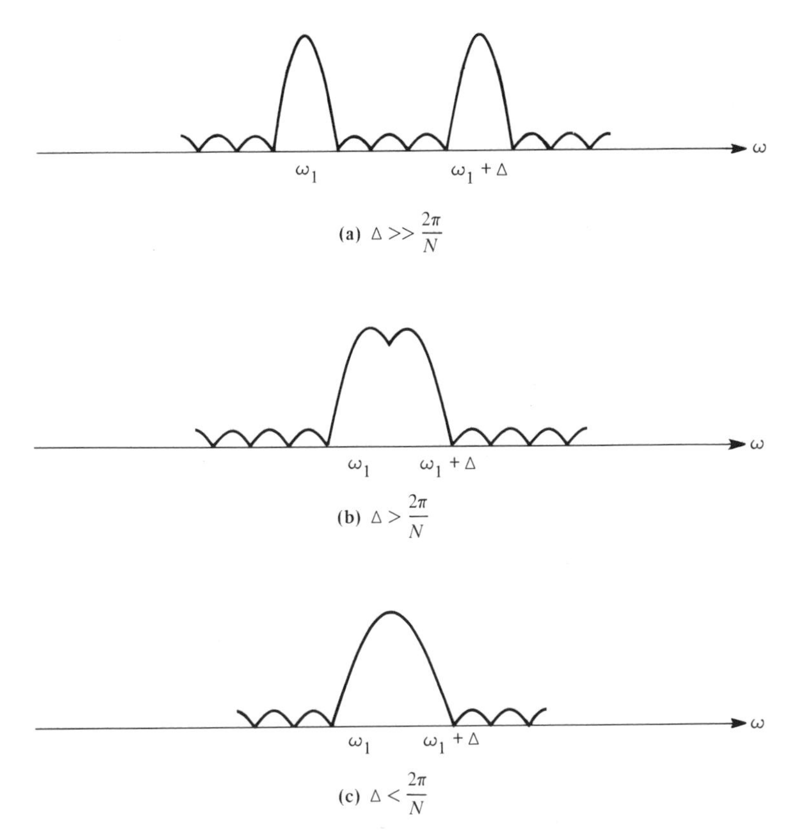

FIGURE 3.12. Plots of the truncated Fourier transform magnitude time series composed of two pure sinusoids whose frequencies are separated by Δ radians: (a) resolution easily seen for $\Delta \gg 2\pi/N$; (b) resolution evident for $\Delta > 2\pi/N$; (c) resolution not achieved for $\Delta < 2\pi/N$.

If the truncated Fourier transform (3.25) associated with the sum of sinusoidal entries given above is used to approximate this underlying Fourier transform, four distinct impulses are not evident. This is made clear by inserting the foregoing expression for $X(e^{j\omega})$ into this convolution integral (3.26). The resultant truncated Fourier transform in a neighborhood of ω_1 and $\omega_1 + \Delta$ is shown in Figure 3.12 for three choices of frequency separation Δ. For Δ relatively large with respect to the rectangular window's main-lobe width $4\pi/N$, two well-defined spectral peaks are evident, indicating the possible presence of two sinusoids. As Δ is decreased, however, the two separate peaks become more difficult to resolve. A value of Δ is eventually reached where these two peaks blend into one and a spectral resolution is not possible. As a rule of thumb, it is possible to resolve two equal-amplitude sinusoids using the truncated Fourier transform (3.25) as long as the frequency difference Δ sufficiently exceeds $2\pi/N$. For unequal-amplitude sinusoids, the frequency difference required for resolution purposes is generally greater than $2\pi/N$.

3.12 DATA WINDOWING

The method presented in Section 3.9 for approximating the Fourier transform from a finite set of time series observations has a more general setting. Let us consider the auxiliary finite-length, *windowed* time series given by

$$x_w(n) = w(n)x(n) \tag{3.34}$$

where $\{w(n)\}$ is a *window sequence* that is zero for $n \notin [0, N-1]$ and it is selected so as to achieve certain signal-processing objectives. Some of the more popularly used window sequences are given in Table 3.4. The rectangular window choice gives rise to the truncated Fourier transform (3.25) discussed previously.

The Fourier transform of the windowed sequence (3.34) is given formally by

$$X_w(e^{j\omega}) = \sum_{n=0}^{N-1} w(n)x(n)e^{-j\omega n} \tag{3.35}$$

To determine how this windowed Fourier transform is related to the underlying Fourier transform $X(e^{j\omega})$, let us again substitute the inverse Fourier integral (3.8) into this expression. After interchanging the order of the integration and summation operations, it is eventually found that

$$X_w(e^{j\omega}) = \frac{1}{2\pi}\int_{-\pi}^{\pi} X(e^{jv})W(e^{j(\omega - v)})\,dv \tag{3.36}$$

where the window transform $W(e^{j\omega})$ is equal to the Fourier transform of the window sequence, that is,

$$W(e^{j\omega}) = \sum_{n=0}^{N-1} w(n)e^{-j\omega n} \tag{3.37}$$

Expression (3.36) indicates that the windowed Fourier transform is equal to a convolution of the underlying Fourier transform $X(e^{j\omega})$ and the window transform $W(e^{j\omega})$. The remarks associated with the rectangular window selection are applicable in this more general setting as well. In particular, the closer $W(e^{j\omega})$ resembles the Dirac delta function $2\pi\delta(\omega)$ in the frequency interval $(-\pi, \pi]$ the more closely $X_w(e^{j\omega})$ will resemble $X(e^{j\omega})$. It is typically found that $X_w(e^{j\omega})$ is a smooth-rippled version of the underlying Fourier transform $X(e^{j\omega})$. For a given value of N, the deleterious smoothing effect is directly proportional to the width of the window's main lobe. Similarly, the effects of ripple decreases as the relative sizes of the amplitudes of the main and largest side lobes diverge. From Table 3.4 the rectangular window is seen to have the smallest main-lobe

ABLE 3.4 Commonly Used Data Windows

Vindow	$w(n)$, $0 \le n \le N - 1$	Main-Lobe Width (rad)	$20 \log \left(\dfrac{\text{Main-Lobe Amplitude}}{\text{Largest Side-Lobe Amplitude}} \right)$ (dB)
ectangular	1	$\frac{4\pi}{N}$	−13
artlett	$\frac{2n}{N-1}$ $0 \le n \le \frac{N-1}{2}$ $2 - \frac{2n}{N-1}$ $\frac{N-1}{2} < n \le N - 1$	$\frac{8\pi}{N}$	−27
anning	$0.5[1 - \cos(2\pi n/N)]$	$\frac{8\pi}{N}$	−32
amming	$0.54 - 0.46 \cos(2\pi n/N)$	$\frac{8\pi}{N}$	−43
lackman	$0.42 - 0.5 \cos(2\pi n/N) +$ $0.8 \cos(4\pi n/N)$	$\frac{12\pi}{N}$	−58

REMARKS

Upon examination of the last three entries of Table 3.4, it is seen that in each of the cosine arguments, the window length parameter N appears as a divisor. The reason for this divisor selection is related to the periodic interpretation that may be given to the associated *discrete Fourier transform*. A justification for this choice is to be found in the excellent tutorial paper by Harris (1978), which is recommended reading for all users of data windows. It must be noted, however, that in much of contemporary literature, this divisor is replaced by $N - 1$ [e.g., the Hamming window appears as $0.54 - 0.46 \cos(2\pi n/(N - 1))$]. Although the divisor choice $N - 1$ does result in the desirable linear phase condition $w(n) = w(N - n - 1)$ for the window, this selection does not have a theoretical justification. Fortunately, for sufficiently moderate values of N, either divisor selection leads to approximately the same windowing effect. For small values of N, however, the user is advised to test each divisor and select that one that is most suitable for the application at hand.

width and it therefore provides the least amount of smoothing. It thus provide typically better performance in resolving closely spaced spectral peaks (i.e. frequency resolution). On the other hand, the other windows listed are seen t have relatively smaller side lobes and are therefore more immune to attendar deleterious ripple effects. It is generally found that windows which possess smal (large) main-lobe widths have attendant large (small) side lobes.

3.13 CONTINUOUS-TIME FOURIER TRANSFORM APPROXIMATION

The continuous-time Fourier transform is used extensively to measure the spec tral (frequency) content of analog signals. This transformation is governed b the integral operation

$$X(\omega_a) = \int_{-\infty}^{\infty} x(t)e^{-j\omega_a t}\, dt \tag{3.38}$$

in which the subscript a has been appended to explicitly designate ω_a as th "analog" frequency variable. As is now shown, the discrete-time Fourier trans form can be utilized to provide an acceptably good approximation to this con tinuous-time Fourier transform.

To begin our development, let the integration time interval $-\infty < t < \infty$ b decomposed into the set of contiguous intervals each of width T specified b $nT \leq t < (n + 1)T$ for $n = 0, \pm 1, \pm 2, \ldots$. The interval width T must b selected small enough so that essentially all of the radian frequency content c $X(\omega_a)$ lies in the interval $-\pi/T \leq \omega_a \leq \pi/T$ (see Section 1.12). With thi restriction in mind, let us equivalently express the Fourier transform integra (3.38) in terms of these time intervals, that is,

$$X(\omega_a) = \sum_{n=-\infty}^{\infty} \int_{nT}^{nT+T} x(t)e^{-j\omega_a t}\, dt$$

Using the rectangular approximation to integration, we may approximate th right side integrals as

$$X(\omega_a) \approx \sum_{n=-\infty}^{\infty} Tx(nT)e^{-j\omega_a Tn} \tag{3.39}$$

This approximation is of good quality, provided that the sampling-time paran eter T is selected suitably small, in accordance with the remarks made above.

Upon comparison of expression (3.39) with the discrete-time Fourier trans form (3.1), it is apparent that they are of identical form. More specifically, th

TABLE 3.5 Approximation of the Continuous-Time Fourier Transform

1. Select the sampling time T so that $X(\omega_a) \approx 0$ for all $|\omega_a| > \pi/T$.
2. Sample the continuous-time signal $x(t)$ being Fourier transformed at the sampling time instants nT to obtain $x(nT)$.
3. Compute the discrete-time Fourier transform (3.1) of the sequence $x(n) = Tx(nT)$.
4. The desired approximation is then given by $X(\omega_a) = X(e^{j\omega_a T})$ for $-\pi/T < \omega_a \leq \pi/T$.

time series elements are identified with *samples* of the continuous-time signal being Fourier transformed, that is,

$$x(n) = Tx(nT) \qquad \text{for } n = 0, \pm 1, \pm 2, \ldots \tag{3.40}$$

while the analog and digital frequencies are related by

$$\omega = \omega_a T \qquad \text{for } -\pi < \omega \leq \pi \tag{3.41}$$

This relationship between the analog and ditigal frequency variables arises again and again in applications concerned with the use of digital techniques in an analog setting. Using correspondence (3.40), we may then apply the discrete-time Fourier transform to the sampled sequence $\{x(n)\}$.

The main steps in this numerical approach are outlined in Table 3.5. In most applications, one does not know a priori the highest-frequency content of $x(t)$. For such situations, good judgment has to be used in selecting T. This parameter must be chosen small enough so that the samples $x(nT)$ adequately represent $x(t)$ but large enough so that an excessive number of samples are not generated (a computational consideration). The resultant discrete-time Fourier transform $X(e^{j\omega})$ will be such that

$$X(\omega_a) \approx X(e^{j\omega})\big|_{\omega = \omega_a T} \qquad \text{for } |\omega_a| < \pi/T \tag{3.42}$$

To depict this behavior visually, we may plot the magnitude and phase of $X(e^{j\omega})$ in a normal fashion and then relabel the frequency axis from $(-\pi, \pi]$ to $(-\pi/T, \pi/T]$.

EXAMPLE 3.6

Using the approach above, compare the continuous-time Fourier transform and its discrete-time Fourier transform approximation for the signal

$$x(t) = e^{-t}u(t)$$

The continuous-time Fourier transform (3.38) associated with this signal i found to be

$$X(\omega_a) = \frac{1}{1 + j\omega}$$

On the other hand, the discrete-time Fourier transform associated with th sampled signal

$$x(n) = Tx(nT) = Te^{-nT}u(nT)$$

is given, according to relationship (3.1), by

$$X(e^{j\omega}) = \sum_{n=0}^{\infty} Te^{-nT}e^{-j\omega n}$$

$$= \frac{T}{1 - e^{-T}e^{-j\omega}} \qquad |\omega| \leq \pi$$

In accordance with relationship (3.42), we therefore have

$$X(\omega_a) \approx \frac{T}{1 - e^{-T}e^{-j\omega_a T}} \qquad |\omega_a| \leq \frac{\pi}{T}$$

Plots of the magnitude of $X(\omega_a)$ and this approximation are given in Figur 3.13 for several choices of sampling time T. Clearly, the approximation i very good for T sufficiently small.

3.14 SUMMARY

Some of the more important properties of the Fourier transform have bee developed. Of note was the observation that the Fourier transform is a specia case of the more general z-transform. This enabled us to construct a Fourie transform pair table and a Fourier transform property table from our previou investigations of the z-transform.

From a practical viewpoint, the concept of the truncated Fourier transform a it relates to finite data sets was of great importance. To evaluate the truncate Fourier transform, the fast Fourier transform algorithm to be developed in th next chapter is shown to provide a computationally efficient procedure.

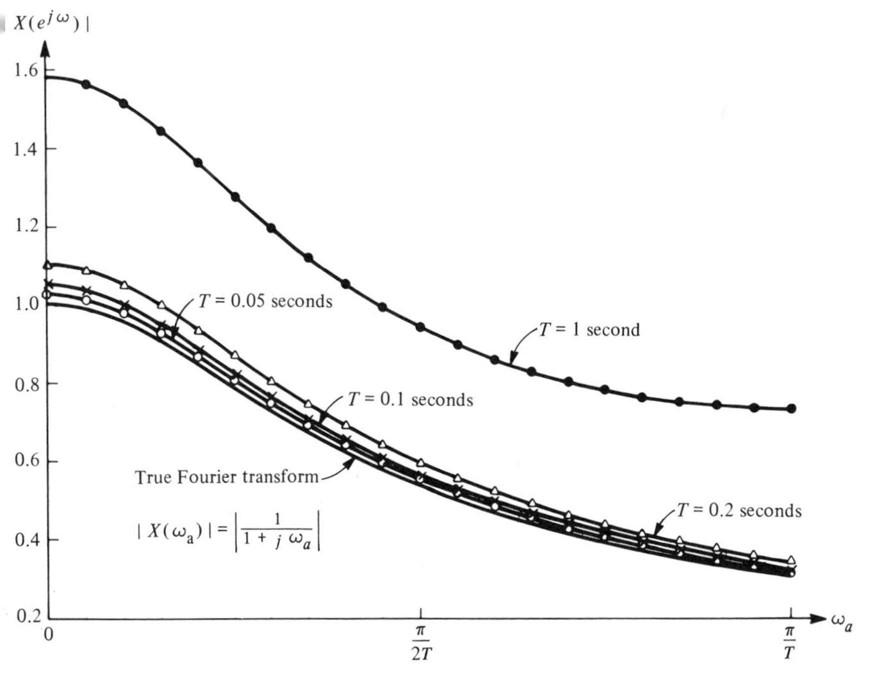

FIGURE 3.13. Digital approximation of Fourier transform of continuous-time signal [$x(t) = e^{-t}u(t)$].

SUGGESTED READINGS

In addition to the books referred to in the Suggested Readings of Chapter 2, the following references are recommended.

BLAHUT, R., *Fast Algorithms for Digital Signal Processing*. Reading, Mass.: Addison-Wesley Publishing Company, Inc., 1985.

BRACEWELL, R. N., *The Fourier Transform and its Applications,* New York: McGraw-Hill Book Company, 1978.

BRIGHAM, B. O. *The Fast Fourier Transform*. Englewood Cliffs, N.J.: Prentice-Hall, Inc., 1974.

COOLEY, J. W., and J. W. TUKEY, "An Algorithm for the Machine Calculation of Complex Fourier Series," *Mathematics of Computation,* Vol. 19, April 1965, pp. 297–301.

HARRIS, F. J., "On the Use of Windows for Harmonic Analysis with the Discrete Fourier Transform," Proceedings of the IEEE, Vol. 66, No. 1, January 1978, pp. 51–83.

RAMIREZ, R., *The FFT: Fundamentals and Concepts*. Englewood Cliffs, N.J.: Prentice-Hall, Inc., 1985.

PROBLEMS

3.1 Determine the Fourier transforms of the time series given in Problem 2.1.

3.2 Determine and sketch the magnitude of the Fourier transform of the time series

$$x(n) = \begin{cases} e^{j\omega_1 n} & -N+1 \le n \le N-1 \\ 0 & \text{otherwise} \end{cases}$$

3.3 Determine the Fourier transforms of the time series

(a) $x_1(n) = 1$ for all n

(b) $x_2(n) = e^{j\omega_1 n}$ for all n

(c) $x_3(n) = u(n+N) - u(n-N)$

3.4 Differentiating the Fourier transform expression (3.1) with respect to ω, prove that

$$\{nx(n)\} = j\frac{dX(e^{j\omega})}{d\omega}$$

3.5 Use the property established in Problem 3.4 to determine the Fourier transform of

(a) $x_1(n) = na^n u(n)$ **(b)** $x_2(n) = n\cos(\omega_1 n)$

3.6 Determine the time series whose associated Fourier transform is given by

(a) $X(e^{j\omega}) = \begin{cases} 1 & \frac{\pi}{4} \le |\omega| \le \frac{\pi}{2} \\ 0 & \text{otherwise} \end{cases}$

(b) $X(e^{j\omega}) = -3e^{j2\omega} + 7e^{-j4\omega}$

(c) $X(e^{j\omega}) = 1 + e^{-j\omega} + e^{-j2\omega} + e^{-j3\omega} + \cdots$

3.7 Determine the unit-impulse response associated with the ideal bandpass and high-pass filter described in Figure 3.5.

3.8 Give a direct prove of the following frequency-shift properties of the Fourier transform.

(a) $\{x(n)\cos(\omega_1 n)\} = 0.5[X(e^{j(\omega-\omega_1)}) + X(e^{j(\omega+\omega_1)})]$

(b) $\{x(n)\sin(\omega_1 n)\} = 0.5j[X(e^{j(\omega+\omega_1)}) - X(e^{j(\omega-\omega_1)})]$

3.9 Prove the convolution and correlation properties of the Fourier transform.

3.10 Prove the time-transpose property of the Fourier transform.

3.11 Let the unit-step signal be applied to the ideal low-pass filter described in Figure 3.5. Sketch the response's Fourier transform magnitude $|Y(e^{j\omega})|$ and find $\{y(n)\}$.

3.12 Compute the truncated Fourier transforms associated with the following time series elements and sketch $|X(e^{j2\pi k/N})|$ versus k for $0 \le k \le N - 1$.

(a) $x(n) = (-1)^n u(n)$ for $0 \le n \le N - 1$

(b) $x(n) = r^n e^{j\omega_0 n}$ for $0 \le n \le N - 1$

where $0 < r < 1$.

3.13 Using the Fourier transform approach of Section 3.13, characterize the approximation of the continuous-time Fourier transform for the signal

(a) $x_a(t) = (\frac{1}{2})^t u(t)$

(b) $x_b(t) = \left(\frac{1}{2}\right)^t \cos\left(\frac{\pi t}{2}\right) u(t)$

CHAPTER 4

Discrete Fourier Transform

4.1

INTRODUCTION

The primary purpose of the Fourier transform is to identify the spectral (i.e. frequency) content of measured data. When the data is of finite length, as it must be in any practical application, this Fourier transform takes the truncated form

$$X_N(e^{j\omega}) = \sum_{n=0}^{N-1} x(n)e^{-j\omega n} \tag{4.1}$$

in which the implicit assumption is made that the data elements $x(n)$ are zero outside the observation interval $0 \leq n \leq N - 1$. The consequences arising when this zero-data condition is violated (as it almost always is) were examined in Chapter 3. Without loss of generality, we have taken $0 \leq n \leq N - 1$ to define the time interval over which the data measurements are made.

If this truncated Fourier transform is to be of any practical value, a systematic procedure for its *numerical evaluation* must be developed. In particular, a computationally feasible method is required for capturing the intrinsic nature of $X_N(e^{j\omega})$ as a function of the continuous frequency variable ω. The *discrete Fourier transform* (DFT) provides a mechanism for achieving this objective.

whereby sampled values of the truncated Fourier transform are computed at a prescribed set of discrete frequencies. A determination of these sampled values is found to entail a finite number of complex addition and multiplication operations and is therefore amenable to digital implementation. Furthermore, the *fast Fourier transform* (FFT) algorithm is found to provide a computationally efficient means for computing these sampled values. In this chapter we explore some of the fundamental issues related to the use of the discrete Fourier transform and its evaluation through the fast Fourier transform algorithm. It is shown that the discrete Fourier transform is a powerful tool both for analyzing data in the frequency domain and for implementing useful signal-processing algorithms.

4.2 DISCRETE FOURIER TRANSFORM

An intuitively appealing procedure for capturing the basic nature of the truncated Fourier transform (4.1) is to compute its value at a discrete set of frequencies. If these discrete frequencies are closely spaced, then the resultant sampled values of $X_N(e^{j\omega})$ may be used to infer its behavior for all other frequencies. With this thought in mind, let the discrete set of frequencies be described by

$$\omega_k = \frac{2\pi k}{N} \qquad \text{for } k = 0, 1, \ldots, N-1 \tag{4.2}$$

It is noted that the number of discrete frequencies in this set (i.e., N) exactly corresponds to the length of the data measurements used in the truncated Fourier transform (4.1). Furthermore, these uniformly spaced frequencies cover one period (i.e., $0 \leq \omega \leq 2\pi$) of the periodic Fourier transform.

Upon evaluating the truncated Fourier transform (4.1) on the discrete frequency set (4.2), the appropriately named *discrete Fourier transform* arises:

$$X_N(e^{j\omega_k}) = \sum_{n=0}^{N-1} x(n)e^{-j2\pi kn/N} \qquad \text{for } k = 0, 1, \ldots, N-1 \tag{4.3}$$

The operation on the right side of this expression is seen to entail N complex number multiplications and $N - 1$ complex number additions for each sampled value determination. Thus a total of N^2 complex number multiplications and $N(N - 1)$ complex number additions are required to compute the N sampled values of the truncated Fourier transform if a direct evaluation of expression (4.3) is made. As we shall see shortly, however, the fast Fourier transform algorithm may be used to compute these sampled values in a much more efficient manner.

The process of computing *discrete* frequency samples of the truncated Fourier transform as specified by relationship (4.3) is commonly referred to as the *discrete Fourier transform* (DFT). It represents a computer programmable method for characterizing the intrinsic nature of the truncated Fourier transform. In wha follows it is preferable to use the more compact notation

$$\begin{aligned} X_N(k) &= X_N(e^{j\omega})\big|_{\omega=2\pi k/N} \\ &= \sum_{n=0}^{N-1} x(n)e^{-j2\pi kn/N} \qquad 0 \le k \le N-1 \end{aligned} \quad (4.4$$

to represent the DFT. Thus the entity $X_N(k)$ denotes the truncated Fourier trans form's value at the discrete frequency $\omega_k = 2\pi k/N$. We shall refer to th elements $X_N(k)$ as being *DFT coefficients* and the computational expression (4.4 as being an *N-point DFT*.

It has previously been established that the Fourier transform may be obtaine by evaluating the z-transform on the unit circle (where $z = e^{j\omega}$) in the comple z-plane. Thus the Fourier transform can be interpreted as being a domain restricted version of the z-transform. In a similar fashion, the discrete Fourie transform constitutes a domain-restricted version of the Fourier transform i which the sample points $e^{j2\pi k/N}$ for $0 \le k \le N-1$ on the unit circle constitut the domain. It therefore follows that the discrete Fourier transform correspond to samples of the z-transform at these N points on the unit circle. This interpla of transforms is depicted in Figure 4.1, where it is implicitly implied that th underlying time series is identically zero outside the interval $0 \le n \le N -$ (or more compactly $[0, N-1]$).

Sampled Magnitude and Phase Plots

To depict the behavior of $X_N(e^{j\omega})$ as a function of frequency, it is first benefici to express the DFT coefficients (4.4) in the polar representation

$$X_N(k) = |X_N(k)|e^{j\phi_N(k)} \qquad 0 \le k \le N-1 \quad (4.5$$

Then plots of the magnitude sequence $|X_N(k)|$ and the phase sequence $\phi_N(k)$ fo $0 \le k \le N-1$ are made in which the separation between samples is equate with the frequency spacing $\Delta\omega = 2\pi/N$ as in expression (4.2). In accordanc with the development above, it must follow that $|X_N(k)|$ and $\phi_N(k)$ correspon to samples of the magnitude and phase functions of $X_N(e^{j\omega})$, respectively, a $\omega_k = 2\pi k/N$ for $0 \le k \le N-1$. This interrelationship is depicted in Figur 4.2.

Periodicity Property

Since the DFT corresponds to sampled values of the truncated Fourier transform it follows immediately that all properties possessed by the latter transform ar shared by the DFT. As an exmaple, the fact that $X_N(e^{j\omega})$ is a periodic functio

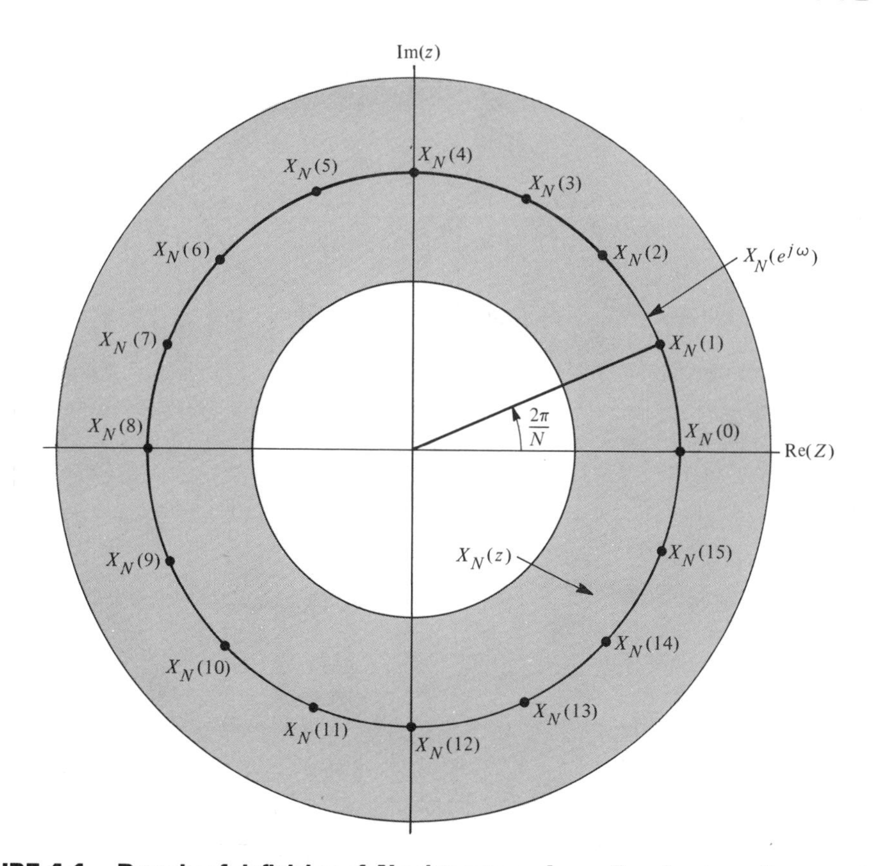

FIGURE 4.1. Domain of definition of *N*-point *z*-transform, Fourier transform, and discrete Fourier transform for *N* = 16.

of ω with fundamental period 2π implies that $X_N(k)$ must itself be a periodic sequence with period N, that is,

$$X_N(k) = X_N(k + N) \qquad \text{for all } k \tag{4.6}$$

This is made apparent upon substitution of $k + N$ for k in expression (4.4) and then simplifying that result. In this approach it is necessary to expand the domain of definition of the DFT from $0 \le k \le N - 1$ to all integer values of k.

EXAMPLE 4.1

To illustrate the mechanics of the DFT, let us consider the length 4 time series

$$x(0) = 2 \qquad x(1) = -1 \qquad x(2) = 1 \qquad x(3) = 4$$

We have purposely selected the time series length to be uncharacteristically small so as to render a paper-and-pencil-computable DFT. Appealing to expression (4.4), the four-point DFT is specified by

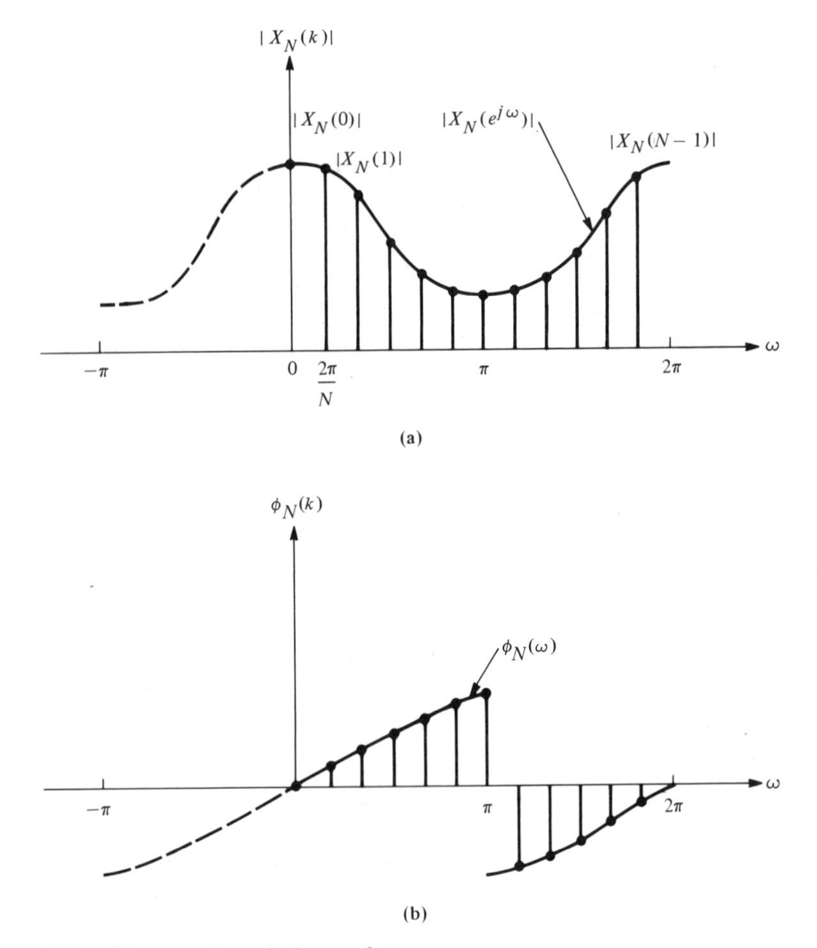

FIGURE 4.2. Magnitude and phase plots.

$$X_4(k) = \sum_{n=0}^{3} x(n)e^{-j2\pi kn/4}$$

$$= 2 - e^{-j\pi k/2} + e^{-j\pi k} + 4e^{-j3\pi k/2} \qquad 0 \leq k \leq 3$$

Upon evaluating this expression at $k = 0, 1, 2, 3$, the following DFT coefficients are obtained:

$$X_4(0) = 6 \qquad X_4(1) = 1 + j5 = \sqrt{26}\, e^{j\tan^{-1}(5)}$$

$$X_4(2) = 0 \qquad X_4(3) = 1 - j5 = \sqrt{26}\, e^{-j\tan^{-1}(5)}$$

As we show in Section 4.3, it is not coincidental that the symmetric DFT coefficient relationship $X_4(1) = \overline{X}_4(3)$ holds for this real-valued time series.

EXAMPLE 4.2

Compute the truncated Fourier transform and the DFT associated with the truncated complex sinusoidal time series

$$x(n) = Ae^{j\omega_1 n} \qquad 0 \le n \le N - 1$$

It was readily shown that the truncated Fourier transform associated with this time series is specified by

$$X_N(e^{j\omega}) = \sum_{n=0}^{N-1} Ae^{j\omega_1 n}e^{-j\omega n}$$

$$= A\, e^{j0.5(\omega_1 - \omega)(N-1)} \frac{\sin\,[0.5\,(\omega - \omega_1)N]}{\sin\,[0.5\,(\omega - \omega_1)]}$$

where ω_1 is a fixed radian frequency in $[0, 2\pi)$. Inserting the above complex sinusoidal time series expression into the DFT expression (4.4), it is found that

$$X_N(k) = \sum_{n=0}^{N-1} Ae^{j\omega_1 n}e^{-j2\pi kn/N}$$

$$= A \sum_{n=0}^{N-1} e^{j(\omega_1 - 2\pi k/N)n} \qquad \text{for } 0 \le k \le N - 1$$

Using the finite geometric series (2.2), we find that this expression simplifies to

$$X_N(k) = A\,\frac{1 - e^{j(\omega_1 - 2\pi k/N)N}}{1 - e^{j(\omega_1 - 2\pi k/N)}}$$

$$= A\, e^{j0.5(\omega_1 - 2\pi k/N)(N-1)} \frac{\sin\,[0.5\,(2\pi k - \omega_1 N)]}{\sin\,[0.5(2\pi k/N - \omega_1)]}$$

$$\text{for } 0 \le k \le N - 1$$

which corresponds to the sampled value of $X_N(e^{j\omega})$ at $\omega = 2\pi k/N$ as previously suggested. A plot of the magnitude of these frequency samples is shown in Figure 4.3 for the choice $N = 10$ and it is seen to correspond to samples of the truncated Fourier transform's magnitude (shown by dashed lines). From the plotted DFT coefficients it is apparent that the underlying time series has a significant amount of spectral energy in a region about ω_1. This observance is possible despite the relatively short data length, $N = 10$, being used here. The truncated Fourier transform is therefore seen again to provide a workable tool for identifying spectral content. In this case, however, several *false* spectral components are indicated through the side lobes of the rectangular window's transform.

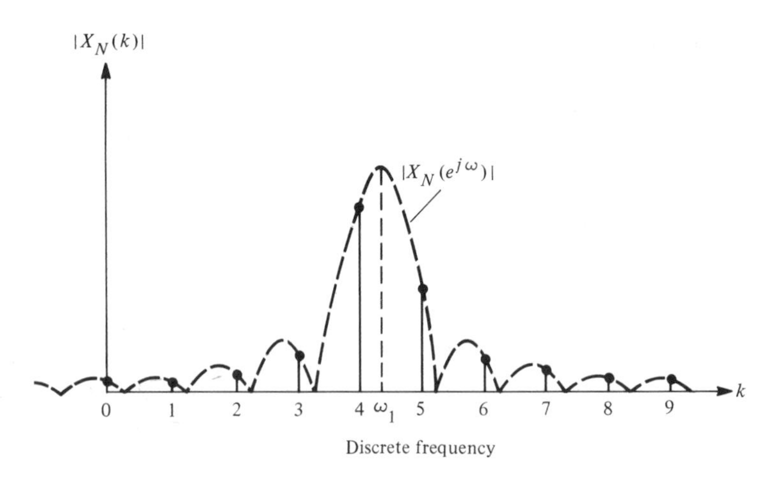

FIGURE 4.3. Plot of the DFT sample magnitude where the dotted line is the sketch of $X_N(e^{j\omega})$ versus ω.

An interesting special case occurs when the sinusoid's radian frequency ω_1 corresponds exactly with one of the discrete sampled frequencies used in the DFT. In particular, if $\omega_1 = 2\pi m/N$, where m is an integer in $[0, N - 1]$, the corresponding DFT coefficients are specified by

$$X_N(k) = \begin{cases} NA & k = m \\ 0 & k \neq m \end{cases}$$

For this case, an exact identification of the sinusoid's frequency is made evident upon examination of the DFT coefficients. It is to be noted, however, that the underlying truncated Fourier transform will still have side lobes. Their effects are here apparently absent, since the discrete frequency samples occur precisely where $X_N(e^{j\omega})$ is either NA or zero. ■

4.3 REAL-VALUED TIME SERIES

When the data being Fourier transformed is real valued, the DFT coefficients possess a useful symmetry. This is made evident upon examining the component $X_N(N - k)$ as given formally by

$$X_N(N - k) = \sum_{n=0}^{N-1} x(n)e^{-j2\pi(N-k)n/N}$$

$$= \sum_{n=0}^{N-1} x(n)e^{j2\pi kn/N}$$

where use of the readily established identity $e^{-j2\pi(k-N)n/N} = e^{j2\pi kn/N}$ has been employed. Upon taking the complex conjugate of this relationship and noting that $\bar{x}(n) = x(n)$, since $x(n)$ is real, it follows that

$$X_N(k) = \overline{X}_N(N - k) \qquad \text{for all } k \tag{4.7}$$

Thus the DFT samples possess an intrinsic complex-conjugate symmetry for the case of real data. One immediate consequence of this property is the fact that only half of the DFT samples need be computed using relationship (4.4) [e.g., $X(0)$, $X(1)$, . . ., $X(N/2)$ when N is even], with the remaining terms being generated using this complex-conjugate symmetry property.

The complex-conjugate symmetry property for the real data case reflects the property previously established for the Fourier transform of real time series. For such time series it was established that the magnitude function $|X(e^{j\omega})|$ and phase function $\phi(\omega)$ are symmetric and antisymmetric functions, respectively. For the DFT, we must then have

$$X_N(k) = \overline{X}_N(-k) \qquad \text{for all } k \tag{4.8}$$

Since $X_N(k)$ is a periodic sequence with period N, it follows that $\overline{X}_N(-k) = \overline{X}_N(N - k)$, which provides another verification of the complex-conjugate symmetry property (4.7).

4.4 ZERO PADDING

The DFT is seen to provide a computationally feasible method for determining sampled values of the truncated Fourier transform. If the data length N is relatively small, however, the frequency sampling interval $\Delta\omega = 2\pi/N$ may be too large to provide an adequate visual interpretation of the frequency behavior of $X_N(e^{j\omega})$. Fortunately, there exists a very simple procedure for overcoming this potential difficulty. It involves sampling the truncated Fourier transform on the more finely spaced set of equally spaced digital frequencies

$$\omega_k = \frac{2\pi}{L} \qquad k = 0, 1, \ldots, L - 1 \tag{4.9}$$

where L is a positive integer selected large enough so as to make the sampling interval $\Delta\omega = 2\pi/L$ sufficiently small for the purposes at hand. Evaluating the truncated Fourier transform (4.1) at these discrete frequencies gives rise to the sampled values

$$X_N(e^{j\omega_k}) = \sum_{n=0}^{N-1} x(n)e^{-j2\pi kn/L} \qquad 0 \le k \le L - 1 \tag{4.10}$$

Thus the data of length N are seen to give rise to L sampled values of the function-truncated Fourier transform, where typically $L > N$.

Relationship (4.10) is of the DFT form, with the notable exception that the number of data points N is not generally equal to the number of frequency samples L. We can rectify this disparity by the simple artifice of appending (*padding*) $L - N$ zeros to the original data to generate the *padded* L-length data sequence

$$x(0), x(1), \ldots, x(N-1), \underbrace{0, 0, \ldots, 0}_{L-N \text{ zeros}} \tag{4.11}$$

The L-point DFT of this auxiliary *padded* sequence results in the sampled values

$$\begin{aligned} X_L(k) &= \sum_{n=0}^{L-1} x(n)e^{-j2\pi kn/L} \\ &= X_N(e^{j\omega})\big|_{\omega=2\pi k/L} \qquad \text{for } 0 \leq k \leq L-1 \end{aligned} \tag{4.12}$$

where use has been made of the fact that $x(n) = 0$ for $N \leq n \leq L - 1$. Thus the L-point DFT of the padded auxiliary sequence (4.11) yields the required refined samples of the truncated Fourier transform.

It has probably occurred to the reader to consider the prudence of going through the seemingly artificial process of padding $L - N$ zeros to force a DFT formulation. The reason for employing this contrivance is strictly computational in nature, namely, we may use the fast Fourier transform algorithm as developed in Section 4.10 to evaluate the L-point DFT of the padded sequence (4.11) in a computationally efficient manner. It is typically found that the computational requirements for evaluating the FFT of the padded data (4.11) are significantly smaller than direct evaluation of expression (4.10).

EXAMPLE 4.3

To illustrate the effects of zero padding, let us consider the time series treated in Example 4.2, in which N zeros have been padded to the original N data elements. The resultant $2N$-length truncated time series is then specified by

$$x(n) = \begin{cases} Ae^{j\omega_1 n} & 0 \leq n \leq N-1 \\ 0 & N \leq n \leq 2N-1 \end{cases}$$

Taking the $2N$-point DFT of this time series, we find that

$$X_{2N}(k) = Ae^{j0.5(\omega_1 - \pi k/N)(N-1)} \frac{\sin[0.5(\pi k - \omega_1 N)]}{\sin[0.5(\pi k/N - \omega_1)]} \quad \text{for } 0 \leq k \leq 2N-1$$

A plot of the magnitude of the padded DFT coefficients $X_{2N}(k)$ is shown in Figure 4.4 for $N = 10$ together with the envelope (in dashed lines) of $X_N(e^{j\omega})$.

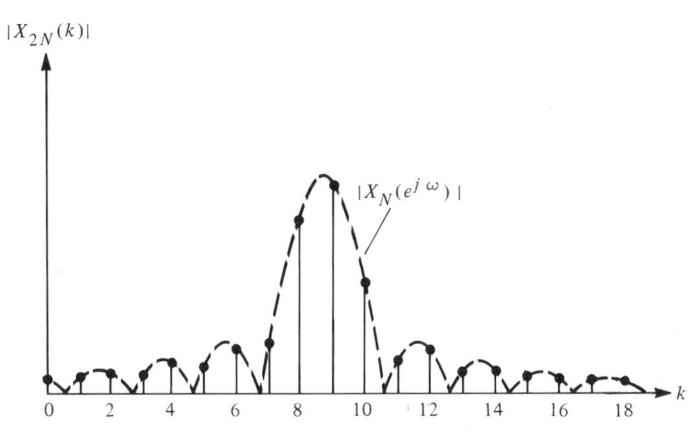

FIGURE 4.4. Plot of the DFT sample magnitudes with N zero padding in which the dashed line is the sketch of $X_N(e^{j\omega})$ versus ω.

Upon comparison of these results with Figure 4.2, it is seen that the padding has provided a better interpolation of the truncated Fourier transform $X_N(e^{j\omega})$. The N-even samples of the DFT $X_{2N}(k)$ are seen to correspond to the N samples of the DFT $X_N(k)$. Clearly, upon padding with more than N zeros, an even better interpolation is achieved. It is important to note, however, that zero padding does not increase the inherent frequency resolution that is limited by the underlying window's main lobe. ■

4.5 INVERSE DISCRETE FOURIER TRANSFORM

As with most transform methods, the true beauty and utility of the DFT is fully realized only after a systematic procedure is available for inverting the DFT operation. Specifically, it is desired to provide a procedure for uniquely recovering the time series elements $x(0), x(1), \ldots, x(N-1)$, which generated the given set of DFT coefficients through the DFT relationship

$$X_N(k) = \sum_{n=0}^{N-1} x(n)e^{-j2\pi kn/N} \qquad 0 \leq k \leq N-1 \tag{4.13}$$

This DFT relationship is seen to be composed of a system of N linear equations relating the N time series elements to the N DFT coefficients. Given the $X_N(k)$ for $0 \leq k \leq N-1$, we may in principle solve this system of linear equations to recover the generating time series elements $x(n)$ for $0 \leq n \leq N-1$. As we now show, this inversion process is manifested by the formula

$$x(n) = \frac{1}{N}\sum_{k=0}^{N-1} X_N(k)e^{j2\pi kn/N} \qquad 0 \leq n \leq N-1 \tag{4.14}$$

This inverse DFT expression is seen to resemble closely the DFT operation, the only distinction being the multiplier $1/N$ and the sign of the term in the exponential.

To prove that the two operations above do in fact constitute operator inverse pairs, let us first substitute DFT coefficient expression (4.13) into equation (4.14), giving the postulated equality

$$x(n) = \frac{1}{N} \sum_{k=0}^{N-1} \left[\sum_{m=0}^{N-1} x(m) e^{-j2\pi km/N} \right] e^{j2\pi kn/N}$$

In this substitution it is necessary to use m as the dummy summation variable in representing $X_N(k)$ since n has been reserved for the time variable in $x(n)$. After interchanging the order of the two summations, this expression becomes

$$x(n) = \frac{1}{N} \sum_{m=0}^{N-1} x(m) \sum_{k=0}^{N-1} [e^{j2\pi(n-m)/N}]^k \tag{4.15}$$

The rightmost summation is recognized as being a finite geometric series, and according to identity (2.2) with $\alpha = e^{j2\pi(n-m)/N}$, it is given by

$$\sum_{k=0}^{N-1} [e^{j2\pi(n-m)/N}]^k = N\delta(n - m) \tag{4.16}$$

where use has been made of the fact that m and n are each restricted to lie in the integer set $0, 1, \ldots, N - 1$. Upon substituting identity (4.16) into equation (4.15), we achieve the desired equivalency $x(n) = x(n)$, thereby verifying the correctness of the hypothesized inverse DFT operation (4.14).

EXAMPLE 4.4

Determine the inverse DFT associated with the four-point DFT coefficients

$$X_4(0) = 6 \qquad X_4(1) = 1 + j5 \qquad X_4(2) = 0 \qquad X_4(3) = 1 - j5$$

In accordance with inverse DFT relationship (4.14), we have

$$\begin{aligned} x(n) &= \tfrac{1}{4} \sum_{k=0}^{3} X_4(k) e^{j2\pi kn/4} \\ &= \tfrac{1}{4} [6 + (1 + j5)e^{j\pi n/2} + (1 - j5)e^{j3\pi n/2}] \end{aligned}$$

Evaluation of this expression for $n = 0, 1, 2, 3$ yields the desired time series that generated the DFT coefficients above, that is,

$$x(0) = 2 \qquad x(1) = -1 \qquad x(2) = 1 \qquad x(3) = 4$$

Not surprisingly, these results are in agreement with those of Example 4.1. Thus the inverse DFT is seen to undo the complex additive and multiplicative operations employed in the DFT, and vice versa. ■

Relationship to the DFT

As noted earlier, the DFT and inverse DFT operations are very similar in structure. We may use this similarity to compute an inverse DFT operation by applying minor changes to an existing DFT algorithm (or computer program). To see how this may be accomplished, let us first multiply each side of expression (4.14) by N and then take complex conjugates to obtain

$$N\overline{x}(n) = \sum_{k=0}^{N-1} \overline{X}_N(k)e^{-j2\pi kn/N}$$

After interchanging the roles of the variables k and n, this relationship may be expressed as

$$N\overline{x}(k) = \sum_{n=0}^{N-1} \overline{X}_N(n)e^{-j2\pi kn/N}$$

The right side of this relationship is now in the form of a DFT in which the "pseudo" time series elements are the conjugated DFT coefficients $\overline{X}_N(n)$. Thus, to generate an inverse DFT operation with a DFT programmed algorithm, we may follow the steps indicated in Table 4.1. By taking this approach, the need to program a separate inverse DFT algorithm is avoided.

TABLE 4.1 Computation of the Inverse DFT with a DFT Algorithm

1. Compute the DFT of the time series specified by $y(n) = \overline{X}_N(n)$ for $0 \leq n \leq N - 1$.
2. If $Y_N(0), Y_N(1), \ldots, Y_N(N - 1)$ denotes the DFT coefficients arising from step 1, the desired inverse DFT time series is specified by $x(n) = \overline{Y}_N(n)/N$ for $0 \leq n \leq N - 1$.

4.6 PERIODIC SIGNALS AND THEIR FOURIER SERIES REPRESENTATION

Although the DFT was developed for characterizing the spectral content of finite-length signals, it is also useful for representing periodic time series. This is made apparent upon examination of the inverse DFT relationship

$$x(n) = \frac{1}{N} \sum_{k=0}^{N-1} X_N(k) e^{j2\pi kn/N} \qquad (4.17)$$

where the domain of definition on n is not restricted to $0 \le n \le N - 1$ as it was in Section 4.5. The time series governed by expression (4.17) is periodic with period N, since it satisfies the readily established property:

$$x(n + N) = x(n) \qquad \text{for all } n$$

The implications of this periodic interpretation are indeed noteworthy and are now briefly discussed.

If the time series $\{x(n)\}$ is periodic with period N, it may always be expressed in the *Fourier series* representation (4.17). In this representation the time series has been decomposed into a sum of complex sinusoids of frequencies $\omega_k = 2\pi k/N$ for $0 \le k \le N - 1$. The amplitudes of these sinusoids are specified by

$$X_N(k) = \sum_{n=0}^{N-1} x(n) e^{-j2\pi kn/N} \qquad 0 \le k \le N - 1 \qquad (4.18)$$

which are recognized as being equal to the DFT coefficients associated with the truncated time series $x(0), x(1), \ldots, x(N - 1)$. This truncated time series is seen to consist of one period of the periodic sequence. Thus the DFT can be a useful mechanism for representing periodic time series.

From these comments and those made in previous sections, it is evident that the DFT coefficients may be used for either one of the following purposes:

1. To approximate the spectral content of a nonperiodic time series from a finite set of observations.
2. To measure the sinusoidal amplitudes in a Fourier series representation of a periodic time series.

If we depend on the specific application at hand, it is generally clear which interpretation is to be made. In most signal-processing applications, the DFT is used to approximate the spectral content of nonperiodic time series. Thus henceforth we emphasize the first interpretation.

4.7 FILTER SYNTHESIS: TIME-DOMAIN IMPLEMENTATION

One of the important dividends of the DFT is its usefulness in synthesizing prototype frequency-discrimination filters. To illustrate this point, we now consider the specific case of low-pass filter synthesis. The method to be followed, however, is readily extended to other standard forms of frequency-discrimination filters (e.g., high-pass, bandpass, and notch filters).

The periodic frequency response associated with an ideal low-pass filter with cutoff frequency ω_c is specified by

$$H_i(e^{j\omega}) = \begin{cases} e^{-j\omega n_0} & 0 \leq \omega \leq \omega_c \text{ and } 2\pi - \omega_c \leq \omega \leq 2\pi \\ 0 & \omega_c < \omega < 2\pi - \omega_c \end{cases} \tag{4.19}$$

where n_0 designates the filter delay constant. The magnitude and phase plots associated with this frequency response are shown in Figure 4.5 by the solid lines. We have identified this frequency response on the unconventional frequency interval $[0, 2\pi)$ rather than on the more conventional interval $(-\pi, \pi]$. This is done in recognition of the fact that the DFT is itself naturally associated with uniformly spaced frequency samples on the frequency interval $[0, 2\pi)$.

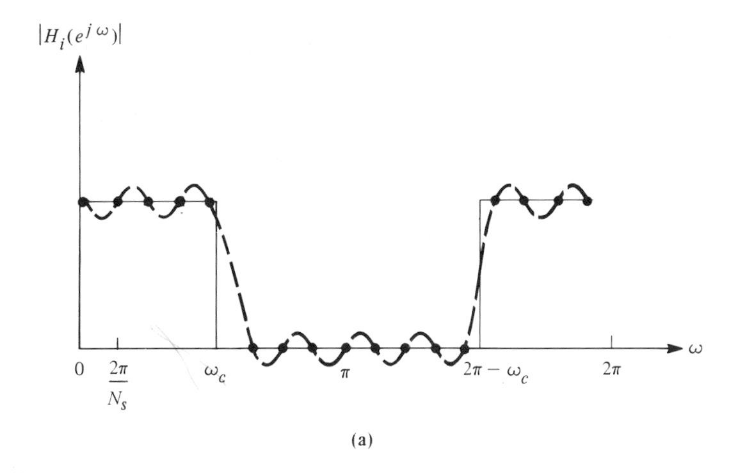

(a)

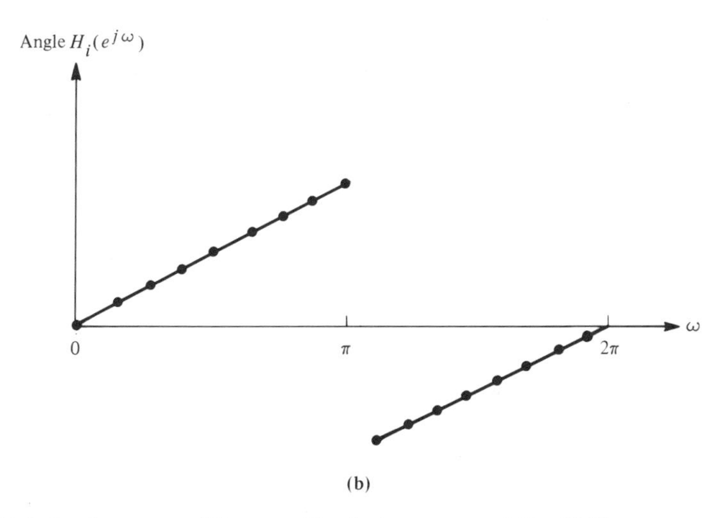

(b)

FIGURE 4.5. Low-pass filter synthesis by means of the DFT.

It will be recalled from Section 4.2 that the DFT coefficients correspond to frequency samples of the associated truncated Fourier transform. Using this property, we now obtain a systematic procedure for selecting the weighting coefficients $h(n)$ of the nonrecursive filter

$$y(n) = \sum_{k=0}^{N-1} h(k)x(n - k) \tag{4.20}$$

so that its associated frequency response

$$H_N(e^{j\omega}) = \sum_{n=0}^{N-1} h(n)e^{-j\omega n} \tag{4.21}$$

agrees exactly with the ideal behavior (4.19) on the discrete frequency set $\omega_k = 2\pi k/N$ for $0 \le k \le N - 1$. This condition requires that the DFT coefficients associated with the $h(n)$ elements satisfy

$$\begin{aligned} H_N(k) &= H_i(e^{j2\pi k/N}) \\ &= \begin{cases} e^{-j2\pi k n_0/N} & 0 \le k \le n_c \text{ and } N - n_c \le k \le N - 1 \\ 0 & n_c < k < N - n_c \end{cases} \end{aligned} \tag{4.22}$$

where

$$n_c = \mathrm{INT}\left(\frac{\omega_c N}{\pi}\right) \tag{4.23}$$

denotes the integer part of the quantity $\omega_c N/\pi$. These sampled values are shown by the dots in Figure 4.5 for the case $N = 16$, $n_0 = 7.5$, and $\omega_c = \pi/4$.

The filter weighting coefficients that give rise to this frequency behavior are obtained directly by taking the inverse DFT of the DFT coefficient sequence (4.22). In particular, incorporation of this sequence into the inverse DFT expression (4.14) gives

$$\begin{aligned} h(n) &= \frac{1}{N}\left[\sum_{k=0}^{n_c} e^{j2\pi k(n-n_0)/N} + \sum_{k=N-n_c}^{N-1} e^{j2\pi k(n-n_0)/N}\right] \\ &= \frac{1}{N}\left\{-1 + \sum_{k=0}^{n_c}\left[e^{j2\pi k(n-n_0)/N} + e^{j2\pi(N-k)(n-n_0)/N}\right]\right\} \end{aligned}$$

If we use the finite geometric series identity (2.2), this expression can eventually be simplifed to

$$h(n) = \frac{\sin\,[\pi(2n_c + 1)(n - n_0)/N]}{N \sin\,[\pi(n - n_0)/N]} \qquad 0 \le n \le N - 1 \tag{4.24}$$

This weighting sequence is symmetric about the point $n = n_0$ where it has a maximum at n_0 (if n_0 is an integer) as illustrated in Figure 4.6.

The nonrecursive filter (4.20) with these weighting coefficient values has a frequency response (4.21) that agrees perfectly with the ideal behavior (4.19) on the discrete frequency set $\omega_k = 2\pi k/N$ for $0 \leq k \leq N - 1$. Unfortunately, it is also found that $H_N(e^{j\omega}) \neq H(e^{j\omega})$ for all other frequencies in $[0, 2\pi)$. This behavior is illustrated in Figure 4.5 by the dashed-line plot. An improved frequency response behavior is achievable by making N suitably large.

Symmetrical Weighting Coefficients

To ensure that $H_N(e^{j\omega})$ provides an adequate approximation to the ideal behavior $H_i(e^{j\omega})$ for all ω, it is necessary to select the delay parameter n_0 judiciously. One important aspect of this selection is that of requiring that the filter weighting coefficients obey the symmetry condition $h(n) = h(N - 1 - n)$ for $0 \leq n \leq N - 1$. This symmetry is desirable, since it ensures that the resultant filter will have a desirable linear phase function, as will be proved in Chapter 5. To effect this symmetry condition, it is necessary to select

$$n_0 = \frac{N - 1}{2} \tag{4.25}$$

which is an integer if N is odd and not an integer for N even. This choice of n_0 is optimal in the sense that the resultant frequency response $H_N(e^{j\omega})$ provides the best integral squared-error match to the ideal behavior (4.19). An additional dividend accrued in using this symmetrical choice for n_0 is that the implementation of nonrecursive expression (4.20) is effected in a more efficient manner.

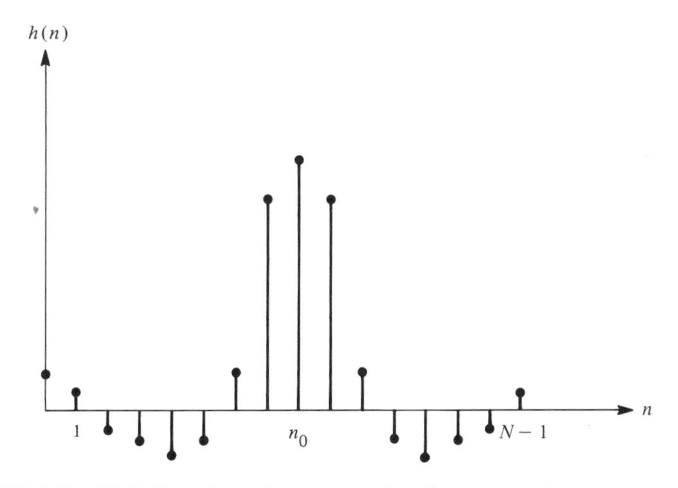

FIGURE 4.6. Unit impulse of nonrecursive low-pass filter.

In particular, due to the fact that $h(n) = h(N - 1 - n)$ for this symmetrical choice of n_0, this nonrecursive expression takes the form

$$y(n) = \sum_{k=0}^{(N/2)-1} h(k)[x(n - k) + x(n - N - 1 - k)] \qquad (4.26a)$$

when N is even and is specified by

$$y(n) = h([N - 1]/2)x(n - [N - 1]/2) + \sum_{k=0}^{(N-1)/2} h(k)[x(n - k) + x(n - N - 1 - k)] \qquad (4.26b)$$

when N is odd. It is apparent that the number of multiplication and additive operations needed to compute the response element $y(n)$ have been reduced by approximately a factor of 2 due to the symmetrical behavior of $h(n)$.

EXAMPLE 4.5

Design a symmetric nonrecursive low-pass filter of length $N = 33$ with cutoff frequency $\omega_c = 0.25\pi$. The unit impulse response of this filter is specified by equation (4.24) with

$$n_c = \text{Int}\left(\frac{0.25\pi\ 33}{\pi}\right) = 8$$

$$n_0 = \frac{32}{2} = 16$$

Namely,

$$h(n) = \frac{\sin[17\pi(n - 16)/33]}{33 \sin[\pi(n - 16)/33]} \qquad 0 \le n \le 32$$

Using these weighting coefficients, a plot of the magnitude of the corresponding frequency response (4.21) in decibels is as shown in Figure 4.7. ■

Other Forms of Frequency-Discrimination Filters

As suggested at the beginning of this section, the philosophy employed here may be extended to other forms of frequency-discrimination filters. In this development, frequent use of the symmetric low-pass N-length prototype sequence

$$h_{\omega_c}(n) = \frac{\sin[\pi(2n_c + 1)(n - n_0)/N]}{N \sin[\pi(n - n_0)/N]} \qquad 0 \le n \le N - 1 \qquad (4.27)$$

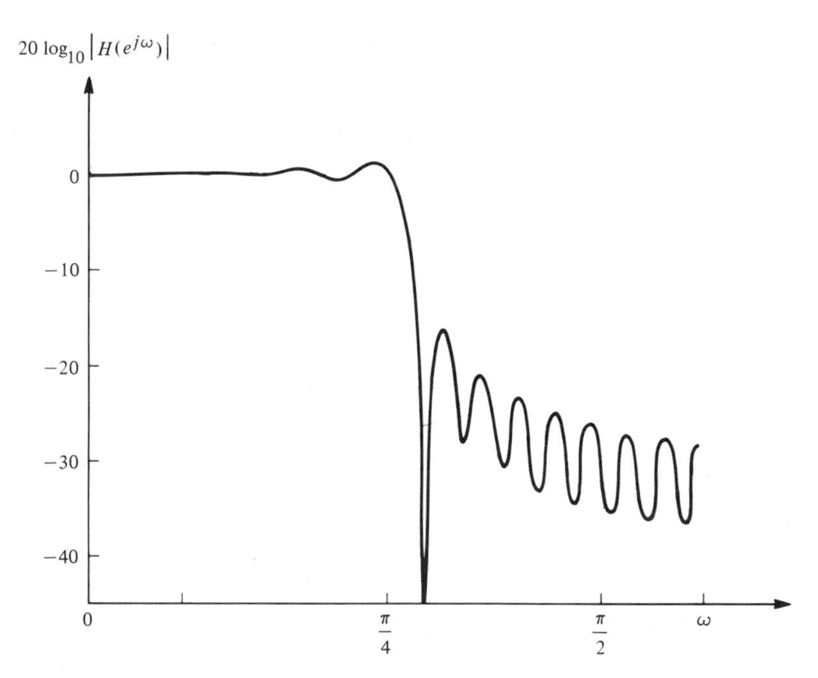

FIGURE 4.7. Magnitude plot of optimal symmetrical low-pass filter with cutoff frequency $\omega_c = \pi/4$ and $N = 33$.

is made where

$$n_0 = \frac{N-1}{2} \quad \text{and} \quad n_c = \text{INT}\left(\frac{\omega_c N}{\pi}\right) \tag{4.28}$$

The subscript ω_c has been appended to $h(n)$ to recognize the associated low-pass filter's cutoff frequency. With these notations in mind, the N-length, unit-impulse responses to be used in convolution summation (4.20) to effect standard forms of frequency-discrimination filters are listed in Table 4.2. The frequency response (4.21) associated with each filter will agree exactly with the idealized behavior on the frequency set $\omega_k = 2\pi k/N$ but will otherwise be different.

4.8 CONVOLUTION OPERATION IMPLEMENTED VIA THE DFT

The convolution summation provides the most direct means for implementing a linear nonrecursive signal-processing algorithm. This convolution summation operation takes the form

$$y(n) = \sum_{k=0}^{M-1} h(k)x(n-k) \tag{4.29}$$

TABLE 4.2 Standard Nonrecursive Filters Using the Prototype Low-Pass Filter Unit Impulse Sequence (4.27)

Filter	Ideal Frequency Response, $H_i(e^{h\omega})$ for $-\pi \le \omega \le \pi$	Unit-Impulse Response, $h(n)$ for $0 \le n \le N-1$
Low pass	$= \begin{cases} e^{-j\omega n_0} & 0 \le \lvert\omega\rvert \le \omega_c \\ 0 & \text{otherwise} \end{cases}$	$h_{\omega c}(n)$
High pass	$= \begin{cases} e^{-j\omega n_0} & \omega_c < \lvert\omega\rvert \le \pi \\ 0 & \text{otherwise} \end{cases}$	$h_\pi(n) - h_{\omega c}(n)$
Bandpass	$= \begin{cases} e^{-j\omega n_0} & \omega_1 < \lvert\omega\rvert \le \omega_2 \\ 0 & \text{otherwise} \end{cases}$	$h_{\omega_2}(n) - h_{\omega_1}(n)$
Band reject	$= \begin{cases} e^{-j\omega n_0} & \text{otherwise} \\ 0 & \omega_1 < \lvert\omega\rvert \le \omega_2 \end{cases}$	$\delta(n - n_0) - h_{\omega_2}(n) + h_{\omega_1}(n)$

in which M designates the filter's length. In all practical applications, there is provided a finite set of data to be processed by this filtering operation. The excitation data are taken here to be of length L, that is,

$$x(0), x(1), \ldots, x(L-1) \tag{4.30}$$

where it is tacitly assumed that $x(n) = 0$ for $n < 0$ and $n \ge L$. Using a direct evaluation of nonrecursive operator (4.29), we may compute the corresponding response elements

$$\underbrace{y(0), y(1), \ldots, y(M-2)}_{\text{transient}}, \underbrace{y(M-1), \ldots, y(L-1)}_{\text{steady state}}, \underbrace{y(L), \ldots, y(L+M-2)}_{\text{transient}} \tag{4.31}$$

where it has been assumed that $M \le L$, as is normally the case. The first and last $M - 1$ terms of the response have been labeled *transient* due to the fact that only a portion of the excitation elements in convolution summation (4.29) are present for $0 \le n \le M - 2$ and $L \le n \le L + M - 2$. The middle $L + 1 - M$ terms correspond to the *steady-state* behavior, in which all the excitation elements are present.

In most signal-processing applications, the number of algebraic computations needed to evaluate each response element is of fundamental importance. This concern arises from a desire to have a computationally fast algorithm for real-time applications and (or) the need to reduce computational costs. With this objective in mind, a standard procedure for measuring the computational requirements necessitated by a direct evaluation of convolution summation (4.29) is now made. Specifically, this entails determining the number of multiplications and additions needed to generate the entire response (4.31). It is readily shown that a total of ML multiplications and $(M - 1)(L - 1)$ additions are required to compute these response elements when a direct evaluation of relationship (4.29) is made.

A reduction in these computational requirements may be achieved by employing the DFT in conjunction with the fast Fourier transform algorithm to be developed later. To see how this may be accomplished, let us take the N-point DFT of the response as generated by relationship (4.29). The integer N must be chosen to at least equal the response's data length, that is,

$$N \geq L + M - 1 \tag{4.32}$$

Under this restriction on N, the DFT of the response sequence is formally given by

$$\begin{aligned} Y_N(k) &= \sum_{n=0}^{N-1} y(n)e^{-j2\pi kn/N} \\ &= \sum_{n=0}^{N-1} \left[\sum_{m=0}^{M-1} h(m)x(n - m) \right] e^{-j2\pi kn/N} \end{aligned}$$

If we interchange the order of summations and then make the change of summation variable substitution $p = n - m$ for the variable n, this expression becomes

$$\begin{aligned} Y_N(k) &= \sum_{m=0}^{M-1} \sum_{p=-m}^{N-1-m} h(m)x(p)e^{-j2\pi k(p+m)/N} \\ &= \sum_{m=0}^{M-1} h(m)e^{-j2\pi km/N} \sum_{p=0}^{L-1} x(p)e^{-j2\pi kp/N} \end{aligned}$$

In going from the first to the second line of this equation, the limits on the summation with respect to p have been changed to $[0, L - 1]$ to reflect the fact that $x(p) = 0$ for $p \notin [0, L - 1]$. The summations relative to m and p are seen to correspond to the N-point DFTs of the padded sequences

$$h(0), h(1), \ldots, h(M - 1), \underbrace{0, 0, \ldots, 0}_{N - M \text{ zeros}} \tag{4.33a}$$

$$x(0), x(1), \ldots, x(L - 1), \underbrace{0, 0, \ldots, 0}_{N - L \text{ zeros}} \tag{4.33b}$$

which are denoted by $H_N(k)$ and $X_N(k)$, respectively. Thus the DFT of the response sequence is simply given by the product form

$$Y_N(k) = H_N(k)X_N(k) \qquad 0 \le k \le N - 1 \tag{4.34}$$

In arriving at this DFT frequency response relationship, it is important to appreciate the need to select N to satisfy (4.32). This always necessitates that the sequences being convolved be adequately padded, in accordance with expression (4.33).

To compute the filtered response sequence using the indirect procedure advocated here, we first generate the N-point DFTs associated with the padded sequences (4.33). Next, the response's DFT is obtained using the product rule (4.34). Finally, the inverse DFT is applied to this response DFT coefficient sequence to generate the required response $y(n)$ for $0 \le n \le N - 1$. This indirect procedure is summarized in Table 4.3.

A casual examination of the steps outlined in Table 4.3 indicates that on the order of $3N^2$ complex multiplications and additions are required to compute the filter response using this indirect approach. Namely, three N-point DFT calculations are required, with each DFT necessitating N^2 complex multiplications and additions. On the other hand, direct evaluation of the convolution operation was found to depend on the order of ML multiplications. Since $3N^2 \ge 3(M + L - 2)^2$, it would appear that the indirect approach is far more demanding computationally than the direct approach. With the introduction of the fast Fourier transform algorithm, however, the computational demands of the DFT are reduced from N^2 to $N \log_2(N)$. In this situation, the indirect approach can lead to significantly smaller computational demands than those of the direct approach.

EXAMPLE 4.6

To illustrate the importance of selecting N in accordance with relationship (4.32) so as to effect convolution, let us consider the two length 2 sequences

$$\{x(0), x(1)\} \qquad \text{and} \qquad \{h(0), h(1)\}$$

Using convolution operator (4.29) with $M = L = 2$, we find the convolved sequence $y = h * x$ to be

TABLE 4.3 Indirect DFT Method for Evaluating a Linear Convolution Operation

1. Compute the N-point DFTs of the padded sequences (4.33) to be convolved in which N is chosen according to expression (4.32). This gives rise to $H_N(k)$ and $X_N(k)$ for $0 \le k \le N - 1$.
2. Determine the response's DFT coefficients according to the product rule $Y_N(k) = H_N(k)\, X_N(k)$ for $0 \le k \le N - 1$.
3. Take the N-point inverse DFT of the $Y_N(k)$ sequence generated in step 2 to determine the filtered response $y(n)$ for $0 \le n \le N - 1$.

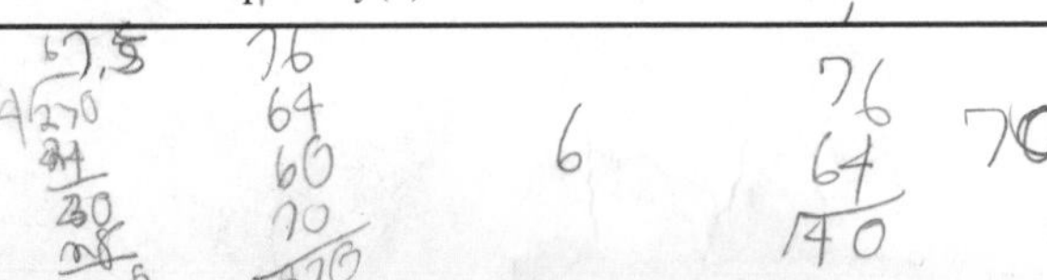

$$y(0) = h(0)x(0) \qquad y(1) = h(0)x(1) + h(1)x(0) \qquad y(2) = h(1)x(1)$$

To achieve the same result using the DFT approach, let us select $N = 3$, which satisfies inequality requirement (4.32). We then have

$$\begin{aligned} Y_3(k) &= H_3(k)X_3(k) \\ &= [h(0) + h(1)e^{-j2\pi k/3}][x(0) + x(1)e^{-j2\pi k/3}] \end{aligned}$$

Carrying out the indicated multiplication, yields

$$Y_3(k) = h(0)x(0) + [h(0)x(1) + h(1)x(0)]e^{-j2\pi k/3} + h(1)x(1)e^{-j4\pi k/3}$$

which is recognized as being the three-point DFT of the convolved sequence $h * x$ generated above. An inverse DFT applied to this sequence then gives rise to the desired convolved sequence.

If the parameter N is inadvertently selected in violation of inequality (4.32), however, a different result arises. For instance, if $N = 2$, we have

$$\begin{aligned} Y_2(k) &= [h(0) + h(1)e^{-j2\pi k/2}][x(0) + x(1)e^{-j2\pi k/2}] \\ &= h(0)x(0) + [h(0)x(1) + h(1)x(0)]e^{-j\pi k} + h(1)x(1)e^{-j2\pi k} \\ &= h(0)x(0) + h(1)x(1) + [h(0)x(1) + h(1)x(0)]e^{-j\pi k} \end{aligned}$$

This is recognized as being a two-point DFT of the sequence

$$y(0) = h(0)x(0) + h(1)x(1) \qquad y(1) = h(0)x(1) + h(1)x(0)$$

which is not in agreement with the convolution result desired. Clearly, the source of the difficulty arises from the fact that the nominal third-time term in the convolution $h(1)x(1)e^{-j2\pi k}$ has been folded back to a zero-time-term component because $e^{-j2\pi k} = 1$. ■

Important Practical Consideration

In many applications it is desired to convolve a relatively short duration time series with a much longer-duration time series. For example, the filtering of an information-bearing signal in which the signal has a long duration and the filter's unit-impulse response is relatively short. Although the DFT method just described could be used for this purpose, it possesses two serious drawbacks: (1) it requires that excessively large-length DFTs be computed, and (2) the computation can begin only after all the long-duration data are available (i.e., off-line computation). Fortunately, it is possible to adapt the DFT method so as to considerably alleviate these shortcomings. A discussion of one such technique is now given.

Let it be desired to convolve (or filter) the long-duration time series $\{x(n)\}$ with a nonrecursive filter whose unit-impulse response $h(n)$ for $0 \le n \le M - 1$ is relatively short. To effect this convolution in view of the comments made above, let us first decompose the time series $\{x(n)\}$ into a set of L-length disjoint segments according to

$$x_m(n) = \begin{cases} x(n) & mL \le n \le (m + 1)L - 1 \\ 0 & \text{otherwise} \end{cases} \tag{4.35}$$

for $m = 0, 1, 2, \ldots$, in which this time series is arbitrarily taken to be zero for negative time. The decomposition is depicted in Figure 4.8. It then follows that the time series $\{x(n)\}$ can be expressed as

$$x(n) = \sum_{m=0}^{\infty} x_m(n) \tag{4.36}$$

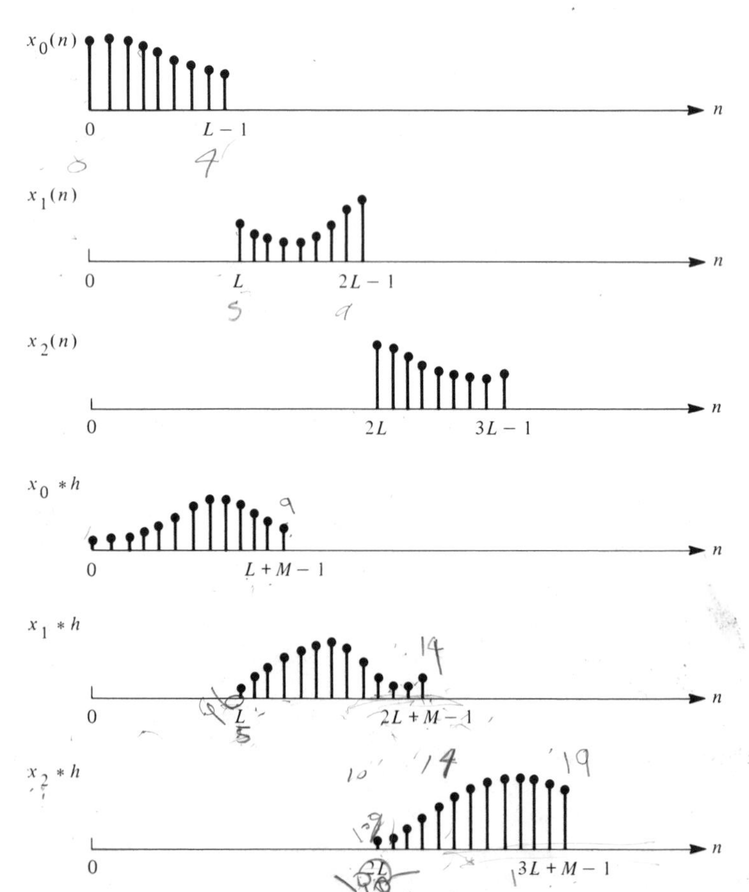

FIGURE 4.8. Convolution using signal decomposition for $L = 9$ and $M = 5$.

Furthermore, the convolution of this time series with the M-length time series $h(n)$ becomes

$$y(n) = \sum_{k=0}^{M-1} h(k)x(n - k)$$

$$= \sum_{k=0}^{M-1} h(k) \sum_{m=0}^{\infty} x_m(n - k)$$

Upon interchanging the order of summations, this response sequence becomes

$$y(n) = \sum_{m=0}^{\infty} \left[\sum_{k=0}^{M-1} h(k)x_m(n - k) \right] \tag{4.37}$$

The term within the nonrecursive filter brackets in this overall response expression is recognized as being the response of the M-length nonrecursive filter to the L-length excitation $x_m(n)$, that is,

$$y_m(n) = \begin{cases} \sum_{k=0}^{M-1} h(k)x_m(n - k) & mL \leq n \leq (m + 1)L + M - 2 \\ 0 & \text{otherwise} \end{cases} \tag{4.38}$$

for $m = 0, 1, 2, \ldots$. This mth response segment of length $L + M - 1$ may be therefore computed by using the indirect DFT approach for $N \geq L + M - 1$. By selecting L appropriately small, the DFT size may be made manageable and the data can be processed on a nearly real-time basis. After computing the response segments (4.38), the overall response is simply

$$y(n) = \sum_{m=0}^{\infty} y_m(n) \tag{4.39}$$

It is to be noted that the $(L + M - 1)$-length contiguous response segments $y_m(n)$ and $y_{m+1}(n)$ will overlap by $M - 1$ points, as shown in Figure 4.8.

4.9 DFT IMPLEMENTATION OF FILTERS

The DFT constitutes a particularly efficient means for implementing frequency-discrimination filtering. To illustrate the reasoning behind this comment, let it be desired to low-pass filter the N-length data sequence $x(n)$ defined for $0 \leq n \leq N - 1$. If the cutoff frequency of this filter is to be ω_c radians, it follows that the idealized frequency-domain behavior

$$Y(e^{j\omega}) = \begin{cases} X_N(e^{j\omega})e^{-j\omega n_0} & 0 \leq |\omega| \leq \omega_c \\ 0 & \omega_c < |\omega| \leq \pi \end{cases} \tag{4.40}$$

is to be effected, where $X_N(e^{j\omega})$ and $Y(e^{j\omega})$ denote the Fourier transforms of the filter's excitation (length N) and response sequences, respectively. The entity n_0 is a delay parameter whose choice is of particular importance from a performance viewpoint. Unfortunately, the response sequence $\{y(n)\}$ so generated is infinite in length and is specified by

$$y(n) = \sum_{k=0}^{N-1} x(k) \frac{\sin[\omega_c(n - n_0 - k)]}{\pi(n - n_0 - k)} \quad \text{for all } n \tag{4.41}$$

This is made evident upon taking the inverse Fourier transform of expression (4.40), in which the truncated Fourier transform expression (4.1) is incorporated.

Although expression (4.41) provides the desired theoretical result, it is of little practical value since the response is of infinite length. To circumvent this difficulty, our previous discussions on the DFT suggest a procedure for approximating frequency relationship (4.40). Specifically, the elements of the N-length sequence

$$y(0), y(1), \ldots, y(N - 1) \tag{4.42}$$

will be selected so that their truncated Fourier transform $Y_N(e^{j\omega})$ agrees with $Y(e^{j\omega})$ on the discrete frequency set $\omega_k = 2\pi k/N$ for $0 \le k \le N - 1$. This is seen to require that the DFT coefficients of the N-length sequence (4.42) satisfy

$$Y_N(k) = \begin{cases} X_N(k)e^{-j\omega n_0} & 0 \le n \le n_c,\ N - n_c \le n \le N - 1 \\ 0 & \text{otherwise in } [0, N - 1] \end{cases} \tag{4.43}$$

where

$$n_c = \text{INT}\ (\omega_c N/\pi) \tag{4.44}$$

The filtered response associated with these DFT coefficients is then obtained by taking an inverse DFT of $Y_N(k)$. A summary of the principal steps in this indirect DFT approach is given in Table 4.4. This approach is made computationally attractive upon use of the fast Fourier transform algorithm implementation of the DFT.

Two precautionary statements must be made concerning this method. First, the desired identity $Y_N(e^{j\omega}) = Y(e^{j\omega})$ holds in general only on the discrete

TABLE 4.4 Low-Pass Filtering via the DFT

1. Compute the N-point DFT of the N-length data $x(n)$ for $0 \le n \le N - 1$. This yields $X_N(0), X_N(1), \ldots, X_N(N - 1)$.
2. Compute the inverse DFT of the related low-pass-filtered DFT coefficient sequence

   $$X_N(0), \ldots, X_N(n_c), 0, 0, \ldots, 0, X_N(N - n_c), \ldots, X_N(N - 1)$$

   This yields the low-pass-filtered response $y(0), \ldots, y(N - 1)$.

frequency set $\omega_k = 2\pi k/N$ for $0 \le k \le N - 1$. Furthermore, the actual goal (often unstated) is that of low-pass filtering the infinite-length data $x(n)$. Because of the finite observations of these data, however, the Fourier transform $X_N(e^{j\omega})$ is a smoothed rippled version of $X(e^{j\omega})$. Each of these factors will result in a distortion of the ideal low-pass-filtered objective.

4.10 FAST FOURIER TRANSFORM ALGORITHM

The widespread use of Fourier transform techniques in signal processing was a direct by-product of the development of computationally efficient algorithms for computing the DFT. Collectively, these procedures are known as *fast Fourier transform* (FFT) algorithms. Whereas direct evaluation of the DFT formula necessitates approximately N^2 complex multiplications and additions, typical FFT algorithms reduce this requirement to $N \log_2 N$ complex multiplications and additions. For relatively modest values of N, this computational savings can be significant (i.e., $\log_2 N < < N$) and makes feasible the utilization of DFT methodology to applications that would not be practical otherwise.

Various versions of FFT algorithms are predicated on the *divide-and-conquer* principle. To illustrate this approach, we now consider the *decimation-in-frequency* FFT algorithm. In this development it is assumed that the data length N is an integer power of 2, that is,

$$N = 2^M \tag{4.45}$$

where M is a positive integer. No loss in generality is incurred by so selecting N, since we may always pad a sufficient number of zeros to a given data set to achieve this requirement. In the initial stage of the divide-and-conquer routine, the DFT expression is decomposed into two separate summations involving the first and last $N/2$ data points, that is,

$$\begin{aligned} X_N(k) &= \sum_{n=0}^{N-1} x(n)e^{-j2\pi kn/N} \\ &= \sum_{n=0}^{(N/2)-1} x(n)e^{-j2\pi kn/N} + \sum_{n=N/2}^{N-1} x(n)e^{-j2\pi kn/N} \qquad 0 \le k \le N-1 \end{aligned}$$

Upon making the change of summation variations $m = n - N/2$ in the second summation, we have

$$\begin{aligned} X_N(k) &= \sum_{n=0}^{(N/2)-1} x(n)e^{-j2\pi kn/N} + \sum_{m=0}^{(N/2)-1} x(m + N/2)e^{-j2\pi k(m+N/2)/N} \\ &= \sum_{n=0}^{(N/2)-1} [x(n) + e^{-j\pi k}x(n + N/2)]e^{-j2\pi kn/N} \qquad 0 \le k \le N-1 \end{aligned} \tag{4.46}$$

Examination of this expression indicates that the number of complex multiplication and addition operations needed to compute the N DFT coefficients has been reduced from N^2 to $N(N/2)$ in this initial divide-and-conquer stage. As we show shortly, a continuation of this approach provides yet further savings.

Let the DFT coefficients (4.46) be separated into even and odd values of k, so that

$$\begin{aligned} X_N(2k) &= \sum_{n=0}^{(N/2)-1} [x(n) + e^{-j2\pi k}x(n + N/2)]e^{-j2\pi(2k)n/N} \\ &= \sum_{n=0}^{(N/2)-1} [x(n) + x(n + N/2)]e^{-j2\pi kn/(N/2)} \end{aligned} \tag{4.47}$$

and

$$\begin{aligned} X_N(2k + 1) &= \sum_{n=0}^{(N/2)-1} [x(n) + e^{-j\pi(2k+1)}x(n + N/2)]e^{-j2\pi(2k+1)n/N} \\ &= \sum_{n=0}^{(N/2)-1} [x(n) - x(n + N/2)]e^{-j2\pi n/N}e^{-j2\pi kn/(N/2)} \end{aligned} \tag{4.48}$$

where each of these expressions holds for $0 \leq k \leq (N/2) - 1$. An important observation is now made: These even and odd DFT coefficient expressions can equivalently be interpreted as being equal to the $(N/2)$-point DFT coefficients of the $N/2$ even-length sequences

$$f(n) = x(n) + x(n + N/2) \qquad 0 \leq n \leq (N/2) - 1 \tag{4.49}$$

$$g(n) = [x(n) - x(n + N/2)]e^{-j2\pi n/N} \qquad 0 \leq n \leq (N/2) - 1 \tag{4.50}$$

respectively. Specifically, it directly follows that

$$X_N(2k) = F_{N/2}(k) \qquad 0 \leq k \leq (N/2) - 1 \tag{4.51}$$

$$X_N(2k + 1) = G_{N/2}(k) \qquad 0 \leq k \leq (N/2) - 1 \tag{4.52}$$

The first stage of decimation-in-frequency, divide-and-conquer routine, in which one N-point DFT computation has been replaced by two $(N/2)$-point DFT computations, has thus been completed. This scheme is shown in Figure 4.9 for the case $N = 8$. Since each $(N/2)$-point DFT computation requires approximately $(N/2)^2$ complex multiplication and addition operations if a direct evaluation is made, this divide-and-conquer routine has produced a factor-of-2 savings in computations [i.e., N^2 versus $2(N/2)^2 = N^2/2$].

Clearly, this divide-and-conquer routine can be continued to achieve yet further computational savings. At the second stage, the DFT of the $N/2$ even-length sequences (4.49) and (4.50) are computed in a similar fashion. This

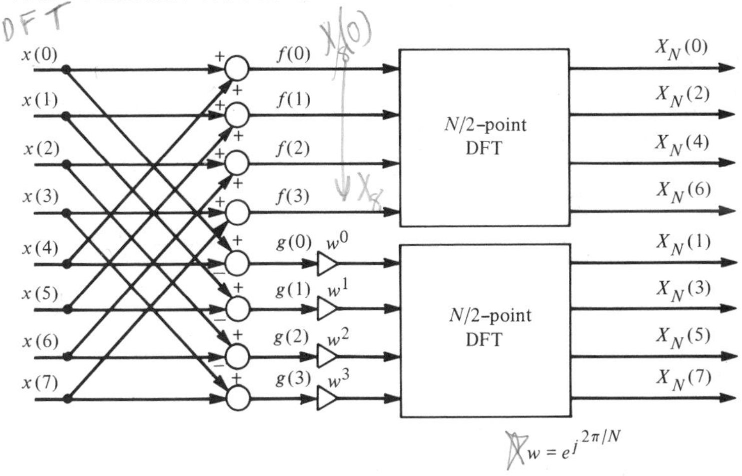

FIGURE 4.9. One stage of a divide-and-conquer routine for converting an N-point DFT into two $N/2$-point DFTs for the case $N = 8$.

entails computing the $(N/4)$-point DFTs of the four $N/4$ even-length sequences, which are composed of the first and last $N/4$ points of the sequences (4.49) and (4.50). This approach is depicted in Figure 4.10 for the case $N = 8$. Upon further analysis it is found that these two applications of the divide-and-conquer routine have reduced the required number of complex multiplication and addition operations to approximately $N + 4(N/4)^2 = N + N^2/4^2$.

We may continue this divide-and-conquer routine to its logical conclusion, whereby $N/2$ two-point DFTs are to be computed. This is illustrated in Figure 4.11 for the case $N = 8$. Upon reflection it is seen that $M = \log_2(N)$ divide-and-conquer stages are required to reach this condition. Furthermore, each of these stages is seen to be composed of a set of $N/2$ *butterfly networks*, a shown in Figure 4.12. These butterfly networks implement the fundamental sub-

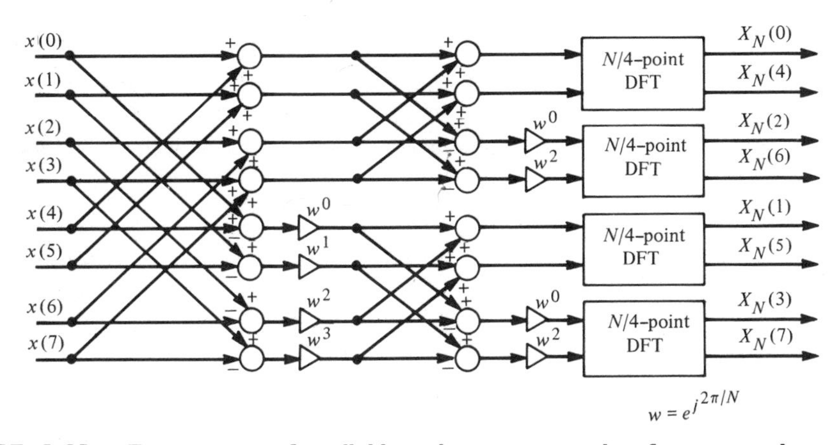

FIGURE 4.10. Two stages of a divide-and-conquer routine for converting an N-point DFT into four $N/4$-point DFTs for the case $N = 8$.

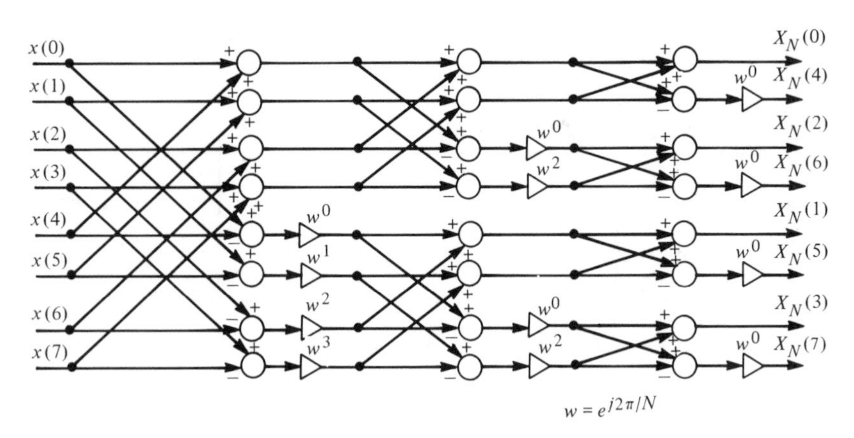

FIGURE 4.11. $M = \log_2 (N)$ stages of a divide-and-conquer routine in which an N-point DFT is converted into $M = \log_2 (N)$ two-point DFTs for the case $N = 8$.

sequence generation operation as represented by expressions (4.49) and (4.50) for the first stage. For the complete divide-and-conquer network we see that the number of complex multiplication and addition operations required to compute the N-point DFT is given approximately by $0.5N \log_2 (N)$ and $N \log_2 (N)$, respectively. Thus the computational complexity of this particular FFT algorithm is said to be of order

$$N \log_2 (N) \tag{4.53}$$

Our development of this FFT algorithm has been purposely made brief. Readers interested in a more detailed treatment are directed to the more specialized literature listed in the suggested reading section at the end of this chapter.

If we employ the divide-and-conquer routine described above, we find that the DFT coefficients appear in scrambled form at the last stage. This is illustrated in Figure 4.11, where these coefficients appear as $X(0)$, $X(4)$, $X(2)$, $X(6)$, $X(1)$, $X(5)$, $X(3)$, and $X(7)$. Due to this unnatural ordering, this particular FFT algorithm is referred to as *decimation in frequency*. To unscramble this ordering, use is made of the fact that the DFT coefficients appear in bit-reversal order at the output stage. That is, the location of the DFT coefficient $X_N(k)$ for $0 \leq k$

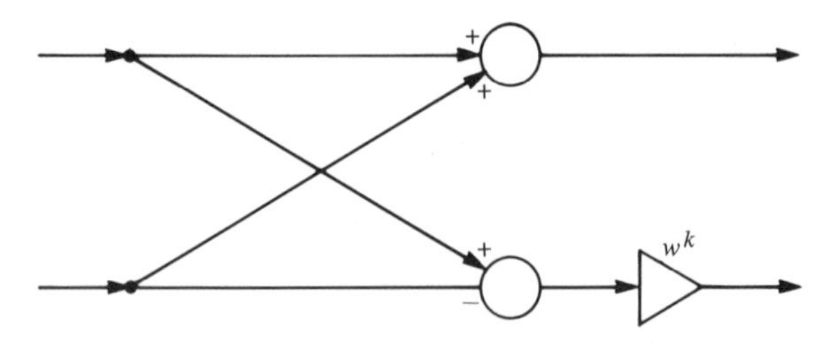

FIGURE 4.12. Butterfly network basic to the FFT algorithm.

```
c
c                  A COOLEY-TUKEY RADIX-2, DIF FFT PROGRAM
c                    COMPLEX INPUT DATA IN ARRAYS X AND Y
c       X AND Y ARE REAL AND IMAGINARY
c       COMPONENTS, RESPECTIVELY.
c ------------------------------------------------------------------------
c
      SUBROUTINE FFT (X,Y,N,M)
      REAL X(1), Y(1)
c
c ---------------------------MAIN FFT LOOPS-----------------------------
c
      N2 = N
      DO 10 K = 1, M
            N1  = N2
            N2  = N2/2
            E   = 6.283185307179586/N1
            A   = 0
            DO 20 J = 1, N2
                  C  = COS (A)
                  S  = SIN (A)
                  A  = J*E
                  DO 30 I = J, N, N1
                        L     = I + N2
                        XT    = X(I) - X(L)
                        X(I)  = X(I) + X(L)
                        YT    = Y(I) - Y(L)
                        Y(I)  = Y(I) + Y(L)
                        X(L)  = C*XT + S*YT
                        Y(L)  = C*YT - S*XT
 30               CONTINUE
 20         CONTINUE
 10   CONTINUE
c
c ------------------------ DIGIT REVERSE COUNTER ------------------------
c
100   J=1
      N1 = N - 1
      DO 104 I=1, N1
            If (I.GE.J) GOTO 101
            XT = X(J)
            X(J) = X(I)
            X(I) = XT
            XT = Y(J)
            Y(J) = Y(I)
            Y(I) = XT
101         K = N/2
102         IF (K.GE.J) GOTO 103
                  J  = J - K
                  K  = K/2
                  GOTO 102
103         J = J + K
104   CONTINUE
      RETURN
      END
```

FIGURE 4.13. Coding of a fast Fourier transform algorithm.

$\leq N - 1$ is obtained by first expressing k as an N-bit number. The location of DFT coefficient $X_N(k)$ then corresponds to the bit reversal of the M-bit representation for k. To illustrate this for the case $N = 8$ or $M = 3$, the coefficient $X_8(6)$ appears in location 3, since the bit reversal of the three-bit representation for 6 as given by (011) is equal to (110), which corresponds to 3.

The decimation-in-frequency FFT algorithm as described above is readily coded. A FORTRAN coding for this algorithm that is relatively short and yet computationally efficient is given in Figure 4.13 (see Burrus and Parks, 1985). We shall not give a detailed description of this program.

4.11 SUMMARY

The utility of the DFT for characterizing the frequency behavior of a finite set of data has been emphasized. In addition, procedures for using DFT methods for indirectly implementing convolution and frequency discrimination operations have been treated. A key requirement for the effective use of the DFT approach in these applications is the FFT algoithm. The FFT algorithm provides a computationally workable method for evaluating the DFT. Its development makes feasible the application of signal-processing methodology to data-processing problems that would otherwise not be possible.

SUGGESTED READINGS

BLAHUT, R., *Fast Algorithms for Digital Signal Processing*. Reading, Mass.: Addison-Wesley Publishing Company, Inc., 1985.

BRIGHAM, B. O., *The Fast Fourier Transform*. Englewood Cliffs, N.J.: Prentice-Hall, Inc., 1974.

BURRUS, C. S., and T. W. PARKS, *DFT/FFT and Convolution Algorithms*. New York: John Wiley & Sons, Inc., 1985.

COOLEY, J. W., and J. W. TUKEY, "An Algorithm for the Machine Calculation of Complex Fourier Series," *Mathematics of Computation*, Vol. 19, April 1965, pp. 297–301.

ELLIOTT, D. F., and K. R. RAO, *Fast Transforms, Algorithms, Analyses, Applications*. New York: Academic Press, 1982.

RAMIREZ, R., *The FFT: Fundamentals and Concepts*. Englewood Cliffs, N.J.: Prentice-Hall, Inc., 1985.

PROBLEMS

4.1 Determine the truncated Fourier transform and DFT associated with the following truncated time series.

(a) $x(0) = 1,\ x(1) = -3$

(b) $x(0) = 2,\ x(1) = -1,\ x(2) = 6$

(c) $x(0) = -4,\ x(1) = 0,\ x(2) = 0,\ x(3) = -1$

Make a superimposed plot of $|X_N(e^{j\omega})|$ versus ω and $|X_N(k)|$ versus k.

4.2 Prove the periodicity property (4.6) of the DFT.

4.3 Determine the DFT of the N-point sequence specified by

$$x(n) = \begin{cases} 1 & 0 \le n \le m \\ 0 & m < n \le N - 1 \end{cases}$$

and evaluate $|X_N(k)|$ and $\phi_N(k)$.

4.4 Show that the following N truncated complex sinusoids sequences

$$x_k(n) = e^{j2\pi kn/N} \qquad 0 \le n \le N - 1$$

are orthogonal in the sense that their inner product satisfies the property

$$\mathbf{x}_k\mathbf{x}_m^* = \sum_{n=0}^{N-1} x_k(n)\bar{x}_m(n) = \begin{cases} N & k = m \\ 0 & k \ne m \end{cases}$$

4.5 The complex numbers $e^{j2\pi k/N}$ for $0 \le k \le N - 1$ play an important role in the DFT. Prove that each of these N numbers corresponds to a Nth root of unity [i.e., $(e^{j2\pi k/N})^N = 1$]. Make a plot of their location on the unit circle in the complex plane.

4.6 The first five DFT coefficients associated with a real truncated time series of length 8 are specified by $X_8(0) = 2$, $X_8(1) = 1 - j2$, $X_8(2) = 0$, $X_8(3) = 3 + j1$, $X_8(4) = -2$. From this information, determine $X_8(5)$, $X_8(6)$, and $X_8(7)$.

4.7 Let there be given the truncated time series $x(n)$ for $0 \le n \le N - 1$ and its associated N-point DFT $X_N(k)$ for $0 \le k \le N - 1$. Consider the auxiliary truncated time series of length $2N$ as specified by

$$y(n) = \begin{cases} x(n/2) & n \text{ even} \\ 0 & n \text{ odd} \end{cases}$$

Establish the relationship between the $2N$-point DFT $Y_{2N}(k)$ and the N-point DFT $X_N(k)$.

4.8 Compute the four-point DFT of the truncated time series $x(0) = 2$, $x(1) = -1$, $x(2) = 1$, $x(3) = 4$. Interpreting the resultant DFT coefficients $X_4(0)$, $X_4(1)$, $X_4(2)$, and $X_4(3)$ as a new truncated time series, compute its four-point DFT. How are these DFT coefficients related to the original truncated time series? Extend this concept to the case of a general N-point truncated time series $x(n)$ for $0 \le n \le N - 1$. Show that this approach provides a means for inverting the DFT coefficient $X_N(k)$ to determine the original N-point time series $x(n)$.

4.9 Prove that *Parseval's relationship*,

$$\sum_{n=0}^{N-1} |x(n)|^2 = \frac{1}{N} \sum_{k=0}^{N-1} |X_N(k)|^2$$

holds for N-point DFTs. *Hint*: Use the inverse DFT relationship. Demonstrate Parseval's relationship for the specific four-point sequence $x(0) = 1$, $x(1) = 2$, $x(2) = 2$, $x(3) = 1$.

4.10 Compute the Fourier series representation for the periodic time series of period N whose first period is specified by:

(a) $x(n = \begin{cases} 1 & 0 \le n \le m \\ 0 & m < 1\ n \le N - 1 \end{cases}$

(b) $x(n) = a^n \qquad 0 \le n \le N - 1$

4.11 Let the time series $\{\tilde{x}(n)\}$ be periodic with period N so that $\tilde{x}(n + N) = \tilde{x}(n)$ for all n. Show that the DFT of the truncated time series specified by

$$x(n) = \begin{cases} \tilde{x}(n) & 0 \le n \le N - 1 \\ 0 & \text{otherwise} \end{cases}$$

gives rise to DFT coefficients $X_N(k)$ for $0 \le k \le N - 1$ such that the identity

$$\tilde{x}(n) = \frac{1}{N} \sum_{k=0}^{N-1} X_N(k) e^{j(2\pi kn/N)}$$

holds for all n. We may think of the latter relationship as constituting a Fourier series representation of the periodic time series $\{\tilde{x}(n)\}$.

4.12 Using the results of Problem 4.11, generate a Fourier series representation for the periodic time series of period N whose first period is specifically by

$$\tilde{x}(n) = \begin{cases} 1 & 0 \le n \le \dfrac{N}{2} \\ 0 & \dfrac{N}{2} < n \le N - 1 \end{cases}$$

in which N is taken to be even.

4.13 The Fourier transform of the generally infinite-length time series $\{x(n)\}$ is specified by

$$X(e^{j\omega}) = \sum_{n=-\infty}^{\infty} x(n) e^{-j\omega n}$$

Show that the values assumed by $X(e^{j\omega})$ at the discrete frequencies $\omega_k = 2\pi k/N$ for $0 \leq k \leq N - 1$ are identical to those obtained by taking the DFT of the truncated time series defined by

$$y(n) = \begin{cases} \sum_{m=-\infty}^{\infty} x(n + mN) & 0 \leq n \leq N - 1 \\ 0 & \text{otherwise} \end{cases}$$

In other words, prove that $Y(k) = X(e^{j2\pi k/N})$.

4.14 Demonstrate the validity of the conjecture made in Problem 4.13 for the time series $x(n) = 3(a)^n u(n)$, where $|a| < 1$.

4.15 Verify the correction of the low-pass unit-impulse response expression (4.24).

4.16 Using the technique described in Section 4.7, design a symmetric nonrecursive low-pass filter of length $N = 16$ and cutoff frequency $\omega_c = 0.1\pi$ radians. Determine the elements $h(n)$ for $0 \leq n \leq 15$ and make a plot of $|H_{16}(e^{j\omega})|$.

4.17 Verify the correctness of the unit-impulse response entries of Table 4.2 for **(a)** a high-pass filter; **(b)** a bandpass filter; **(c)** a band-reject filter.

4.18 Using a direct application of the convolution operation, convolve the sequences

$$\begin{array}{lll} x(0) = 1 & x(1) = -3 & x(2) = 2 \\ h(0) = 2 & h(1) = 1 & \end{array}$$

Obtain the same results using the indirect approach of Section 4.8 when N is set equal to 4. In a similar fashion, show that a choice of $N = 3$ yields an incorrect result.

4.19 The *Hilbert transform* is useful for processing narrow-band signals, where quadrature components (i.e., a $\pi/2$ phase shift) are required. Let $\{x(n)\}$ be a real signal that excites the linear system with frequency response

$$H(e^{j\omega}) = \begin{cases} -1 & 0 \leq \omega \leq \pi \\ 1 & -\pi < \omega < 0 \end{cases}$$

This system's response $\{y(n)\}$ is referred to as the Hilbert transform of $\{x(n)\}$. Compute this system's unit-impulse response. If $x(n) = A \cos(\omega_0 n + \theta)$, show that $y(n) = A \sin(\omega_0 n + \theta)$.

CHAPTER 5

Theory of Frequency-Discrimination Digital Filters

5.1

INTRODUCTION

One of the primary benefits of the Fourier transform arises from its ability to identify salient frequency-domain features of a signal. This possibility is made evident upon examination of the inverse Fourier transform relationship

$$x(n) = \frac{1}{2\pi} \int_{-\pi}^{\pi} X(e^{j\omega})e^{-j\omega n}\, d\omega$$

in which $X(e^{j\omega})$ designates the Fourier transform corresponding to signal $\{x(n)\}$. Specifically, a signal's time-domain description is seen to be directly related to its frequency-domain behavior through this integral relationship. This linkage was demonstrated in Chapter 3, where a sinusoidal time series was shown to have a Fourier transform that consisted of Dirac delta impulses. We shall now use the frequency-domain description to study the very important application of frequency-discrimination filtering. In particular, many of the more theoretical aspects of filtering will be examined. Procedures for synthesizing such frequency-discrimination filters are taken up in Chapter 6.

5.2

CONCEPT OF FREQUENCY DISCRIMINATION

Frequency-discrimination filtering provides one of the more important applications of linear systems. To describe this concept mathematically, let us consider a general linear system as governed by the convolution summation

$$y(n) = \sum_{k=-\infty}^{\infty} h(k)x(n - k) \tag{5.1}$$

in which $h(n)$ designates the system's unit-impulse response (or weighting coefficients). In the frequency domain, this convolution operation takes on the equivalent Fourier transform product form

$$Y(e^{j\omega}) = H(e^{j\omega})X(e^{j\omega}) \tag{5.2}$$

in which the entity $H(e^{j\omega})$ denotes the system's *frequency response* as specified by

$$H(e^{j\omega}) = \sum_{n=-\infty}^{\infty} h(n)e^{-j\omega n} \tag{5.3}$$

When synthesizing a linear system, it is generally required to select values for the weighting coefficients $h(n)$ so that the resulting linear system effects a given signal-processing objective. For instance, in a frequency-discrimination filtering application, the linear operator (5.1) is required to selectively transmit certain excitation sinusoidal components while rejecting other sinusoidal components. With this objective in mind, the following terminology relative to this frequency-discrimination objective is introduced.

1. *Passband:* set of sinusoidal frequencies ω in the interval $(-\pi, \pi]$ that the filter is to transmit.
2. *Stopband:* set of sinusoidal frequencies ω in the interval $(-\pi, \pi]$ that the filter is to reject.
3. *Transition band:* set of frequencies ω in the interval $(-\pi, \pi]$ separating the passbands and the stopbands.

The *ideal frequency response* associated with a digital filter that achieves the implied frequency discrimination is then characterized by

$$H(e^{j\omega}) = \begin{cases} 1 & \text{for all } \omega \text{ in passband} \\ 0 & \text{for all } \omega \text{ in stopband} \end{cases} \tag{5.4}$$

It is to be noted that the filter's behavior in the transition band is not stipulated

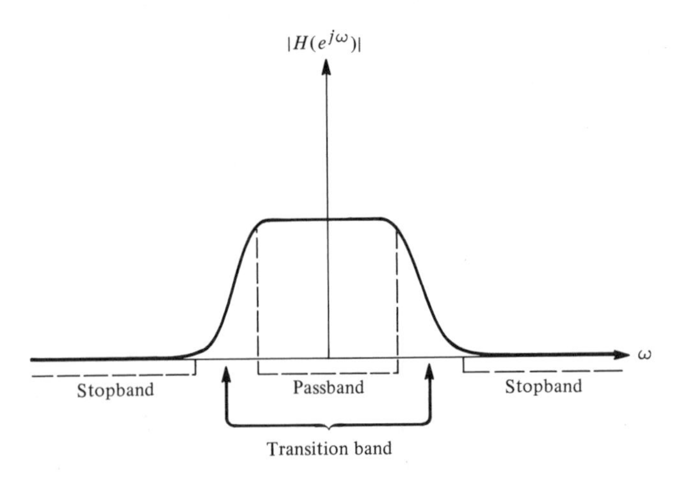

FIGURE 5.1. Discrimination bands for a filter.

here and is in a sense arbitrary. A visual depiction of this frequency-discrimi nation characteristic is shown in Figure 5.1.

The exact nature of a frequency-discrimination filter is then determined b how one selects the passband, stopband, and transition band. To illustrate typical application where this selection is readily made, let there be available set of data measurements that is composed of an information-bearing signa $\{s(n)\}$ and an additive-noise component $\{w(n)\}$, namely

$$x(n) = s(n) + w(n) \tag{5.5}$$

It is now desired to apply these data measurements to a linear operator whos primary function is that of transmitting the information signal while simulta neously rejecting (removing) the additive-noise component.

To determine the conditions under which this ideal filtering operation is pos sible, a frequency-domain analysis is particularly revealing. Upon taking th Fourier transform of expression (5.5) and then substituting that result into expressio (5.2), we have

$$\begin{aligned} Y(e^{j\omega}) &= H(e^{j\omega})S(e^{j\omega}) + H(e^{j\omega})W(e^{j\omega}) \\ &= Y_s(e^{j\omega}) + Y_w(e^{j\omega}) \end{aligned}$$

where $S(e^{j\omega})$ and $W(e^{j\omega})$ denote the Fourier transforms of the information sign and additive noise, respectively. If the filter's response is to equal the info mation-bearing signal [i.e., $Y(e^{j\omega}) = S(e^{j\omega})$], it is seen that two frequency domain conditions must be satisfied; namely, the spectral contents of the sign and noise must be disjoint in the sense that

$$S(e^{j\omega})W(e^{j\omega}) = 0 \qquad \text{for all } \omega \tag{5.6}$$

and the filter's frequency response must satisfy

$$H(e^{j\omega}) = \begin{cases} 1 & \text{for all } \omega \text{ such that } S(e^{j\omega}) \neq 0 \\ 0 & \text{for all } \omega \text{ such that } W(e^{j\omega}) \neq 0 \end{cases} \tag{5.7}$$

Under these idealized conditions it is conceptually possible to achieve perfect additive-noise suppression and signal transmission. Moreover, the frequency response specification (5.7) provides a natural means for selecting the filter's passband and stopband.

In any practical application, the disjoint condition (5.6) is never perfectly satisfied. Fortunately, it commonly happens that the signal and noise will contain *most* of their energy in readily identifiable disjoint frequency bands. Using these disjoint bands as pass and stop bands, we find that although the resultant frequency-discrimination filter will not perfectly separate the information signal from the additive noise, it does give a reasonably noise-free version of the information signal.

5.3 SINUSOIDAL RESPONSE AND THE FILTER FREQUENCY RESPONSE

The discussion in Section 5.2 emphasizes the importance attached to the frequency response characteristics of linear systems. In this chapter we examine the basic attributes of linear digital signal-processing algorithms (digital filters) that enable them to discriminate between different frequency components. To begin this study, we examine the ability of a linear system to modify the amplitude and phase of a pure sinusoid excitation. The approach here parallels the development in Section 1.12, which was concerned with the sinusoidal transmission characteristics of linear systems. In particular, let us consider the case when the general stable time-invariant linear system

$$y(n) = \sum_{k=-\infty}^{\infty} h(k)x(n-k) \tag{5.8}$$

is excited by the real two-sided sinusoid

$$x(n) = A\cos(\omega n + \theta) \tag{5.9}$$

where the real parameters A, ω, and θ denote the sinusoid's amplitude, frequency, and phase, respectively. In the analysis that follows it is assumed that the linear system's weighting sequence $\{h(n)\}$ is real.

To compute the sinusoidal response, the Euler representation for the excitation (5.9) is inserted into expression (5.8) to yield

$$y(n) = \sum_{k=-\infty}^{\infty} h(k)A\,\frac{e^{j(\omega n-\omega k+\theta)} + e^{-j(\omega n-\omega k+\theta)}}{2}$$

Upon taking all multiplicative terms not depending on the index k outside the summation, it follows that

$$y(n) = \frac{A}{2}\left[e^{j(\omega n+\theta)} \sum_{k=-\infty}^{\infty} h(k)e^{-j\omega k} + e^{-j(\omega n+\theta)} \sum_{k=-\infty}^{\infty} h(k)e^{j\omega k}\right]$$

The first summation on the right-hand side is recognized as being the Fourier transform of the linear system's weighting sequence (or its *frequency response*), that is,

$$\begin{aligned} H(e^{j\omega}) &= \sum_{k=-\infty}^{\infty} h(k)e^{-j\omega k} \\ &= |H(e^{j\omega})|e^{j\phi(\omega)} \end{aligned} \tag{5.10}$$

We have elected here to express the frequency response $H(e^{j\omega})$ in its polar form, where $|H(e^{j\omega})|$ and $\phi(\omega)$ denote the transform's real-valued magnitude and phase functions, respectively. Similarly, the second summation on the right-hand side is seen to equal the complex conjugate of $H(e^{j\omega})$, since the $h(k)$ are taken to be real. The response is therefore expressible as

$$y(n) = \frac{A}{2}\left[e^{j(\omega n+\theta)}|H(e^{j\omega})|e^{j\phi(\omega)} + e^{-j(\omega n+\theta)}|H(e^{j\omega})|e^{-j\phi(\omega)}\right]$$

where use of the real valuedness of the $\{h(n)\}$ sequence has been implicitly made. Upon combining exponential terms and using the Euler identity once more, the desired form for the sinusoidal response is found to be

$$y(n) = |H(e^{j\omega})|\, A \cos\,[\omega n + \theta + \phi(\omega)] \tag{5.11}$$

This is a rather remarkable result in that it indicates that the response of a real stable time-invariant linear system to a pure sinusoid of frequency ω is itself a pure sinusoid of frequency ω. The only effect that the linear system has imparted is that of changing the sinusoid's amplitude by the magnitude factor $|H(e^{j\omega})|$ and increasing its phase by the phase factor $\phi(\omega)$. This feature is shown in Figure 5.2. It is to be noted that if a one-sided sinusoid had been applied to the same system, the steady-state behavior would be also governed by expression (5.11) for large positive n. In this case a transient response term caused by the

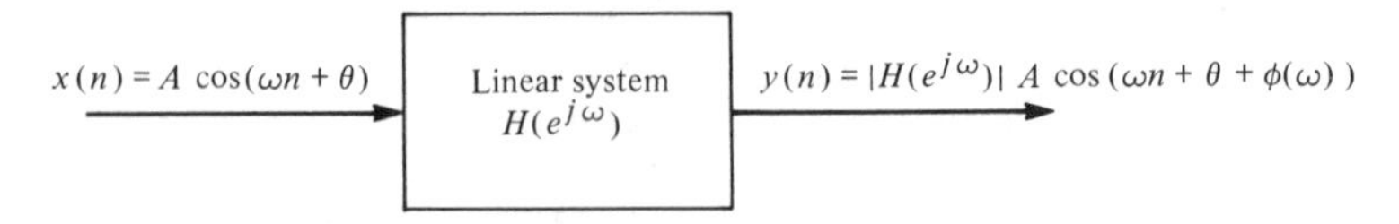

FIGURE 5.2. Sinusoidal response characteristics of a linear system.

sudden application of the excitation at $n = 0$ would be present and this term would eventually decay to zero due to the assumed stability of the system.

From the analysis above, the importance of the *frequency response function* (5.10) is apparent relative to the sinusoidal transmission characteristics of a linear system. To visualize this behavior, plots of the frequency response's magnitude $|H(e^{j\omega})|$ and phase $\phi(\omega)$ as a function of frequency are revealing. Because of the requirement that the filter's impulse response $\{h(n)\}$ be *real*, it follows from observations made in Chapter 3 that the magnitude function is an even function of ω,

$$|H(e^{j\omega})| = |H(e^{-j\omega})| \tag{5.12a}$$

while the phase function is an odd function of ω,

$$\phi(\omega) = -\phi(-\omega) \tag{5.12b}$$

With this in mind, we need only plot the functions $|H(e^{j\omega})|$ and $\phi(\omega)$ over the principal frequency range $0 \leq \omega \leq \pi$ to depict completely the linear system's frequency discrimination behavior. Its behavior for frequencies outside this principal frequency range is obtained directly by using the evenness and oddness

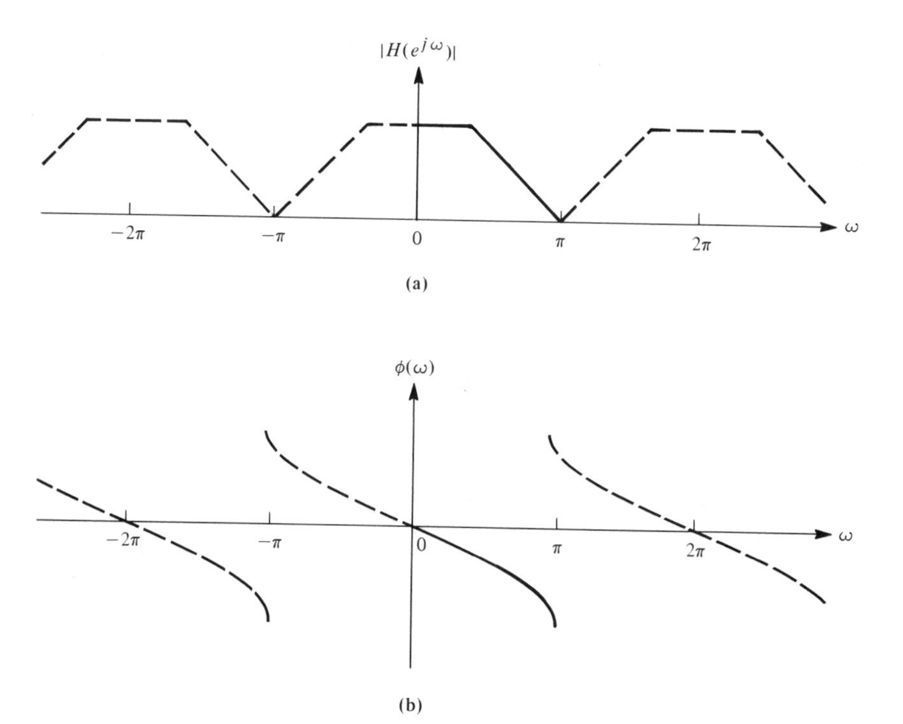

FIGURE 5.3. Magnitude and phase functions associated with a real-valued linear system.

properties (5.12) and the additional property that the magnitude and phase functions are periodic functions of ω with period 2π. For example, these properties may be used to identify the transfer function behavior outside the principal frequency range, as shown in Figure 5.3. The behavior plotted within the principal frequency range is shown as a continuous line; the corresponding inferred behavior outside that range is shown by the dashed lines.

5.4 DISTORTIONLESS SIGNAL TRANSMISSION

It was shown in Section 5.3 that a linear system's frequency response function provides a direct measure of its sinusoidal transmission characteristics. As is now demonstrated, the frequency response may also be used for evaluating a linear system's transmission properties for more general types of excitation. To begin this discussion, let us determine the conditions under which a linear system will transmit an excitation without distortion. This has very important implications when designing frequency discrimination filters. A distortionless signal transmission is said to have occurred if the following excitation–response pair prevails:

$$x(n) \Longrightarrow y(n) = Ax(n - n_0) \tag{5.13}$$

where A is a constant and n_0 is an integer. That is, the linear system's response to the given excitation is simply a scaled-by-A and shifted-by-n_0 version of that excitation in the distortionless case being considered here.

To determine the linear system requirements needed for distortionless signal transmission, let us take the Fourier transform of the response term in the excitation-response pair relation (5.13). Upon using the Fourier transform shifting property, it follows that this transform is given by

$$Y(e^{j\omega}) = AX(e^{j\omega})e^{-j\omega n_0} \tag{5.14}$$

Next we use the linear system property $Y(e^{j\omega}) = H(e^{j\omega})X(e^{j\omega})$ to conclude that a distortionless signal transmission will occur whenever the system's frequency response satisfies the following magnitude and phase requirements:

$$\begin{aligned} |H(e^{j\omega})| &= A \\ \arg\,[H(e^{j\omega})] &= -\omega n_0 \end{aligned} \qquad \text{for all } \omega \text{ for which } X(e^{j\omega}) \neq 0 \tag{5.15}$$

Specifically, the signal $\{x(n)\}$ will be transmitted undistorted by a linear system, provided that the system's associated frequency response function has a constant magnitude A and linear phase $-\omega n_0$ over those frequencies where $X(e^{j\omega})$ is nonzero. If either of these two conditions is not met, the response signal will be distorted. It is important to note that the distortionless transmission requirement (5.15) need hold only over those frequencies where the signal's Fourier transform is nonzero. For frequencies in which $X(e^{j\omega})$ is zero, the frequency response function can have a nonconstant magnitude or a nonlinear phase while maintaining distortionless transmission for that particular signal.

The linear phase requirement mandated for distortionless transmission can also be specified in terms of the *group delay* function, defined by

$$\tau(\omega) = -\frac{d\phi(\omega)}{d\omega} \tag{5.16}$$

That is, a linear system will give rise to a distortionless transmission if its associated group delay function is constant [i.e., $\tau(\omega) = n_0$] over frequencies where $X(e^{j\omega})$ is nonzero. For example, in synthesizing low-pass, high-pass, and bandpass filters, it is typically required to maintain a nearly constant group delay in the filter's passband.

5.5 PROTOTYPE LOW-PASS FILTERS

A prototype ideal low-pass filter has a frequency response magnitude as depicted in Figure 5.4. Such a filter transmits any sinusoid whose frequency lies within its passband $-\Delta \leq \omega \leq \Delta$ with a gain of one, while perfectly rejecting all other sinusoids. Also shown in this figure is a plausible phase behavior associated with a prototype ideal low-pass filter. Although any phase function could be used without affecting the ideal magnitude property of the low-pass filter, there are certain phase property constraints that must be imposed if a useful filter is to result. We shall now address this important issue.

The unit-impulse response associated with the prototype ideal low-pass filter is formally given by the inverse Fourier transform integral

$$h_\Delta(n) = \frac{1}{2\pi}\int_{-\Delta}^{\Delta} e^{\phi(\omega)}e^{j\omega n}\, d\omega \tag{5.17}$$

where the magnitude behavior shown in Figure 5.4 has been used in arriving at this result. The subscript Δ has been appended to the unit-impulse response to take note of the ideal low-pass filter's bandwidth.

Upon examination of expression (5.17), it is apparent that there exists an infinite set of digital filters which have the same ideal magnitude behavior as that shown in Figure 5.4 [i.e., one filter for each distinctive choice of $\phi(\omega)$]. Clearly, the basic nature of a specific filter's unit-impulse response is linked directly to its associated phase function. For typical applications, it is generally required that the filter operation being employed possess certain properties. Depending on the particular application, one or more of the following practical conditions often must be satisfied:

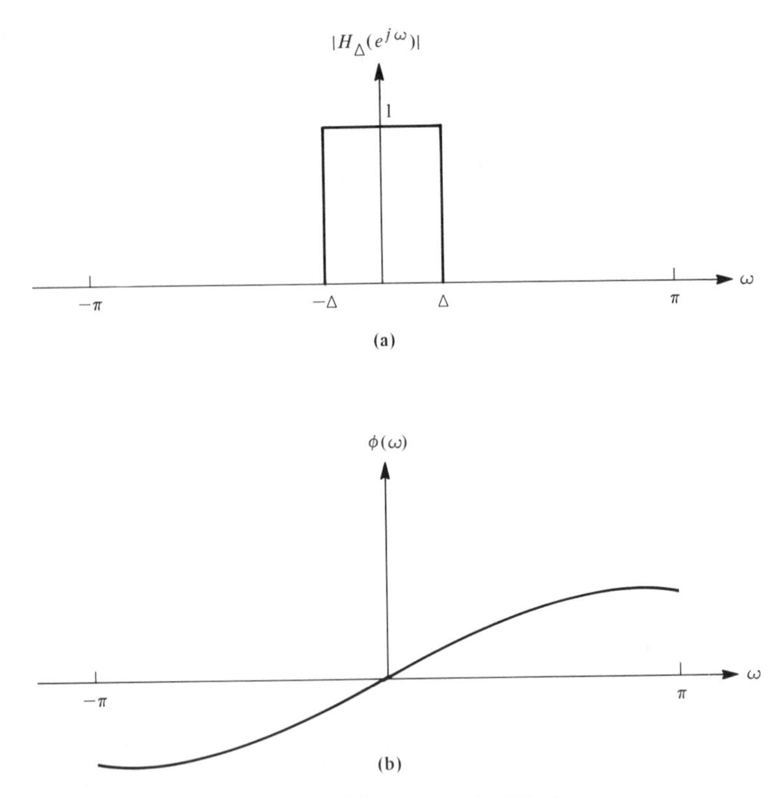

FIGURE 5.4. Ideal low-pass filter: (a) magnitude; (b) phase.

1. The filtering operation is real valued [i.e., $h(n)$ is real], which requires that $\phi(\omega)$ be antisymmetric so that $\phi(\omega) = -\phi(-\omega)$.
2. The filtering operation is causal, necessitating that $h_\Delta(n) = 0$ for $n < 0$.
3. The phase function is reasonably linear in the passband, $-\Delta \leq \omega \leq \Delta$.
4. The filter's *time constant* is relatively small (see Problem 5.21).

Unfortunately, it is not possible to satisfy all these conditions simultaneously while retaining the postulated ideal frequency response magnitude behavior. If we are to synthesize a practical low-pass filter that emulates the idealized filter, it is then necessary to compromise on the filter's magnitude behavior. Namely, we shall seek an implementable filter that possesses desirable properties such as those listed above and has a frequency response magnitude which closely approximates that of the ideal behavior. A few selected synthesis procedures that accomplish these objectives are presented in Chapter 6.

EXAMPLE 5.1

Determine the unit-impulse response associated with the ideal low-pass frequency response

$$H_\Delta(e^{j\omega}) = \begin{cases} e^{-jn_0\omega} & |\omega| \le \Delta \\ 0 & \Delta < |\omega| \le \pi \end{cases}$$

in which the phase function has been selected here to have the linear behavior $\phi(\omega) = -n_0\omega$, where n_0 is a fixed scalar (not necessarily an integer). Taking the inverse Fourier transform of this frequency response, we have

$$\begin{aligned} h_\Delta(n) &= \frac{1}{2\pi}\int_{-\Delta}^{\Delta} e^{-jn_0\omega}e^{j\omega n}\,d\omega \\ &= \frac{\sin[\Delta(n - n_0)]}{\pi(n - n_0)} \end{aligned} \tag{5.18}$$

This unit-impulse response is seen to have an infinite length, to be noncausal in nature, and to have a symmetry about $n = n_0$ (if n_0 is an integer). Unfortunately, the first two conditions are often undesirable from a practical filtering viewpoint. ■

5.6 PROTOTYPE BANDPASS FILTERS

The frequency response magnitude associated with an idealized bandpass filter is shown in Figure 5.5, where ω_c denotes the filter's center frequency and 2Δ its bandwidth. We now demonstrate how such a bandpass filter may be synthesized by using an appropriately designed low-pass filter. Specifically, it is assumed that a low-pass filter has been synthesized which has a frequency response magnitude that closely approximates the ideal behavior depicted in Figure 5.4. Let the unit-impulse response of this low-pass filter be denoted by $h_\Delta(n)$.

We next consider a related auxiliary filter whose unit-impulse response is specified by

$$h_{\text{bp}}(n) = 2h_\Delta(n)\cos(\omega_c n) \tag{5.19}$$

It is now shown that this auxiliary unit-impulse response corresponds to a bandpass filter which has a frequency response magnitude that closely approximates the ideal behavior shown in Figure 5.5. This is readily proved by first taking Fourier transform of $h_{\text{bp}}(n)$, that is,

$$\begin{aligned} H_{\text{bp}}(e^{j\omega}) &= \sum_{n=-\infty}^{\infty} 2h_\Delta(n)\cos(\omega_c n)e^{-j\omega n} \\ &= \sum_{n=-\infty}^{\infty} [h_\Delta(n)e^{-j(\omega-\omega_c)n} + h_\Delta(n)e^{-j(\omega+\omega_c)n}] \end{aligned}$$

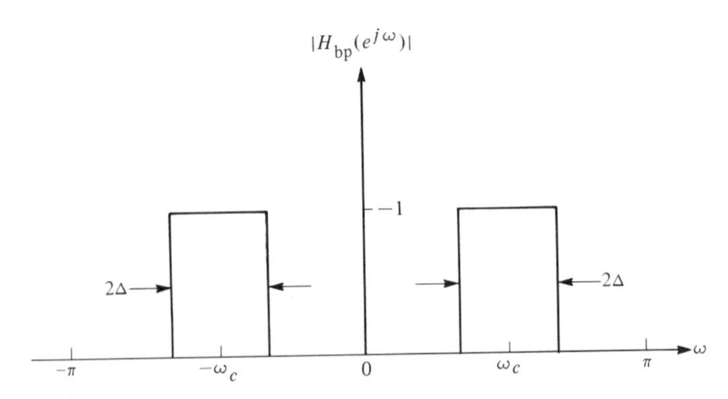

FIGURE 5.5. Prototype bandpass filter.

The Euler identity for cos $(\omega_c n)$ has been used here. Upon examination of this expression, it is seen that the first summation term is equal to a right-shifted-by-ω_c version of the low-pass filter frequency response, while the second term corresponds to a left-shifted-by-ω_c version of the low-pass filter's frequency response. Thus the frequency response associated with the postulated unit-impulse response (5.19) is simply

$$H_{\text{bp}}(e^{j\omega}) = H_\Delta(e^{j(\omega-\omega_c)}) + H_\Delta(e^{j(\omega+\omega_c)}) \tag{5.20}$$

The magnitude of the first component in this expression is seen to equal the magnitude of the given low-pass filter [i.e., $H_\Delta(e^{j\omega})$] shifted to the right by ω_c radians. That is, it corresponds to the rightmost passband centered at ω_c, as shown in Figure 5.5. Similarly, the magnitude of the second term corresponds to the leftmost passband centered at $-\omega_c$. We therefore conclude that the postulated unit-impulse response (5.19) corresponds to a bandpass filter whose frequency response magnitude closely approximates that shown in Figure 5.5. This presumes, of course, that the given low-pass filter with unit-impulse response $\{h_\Delta(n)\}$ has a frequency response magnitude that itself closely approximates the ideal behavior illustrated in Figure 5.4.

EXAMPLE 5.2

Consider the ideal bandpass filter whose magnitude is shown in Figure 5.5 and whose phase we take to be zero [i.e., $\phi(\omega) = 0$]. Using the inverse Fourier transform, we find that the corresponding unit-impulse response is given by

$$\begin{aligned} h(n) &= \frac{1}{2\pi}\int_{-\omega_c-\Delta}^{-\omega_c+\Delta} e^{j\omega n}\,d\omega + \frac{1}{2\pi}\int_{\omega_c-\Delta}^{\omega_c+\Delta} e^{j\omega n}\,d\omega \\ &= \frac{1}{\pi n}\{\sin[(\omega_c - \Delta)n] - \sin[(\omega_c + \Delta)n]\} \end{aligned}$$

Upon applying standard trigonometric identities to the constituent sinusoid terms here, this expression simplifies to

$$h(n) = 2h_\Delta(n)\cos(\omega_c n)$$

where $h_\Delta(n) = \sin(\Delta n)/\pi n$ corresponds to the unit-impulse response associated with the zero-phase ideal low-pass filter with cutoff frequency Δ radians. Not surprisingly, this result is in exact accordance with the postulated bandpass expression (5.19). ■

A procedure for implementing the bandpass filter as characterized by weighting sequence (5.19) is readily determined by appealing to its convolution summation representation

$$\begin{aligned} y(n) &= \sum_{k=-\infty}^{\infty} h_{\text{bp}}(k)x(n-k) \\ &= 2\sum_{k=-\infty}^{\infty} h_\Delta(k)\cos(\omega_c k)x(n-k) \\ &= 2\sum_{k=-\infty}^{\infty} h_\Delta(k)\cos[\omega_c(k-n)+\omega_c n]x(n-k) \end{aligned}$$

Upon using a standard trigonometric identity on the cosine term, this relationship can be simplified to

$$\begin{aligned} y(n) = {} & 2\cos(n\omega_c)\sum_{k=-\infty}^{\infty} h_\Delta(k)\cos[\omega_c(n-k)]x(n-k) \\ & + 2\sin(n\omega_c)\sum_{k=-\infty}^{\infty} h_\Delta(k)\sin[\omega_c(n-k)]x(n-k) \end{aligned} \tag{5.21}$$

The first summation in this expression can be interpreted as being the response of an underlying low-pass filter with unit-impulse response $\{h_\Delta(n)\}$ to the excitation $\{\cos(\omega_c n)x(n)\}$. A similar interpretation can be given to the second summation, with the excitation being replaced by $\{\sin(\omega_c n)x(n)\}$. An implementation of bandpass filter relationship (5.21) is depicted in Figure 5.6. It is seen to require two identically designed low-pass filters and four multipliers. It is noteworthy that the filter's center frequency is controlled exclusively by the frequency of the sinusoids appearing at the multiplier terminals, while the filter's bandwidth is governed by the bandwidth of the low-pass filter. An alternative method for synthesizing the bandpass filter as represented by expression (5.19) is presented in Section 5.9. The latter method does not explicitly utilize low-pass filters or sinusoidal multiplications in its implementation.

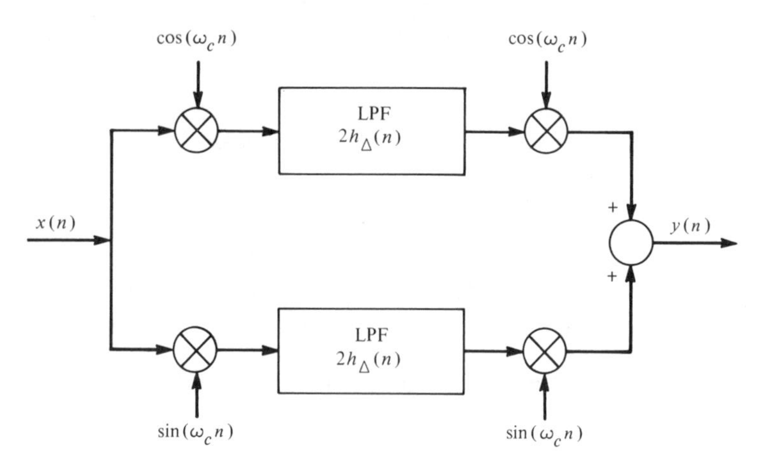

FIGURE 5.6. Bandpass filter implementation.

5.7 PROTOTYPE HIGH-PASS FILTERS

A prototype high-pass filter has the frequency response magnitude shown in Figure 5.7. To synthesize such a filter, let there be available a low-pass filter whose unit-impulse response $h_\Delta(n)$ has an associated frequency response magnitude that closely resembles that shown in Figure 5.4. It is now shown that the auxiliary unit-impulse response as specified by

$$h_{\rm hp}(n) = (-1)^n h_\Delta(n) \tag{5.22}$$

has the required high-pass filter behavior. This follows directly upon taking the Fourier transform of this sequence, which results in

$$H_{\rm hp}(e^{j\omega}) = H_\Delta(e^{j(\omega-\pi)})$$

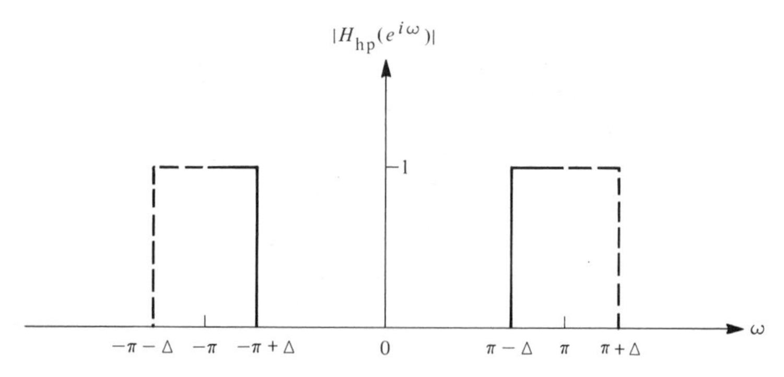

FIGURE 5.7. Prototype high-pass filter.

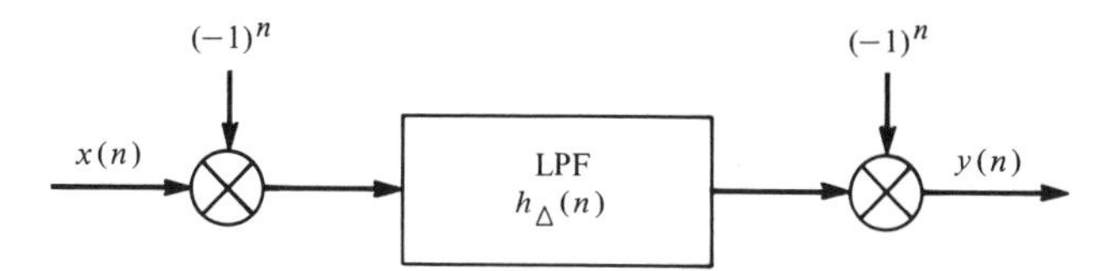

FIGURE 5.8. High-pass filter implementation.

Thus the magnitude of the frequency response is identical to that of the low-pass filter shifted by π radians. However, this corresponds precisely to the high-pass filter behavior desired.

An implementation of the high-pass filtering operation as characterized by expression (5.22) is given formally by the associated convolution summation

$$y(n) = \sum_{k=-\infty}^{\infty} (-1)^k h_\Delta(k) x(n-k)$$

$$= (-1)^n \sum_{k=-\infty}^{\infty} h_\Delta(k)(-1)^{n-k} x(n-k)$$

Similar to the approach taken in the bandpass filter case, we may interpret the summation as representing the response of the given low-pass filter with unit-impulse response $\{h_\Delta(n)\}$ to the excitation $\{(-1)^n x(n)\}$. This filtering operation may therefore be realized by the simple configuration given in Figure 5.8. An alternative implementation of this high-pass filter is presented in Section 5.9.

5.8 PROTOTYPE BAND-REJECT FILTERS

In various applications it is required to remove a band of frequency components while transmitting all other frequency components undistorted. A prototype band-reject filter that accomplishes this task would have the idealized frequency re-

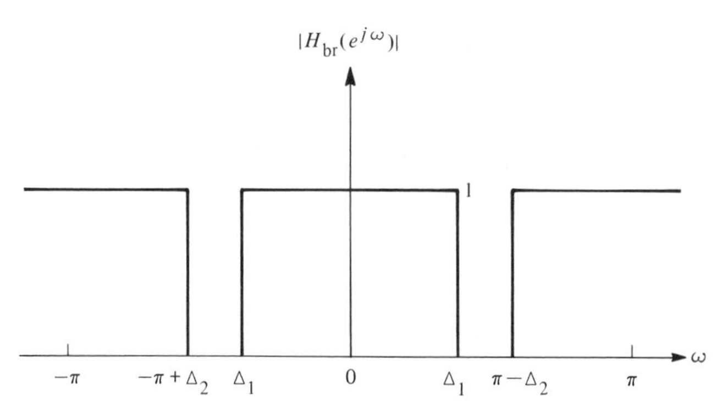

FIGURE 5.9. Prototype band-reject filter.

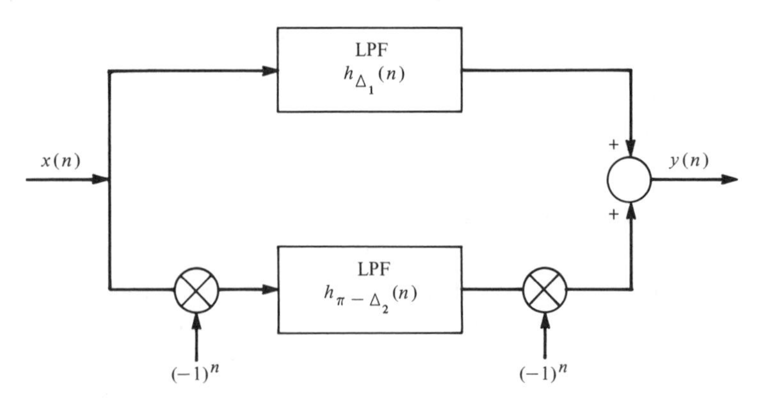

FIGURE 5.10. Band-reject filter implementation.

sponse magnitude illustrated in Figure 5.9. A little thought will convince the reader that this filtering operation may be realized by connecting a low-pass filter with bandwidth Δ_1 in parallel with a high-pass filter with bandwidth $\pi - \Delta_2$. The unit-impulse response associated with such a configuration is therefore given by

$$h_{br}(n) = h_{\Delta 1}(n) + (-1)^n h_{\pi - \Delta_2}(n) \tag{5.23}$$

In accordance with the developments in Section 5.7, an implementation of this band-reject filter is made using the configuration shown in Figure 5.10. It is seen that two appropriately designed low-pass filters are required.

5.9 DIRECT IMPLEMENTATION OF FREQUENCY-DISCRIMINATION FILTERS

In this section an alternative method is presented for implementing frequency-discrimination filters as obtained using the philosophy of Sections 5.6 to 5.8. For illustrative purposes we use the bandpass filter expression

$$h_{bp}(n) = 2h_{\Delta}(n) \cos(\omega_c n) \tag{5.24}$$

in which $\{h_{\Delta}(n)\}$ corresponds to the unit-impulse response of a given low-pass filter with bandwidth Δ. This alternative method is predicated on first determining the transfer function of the filter under investigation. For the unit-impulse response above, this is seen to give

$$H_{bp}(z) = \sum_{n=-\infty}^{\infty} 2h_{\Delta}(n) \cos(\omega_c n) z^{-n}$$

$$= \sum_{n=-\infty}^{\infty} h_{\Delta}(n)[(e^{-j\omega_c}z)^{-n} + (e^{j\omega_c}z)^{-n}] \quad (5.25)$$

$$= H_{\Delta}(e^{-j\omega_c}z) + H_{\Delta}(e^{j\omega_c}z)$$

In this expression $H_{\Delta}(e^{\pm j\omega_c}z)$ is simply obtained by replacing z with $e^{\pm j\omega_c}z$ everywhere that it appears in $H_{\Delta}(z)$.

The effect of transfer function relationship (5.25) on the bandpass filter's zero and pole locations is readily determined. In particular, let the low-pass filter being used in this bandpass filter operation have the following factored transfer function representation:

$$H_{\Delta}(z) = \alpha \frac{\prod_{k=1}^{q} (1 - z_k z^{-1})}{\prod_{m=1}^{p} (1 - p_m z^{-1})} \quad (5.26)$$

where the z_k and p_m correspond to the low-pass filter's zeros and poles, respectively. The transfer function of the associated bandpass filter, from expression (5.25), is given by

$$H_{bp}(z) = \alpha \frac{\prod_{k=1}^{q} (1 - z_k e^{j\omega_c} z^{-1})}{\prod_{m=1}^{p} (1 - p_m e^{j\omega_c} z^{-1})} + \alpha \frac{\prod_{k=1}^{q} (1 - z_k e^{-j\omega_c} z^{-1})}{\prod_{m=1}^{p} (1 - p_m e^{-j\omega_c} z^{-1})}$$

$$= \alpha \frac{\prod_{k=1}^{q} \prod_{m=1}^{p} [(1 - z_k e^{j\omega_c} z^{-1})(1 - p_m e^{-j\omega_c} z^{-1}) + (1 - z_k e^{-j\omega_c} z^{-1})(1 - p_m e^{j\omega_c} z^{-1})]}{\prod_{m=1}^{p} (1 - p_m e^{j\omega_c} z^{-1})(1 - p_m e^{-j\omega_c} z^{-1})} \quad (5.27)$$

Upon examination of transfer functions (5.26) and (5.27), it is apparent that each low-pass filter pole p_m gives rise to the two (rotated) bandpass filter poles $e^{j\omega_c}p_m$ and $e^{-j\omega_c}p_m$. The bandpass filter $q + p$ zeros are dependent on the low-pass filter's zeros and poles in a more complex manner.

After carrying out the multiplications indicated in expression (5.27), the following equivalent expression is eventually obtained:

$$H_{bp}(z) = \frac{b_0 + b_1 z^{-1} + \cdots + b_{p+q} z^{-p-q}}{1 + a_1 z^{-1} + \cdots + a_{2p} z^{-2p}}$$

Adopting a causal interpretation to this filter transfer function (this assumes that all the poles satisfy $|p_m| < 1$ for stability), we see that the recursive relationship

$$y(n) = \sum_{k=0}^{p+q} b_k x(n-k) - \sum_{k=1}^{2p} a_k y(n-k) \quad (5.28)$$

TABLE 5.1 Transfer Functions of Prototype Filters

Filter	Impulse	Transfer Function
Low-pass	$h_\Delta(n)$	$H_\Delta(z)$
Bandpass	$2h_\Delta(n)\cos(\omega_c n)$	$H_\Delta(e^{-j\omega_c}z) + H_\Delta(e^{j\omega_c}z)$
High-pass	$(-1)^n h_\Delta(n)$	$H_\Delta(-z)$
Band-reject	$h_{\Delta_1}(n) + (-1)^n h_{\pi-\Delta_2}(n)$	$H_{\Delta_1}(z) + H_{\pi-\Delta_2}(-z)$

implements the given bandpass filtering operation. If the original low-pass filter is a real-valued operation [i.e., the $h_\Delta(n)$ are real], it follows that this bandpass recursive relationship is also real valued (i.e., the a_k and b_k coefficients are real).

To implement the bandpass filter operation as characterized by expression (5.24), we can utilize either the scheme shown in Figure 5.6 or recursive relationship (5.28). Although these implementations are equivalent, one may be preferable to the other in a given application. For example, the recursive relationship is somewhat more efficient computationally (i.e., fewer multiplications per output element are required). On the other hand, the Figure 5.6 implementation is superior in those applications requiring several bandpass filtering operations using the same bandwidth but different center frequencies. For such applications the same low-pass filter may be used, thus giving rise to an economic implementation.

The design of high-pass and band-reject filters can be accomplished in a similar fashion. With reference to the specifications given in Sections 5.5 to 5.8, the transfer function associated with these filters are listed in Table 5.1. In each case, the overall transfer function can be implemented by a recursive expresison.

EXAMPLE 5.3

Let the low-pass filter being used have the transfer function

$$H_\Delta(z) = \frac{1-\alpha}{1-\alpha z^{-1}}$$

From expression (5.27) the corresponding bandpass filter is then given by

$$\begin{aligned} H_{\text{bp}}(z) &= \frac{1-\alpha}{1-\alpha e^{j\omega_c} z^{-1}} + \frac{1-\alpha}{1-\alpha e^{-j\omega_c} z^{-1}} \\ &= 2(1-\alpha)\frac{1-\alpha\cos(\omega_c)z^{-1}}{1-2\alpha\cos(\omega_c)z^{-1}+\alpha^2 z^{-2}} \end{aligned}$$

It is apparent that the recursive implementation of this bandpass filter is real valued in nature provided that the associated low-pass filter operation is similarly characterized (i.e., α is real). ■

5.10 ALL-PASS NETWORKS

It often happens that the frequency response function associated with a linear system has a satisfactory magnitude function, but its associated phase function may not be suitable (e.g., it is highly nonlinear in the passband). In such situations it is possible to modify the phase function properly while maintaining the magnitude function with the use of an all-pass network.† An *all-pass network* is defined to be a linear system whose magnitude function is 1 for all frequencies, that is,

$$H_{\text{ap}}(e^{j\omega}) = e^{j\phi_{\text{ap}}(\omega)} \tag{5.29}$$

where $\phi_{\text{ap}}(\omega)$ designates the all-pass network's phase function. As is now demonstrated, the aforementioned phase compensation may be achieved by cascading this all-pass network with the given linear system as depicted in Figure 5.11. By using the cascading property of linear systems established previously, it follows that the transfer function associated with this cascaded configuration is given by

$$\begin{aligned} H_c(e^{j\omega}) &= H(e^{j\omega})H_{\text{ap}}(e^{j\omega}) \\ &= |H(e^{j\omega})|e^{j\phi(\omega)}|H_{\text{ap}}(e^{j\omega})|e^{j\phi_{\text{ap}}(\omega)} \\ &= |H(e^{j\omega})|e^{j[\phi(\omega)+\phi_{\text{ap}}(\omega)]} \end{aligned} \tag{5.30}$$

Conceptually, it is possible to choose the all-pass network's phase function $\phi_{\text{ap}}(\omega)$ so as to correct (or compensate) for the unsatisfactory behavior in the given linear system's phase function $\phi(\omega)$ while not altering its magnitude behavior, since $|H_c(e^{j\omega})| = |H(e^{j\omega})|$.

Prototype First-Order, All-Pass Network

Every all-pass network can be generated by a proper cascading of prototype first-order all-pass networks whose transfer functions are of the form

$$H_{\text{ap}}(z) = r\frac{1 - r^{-1}e^{j\theta}z^{-1}}{1 - re^{j\theta}z^{-1}} \tag{5.31}$$

†As shown in Section 5.12, an all-pass network may also be used for stabilizing linear systems.

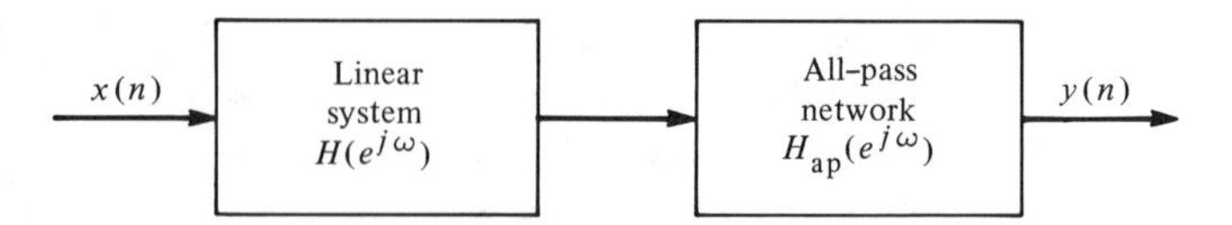

FIGURE 5.11. Phase compensation using all-pass networks.

where r is a positive parameter and θ a phase angle. This transfer function ha[s] a zero at $r^{-1}e^{j\theta}$ and a pole at $re^{j\theta}$. If this all-pass network is taken to be causal the parameter r must be selected so that $0 \le r < 1$, to ensure its stability. Upo[n] setting $z = e^{j\omega}$ in this expression, it is readily found that this network has th[e] following magnitude and phase functions:

$$|H_{ap}(e^{j\omega})| = 1$$
$$\tan[\phi_{ap}(\omega)] = \frac{-(1 - r^2)\sin(\omega - \theta)}{(1 + r^2)\cos(\omega - \theta) - 2r} \tag{5.32}$$

Thus transfer function (5.31) does indeed constitute a first-order all-pass net work, since its associated magnitude function equals 1 for all frequencies.

The behavior of the phase function $\phi_{ap}(\omega)$ as specified by expression (5.32) is now examined. In particular, this function is seen to take on the values $\pm\pi/2$ at the two radian frequencies where the denominator term in relationship (5.32) equals zero, that is, at

$$\omega_1 = \theta + \Delta \qquad \text{and} \qquad \omega_2 = \theta - \Delta \tag{5.33a}$$

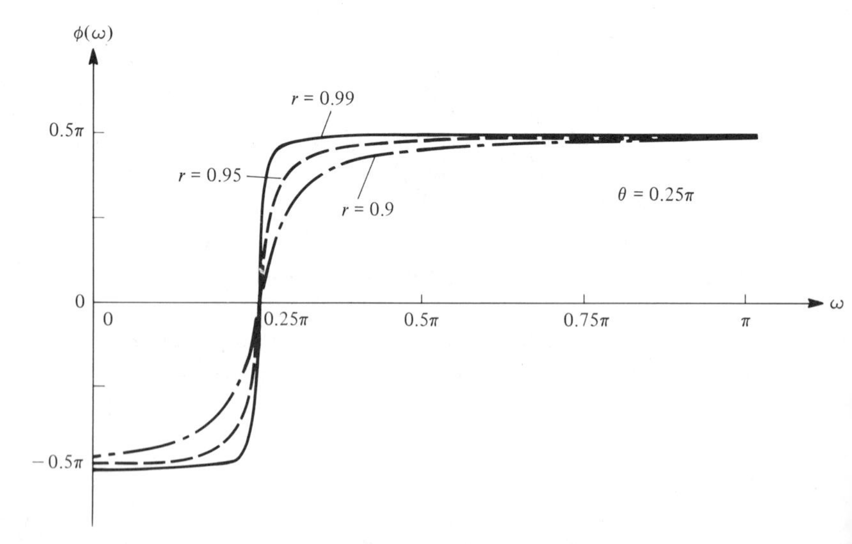

FIGURE 5.12. Phase plots for a first-order all-pass network with r = 0.9, 0.95, and 0.99.

respectively, where the parameter Δ is specified by

$$\Delta = \cos^{-1}\left(\frac{2r}{1 + r^2}\right) \tag{5.33b}$$

The phase function therefore approximately increases by π radians over the frequency interval (ω_1, ω_2), this increase being most rapid for r close to 1. This behavior is depicted in Figure 5.12 for $\theta = 0.25\pi$ and for different values of r. The first-order all-pass network (5.31) therefore affects primarily the phase properties in a neighborhood about θ. From this phase plot, it is apparent that if we seek to modify the phase characteristics of a given linear system over a range of frequencies, it will generally be necessary to employ several cascaded all-pass networks of form (5.31), each with a different selection of θ and r.

Prototype Second-Order All-Pass Network

The difference equation associated with all-pass network (5.30) will have complex coefficients if the parameter θ is not equal to an integer multiple of π^1.† To obtain an all-pass network with real coefficients, it is then necessary to cascade to the first-order all-pass network its complex-conjugate mate with zero at $r^{-1}e^{-j\theta}$ and pole at $re^{-j\theta}$. The resultant second-order all-pass network is then specified by

$$\begin{aligned} H_{\text{ap}}(z) &= r\,\frac{1 - r^{-1}e^{j\theta}z^{-1}}{1 - re^{j\theta}z^{-1}}\; r\,\frac{1 - r^{-1}e^{-j\theta}z^{-1}}{1 - re^{-j\theta}z^{-1}} \\ &= r^2\,\frac{1 - 2\cos(\theta)r^{-1}z^{-1} + r^{-2}z^{-2}}{1 - 2\cos(\theta)rz^{-1} + r^2 z^{-2}} \end{aligned} \tag{5.34}$$

and is seen to have real coefficients, as required. Moreover, its associated magnitude function is identically 1 and its phase function is equal to the sum of the individual first-order all-pass phase functions, that is,

$$\begin{aligned} \phi_{\text{ap}}(\omega) &= \tan^{-1}\left[\frac{-(1 - r^2)\sin(\omega - \theta)}{(1 + r^2)\cos(\omega - \theta) - 2r}\right] \\ &\quad + \tan^{-1}\left[\frac{-(1 - r^2)\sin(\omega + \theta)}{(1 + r^2)\cos(\omega + \theta) - 2r}\right] \end{aligned} \tag{5.35}$$

In summary, to synthesize an all-pass network governed by a difference equation with real coefficients, it is necessary to include the associated complex-conjugate mate of network (5.31).

†This implementing difference equation is specified by $y(n) = rx(n) - e^{j\theta}x(n-1) + re^{j\theta}y(n-1)$

5.11

MINIMUM-PHASE–MAXIMUM-PHASE FILTERS

The concept of all-pass networks has an interesting and relevant utilization in linear nonrecursive filtering. In particular, we shall consider the qth-order nonrecursive expression

$$y(n) = \sum_{k=0}^{q} b_k x(n - k) \tag{5.36}$$

and its associated transfer function

$$H(z) = \sum_{k=0}^{q} b_k z^{-k} \tag{5.37}$$

$$= b_0 \prod_{k=1}^{q} (1 - z_k z^{-1}) \tag{5.38}$$

It is assumed here that the zeros z_k are distinct and that none lie on the unit circle. Let us now cascade a specific first-order all-pass network with pole at z_1 to this nonrecursive filter. The overall transfer function associated with this cascaded configuration is then

$$\begin{aligned} H_c(z) &= H(z)|z_1| \frac{1 - \bar{z}_1^{-1} z^{-1}}{1 - z_1 z^{-1}} \\ &= b_0 |z_1| (1 - \bar{z}_1^{-1} z^{-1}) \prod_{k=2}^{q} (1 - z_k z^{-1}) \end{aligned} \tag{5.39}$$

where $\bar{z}_1$ denotes the complex conjugate of z_1. In effect, this system cascading has produced another qth-order nonrecursive (all-zero) filter in which the original zero at $z = z_1$ has been replaced by $\bar{z}_1^{-1}$ while the remaining zeros have been unaltered. Of particular importance, it is noted that the magnitude functions associated with the two transfer functions $H(z)$ and $H_c(z)$ are identical due to the all-pass network behavior. However, their respective phase functions are different.

Clearly, in using the foregoing approach we could have interchanged any subset of the q zeros by an appropriated cascading of first-order all-pass networks without changing the overall filter's magnitude function. Since there were assumed to be q distinct zeros in the original transfer function (5.38), it follows that there will be 2^q different such configurations generated in this manner. Each of these transfer functions has its own unique qth-order nonrecursive system description; however, they share the *same* common magnitude function but have different phase functions. The zeros of these equivalent-magnitude nonrecursive filters will be at either z_k or $\bar{z}_k^{-1}$ for $k = 1, 2, \ldots, q$. We are assuming here, of course, that the q zeros $z_1, z_2, \ldots, z_q$ are distinct and located off the unit circle. If this is not the case, it can be argued that there will be fewer than 2^q members in this "equivalence" class of nonrecursive filters.

The natural question arises as to whether there exists a preferred qth-order nonrecursive system to use from this class of equivalent-magnitude functions. There are two members of this class that are readily identified, as noted in the following definition.

Definition 5.1. Consider the class of equivalent qth-order linear nonrecursive systems that have the same magnitude function and are identified by the roots z_k for $1 \le k \le q$. If these zeros are distinct and do not lie on the unit circle, there will exist 2^q different members in this class. Moreover, that particular choice which has all its q zeros inside (outside) the unit circle is called the *minimum* (*maximum*)*-phase* realization filter.

The minimum- and maximum-phase filters of the equivalent class of linear nonrecursive filters each possess a salient property. Specifically, if $h(n)$ for $0 \le n \le q$ denotes the weighting sequence of any member of this class, it follows that the minimum-phase member will maximize each of the sums

$$\sum_{k=0}^{n} h^2(k) \qquad \text{for all } 0 \le n \le q \tag{5.40}$$

over all members in the class. Similarly, the maximum-phase member will minimize these sums. In words, the minimum-phase system has a weighting sequence that is most active (has its largest values) for small values of n. It thus possesses the fastest time constant in the equivalent class of magnitude functions. On the other hand, the maximum-phase system has the slowest time constant in this class.

EXAMPLE 5.4

Consider the linear second-order nonrecursive system with transfer function

$$\begin{aligned} H_1(z) &= 3 + 5.25z^{-1} - 1.5z^{-2} \\ &= 3(1 - 0.25z^{-1})(1 + 2z^{-1}) \end{aligned}$$

Construct the class of equivalent-magnitude functions associated with this transfer function. Using expression (5.39), we find that the $2^2 = 4$ members of this class are given by

$$\begin{aligned} H_2(z) &= 3(1 - 0.25z^{-1})(1 + 2z^{-1})\, 0.25\left(\frac{1 - 4z^{-1}}{1 - 0.25z^{-1}}\right) \\ &= 0.75(1 + 2z^{-1})(1 - 4z^{-1}) = 0.75 - 1.5z^{-1} - 6z^{-2} \\ H_3(z) &= 3(1 - 0.25z^{-1})(1 + 2z^{-1})\, 2\left(\frac{1 + 0.5z^{-1}}{1 + 2z^{-1}}\right) \end{aligned}$$

$$= 6(1 - 0.25z^{-1})(1 + 0.5z^{-1}) = 6 + 1.5z^{-1} - 0.75z^{-2}$$

$$H_4(z) = 3(1 - 0.25z^{-1})(1 + 2z^{-1})\, 0.25\left(\frac{1 - 4z^{-1}}{1 - 0.25z^{-1}}\right) 2\left(\frac{1 + 0.5z^{-1}}{1 + 2z^{-1}}\right)$$

$$= 1.5(1 - 4z^{-1})(1 + 0.5z^{-1}) = 1.5 - 5.25z^{-1} - 3z^{-2}$$

The transfer functions $H_3(z)$ and $H_2(z)$ are seen to correspond to the minimum-phase and maximum-phase members, respectively. Plots of the weighting sequence corresponding to each of these equivalent magnitude functions are shown in Figure 5.13. It is apparent that the minimum- and maximum-phase filters are seen to maximize and minimize, respectively, the summations (5.40). ■

5.12 STABILIZATION OF A CAUSAL RECURSIVE FILTER

Upon synthesizing a recursive frequency-discrimination filter, it may happen that although the resultant filter will have a satisfactory frequency response magnitude behavior, the associated transfer function will have some of its poles located outside the unit circle. For example, this situation arises when modeling an ideal unit-impulse response that has a noncausal component. Whatever the cause, however, this pole condition is to be avoided if it is desired to implement the transfer function by a real-time (i.e., causal) operation, as is normally the case. This is a direct consequence of an observation made previously that transfer function poles located outside the unit circle give rise to unstable causal operations (or stable anticausal operations). As we now show, this apparent dilemma is readily resolved by employing the all-pass network concept.

It is now assumed that a recursive frequency-discrimination filter has been synthesized which has a satisfactory frequency response magnitude behavior. Let the transfer function associated with this filter be expressed in the factored form

$$H(z) = \frac{b_0 + b_1 z^{-1} + \cdots + b_q z^{-q}}{\prod_{k=1}^{s} (1 - p_k z^{-1}) \prod_{k=s+1}^{p} (1 - p_k z^{-1})} \tag{5.41}$$

where the poles p_k for $1 \le k \le s$ each have a magnitude of *less* than one while the remaining poles for $s + 1 \le k \le p$ have magnitudes greater than one. This transfer function is seen to have s poles located inside the unit circle and its remaining $p - s$ poles located outside the unit circle. To remove the destabilizing effect (assuming that a causal implementation is desired) of the $p - s$ poles located outside the unit circle while retaining the satisfactory magnitude response behavior, let us cascade this transfer function with the all-pass network

$$H_{ap}(z) = \prod_{k=s+1}^{p} |p_k|^{-1} \frac{1 - p_k z^{-1}}{1 - \bar{p}_k^{-1} z^{-1}} \tag{5.42}$$

After carrying out the indicated $p - s$ pole–zero cancellations in the cascaded system, the resultant transfer function is specified by

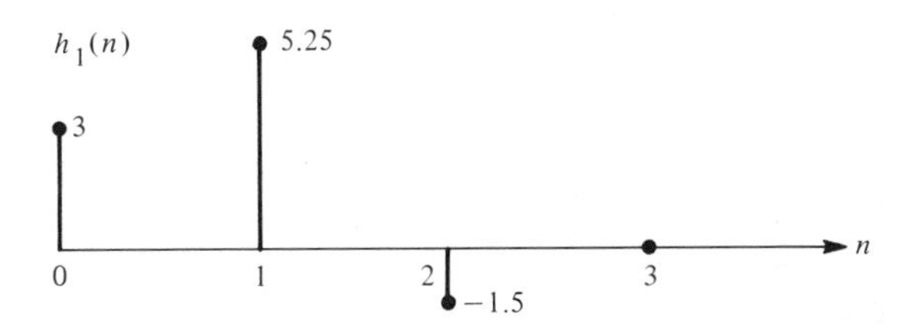

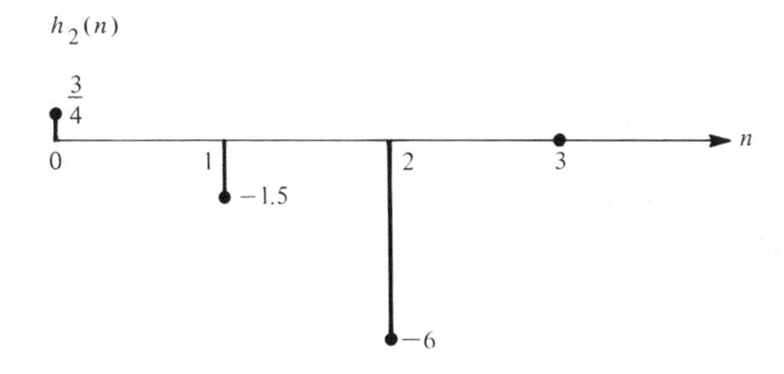

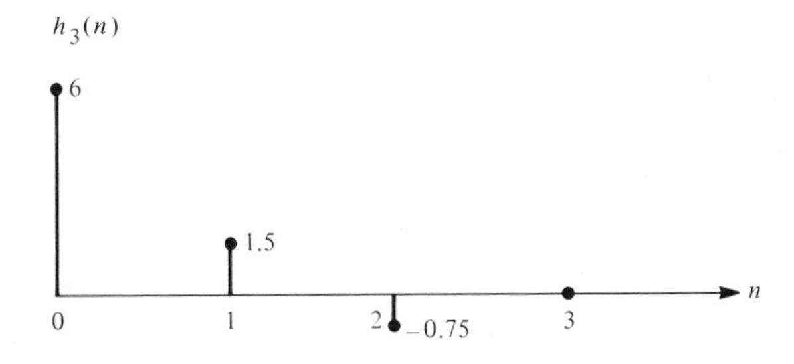

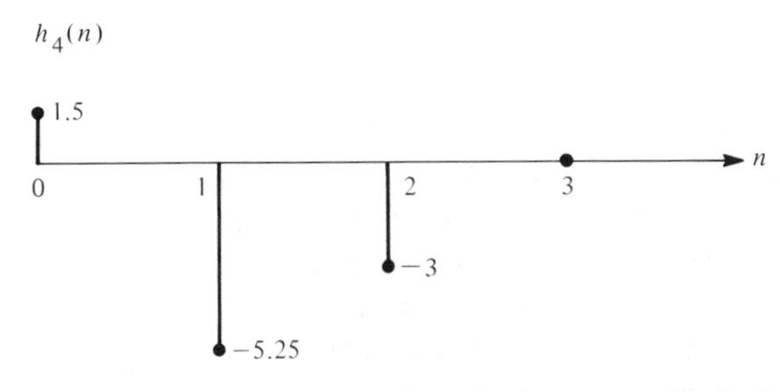

FIGURE 5.13. Weighting sequence of equivalent magnitude functions in Example 5.4.

$$
\begin{aligned}
H_c(z) &= H(z)H_{ap}(z) \\
&= \frac{(b_0 + b_1 z^{-1} + \cdots + b_q z^{-q}) \prod_{k=s+1}^{p} |p_k^{-1}|}{\prod_{k=1}^{s} (1 - p_k z^{-1}) \prod_{k=s+1}^{p} (1 - \bar{p}_k^{-1} z^{-1})}
\end{aligned} \tag{5.43}
$$

This related transfer function has all of its poles located inside the unit circle. Moreover, due to the all-pass nature of network (5.42), it follows directly that

$$|H_c(e^{j\omega})| = |H(e^{j\omega})|$$

Thus the associated transfer function $H_c(z)$ has the desired magnitude function behavior as represented in the original synthesized filter $H(z)$. More important, this associated transfer function may be implemented by a causal stable operation. As a precautionary note, the phase characteristics of this stabilized filter can be an important consideration in applications requiring nearly distortionless behavior (i.e., linear phase). In such applications the user must carefully examine the phase behavior of the stabilized filter.

EXAMPLE 5.5

Using the concept of all-pass network as described above, determine a stabilized causal system that has the same frequency response as that of the system governed by

$$H(z) = \frac{2 + z^{-1}}{(1 - 0.8z^{-1})(1 - 2z^{-1})}$$

If a causal filtering operation is effected, the pole at $z = 2$ creates an unstable implementation. To compensate for this instability, let us cascade an all-pass network of form (5.31) with $r = 0.5$ and $\theta = 0$ to this system. The resulting cascaded system's overall transfer function is then given by

$$
\begin{aligned}
H_c(z) = H(z)H_{ap}(z) &= \frac{2 + z^{-1}}{(1 - 0.8z^{-1})(1 - 2z^{-1})} \frac{1}{2} \left(\frac{1 - 2z^{-1}}{1 - 0.5z^{-1}} \right) \\
&= \frac{1 + 0.5z^{-1}}{(1 - 0.8z^{-1})(1 - 0.5z^{-1})}
\end{aligned}
$$

after the common zero–pole concellation has been made. Due to the magnitude behavior of the all-pass network, it follows that the magnitude functions associated with transfer functions $H(z)$ and $H_c(z)$ are identical. The causal transfer function $H_c(z)$ that arises through this procedure, however, is seen to be stable. Clearly, this concept can be applied to more general situations where all unstable poles are replaced by their conjugate reciprocal stable poles. ■

5.13

FILTERING OF UNIFORMLY SAMPLED CONTINUOUS-TIME SIGNALS

One of the most important practical signal-processing applications is concerned with the digital filtering of uniformly sampled continuous-time signals. There are many delicate issues that must be addressed if this task is to be achieved successfully. They center around the selection of the sampling time as it relates to the frequency content of the analog signal being sampled. If proper precautions are not exercised, one may obtain unexpected (and undesired) signal-processing results.

To gain an insight into these sampling-operation considerations, let us examine the specific continuous-time sinusoidal signal

$$x_a(t) = A \sin(\omega_a t + \theta) \tag{5.44}$$

whose analog frequency is ω_a radians per second. Upon uniformly sampling this signal at the sampling rate $f_s = 1/T$ hertz (or "samples/second"), there arises the discrete-time sinusoidal signal

$$x(nT) = A \sin(\omega_a nT + \theta) \tag{5.45}$$

Thus the process of uniform sampling an analog sinusoidal signal has produced a digital sinusoid in which the analog and digital frequencies are related according to

$$\underset{\omega_a \text{ rad/sec}}{\text{analog}} \Longleftrightarrow \underset{\omega = \omega_a T \text{ rad/sample}}{\text{digital}} \tag{5.46}$$

We shall now carefully analyze this relationship so as to comprehend the potential pitfalls that can arise in the sampling process.

Nyquist Sampling Rate

It is recalled from Chapter 1 that if the sampling operation is to be reversible, the sampling rate selected must exceed twice the analog signal's highest frequency content. Relative to the specific analog sinusoidal signal (5.44), this requirement is equivalent to selecting the sample period T to satisfy

$$\frac{1}{T} > 2 f_a = \frac{\omega_a}{\pi}$$

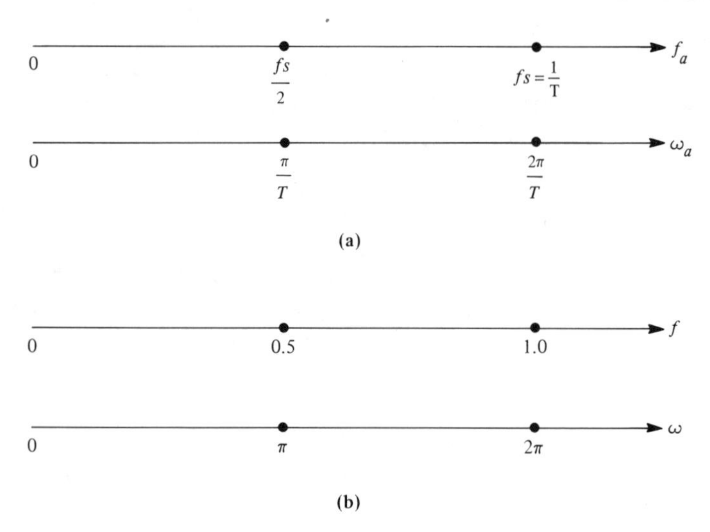

FIGURE 5.14. Relationship between analog and digital frequencies.

Under the assumption that T has been so selected, it follows that the corresponding digital frequency (5.46) will satisfy

$$\omega = \omega_a T < \pi$$

In summary, when the Nyquist rate criterion is satisfied, the digital radian frequency is always less than π radians. The interrelationship between the analog and digital frequencies for a proper choice of T is shown in Figure 5.14, where $f_s = 1/T$ denotes the sampling frequency.

Frequency Aliasing (Folding)

With the foregoing thoughts in mind, let us now consider the consequences when the Nyquist sample rate criterion is violated. For the analog sinusoidal signal (5.45), this is equivalent to selecting T so that it yields the *undersampled* condition

$$\omega_a T > \pi$$

To see the effects of this undersampling, let us express the digital frequency quantity $\omega_a T$ appearing in relationship (5.45) as

$$\omega_a T = m\pi + \overline{\omega}_a T \tag{5.47}$$

where m is a nonzero integer and $0 \leq \overline{\omega}_a T < \pi$. With this unique decomposition, the uniformly sampled sinusoid (5.45) is equivalently expressed as

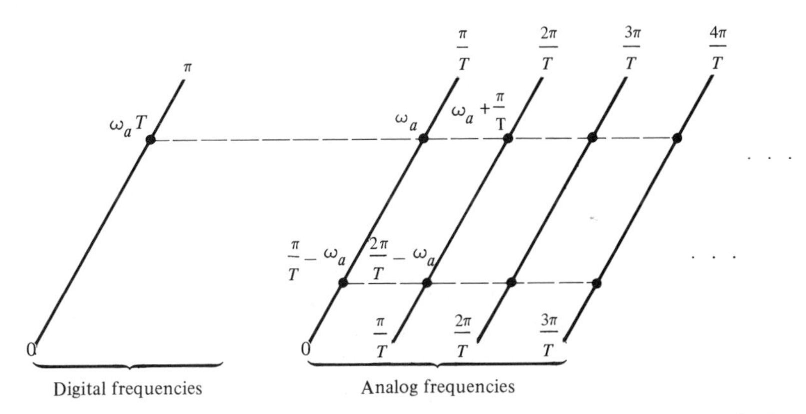

FIGURE 5.15. Frequency aliasing (folding) and analog frequencies equivalent to digital frequency $\omega_a T$.

$$\begin{aligned} A \sin(\omega_a nT + \theta) &= A \sin(m\pi + \overline{\omega}_a T + \theta) \\ &= (-1)^m A \sin(\overline{\omega}_a T + \theta) \end{aligned}$$

Thus the process of undersampling has produced a discrete-time sinusoid of digital frequency $\overline{\omega}_a T$.

Upon careful reflection on this result, a troublesome conclusion is reached: namely, a discrete-time sinusoid with frequency $\omega_a T$ that arises through the process of uniform sampling can be generated from any of the family of analog sinusoids with radian frequencies

$$\omega_a, \quad \frac{\pi}{T} \pm \omega_a, \quad \frac{2\pi}{T} \pm \omega_a, \quad \frac{3\pi}{T} \pm \omega_a, \quad \ldots$$

where ω_a is such that $\omega_a T < \pi$. Thus the process of uniform sampling has made these analog frequencies be, in a sense, equivalent. This phenomenon is known as *frequency aliasing* (or folding) and is depicted in Figure 5.15, where all analog frequencies on the horizontal line are *equivalent* and transform into the same digital frequency, $\omega_a T$.

Guard Band Filtering

To remove the undesirable frequency ambiguity caused by uniform sampling, it is therefore critical to preprocess the analog signal being sampled so as not to create the foregoing frequency ambiguity. This is readily achieved by applying the analog signal to an *analog* low-pass filter that passes only frequencies in the interval $0 \leq \omega_a < \pi/T$. The filter's response signal is then uniformly sampled at the rate $1/T$ or faster, with no resultant frequency ambiguity, due to the lack

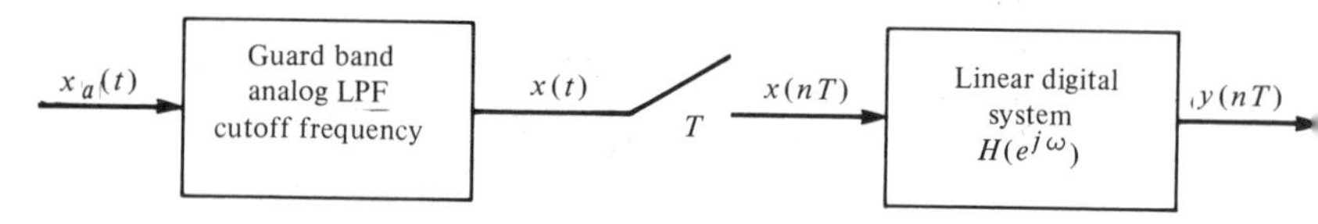

FIGURE 5.16. Digital signal processing of analog signals.

of analog frequency components exceeding π/T rad/sec. This "guard band" filtering concept is depicted in Figure 5.16, where the overall digital signal processor is to effect a desired objective on the sampled signal. In virtually all applications entailing sampling, the use of guard band filtering is absolutely essential. If the frequency response of the linear digital system is $H(e^{j\omega})$, it follows from the comments above that the frequency response of the *effective* analog system must then be

$$H(e^{j\omega_a T}) \qquad \text{for } -\pi/T < \omega_a \leq \pi/T \tag{5.48}$$

The behavior of this frequency response may readily be displayed by first plotting the digital frequency response functions $|H(e^{j\omega})|$ and $\phi(\omega)$ versus ω and then rescaling the abscissa by a factor of $1/T$. Thus the digital frequency π radians becomes the analog frequency π/T radians per second.

EXAMPLE 5.6

Let it be desired to low-pass filter an analog signal in which the filter's cutoff frequency is to be ω_c radians per second. If this filtering operation is to be achieved with the scheme depicted in Figure 5.16, the digital signal processor would correspond to a digital low-pass filter with cutoff frequency of $\omega_c T$ radians.

In a similar fashion, an analog bandpass filtering operation over the passband $\omega_1 \leq \omega_a \leq \omega_2$ would necessitate the use of a digital bandpass filter whose passband is $\omega_1 T \leq \omega \leq \omega_2 T$. Clearly, a wide variety of *analog* frequency-discrimination filtering operations can be implemented in this manner by appropriately selecting the associated digital signal-processing operator. ■

5.14 SUMMARY

An examination of the fundamental issues involved in frequency-discrimination filtering has been presented. This entailed the introduction of the concept of ideal low-pass, high-pass, and band-reject filters. It was shown that each of these important filter classes could be synthesized with the exclusive use of low-pass filters.

SUGGESTED READINGS

ANTONIOU, A., *Digital Filters, Analysis and Synthesis*. New York: McGraw-Hill Book Company, 1979.

CADZOW, J. A., *Discrete-Time Systems*. Englewood Cliffs, N.J.: Prentice-Hall, Inc., 1973.

CADZOW, J. A., "Recursive Filter Synthesis via Gradient Based Algorithms," *IEEE Transactions on Acoustics, Speech, and Signal Processing*, Vol. ASSP-24, No. 5, October 1976, pp. 349–355.

CHEN, C. T., *One-Dimensional Digital Signal Processing*. New York: Marcel Dekker, Inc., 1979.

GOLD, B., and C. M. RADER, *Digital Processing of Signals*. New York: McGraw-Hill Book Company, 1969.

OPPENHEIM, A. V., and R. W. SCHAFER, *Digital Signal Processing*. Englewood Cliffs, N.J.: Prentice-Hall, Inc., 1975.

PELED, A., and B. LIU, *Digital Signal Processing*. New York: John Wiley & Sons, Inc., 1976.

STEARS, S. D., *Digital Signal Analysis*. Rochelle Park, N.J.: Hayden Book Company, Inc., 1975.

TRETTER, S. A., *Discrete-Time Signal Processing*. New York: John Wiley & Sons, Inc., 1976.

PROBLEMS

5.1 Determine the transfer function associated with the recursive linear operator

$$y(n) = x(n) - 0.5x(n - 1) + 0.81y(n - 2)$$

Find this transfer function's zeros and poles, and sketch their location in the complex z-plane. Sketch the behavior of the associated magnitude and phase functions in the interval $-\pi < \omega \leq \pi$. Indicate how the zero–pole location influences the location of the peaks and nulls in the magnitude function.

5.2 Repeat Problem 5.1 for the linear nonrecursive operator

$$y(n) = x(n) - \tfrac{1}{2}x(n - 1) + x(n - 2) - \tfrac{1}{2}x(n - 3)$$

5.3 Given a stable linear operator with weighting sequence $\{h(n)\}$, show that its response to the excitation

$$x(n) = A_1 \cos(\omega_1 n + \theta_1) + A_2 \cos(\omega_2 n + \theta_2)$$

is given by

$$y(n) = A_1|H(e^{j\omega_1})|\cos[\omega_1 n + \theta_1 + \phi(\omega_1)] + A_2|H(e^{j\omega_2})|\cos[\omega_2 n + \theta_2 + \phi(\omega_2)]$$

where $H(e^{j\omega}) = |H(e^{j\omega})|e^{j\phi(\omega)}$ is the linear operator's frequency response.

5.4 Extend the results of Problem 5.3 to show that when the excitation is specifie by

$$x(n) = \sum_{k=1}^{q} A_k \cos(\omega_k n + \theta_k)$$

the corresponding linear system response is

$$y(n) = \sum_{k=1}^{q} A_k|H(e^{j\omega_k})| \cos[\omega_k n + \theta_k + \phi(\omega_k)]$$

5.5 Determine the group delay function $\tau(\omega)$ associated with the nonrecursive linea operator

$$y(n) = \tfrac{1}{3}[x(n - 1) + x(n) + x(n + 1)]$$

This operator provides a crude approximation of a low-pass filter. Make a plo of the group delay function $\tau(\omega)$ and the magnitude function $|H(e^{j\omega})|$ fo $-\pi < \omega \leq \pi$.

5.6 Show that the magnitude function associated with the frequency response of the linear operator

$$y(n) = rx(n) - e^{j\theta}x(n - 1) + re^{j\theta}y(n - 1)$$

has the value 1 for all frequencies. This operator is called an all-pass network and it is implicitly assumed that $0 \leq r < 1$. Determine the group delay function associated with this all-pass network.

5.7 Demonstrate that the samples which arise from uniformly sampling the two real-valued, continuous-time sinusoids

$$x_1(t) = 2\sin(2\pi f_0 t) \qquad \text{and} \qquad x_2(t) = 2\sin[2\pi(f_0 + 10)t]$$

are identical when the sample period is $T = \frac{1}{2}$ sec. For such sampling periods, the process of uniform sampling thus views these two different continuous-time sinusoids as being identical.

5.8 Develop a procedure for estimating the parameters, A_1, ω_1, and ϕ_1 from the time series elements $x(n)$ for $1 \leq n \leq N$ if it is known that $x(n)$ is governed by

$$x(n) = A_1 \cos(\omega_1 n + \phi_1)$$

5.9 In many applications it is essential that the frequencey-discrimination filter have a linear-phase characteristic. Consider the nonrecursive filter specified by

$$y(n) = \sum_{k=0}^{q} b_k x(n - k)$$

Show that this filter has a linear phase, provided that the real-valued b_k coefficients are selected to be
(a) Symmetric so that $b_k = b_{q-k}$ for $0 \leq k \leq q$
(b) Skew symmetric so that $b_k = -b_{q-k}$ for $0 \leq k \leq q$

For each of these selections, determine the phase function $\phi(\omega)$ for $-\pi < \omega \leq \pi$. In your analysis, treat separately the cases where q is even and odd.

5.10 In synthesizing frequency-discrimination filters, it is often possible to generate a good approximation to an ideal magnitude characteristic, but the corresponding phase function is not satisfactory. It is possible to retain the filter's magnitude characteristic while achieving a desirable zero-phase function by using a straightforward procedure to be now described. Specifically, let $\{h(n)\}$ denote the weighting sequence associated with a given filter with a nonlinear phase function. Let us first process the data signal $\{x(n)\}$ according to the convolution summation

$$v(n) = \sum_{k=-\infty}^{\infty} h(k)x(n - k) \qquad \text{normal excitation} \tag{1}$$

in which the weighting sequence is not restricted to be causal. This weighting sequence can be associated with a finite nonrecursive relationship or a finite-order recursive relationship. Next, the *time transpose* of this response signal [i.e., $v(-n)$] is applied to a copy of the original filter to generate the auxiliary response

$$w(n) = \sum_{k=-\infty}^{\infty} h(k)v(k - n) \qquad \text{time-transposed response} \tag{2}$$

Finally, the overall system response is defined to equal the time transpose of $w(n)$, that is,

$$y(n) = w(-n) \tag{3}$$

(a) Determine the overall weighting sequence $\{h_0(n)\}$ corresponding to this *linear operation* which produces $\{y(n)\}$ from $\{x(n)\}$ in terms of the original filter's weighting sequence $\{h(n)\}$. Show that $\{h_0(n)\}$ is an even function of n and therefore its associated phase function is identically zero.

(b) Show that the magnitude function associated with this three-stage operation is given by

$$H_0(e^{j\omega}) = |H(e^{j\omega})|^2$$

Thus at the expense of loss in causality, we retain the original magnitude characteristics (squared) and achieve a zero-phase characteristic.

5.11 With the objectives outlined in Problem 5.10 in mind, consider the three-stage linear operation

$$v(n) = \sum_{k=-\infty}^{\infty} h(k)x(n-k) \qquad \text{normal excitation} \tag{1}$$

$$w(n) = \sum_{k=-\infty}^{\infty} h(k)x(k-n) \qquad \text{time-transposed excitation} \tag{2}$$

$$y(n) = v(n) + w(-n) \tag{3}$$

where $\{h(n)\}$ is an arbitrary weighting sequence. Show that the overall linear operation that generates $\{y(n)\}$ form $\{x(n)\}$ has its frequency response specified by

$$\begin{aligned} H_2(e^{j\omega}) &= H(e^{j\omega}) + \overline{H}(e^{j\omega}) \\ &= 2|H(e^{j\omega})|\cos(\phi(\omega)) \end{aligned}$$

where $|H(e^{j\omega})|$ and $\phi(\omega$ are the magnitude and phase functions associated with the original filter $\{h(n)\}$. Thus the overall linear operation has a zero-type phase characteristic.

5.12 The filtering of continuous-time signals is often effected by using the configuration depicted in the accompanying figure, whereby the digital-to-analog (D/A) converter changes the digital filter's response $y(n)$ into a continuous-time signal according to the rule $\tilde{y}(t) = y(nT)$ for $nT \le t < (n+1)T$.

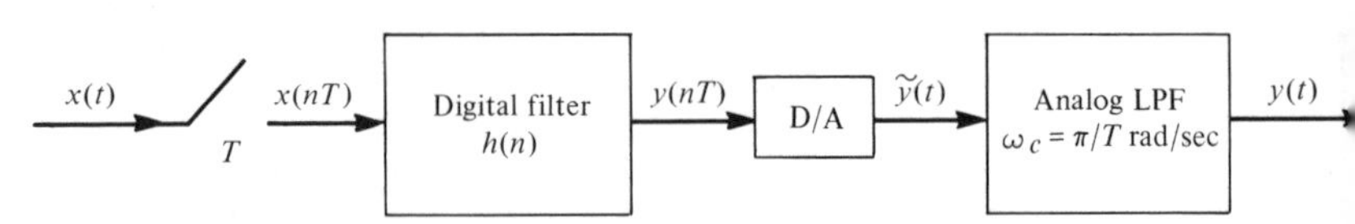

Consider the case in which the digital filter is

(a) A low-pass filter with cutoff frequency $\omega_1 = 0.15\pi$ radians

(b) A bandpass filter with passband $0.2\pi \le \omega \le 0.25\pi$

(c) A high-pass filter with cutoff frequency $\omega_2 = 0.35\pi$ radians

Determine the passbands of the overall equivalent analog filter for each of these cases when the sampling period is $T = 0.001$ sec and $T = 0.1$ sec.

13 Show that the prototype order 1 all-pass system (5.31) has a unity magnitude function and that its phase function is specified by relationship (5.32).

14 Consider the nonrecursive system with transfer function

$$H_1(z) = 8 - 6z^{-1} + z^{-2}$$

Determine its magnitude function and find three other nonrecursive systems that have the same magnitude behavior. Which of these four magnitude-equivalent systems is the minimum-phase system and which the maximum-phase system? Evaluate

$$\sum_{k=0}^{n} h^2(k) \qquad \text{for } 0 \leq n \leq 2$$

for each of the four linear systems belonging to this class.

15 Consider the nonrecursive systems with transfer function

$$H(z) = \alpha \prod_{k=1}^{q} (1 - z_k z^{-1})$$

in which the roots are all distinct and none have magnitude one. Construct a proof to show that there are 2^q members in the class of linear nonrecursive systems that have the same magnitude function as that of $H(z)$. Moreover, the minimum-phase selection is such as to maximize the sum

$$\sum_{k=0}^{n} h^2(k)$$

for all $0 \leq n \leq q$. Here $\{h(n)\}$ represents the weighting sequence associated with any member in the class being investigated.

16 Consider the two linear recursive systems with transfer functions

$$\textbf{(a)}\ H(z) = \frac{z + 2}{(z + 0.9)(z - 0.9)}$$

$$\textbf{(b)}\ H(z) = \frac{z + 2}{0.81(z + \frac{10}{9})(z - \frac{10}{9})}$$

The first represents a stable causal system, and the second represents a stable anticausal system. Plot their respective magnitude functions. Are they the same If so, why? *Hint*: Use the concept of all-pass networks.

5.17 Determine the stable causal system that has the same magnitude function as that of the unstable causal system with transfer function

$$H(z) = \frac{z^3 + 2z + 5}{(z - 0.5)(z - 3)}$$

5.18 Stabilize the following causal systems using the all-pass network compensation concept.

$$\textbf{(a)}\ H_1(z) = \frac{2z^2 - 5z + 2}{z^2 - 4z + 4}$$

$$\textbf{(b)}\ H_2(z) = \frac{z}{(z - 5)(z - \frac{1}{3})}$$

5.19 How many systems have the same magnitude function as the following system

$$H(z) = \frac{z - 3}{z^3 - 2z^2 + \frac{5}{4}z - \frac{1}{4}}$$

5.20 Show that if z_0 is a root of the polynomial

$$A(z) = a_0 z^n + a_1 z^{n-1} + \cdots + a_{n-1} z + a_n$$

so that $A(z_0) = 0$, then z_0^{-1} is a root of its reversed image polynomial

$$A_r(z) = a_n z^n + a_{n-1} z^{n-1} + \cdots + a_1 z + a_0$$

Thus the roots of $A(z)$ and its reversed order image $A_r(z)$ are reciprocals of one another.

5.21 The time constant of a linear causal system $\{h(n)\}$ can be defined as that positive number n_0 for which

$$\sum_{n=0}^{n_0} h^2(n) = 0.95 \sum_{n=0}^{\infty} h^2(n)$$

Using this definition, determine the time constant of the system

$$y(n) = x(n) - 0.8\, y(n - 1)$$

CHAPTER 6

Design of Frequency-Discrimination Digital Filters

6.1 INTRODUCTION

Fundamental issues involved in the theoretical study of frequency-discrimination filters were examined in Chapter 5. Of particular note was the observation that an important class of frequency-discrimination filters (e.g., bandpass, high pass, band rejection) can be analyzed in the guise of low-pass filters. With these developments we are now in a position to address the practical task of filter design (or synthesis), namely, systematic procedures for assigning values to the parameters of linear nonrecursive and recursive operators so that their associated frequency responses take on desired behaviors. In accordance with the above remarks and the developments of Chapter 5, our efforts here are directed toward low-pass filter synthesis.

The presentation in this chapter is to a certain extent philosophical in nature in that basic principles are emphasized. An exhaustive coverage of the several methods of filter synthesis methods is avoided. From a pedagogical viewpoint, it is strongly felt that a thorough understanding of basic principles is the most expedient and motivating means for mastering the intricate aspects of filter synthesis. With such a background, we are better able to adapt available techniques to meet specialized requirements that invariably arise in applications. Moreover, a person interested in other synthesis techniques may use this background to read the extensive filter synthesis literature with more comprehension.

6.2 SYNTHESIS OF NONRECURSIVE LOW-PASS FILTERS

The discussion of frequency-discrimination filters in Chapter 5 was theoretical in nature. If those concepts are to be of practical importance, systematic procedures for synthesizing low-pass filters must be developed. One such procedure was presented in Section 4.7, which made use of the discrete Fourier transform. In this section we examine a more direct time-domain approach. In particular, a procedure is presented for assigning values to the b_k parameters of the causal nonrecursive filter

$$y(n) = \sum_{k=0}^{q} b_k x(n - k) \tag{6.1}$$

so that its associated frequency response

$$H(e^{j\omega}) = \sum_{n=0}^{q} b_n e^{-j\omega n} \tag{6.2}$$

takes on a prescribed low-pass-filter behavior. This nonrecursive filter is said to be of *order* q because its response is exclusively dependent on the present and q most recent excitation elements. Nonrecursive filters are often preferred over their recursive counterpart because of their guaranteed stability, their inherent finite data window, and the relative ease with which they are synthesized.

Ideal Low-Pass Filter

In the standard approach to filter synthesis, there is typically postulated an ideal behavior that is to be emulated by a realizable filter algorithm. For the prototype low-pass filter, this behavior is specified here by the ideal frequency response characteristic

$$H_\Delta(e^{j\omega}) = \begin{cases} 1 & |\omega| \leq \Delta \\ 0 & \Delta < |\omega| \leq \pi \end{cases} \tag{6.3}$$

This zero-phase idealized filter is seen to transmit all frequency components up to Δ radians perfectly while totally rejecting all frequency components between Δ and π radians. As such, this filter is said to have a *bandwith* (or alternatively, a *cutoff frequency*) of Δ radians. The subscript Δ has been appended to the frequency response function (6.3) to identify the filter's bandwidth explicitly.

The unit-impulse response associated with the postulated ideal low-pass filter is obtained directly by taking the inverse Fourier transform of its frequency response (6.3). This results in the classical sinc function behavior as specified by

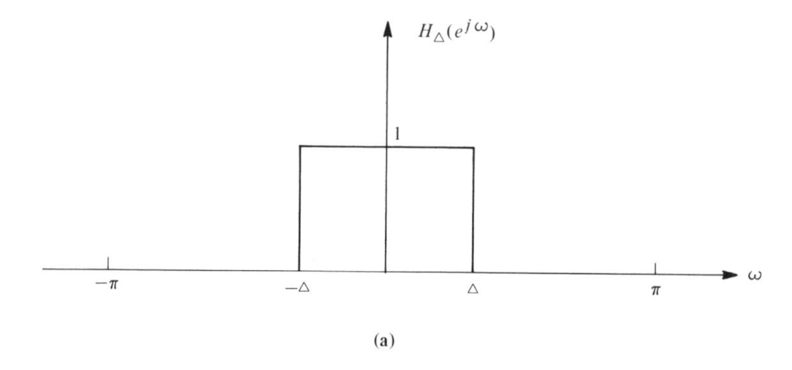

(a)

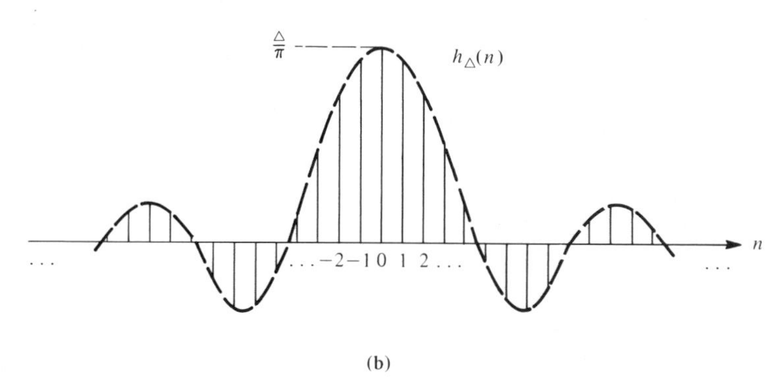

(b)

FIGURE 6.1. Ideal low-pass filter: (a) frequency response; (b) unit weighting sequence.

$$h_\Delta(n) = \frac{\sin(\Delta n)}{\pi n} \qquad \text{for all } n \tag{6.4}$$

Plots of this ideal unit-impulse response and its corresponding ideal frequency response are given in Figure 6.1. It is seen that $h_\Delta(n)$ is an even function of time which decays to zero as the magnitude of n gets large.

Truncated Filter

To implement the ideal low-pass filter as mathematically characterized by unit-impulse response expression (6.4), we could appeal to the associated convolution summation representation; namely, the ideal filter's output as time n is computed using the algorithm

$$y(n) = \sum_{k=-\infty}^{\infty} \frac{\sin(\Delta k)}{\pi k} x(n - k) \tag{6.5}$$

An examination of this expression, however, reveals that an infinite number o multiplication and addition operations are required to compute each respons element. Fortunately, a very simple procedure exists for circumventing thi undesirable feature. In particular, from the observation that $h_\Delta(n)$ goes to zero as the magnitude of n gets large, it is intuitively apparent that the related *truncated nonrecursive filter* of order q,

$$y(n) = \sum_{k=-q/2}^{q/2} \frac{\sin(\Delta k)}{\pi k} x(n-k) \tag{6.6}$$

should yield a reasonably good approximation to the ideal filter operation (6.5). This, of course, presumes that the positive even-valued integer q is selected suitably large. It is to be noted that this modified filter's unit-impulse response has length $q + 1$ and is noncausal, since $h_\Delta(n)$ is generally nonzero for $-1 \geq n \geq -q/2$. In Section 6.3 we examine a method for selecting th "truncation" (or filter-order) parameter q so that this approximation to the idea low-pass-filter behavior is suitably accurate.

Truncated–Causal Filter

In applications requiring causal filtering operations, the noncausal filter (6.6 may be readily modified to yield a causal operation. This simply entails right shifting the unit-impulse response by $q/2$ units, which is seen to result in th associated causal filtering algorithm

$$y(n) = \sum_{k=0}^{q} \frac{\sin[\Delta(k - q/2)]}{\pi(k - q/2)} x(n-k) \tag{6.7}$$

This expression is in the form of the traditional nonrecursive causal filterin operation (6.1). It is a simple matter to show that the frequency response magnitudes of filters (6.6) and (6.7) are identical. It is important to note, howeve that the shifted low-pass filter (6.7) has an inherent delay of $q/2$ units. As such the filter will delay by $q/2$ units any excitation component that lies within th filter's passband. This can be an important consideration in those application where such delays are to be avoided.

In summary, a nonrecursive approximation of the ideal low-pass frequenc response behavior (6.3) can be affected by the truncated algorithm (6.6). Th degree of approximation is determined by the choice of q (exact at $q = +\infty$). A method for selecting the filter-order parameter q is addressed in Section 6.3 In situations necessitating causal operations, the modified filtering algorithr (6.7) provides an equivalent low-pass-filtering behavior in which an inherer delay of $q/2$ units is introduced.

6.3

CHOICE OF THE NONRECURSIVE FILTER'S ORDER

The central issue to be addressed here is that of measuring how accurately the truncated nonrecursive low-pass filter's (6.6) frequency response matches that of the ideal untruncated behavior (6.5). This is readily answered by noting that the ideal frequency response can be decomposed as

$$\begin{aligned} H_\Delta(e^{j\omega}) &= \sum_{n=-\infty}^{\infty} \frac{\sin(\Delta n)}{\pi n} e^{-j\omega n} \\ &= \sum_{n=-q/2}^{q/2} \frac{\sin(\Delta n)}{\pi n} e^{-j\omega n} + \sum_{n=1+q/2}^{\infty} \frac{\sin(\Delta n)}{\pi n} (e^{-j\omega n} + e^{j\omega n}) \end{aligned} \tag{6.8}$$

The first summation on the right side of expression (6.8) is recognized as being the frequency response of the truncated nonrecursive filter (6.6), that is,

$$H_\Delta^{(q)}(e^{j\omega}) = \sum_{n=-q/2}^{q/2} \frac{\sin(\Delta n)}{\pi n} e^{-j\omega n} \tag{6.9}$$

The superscript (q) has been appended so as to explicitly denote the truncated nonrecursive filter's order. From expression (6.8), it is seen that the ideal and truncated frequency responses differ by

$$H_\Delta(e^{j\omega}) - H_\Delta^{(q)}(e^{j\omega}) = \frac{2}{\pi} \sum_{n=1+q/2}^{\infty} \frac{\sin(\Delta n)\cos(\omega n)}{n} \tag{6.10}$$

in which the Euler identity for $\cos(\omega n)$ has been used in arriving at this result. This relationship will be referred to as the *filter frequency response error*.

Examination of relationship (6.10) indicates that the filter frequency response error is a periodic function of ω of period 2π whose first $q/2$ harmonics are absent. This behavior is manifested in a high harmonic ripple behavior, which represents the amount by which the truncated nonrecursive filter's frequency response (6.9) deviates from the ideal behavior (6.3). Since the $\sin(\Delta n)/\pi n$ term decays to zero for large n, it is intuitively clear that the filter frequency response error can be made as small as desired by selecting q adequately large. We shall now verify this conjecture mathematically.

Normalized Frequency Response Error Criterion

A standard method for measuring the degree to which the ideal and truncated frequency responses differ is given by the integral squared-error criterion

$$\int_{-\pi}^{\pi} |H_\Delta(e^{j\omega}) - H_\Delta^{(q)}(e^{j\omega})|^2 \, d\omega = \int_{-\pi}^{\pi} \left| \frac{2}{\pi} \sum_{n=1+q/2}^{\infty} \frac{\sin(\Delta n)\cos(\omega n)}{n} \right|^2 d\omega$$

$$= \frac{4}{\pi} \sum_{n=1+q/2}^{\infty} \frac{\sin^2(\Delta n)}{n^2} \tag{6.11}$$

In arriving at this result, standard integration properties of the sinusoidal product terms appearing on the right-hand integral are employed. Clearly, the right-side expression progressively approaches zero as the truncation parameter q is made larger. This in turn indicates that the truncated filter's frequency response better approximates the ideal behavior in the integral squared-error sense as the filter order q increases. We may therefore use this integral squared-error criterion's behavior for selecting an appropriate value for q. To render expression (6.11) of more computational value, we shall shortly use the readily established identity

$$2\Delta = \int_{-\pi}^{\pi} |H_\Delta(e^{j\omega})|^2 \, d\omega = \frac{2\Delta^2}{\pi} + \frac{4}{\pi} \sum_{n=1}^{\infty} \frac{\sin^2(\Delta n)}{n^2} \tag{6.12}$$

where the first equality follows directly upon integrating the postulated ideal behavior (6.3) and the second equality is obtained along the same lines that lead to expression (6.11).

The fidelity with which the truncated filter's frequency response matches the ideal objective is conveniently measured by the *normalized frequency response error criterion* as defined by

$$\rho(q) = \frac{\int_{-\pi}^{\pi} |H_\Delta(e^{j\omega}) - H_\Delta^{(q)}(e^{j\omega})|^2 \, d\omega}{\int_{-\pi}^{\pi} |H_\Delta(e^{j\omega})|^2 \, d\omega}$$

$$= 1 - \frac{\Delta}{\pi} - \frac{2}{\Delta\pi} \sum_{n=1}^{q/2} \frac{\sin^2(\Delta n)}{n^2} \qquad \text{for } q = 2, 4, 6, \ldots \tag{6.13}$$

in which relationships (6.11) and (6.12) have been used in establishing this equality. The argument q has been incorporated in the criterion $\rho(q)$ so as to recognize its dependency on the nonrecursive filter order. At $q = -1$, the frequency response match criterion is defined to be one, which reflects the fact that the corresponding truncated filter's frequency response is identically zero. As q takes on the values $0, 2, 4, 6, \ldots$, this criterion's value monotonically decreases from 1 and approaches zero as q approaches plus infinity. The filter designer's primary task is then to find the *smallest* value of q so that this normalized error criterion is made reasonably close to the ideal value of zero. A reasonably small value is desired because q gives a direct measure of the filter's memory requirement and computational complexity.

For a given choice of the low-pass filter's bandwith Δ, we can use expression (6.13) to compute the behavior of the normalized frequency response error criterion as a function of q. This behavior can then be utilized for selecting the truncation parameter q so that a prescribed frequency response approximation

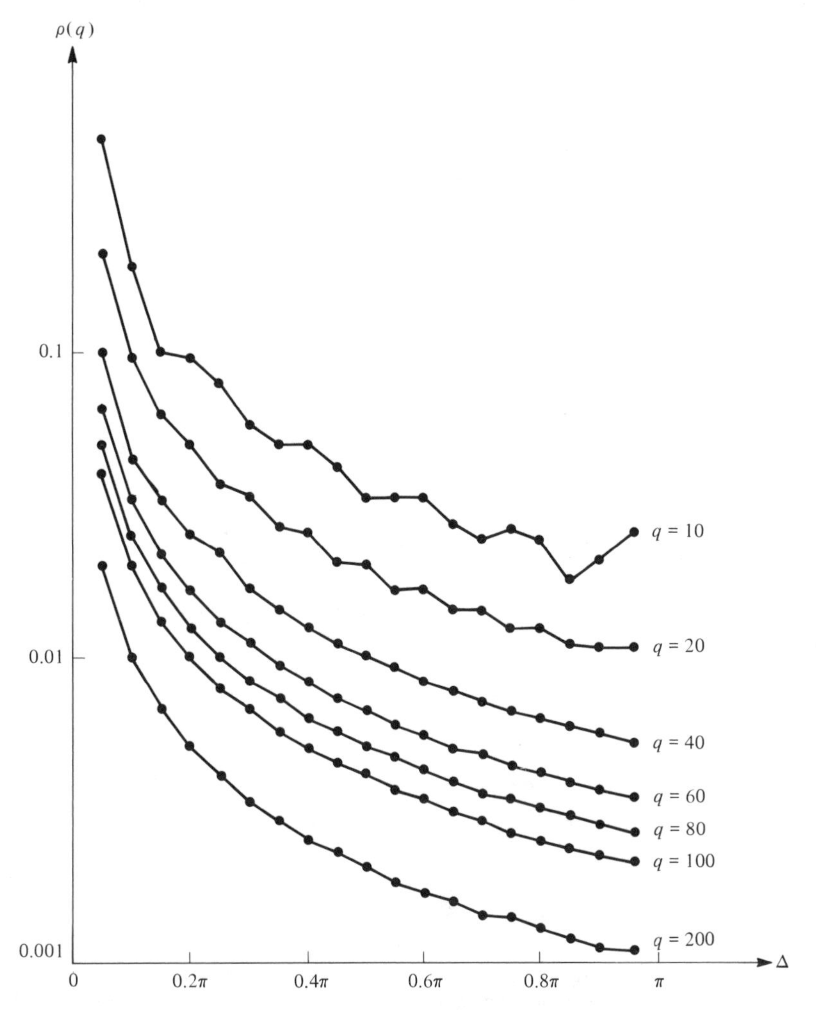

FIGURE 6.2. Plot of normalized frequency response error criterion ρ(*q*) versus filter cutoff frequency Δ for several choices of nonrecursive filter order *q*.

is achieved. In this approach it is beneficial to use the following equivalent iterative expression for (6.13):

$$\rho(q) = \rho(q - 2) - \frac{8 \sin^2(\Delta q/2)}{\Delta \pi q^2} \qquad \text{for } q = 2, 4, 6, \ldots \tag{6.14}$$

with initial condition

$$\rho(0) = 1 - \frac{\Delta}{\pi} \tag{6.15}$$

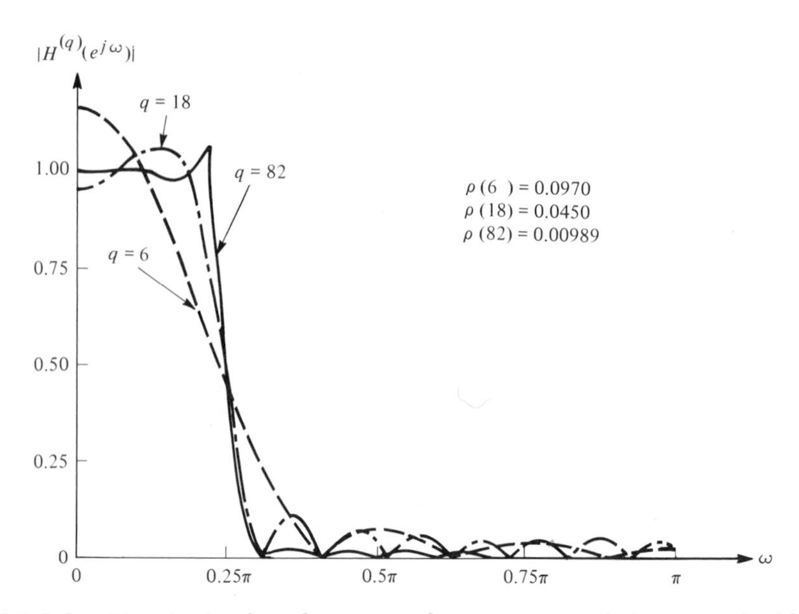

FIGURE 6.3. Magnitude plots for truncation parameter choices $q = 6$, 18, and 82.

This iterative expression's validity is readily established by substituting relationship (6.13) into (6.14).

To gain an appreciation for how the low-pass filter's bandwidth Δ and order q are interrelated, expression (6.13) was examined for several choices of cutoff frequencies Δ (i.e., $\Delta = 0.05k\pi$ for $k = 1, 2, \ldots, 19$) over the principal frequency range $[0, \pi)$. At each of these cutoff frequencies, the value of $\rho(q)$ was computed for q taking on the values 10, 20, 40, 60, 80, 100, and 200. The results are plotted in Figure 6.2. From these plots it is apparent that the filter order needed to achieve a given transfer function error level generally decreases as the low-pass filter cutoff frequency Δ increases. Moreover, a relatively significant increase in filter order is required to improve slightly the normalized frequency response error criterion.

EXAMPLE 6.1

Let it be desired to design a nonrecursive filter with a cutoff frequency of $\Delta = 0.25\pi$ radians. To gain an insight into the use of the normalized frequency response error criterion, we shall now determine the required filter order q needed to cause $\rho(q)$ to be first equal to or less than 0.10, 0.05, and 0.01. Moreover, to gain an insight into the use of this $\rho(q)$ measure, plots of the magnitude of the frequency response (6.9) associated with each of these filtering orders are given. From Figure 6.2, the required filter orders required are found by interpolation to be as follows:

At $q = 6$: $\rho(6) = 0.0970.$
At $q = 18$: $\rho(18) = 0.0450.$
At $q = 82$: $\rho(82) = 0.00989.$

Plots of the magnitudes of $H^{(q)}(e^{j\omega})$ for these three order choices are shown in Figure 6.3. Clearly, the choice of $\rho = 0.1$ provides a relatively crude approximation to the ideal frequency response behavior, while the selection $\rho = 0.01$ gives a reasonably satisfactory approximation. ■

6.4 NONRECURSIVE FILTERS: GENERAL WINDOW APPROACH

In Section 6.3 a straightforward procedure for synthesizing a nonrecursive low-pass filter with prescribed frequency response characteristics was developed. This procedure employed the intuitively appealing concepts of a truncation of the ideal low-pass filter's unit-impulse response. As we now demonstrate, this method is a special case of the more general windowing technique of filter synthesis. In this approach we consider the finite-length weighting sequence as generated according to the sequence product rule

$$h(n) = w(n)h_d(n) \tag{6.16}$$

where $\{h_d(n)\}$ represents the weighting sequence corresponding to the underlying *ideal* filter and $\{w(n)\}$ denotes a *window* sequence. The window sequence is here taken to have a finite length and to be nonzero only over those values of n for which $h_d(n)$ is deemed to significantly differ from zero. From the window's finite-length requirement it follows that the synthesized filter will also have a finite length and is thereby implementable in a nonrecursive manner.

Since the synthesized filter's unit impulse response is specified by the sequence product (6.16), we may use the Fourier transform property (3.19) to obtain its associated frequency response: namely, the windowed filter (6.16) has its frequency response given by the convolution integral operation

$$H(e^{j\omega}) = \frac{1}{2\pi} \int_{-\pi}^{\pi} H_d(e^{j\nu})W(e^{j(\omega-\nu)})\, d\nu \tag{6.17}$$

In this representation, $W(e^{j\omega})$ denotes the Fourier transform of the window sequence $\{w(n)\}$ as specified by

$$W(e^{j\omega}) = \sum_{n=-\infty}^{\infty} w(n)e^{-j\omega n}$$

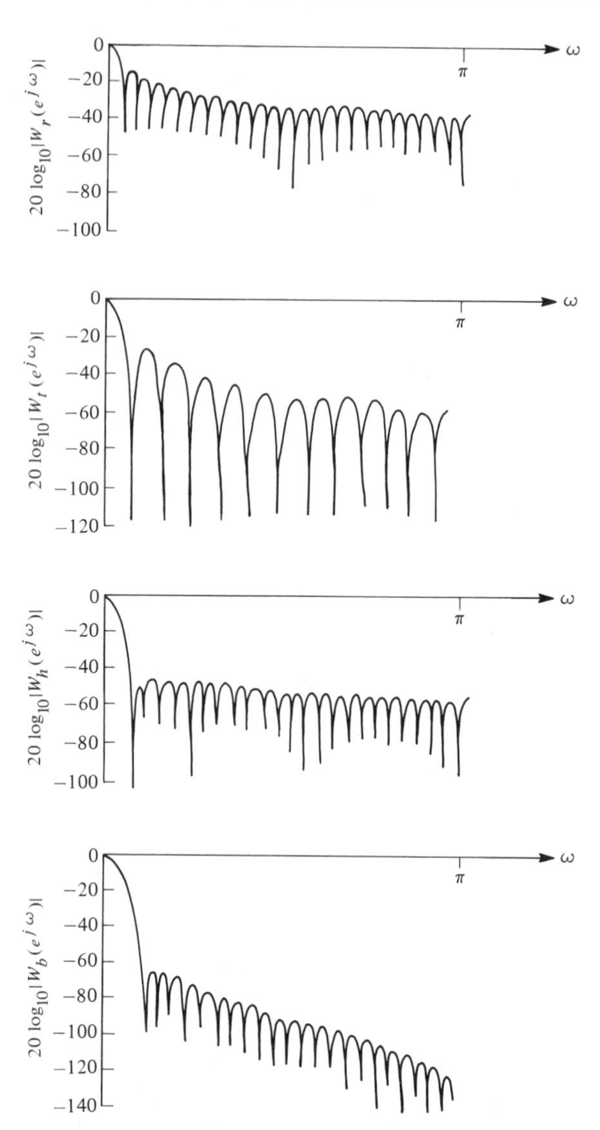

FIGURE 6.4. Magnitude of window's Fourier transform in decibels for $q = 50$: (a) rectangular window; (b) triangular window; (c) Hamming window; (d) Blackman window.

where typically only a contiguous finite number of the $w(n)$ terms are nonzero. Thus the synthesized filter's frequency response is obtained simply by convolving the ideal frequency response with the window's Fourier transform. From expression (6.17) it is apparent that if the window's Fourier transform closely approximates the Dirac delta function $2\pi\delta(\omega)$ for $\omega \in (-\pi, \pi]$, the synthesized filter will have a frequency response characteristic that closely matches that of the desired behavior $H_d(e^{j\omega})$.

TABLE 6.1 Some Commonly Used Windows of Width $q + 1$

Window	$w(n)$ for $-q/2 \leq n \leq q/2$
Rectangular	1
Triangular (Bartlett)	$1 - 2\|n\|/q$
Hamming	$0.54 + 0.46 \cos(2n\pi/q)$
Blackman	$0.42 + 0.50 \cos(2n\pi/q) + 0.08 \cos(4n\pi/q)$

The symmetric windows listed in Table 6.1 possess the desired impulse-type behavior provided that the window-length parameter q is selected suitably large. This is made evident in Figure 6.4, where plots of each of the window's associated Fourier transform magnitudes as measured in decibels [i.e., $20 \log |W(e^{j\omega})|$]

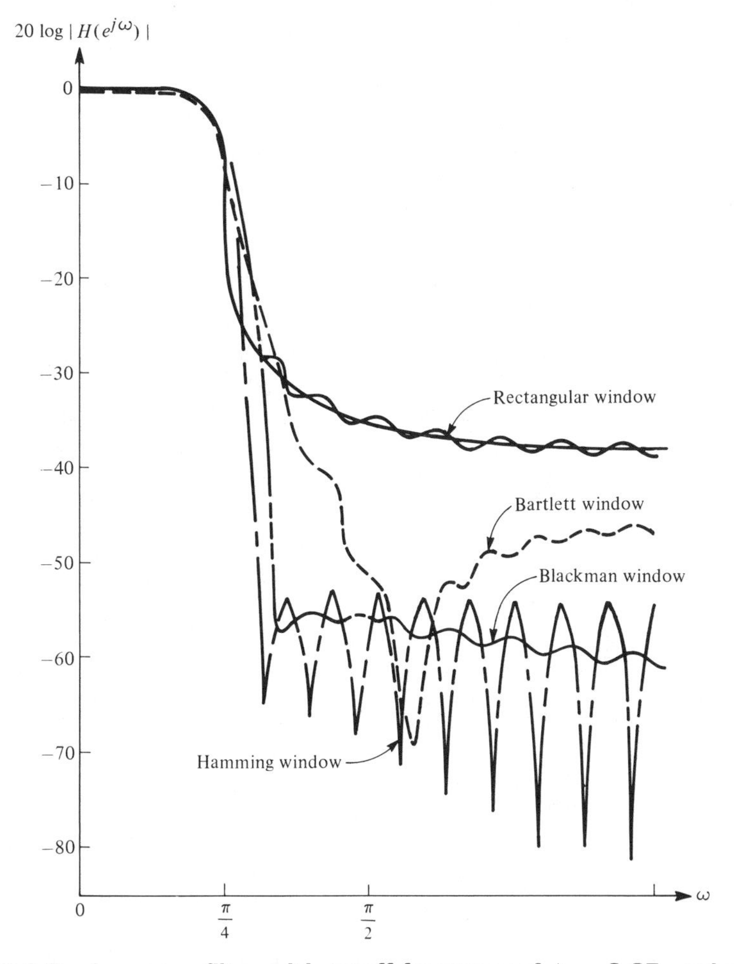

FIGURE 6.5. Low-pass filter with cutoff frequency of $\Delta = 0.25\pi$ using the windows given in Table 6.1.

are made for the case $q = 50$. A number of general conclusions can be inferred from these specific window selections. Of primary importance are the following:

1. The Fourier transform of the rectangular window has the narrowest main lobe and therefore provides the best replication of sharp transitions in the desired magnitude behavior.
2. The first side lobe associated with the rectangular window's Fourier transform has the highest level, being only 13 dB below that of its main lobe. This indicates that the associated windowed frequency response (6.17) will have the highest ripple level of the windows listed in Table 6.1.

Thus the performance trade-off between a sharp transition replication and a low ripple level is of prime consideration when selecting the data window. A gradual tapering of the window to zero at $\pm q/2$ is seen to generate a wider main lobe but also a lower side-band level.

To depict the effects of data windowing, let us consider an ideal low-pass filter with a cutoff frequency $\Delta = 0.25\pi$. Using each of the windows given in Table 6.1 for filter order $q = 50$, the associated transfer function's (6.17) frequency response magnitude is sketched in Figure 6.5. The transition replication and rippling effects mentioned above are apparent from these plots.

6.5 RECURSIVE BUTTERWORTH FREQUENCY-DISCRIMINATION FILTERS

To achieve a sufficiently sharp passband–stopband transition, it is generally necessary to select the nonrecursive filter's order to be rather large. Unfortunately, this in turn leads to a computationally intensive filtering algorithm, since the number of multiplication and summation operations required for each response element evaluation is proportional to this order. If our primary objective is that of closely approximating an ideal frequency-discrimination behavior in a computationally efficient manner, it is often beneficial to consider linear recursive filters. A causal linear recursive filter of order (p, q) is governed by

$$y(n) = \sum_{k=0}^{q} b_k x(n - k) - \sum_{k=1}^{p} a_k y(n - k) \tag{6.18}$$

This filtering algorithm is said to be recursive, since previously computed response elements [i.e., the $y(n - k)$ for $1 \leq k \leq p$] are used in evaluating each current response element. Implementation of this recursive algorithm is seen to entail $p + q + 1$ multiplication and $p + q$ addition operations to evaluate each response element $y(n)$. It is widely recognized that a prescribed frequency-discrimination approximation is typically achievable with a much lower order recursive filter than its nonrecursive filter counterpart.

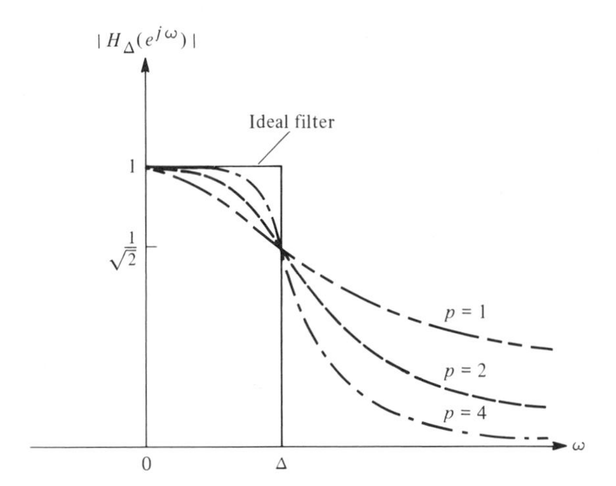

FIGURE 6.6. Magnitude function associated with the low-pass Butterworth function.

As shown in Chapter 5, the classical high-pass, bandpass, and band-reject frequency-discrimination filters can be synthesized with an appropriate interconnection of low-pass filters. With this in mind, we need only consider the synthesis of recursive low-pass filters. To effect one such synthesis, we now appeal to a trigonometric-based procedure. In particular, consider a low-pass filter whose frequency-response's squared magnitude behavior is governed by the *Butterworth function*

$$|H_\Delta(e^{j\omega})|^2 = \frac{1}{1 + \tan^{2p}(\omega/2)/\tan^{2p}(\Delta/2)} \tag{6.19}$$

where the positive integer p is subsequently identified with the filter order parameter p appearing in recursive relationship (6.18), and Δ is the desired recursive filter's *cutoff* frequency. In this section it is shown that there exists a linear recursive filter whose associated frequency response magnitude is given by the above Butterworth function. This realization provides the motivation for studying Butterworth filters.

Plots of the Butterworth frequency response function (6.19) for different choices of the filter order parameter p are shown in Figure 6.6. From these plots it is clear that the magnitude function monotonically decreases from 1 at $\omega = 0$ and that the filter's half-power point [(i.e., where $|H(e^{j\omega})|^2 = 0.5$] is located at $\omega = \Delta$. Moreover, the transition from the nominal passband $[0, \Delta]$ to the nominal stopband $(\Delta, \pi]$ becomes more rapid as the filter-order parameter p is increased. Namely, the filter's magnitude function uniformly becomes closer to one (zero) in the nominal passband (stopband) as the filter order increases. As a point of fact, as p increases to plus infinity, this magnitude behavior approaches that of the ideal low-pass filter illustrated in Figure 6.1.

Filter-Order-Selection Procedure

To provide a systematic procedure for selecting the Butterworth filter-order parameter p to meet prescribed frequency-discrimination behavior, let us consider the filtering requirements as depicted in Figure 6.7. Specifically, let it be desired to synthesize a Butterworth filter whose associated frequency response's magnitude is simultaneously equal to or larger than $1 - \delta_a$ for all frequencies $0 \leq \omega \leq \omega_a$ and is smaller than δ_b for all frequencies $\omega_b \leq \omega \leq \pi$. In this approach, the frequency range (ω_a, ω_b) is identified as the filter transition band while the positive parameters δ_a and δ_b control how close the filter's magnitude is near one in the nominal passband $[0, \omega_a]$ and zero in the nominal stopband $(\omega_b, \pi]$, respectively. The filter designer normally selects the parameters ω_a, ω_b, δ_a, and δ_b to reflect frequency discrimination requirements that are needed in a specific application.

To determine how these parameters affect the required filter order, let us first solve expression (6.19) for the parameter p. This is readily shown to yield

$$p = \frac{1}{2} \frac{\log [|H_\Delta(e^{j\omega})|^{-2} - 1]}{\log [\tan (\omega/2)/\tan (\Delta/2)]} \tag{6.20}$$

Upon evaluating this expression at the frequencies ω_a and ω_b, and setting $|H_\Delta(e^{j\omega_a})| = 1 - \delta_a$ and $|H_\Delta(e^{j\omega_b})| = \delta_b$, respectively, the individual filter orders needed to satisfy these two magnitude requirements are determined to be

$$p_a = \frac{1}{2} \frac{\log [(2\delta_a - \delta_a^2)/(1 - \delta_a)^2]}{\log [\tan(\omega_a/2)/\tan (\Delta/2)]} \tag{6.21}$$

$$p_b = \frac{1}{2} \frac{\log [-1 + \delta_b^{-2}]}{\log [\tan (\omega_b/2)/\tan (\Delta/2)]} \tag{6.22}$$

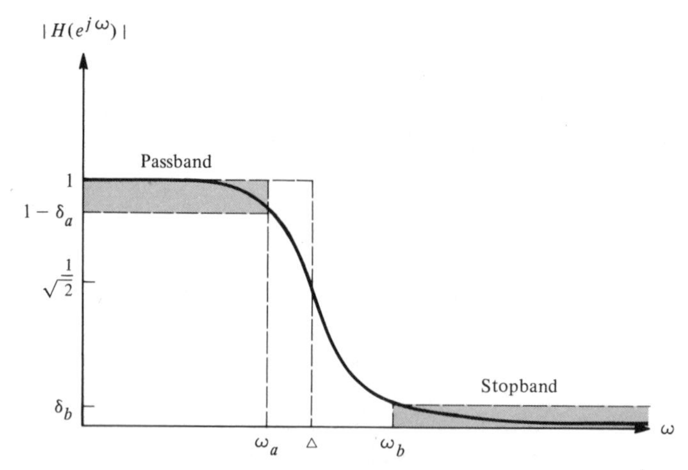

FIGURE 6.7. Low-pass filter specifications.

It is generally found that entities p_a and p_b are not integers due to the fact that the parameters ω_a, ω_b, δ_a, and δ_b will not correspond exactly with the Butterworth magnitude function (6.19). Because of the monotonic nature of this magnitude function, however, it immediately follows that any *integer*-valued filter-order selection that satisfies

$$p \geq \max\,(p_a, p_b) \tag{6.23}$$

will cause the magnitude function (6.19) to at least equal the required behavior shown in Figure 6.7. In summary, once the parameters ω_a, ω_b, δ_a, and δ_b have been specified, expression (6.23) may be used to identify the recursive filter order needed to meet or exceed the desired magnitude frequency response behavior. It is possible to use other procedures for selecting the filter order p.

EXAMPLE 6.2

Let it be desired to design a low-pass filter with a nominal cutoff frequency $\Delta = 0.25\pi$ radians. Due to specific filtering requirements, the following parameter specifications are made: $\omega_a = 0.20\pi$ radians, $\omega_b = 0.41\pi$ radians, $\delta_a = \delta_b = 0.1$. Inserting these parameters into expressions (6.21) and (6.22), it is found that

$$p_a = 2.986 \qquad p_b = 3.863$$

Thus, in accordance with expression (6.23), the required filtering specification are met (actually exceeded) provided that the filter order is selected to be at least equal to 4 (i.e., $p \geq 4$). ■

6.6 BUTTERWORTH FILTER SYNTHESIS

To determine the recursive filter that has the squared magnitude behavior (6.19), let us now examine the related rational function of z:

$$H_\Delta(z)H_\Delta(z^{-1}) = \frac{1}{1 + \dfrac{(-1)^p}{\tan^{2p}(\Delta/2)}\left(\dfrac{z-1}{z+1}\right)^{2p}} \tag{6.24}$$

We have suggestively expressed the left-hand side as the product of a function $H_\Delta(z)$ and its image $H_\Delta(z^{-1})$, where the image function is obtained by replacing z by z^{-1} everywhere it appears in $H_\Delta(z)$. This particular factorization is justified on the basis that the right-hand-side function has the same value at z as it does at z^{-1} for all choices of z. It is to be noted that the functions $H_\Delta(z)$ and $H_\Delta(z^{-1})$ constituting this factorization of necessity will have zeros and poles that are the reciprocals of one another. For reasons that are to be made clear shortly, the poles of the rational function (6.24) that lie inside the unit circle are assigned to $H_\Delta(z)$, while their reciprocal mates are associated with the image function $H(z^{-1})$.

Upon using the Euler identity, it is readily established that upon setting $z = e^{j\omega}$ in function (6.24), the required Butterworth squared magnitude behavior (6.19) results. We may therefore identify $H_\Delta(z)$ as being the transfer function of the recursive filter that has the prescribed Butterworth magnitude behavior [since $|H_\Delta(e^{j\omega})|^2 = H_\Delta(e^{j\omega})\overline{H}_\Delta(e^{j\omega}) = H_\Delta(e^{j\omega})H_\Delta(e^{-j\omega})$]. Moreover, this recursive filter must be stable, since only those poles of rational function (6.24) that lie inside the unit circle have been assigned to $H_\Delta(z)$.

To obtain the desired recursive filter transfer function, it is then necessary to perform a factorization of the numerator and denominator polynomials appearing on the right-hand side of expression (6.24). This is straightforwardly accomplished by first multiplying the numerator and denominator of this expression by $\tan^{2p}(\Delta/2)(z + 1)^{2p}$ to obtain

$$H_\Delta(z)H_\Delta(z^{-1}) = \frac{\tan^{2p}(\Delta/2)(z + 1)^{2p}}{\tan^{2p}(\Delta/2)(z + 1)^{2p} + (-1)^p(z - 1)^{2p}}$$

After factoring the denominator polynomial, the causal stable recursive filter associated with the Butterworth frequency response behavior is specified by (Cadzow, 1973; Gold and Rader, 1969)

$$H_\Delta(z) = \left(\prod_{m=1}^{p} \frac{1 - p_m}{2}\right) \frac{(1 + z^{-1})^p}{\prod_{k=1}^{p} (1 - p_k z^{-1})} \tag{6.25}$$

This transfer function is seen to have its p zeros located at $z = -1$, while its p poles are found to be given by

$$p_k = \frac{1 - \tan^2(\Delta/2) + j2\tan(\Delta/2)\cos[(2k - 1)\pi/2p]}{1 + 2\tan(\Delta/2)\sin[(2k - 1)\pi/2p] + \tan^2(\Delta/2)} \quad \text{for } 1 \le k \le p \tag{6.26}$$

It can be shown that these poles all lie on a circle whose radius is specified by $r = 2\tan(\Delta/2)/[1 - \tan(\Delta/2)]$ and whose center is located on the real axis

TABLE 6.2 Method for Synthesizing a Recursive Butterworth Low-Pass Filter

1. Determine the required filter-order parameter p. This might be effected by requiring that the frequency-discrimination behavior be as depicted in Figure 6.7. The required order is then specified by expressions (6.21) to (6.23).
2. The required filter transfer function is then given by expression (6.25), in which the poles are given in equation (6.26). Upon carrying out the indicated products, we find that the coefficients of the corresponding recursive filter algorithm (6.18) are generated, in which $q = p$.

with abscissa value equal to $[1 + \tan^2(\Delta/2)]/[1 - \tan^2(\Delta/2)]$. The basic steps of the Butterworth low-pass filter synthesis are outlined in Table 6.2.

Practical Consideration

The pth-order linear recursive equation, which implements the low-pass transfer function (6.25), is obtained by first carrying out the products indicated in the numerator and denominator. This operation will result in the real-valued coefficients that identify the associated filter recursive expression. These coefficients are guaranteed to be real due to the fact that the filter's poles occur in complex conjugate pairs, whereby $p_k = \overline{p}_{p+1-k}$ for $1 \leq k \leq p$. From a practical viewpoint, however, it is preferable to implement the resultant pth-order linear recursive expression as a cascade of second-order systems (and one first-order system if p is odd). If the order p is even, this cascaded configuration would be specified by

$$H_\Delta(z) = b \prod_{k=1}^{p/2} \frac{1 + 2z^{-1} + z^{-2}}{1 - 2\,\mathrm{Re}\,(p_k)z^{-1} + |p_k|^2 z^{-2}} \tag{6.27a}$$

while for p odd we have

$$H_\Delta(z) = b \frac{1 + z^{-1}}{1 - p_{(p+1)/2}z^{-1}} \prod_{k=1}^{(p-1)/2} \frac{1 + 2z^{-1} + z^{-2}}{1 - 2\,\mathrm{Re}\,(p_k)z^{-1} + |p_k|^2 z^{-2}} \tag{6.27b}$$

in which

$$b = 2^{-p} \prod_{k=1}^{p} (1 - p_k) \tag{6.27c}$$

This second-order cascade equivalency is superior because it desensitizes the overall filtering operation to the inevitable coefficient roundoff encountered in a digital implementation. Such coefficient roundoff can lead to an unstable filter realization even though the untruncated coefficients are guaranteed to give a stable filter.

Normalized Frequency Response Magnitude Error Criterion

To measure how closely the synthesized Butterworth filter's frequency response magnitude (6.19) approximates the ideal objective the *normalized frequency response magnitude error criterion* specified by

$$\rho(p) = \frac{\int_{-\pi}^{\pi} [|H_d(e^{j\omega})| - |H_\Delta(e^{j\omega})|]^2 \, d\omega}{\int_{-\pi}^{\pi} |H_d(e^{j\omega})|^2 \, d\omega}$$

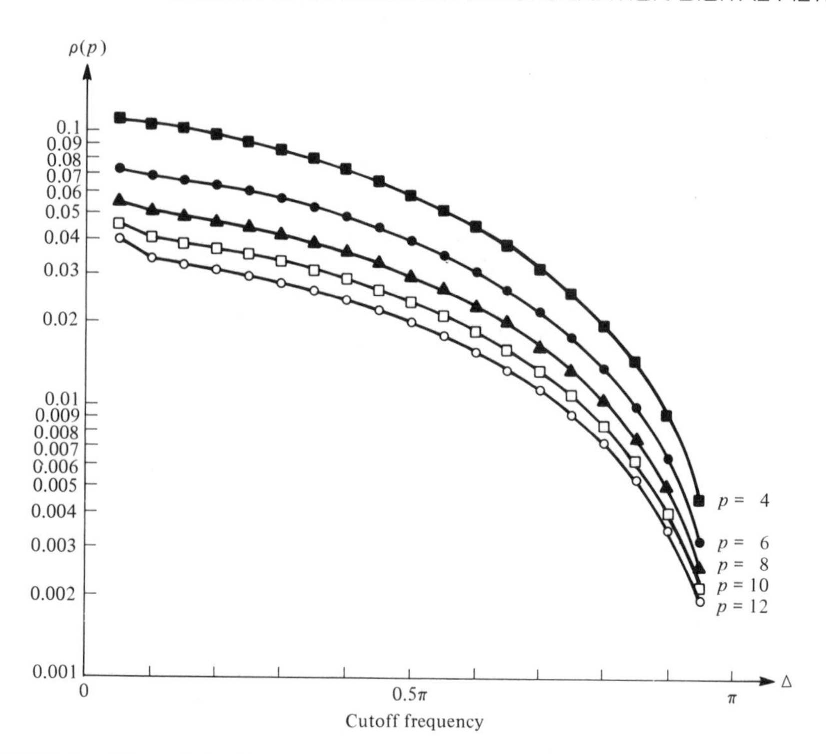

FIGURE 6.8. Plot of the Butterworth filter's normalized frequency response error criterion ρ(p) versus filter cutoff frequency Δ for several choices of filter order p

is used. The argument p specifies the recursive filter's order, while the ideal frequency response magnitude used in this criterion is governed by

$$|H_d(e^{j\omega})| = \begin{cases} 1 & |\omega| \le \Delta \\ 0 & \Delta < |\omega| \le \pi \end{cases}$$

It can be shown that this normalized criterion takes on values exclusively in the interval $0 \le \rho(p) \le 1$. To gain a feeling for the performance of recursive Butterworth filters, plots of $\rho(p)$ versus Δ are shown in Figure 6.8 for choices of filter order $p = 4, 6, 8, 10$, and 12. These plots will enable us to make an informed selection of filter order. Upon comparison of these plots with those given in Figure 6.2, it is apparent that recursive filters are generally more effective than their nonrecursive counterpart in achieving a prescribed frequency response behavior.

EXAMPLE 6.3

Let us continue with the recursive filtering synthesis begun in Example 6.2, in which $\Delta = 0.25\pi$ radians and $p = 4$. Using step 2 of Table 6.2, we determine the polynomial roots using expression (6.26) with $\omega_1 = 0.25\pi$ and $p = 4$. This is found to yield the following four stable roots:

$$p_1, p_2 = 0.427692 \pm j0.163680 \qquad p_3, p_4 = 0.556520 \pm j0.514171$$

while the constant b is specified by relationship (6.27c):

$$b = 0.010210$$

The filter transfer function arising from this design procedure is then obtained using expression (6.27a) with these parameters. After carrying out the indicated multiplications, we have

$$H_\Delta(z) = \frac{0.010210(1 + z^{-1})^4}{(1 - 0.85538z^{-1} + 0.209712z^{-2})(1 - 1.113040z^{-1} + 0.574087z^{-2})}$$

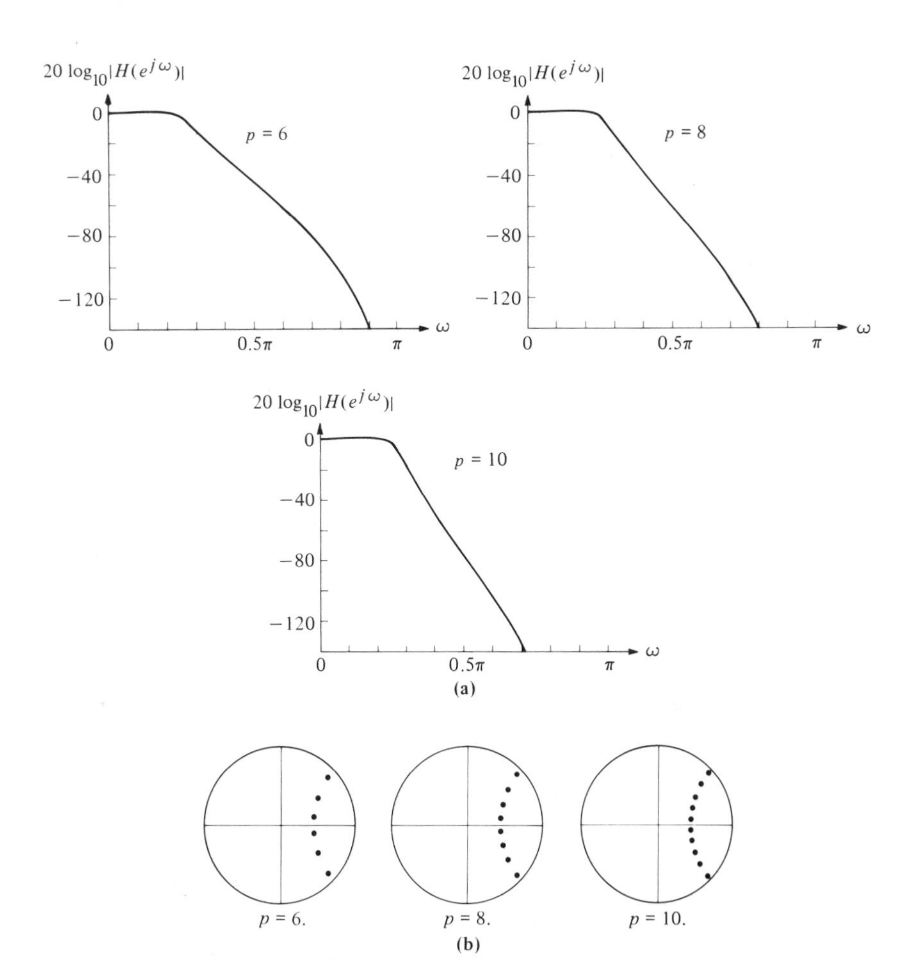

FIGURE 6.9. Location of Butterworth filter poles relative to the unit circle for cutoff frequency $\Delta = 0.25\pi$ and filter orders p = 6, 8, and 10.

This filtering operation may be implemented by a fourth-order linear difference equation or alternatively, as the cascade of two second-order linear difference equations. ■

EXAMPLE 6.4

It is insightful to examine the pole locations of a typical Butterworth filter. With this in mind, let us consider three Butterworth filters of order $p = 6$, 8, and 10, in which the cutoff frequencies is $\Delta = 0.25\pi$ radians. For this cutoff frequency, the corresponding Butterworth frequency response magnitudes as specified by relationship (6.19) are plotted in Figure 6.9a for $p = 6$, 8, and 10. By using the Butterworth filter synthesis procedure outlined in Table 6.2, the poles associated with each of the three filters are determined and also plotted in Figure 6.9b. The distribution of the poles on a circle is evident. ■

6.7 POSITIVE-SEMIDEFINITE FOURIER TRANSFORMS

In various signal-processing applications, the appearance of Fourier transforms that are positive-semidefinite functions of frequency arise in a natural manner. As examples, the frequency response of an ideal zero-phase frequency-discrimination filter, the spectral density function of a wide-sense-stationary time series, and the coherence function used in multiple signal analysis each fall into this category. It will be shown in Sections 6.8 and 6.9 that the concept of positive-definite Fourier transforms can be used to effectively synthesize frequency-discrimination filters. With this serving as motivation, we now examine some properties possessed by *positive-semidefinite Fourier transforms*. The Fourier transform $G(e^{j\omega})$ is said to be positive-semidefinite if

$$G(e^{j\omega}) = \sum_{n=-\infty}^{\infty} g(n)\, e^{-j\omega n} \geq 0 \qquad \text{for all } \omega \tag{6.28}$$

A more thorough treatment of positive-semidefinite Fourier transforms is given in Cadzow and Sun (1986). For obvious reasons, we refer to the generating sequence (i.e., its inverse Fourier transform)

$$g(n) = \frac{1}{2\pi} \int_{-\pi}^{\pi} G(e^{j\omega}) e^{j\omega n}\, d\omega \tag{6.29}$$

as being a *positive-semidefinite sequence*. Clearly, only a small subclass of sequences is positive semidefinite. As noted above, however, this subclass plays an important role in signal-processing theory and applications.

Since a positive-semidefinite Fourier transform is by its very definition real valued, this imposes a constraint on its generating sequence (6.29). A necessary condition for a sequence to have a positive-semidefinite Fourier transform is that it be a complex conjugate symmetric sequence, that is,

$$g(n) = \overline{g}(-n) \tag{6.30}$$

where the overbar symbol denotes the complex-conjugation operation. It is important to appreciate the fact that all positive-semidefinite Fourier transforms have conjugate symmetric generating sequences, but the converse need not follow. Thus only a (small) subset of conjugate symmetric sequences are positive semidefinite.

EXAMPLE 6.5

Consider the frequency response associated with the ideal low-pass filter as specified by

$$H_\Delta(e^{j\omega}) = \begin{cases} 1 & |\omega| \leq \Delta \\ 0 & \Delta < |\omega| \leq \pi \end{cases}$$

This Fourier transform is seen to be a positive-semidefinite function. Moreover, the sequence that gives rise to it,

$$h_\Delta(n) = \frac{\sin(\Delta n)}{\pi n}$$

satisfies the required symmetrical condition (6.30). It is important to note that this sequence takes on negative values even though its Fourier transform is a positive-semidefinite function of frequency. ■

Spectral Factorization Representation

If the Fourier transform $G(e^{j\omega})$ is positive semidefinite, a well-known theorem indicates that it is always possible to find another Fourier transform $H(e^{j\omega})$ such that the *spectral factorization representation*

$$\begin{aligned} G(e^{j\omega}) &= H(e^{j\omega})\overline{H}(e^{j\omega}) \\ &= |H(e^{j\omega})|^2 \end{aligned}$$

holds for all ω (e.g., see Papoulis, 1977). This is a most significant result, since it indicates that the analysis of positive-semidefinite functions can be equivalently made through general Fourier transforms $H(e^{j\omega})$ in which constraints such as positive semidefiniteness need not be imposed. It should be noted that no claim for uniqueness of the spectral factorization has been here made. That is, if $H(e^{j\omega})$ is a Fourier transform for which such a representation holds, the related Fourier transform $H(e^{j\omega})e^{j\phi(\omega)}$ will also satisfy this representation for any choice of the phase function $\phi(\omega)$. We therefore conclude that an infinite number of different spectral factorization representations exist. The magnitude of each of the factor terms [i.e., the $H(e^{j\omega})$] contained in this infinite set, however, are all identical and equal to $\sqrt{G(e^{j\omega})}$. We now summarize this fundamental result.

Theorem 6.1 (Spectral Factorization Theorem). Let the conjugate symmetric sequence $\{g(n)\}$ be square summable and its Fourier transform $G(e^{j\omega})$ be positive semidefinite. It then follows that this positive-semidefinite Fourier transform may be factorized as

$$G(e^{j\omega}) = |H(e^{j\omega})|^2 \tag{6.31}$$

in which the function $H(e^{j\omega})$ has a square summable inverse Fourier transform $\{h(n)\}$. Furthermore, if $\{g(n)\}$ happens to be real valued, then $G(e^{j\omega})$ is a symmetric positive-semidefinite function of ω and $\overline{H}(e^{j\omega}) = H(e^{-j\omega})$.

6.8 RATIONAL APPROXIMATION OF POSITIVE-SEMIDEFINITE SYMMETRIC FOURIER TRANSFORMS

In this section a straightforward procedure is developed for approximating a positive-semidefinite Fourier transform by a rational function. Rational approximations are useful in those applications related to positive-semidefinite functions. For instance, such an approximation can be used for implementing a linear recursive filter whose frequency response closely matches that of an ideal behavior. Similarly, rational approximations play a prominent role in spectral analysis and system identification.

The spectral factorization theorem is applicable to all positive-semidefinite Fourier transforms. To simplify the analysis to follow, we require further that the positive-semidefinite function under analysis be a symmetric function of frequency. Fortunately, in applications related to real-valued signal processing, this symmetrical condition is automatically satisfied. Typically, the positive-semidefinite symmetric function $G(e^{j\omega})$ being factorized will have a relatively complex structure that makes its interpretation (or implementation) very difficult. This being the case, it is often beneficial to approximate the function $G(e^{j\omega})$ by the simpler rational symmetric function

$$G_{2p}(e^{j\omega}) = \frac{\beta_0 + \sum_{k=1}^{p} \beta_k(e^{j\omega k} + e^{-j\omega k})}{1 + \sum_{k=1}^{p} \alpha_k(e^{j\omega k} + e^{-j\omega k})} \tag{6.32}$$

in which the symmetry of the polynomial coefficients is to be noted. The real-valued α_k and β_k parameters constituting this approximating rational model are to be selected so that this model provides a suitably accurate representation for $G(e^{j\omega})$ and the model is also a positive semidefinite function. The latter condition is imposed so as to ensure that the rational model (6.32) can be represented by

$$G_{2p}(e^{j\omega}) = \left| \frac{b_0 + b_1 e^{-j\omega} + \cdots + b_p e^{-jp\omega}}{1 + a_1 e^{-j\omega} + \cdots + a_p e^{-jp\omega}} \right|^2 \tag{6.33}$$

as is ensured by the spectral factorization theorem. The a_k and b_k parameters of this factored form are also to be real valued. We shall now discuss the ramifications of these two imposed conditions.

Since the positive-semidefinite symmetric function $G(e^{j\omega})$ being approximated is generally nonrational, it is apparent that an error is incurred when using the rational model (6.32) to represent $G(e^{j\omega})$. For our purposes, the goodness of the rational-model representation will be measured by the standard integral of the squared-error criterion,

$$\rho(\alpha_k, \beta_k) = \frac{1}{2\pi} \int_{-\pi}^{\pi} |G(e^{j\omega}) - G_{2p}(e^{j\omega})|^2 \, d\omega \tag{6.34}$$

The closer to zero this criterion can be made, the better the rational model is said to represent the positive-semidefinite symmetric function being approximated. Ideally, it is desirable to select the model's α_k and β_k parameters so as to render this criterion its minimum value. Unfortunately, this minimization is intractable due to the nonlinear dependency of the integrand term $G_{2p}(e^{j\omega})$ on the α_k and β_k parameters. If we are to achieve an optimum solution, it is generally necessary to employ nonlinear programming techniques.

An approximating method that circumvents the computationally demanding nonlinear programming solution approach is now described. It is predicated on applying Parseval's relationship (3.21) to equivalently express the integral squared-error criterion (6.34) as the *sum-of-squared-error* criterion

$$\rho(\alpha_k, \beta_k) = \sum_{n=-\infty}^{\infty} |g(n) - g_{2p}(n)|^2 \tag{6.35}$$

In this time-domain equivalency, the real-valued symmetric sequences $g(n)$ and $g_{2p}(n)$ correspond to the inverse Fourier transforms of $G(e^{j\omega})$ and $G_{2p}(e^{j\omega})$, respectively. From these equivalent criteria, it is seen that the degree to which $G_{2p}(e^{j\omega})$ approximates $G(e^{j\omega})$ is directly related to the degree to which $g_{2p}(n)$ approximates $g(n)$. With this correspondence in mind, a straightforward procedure is now presented for generating a sequence $g_{2p}(n)$, which yields a good approximation to $g(n)$.

Method of Rational-Model Approximation

The rational-model approximation method to be presented is based on a prudent selection of the causal linear recursive system as governed by the transfer function

$$\tilde{G}_p(z) = \frac{\tilde{b}_0 + \tilde{b}_1 z^{-1} + \cdots + \tilde{b}_p z^{-p}}{1 + a_1 z^{-1} + \cdots + a_p z^{-p}} = \frac{\tilde{B}_p(z)}{A_p(z)} \tag{6.36}$$

In particular, the real-valued a_k and $\tilde{b}_k$ parameters of this transfer function are to be chosen so that its unit-impulse response $\{\tilde{g}_p(n)\}$ provides a suitably good approximation to the causal component of the symmetric positive-semidefinite sequence $g(n)$. This *causal component* is defined by

$$g^c(n) = g(n)u(n) - 0.5g(0)\delta(n) \tag{6.37}$$

Various methods exist for this parameter selection. A particularly simple and effective method is described in Section 6.9. Independent of the method used, however, the recursive system's order p must be chosen adequately large if $\{\tilde{g}_p(n)\}$ is to provide a suitably good approximation of $\{g^c(n)\}$. This order choice typically entails empirical experimentation on the user's part.

Once the linear system (6.36) has been synthesized so as to achieve the objective described above, it follows that the related symmetrical sequence

$$\tilde{g}_{2p}(n) = \tilde{g}_p(n) + \tilde{g}_p(-n) \tag{6.38}$$

will provide a similarly good approximation of $g(n)$. The Fourier transform of this symmetrical sequence as given by

$$\tilde{G}_{2p}(e^{j\omega}) = \frac{\tilde{B}_p(e^{j\omega})}{A_p(e^{j\omega})} + \left[\frac{\tilde{B}_p(e^{-j\omega})}{A_p(e^{-j\omega})}\right] = \frac{A_p(e^{j\omega})\tilde{B}_p(e^{-\omega}) + A_p(e^{-j\omega})\tilde{B}_p(e^{j\omega})}{A_p(e^{j\omega})A_p(e^{-j\omega})} \tag{6.39}$$

must also be a good approximation to $G(e^{j\omega})$, however, due to the equivalences (6.34) and (6.35). Upon examination of the right-hand side of the real-valued symmetric transform (6.39), it is seen to be of the required rational form (6.32). It therefore yields the desired rational approximation of $G(e^{j\omega})$, provided that it satisfies the additional property of itself being positive semidefinite. This condition, is, however, often violated. A means for overcoming this potential shortcoming will be described shortly.

To implement the conjugate symmetric system described by expression (6.38), the $2p$th-order linear recursive configuration depicted in Figure 6.10 can be employed. This implementation is seen to consist of the pth-order causal system (6.36) connected in parallel with its anticausal system image. As such, this configuration is of utility only in those applications in which non-real-time signal processing can be employed (i.e., prerecorded data). The frequency response of this system is seen to be $G_{2p}(e^{j\omega})$ and it has a desirable zero-phase charac-

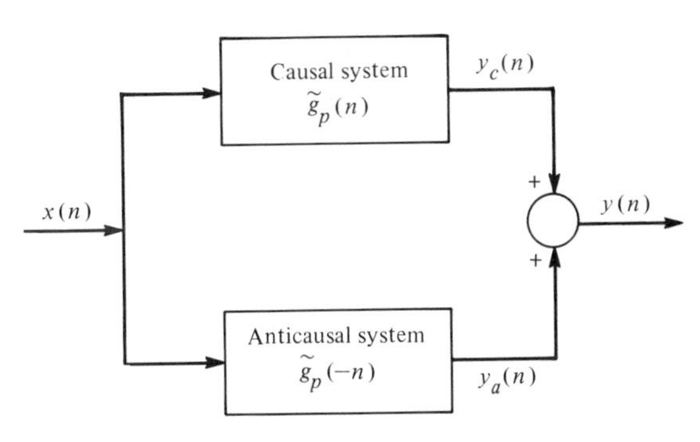

FIGURE 6.10. Implementation of symmetrical unit-impulse-response system.

teristic. This zero-phase characteristic can be of vital importance in applications requiring distortionless transmission.

Positive-Semidefinite Condition

In filter synthesis it is desirable that the rational form (6.32) be expressible in the factored form

$$G_{2p}(z) = \left[\frac{B_p(z)}{A_p(z)}\right]\left[\frac{B_p(z^{-1})}{A_p(z^{-1})}\right] \tag{6.40}$$

where a z-domain representation has been used. The pth-order polynomials $A(z)$ and $B(z)$ constituting this product decomposition are to have real coefficients. The spectral factorization theorem indicates that this factorization is always achievable provided that the numerator term in expression (6.39) is positive semidefinite. This numerator term can be expressed in the z-domain as

$$\begin{aligned} C(z) &= A_p(z)\tilde{B}_p(z^{-1}) + A_p(z^{-1})\tilde{B}_p(z) \\ &= c_0 + \sum_{n=1}^{p} c_k(z^{-k} + z^k) \end{aligned} \tag{6.41}$$

Unfortunately, it often happens that this numerator polynomial is not positive semidefinite. To maintain the satisfactory approximation of $G(e^{j\omega})$ by $G_{2p}(e^{j\omega})$ while effecting a positive-semidefinite numerator, it may therefore be necessary to perturb the coefficients c_k *slightly* so as to cause a resultant positive-semidefinite numerator. That is, it is desired to select suitably small values for the perturbation elements δc_k so that the approximating numerator term

$$\hat{C}(z) = (c_0 + \delta_0) + \sum_{n=1}^{p} (c_n + \delta_n)(z^{-k} + z^k) \tag{6.42}$$

is positive semidefinite and has a Fourier transform that closely approximates $C(e^{j\omega})$. One such procedure is that of selecting the real-valued b_n parameters governing the polynomial $B_p(z)$ appearing in expression (6.40) to minimize the functional

$$2 \sum_{n=1}^{p} \left(c_n - \sum_{k=0}^{p-n} b_k b_{k+n} \right)^2 + \left(c_0 - \sum_{k=0}^{p} b_k^2 \right)^2$$

This minimization will entail the use of a nonlinear programming approach, since the b_n coefficient appear up to powers of 4.

6.9 RECURSIVE SYSTEM SYNTHESIS: SQUARED-ERROR CRITERION APPROACH

In various applications it is desired to synthesize a causal real linear recursive system so that its dynamic characteristics best match those of an ideal objective. This need was illustrated in Sections 6.5 and 6.6. The motivation for seeking a recursive instead of a nonrecursive implementation generally arises from computational considerations: Recursive operators tend to be more efficient in terms of model order needed to achieve a prescribed behavior. Unfortunately, the synthesis of recursive operators also tends to be relatively difficult. In this section we present a straightforward procedure for recursive operator synthesis that typically results in satisfactory dynamic performance. Although our immediate objective is that of frequency-discrimination filtering, the approach to be given is applicable to many other forms of signal processing as well.

To begin our development, it is first assumed that there is given a real-valued causal unit-impulse response $\{h_d(n)\}$ which represents an idealized signal-processing operation. For example, $\{h_d(n)\}$ might correspond to the causal component of an ideal low-pass filter's unit impulse response. Whatever the case, it is desired to select the real-valued a_k and b_k parameters of the causal linear recursive system

$$y(n) + \sum_{k=1}^{p} a_k y(n-k) = \sum_{k=0}^{p} b_k x(n-k) \tag{6.43}$$

so that its associated unit-impulse response $\{h(n)\}$ *best approximates* the prescribed ideal behavior $\{h_d(n)\}$. To measure the quality of this approximation, the standard *sum-of-squared-error criterion*

$$f(a_k, b_k) = \sum_{n=0}^{\infty} [h_d(n) - h(n)]^2 \tag{6.44}$$

is often employed. This criterion has been suggestively expressed as a function of the foregoing recursive system's a_k and b_k parameters, which uniquely identify $\{h(n)\}$ in a highly nonlinear fashion.

The task at hand is then to select the recursive system's a_k and b_k parameters so as to minimize the squared-error criterion (6.44). For such a selection, the approximation $h(n) \approx h_d(n)$ for $n \geq 0$ typically follows provided that the recursive system's order parameter p has been selected adequately large. This then implies that the corresponding optimum (i.e., minimizing) linear recursive system will possess dynamics that nearly replicate those of the ideal model. Unfortunately, the minimization of squared-error criterion (6.44) is most difficult and an analytical solution is generally not feasible. We must therefore resort to iterative numerical techniques for a solution (e.g., descent algorithms). An important aspect of these iterative techniques is that of getting a good set of *initial* parameter values to start the iteration. In what is to follow, an analytically feasible method for obtaining these initial parameter values is presented. Frequently, these initial values are entirely adequate for the application at hand.

The unit-impulse response of linear recursive system (6.43) is formally obtained by letting $x(n) = \delta(n)$ and setting the initial conditions $y(n) = 0$ for all $n < 0$. This is found to give rise to the expression

$$h(n) = \begin{cases} b_0 & n = 0 \\ b_n - \sum_{k=1}^{n} a_k h(n-k) & 1 \leq n \leq p \\ -\sum_{k=1}^{p} a_k h(n-k) & n \geq p+1 \end{cases} \tag{6.45}$$

for the recursive system's unit-impulse response elements. After substituting this relationship into the squared-error criterion, we have

$$\begin{aligned} f(a_k, b_k) = [h_d(0) - b_0]^2 &+ \sum_{n=1}^{p} \left[h_d(n) - b_n + \sum_{k=1}^{n} a_k h(n-k) \right]^2 \\ &+ \sum_{n=p+1}^{\infty} \left[h_d(n) + \sum_{k=1}^{p} a_k h(n-k) \right]^2 \end{aligned} \tag{6.46}$$

As indicated earlier, the nonlinear behavior of the $h(n)$ terms on the a_k and b_k parameters precludes any analytical procedure for selecting the optimum parameters. It is possible, however, to modify this criterion suitably so as to render a tractable procedure for obtaining a *pseudo-optimum* parameter selection.

The modification entails the implicit assumption that the recursive system that eventually arises from the synthesis is such that the objective $h(n) = h_d(n)$ is closely approximated. Under this hypothesis, a replacement of $h(n)$ by $h_d(n)$ in the squared-error criterion (6.46) gives rise to the *auxiliary squared-error criterion*

$$f_1(a_k, b_k) = [h_d(0) - b_0]^2 + \sum_{n=1}^{p} \left[h_d(n) - b_n + \sum_{k=1}^{n} a_k h_d(n-k) \right]^2 + \sum_{n=p+1}^{\infty} \left[h_d(n) + \sum_{k=1}^{p} a_k h_d(n-k) \right]^2 \quad (6.47)$$

This auxiliary criterion serves as a good approximation to the original squared error criterion as long as the a_k and b_k parameters are such that $h(n) \approx h_d(n)$. As such, the minimization of this related auxiliary criterion tends to render the original criterion a near-minimum value. Fortunately, the latter minimization is *tractable,* since the a_k and b_k parameters here appear in a *quadratic* manner due to the replacement of the nonlinear $h(n)$ terms by the known elements $h_d(n)$.

To minimize auxiliary criterion (6.47), we then simply apply the standard calculus technique of setting to zero the partials of $f_1(a_k, b_k)$ with respect to each of the recursive system parameters. In particular, the necessary condition for a minimum $\partial f_1(a_k, b_k)/\partial b_m = 0$ is found to result in

$$b_n^{o} = \begin{cases} h_d(0) & n = 0 \\ h_d(n) + \sum_{k=1}^{n} a_k^{o} h_d(n-k) & 1 \leq n \leq p \end{cases} \quad (6.48)$$

where the superscript ^{o} is used to denote an optimum parameter choice insofar as minimizing $f_1(a_k, b_k)$. This selection is seen to cause the first $p + 1$ error squared terms in criterion (6.47) to be identically zero. In a similar fashion, the necessary conditions $\partial f_1(a_k, b_k)/\partial a_m = 0$ is found to result in the following linear system of p equations in the p unknown a_k^{o} parameters

$$\sum_{k=1}^{p} c_{mk} a_k^{o} = -c_{mo} \qquad 1 \leq m \leq p \quad (6.49)$$

The coefficients characterizing this system of equations are specified as

$$c_{mk} = \sum_{n=p+1}^{\infty} h_d(n-m) h_d(n-k) \qquad 0 \leq k \leq p, \quad 1 \leq m \leq p \quad (6.50)$$

In summary, the optimum selection of the recursive system parameters with regard to minimizing auxiliary criterion (6.47) are obtained by following the

TABLE 6.3 Linear Recursive System Synthesis Method

1. Solve the linear system of equations (6.49) for the a_k^{o} parameters.
2. Using the a_k^{o} parameters obtained at step 1, compute the b_k^{o} according to relationship (6.48).

two-step procedure outlined in Table 6.3. In this solution procedure it is necessary to evaluate the c_{mk} coefficients (6.50), which each involve an infinite number of summation and multiplication operations. To achieve a computationally workable evaluation of the c_{mk} coefficients, we could change the upper summation limit to L, where L is selected suitably large so that $h_d(n)$ is approximately zero for all $n > L$. Alternatively, it is possible to employ Parseval's relationship (3.21) for the choice $w(n) = h_d(n - m)$ and $x(n) = h_d(n - k)$. This is found to result in

$$c_{mk} = \frac{1}{2\pi}\int_{-\pi}^{\pi} |H_d(e^{j\omega})|^2 e^{j\omega(m-k)}\, d\omega - \sum_{n=\max(k,m)}^{p} h_d(n - m)h_d(n - k)$$

$$\text{for } 0 \leq k \leq p \quad \text{and} \quad 1 \leq m \leq p \tag{6.51}$$

where $H_d(e^{j\omega})$ denotes the Fourier transform of the causal ideal unit impulse response $\{h_d(n)\}$. Upon evaluating the constituent integral (this may entail numerical integration) and carrying out the $[p - \max(k, m)]$ multiplications and $[p + 1 - \max(k, m)]$ summations, the required coefficients are readily computed.

In solving relationship (6.49) for the a_k^o parameters, it is advisable to use linear-equation-solving software routines that take advantage of the linear system's coefficient symmetry (i.e., $c_{mk} = c_{km}$). Such routines are typically more efficient computationally as well as being more accurate than solution routines that do not invoke a symmetric coefficient condition. In Section 6.10 it is seen that in a low-pass filtering application using this approach, the consistent system of linear equations (6.49) are ill-posed even for moderate values of the order integer p. This necessitates our having to use solution techniques that are less sensitive to such ill-posed conditions. We develop such techniques in the companion textbook. Computational considerations of this nature are important and should be kept in mind when addressing various signal-processing problems.

Because of the manner in which the b_n^o parameters are selected through expression (6.48), it is apparent that the resultant linear recursive system's unit impulse response is such that

$$h^o(n) = h_d(n) \qquad 0 \leq n \leq p \tag{6.52}$$

namely, there is a perfect agreement with the ideal behavior over the time span $0 \leq n \leq p$. The behavior of $h^o(n)$ for $n \geq p + 1$ then dictates the fidelity of the synthesized recursive system. If the match over this time span is deemed unsatisfactory, a better approximation can be obtained either by (1) increasing the filter order p or (2) using a numerical iterative algorithm for finding the a_k and b_k parameters that minimize the original squared-error criterion (6.44). The parameters obtained from expressions (6.48) and (6.49) would serve as appropriate initial conditions for such algorithms.

One noteworthy feature of this synthesis method is that if the ideal objective $\{h_d(n)\}$ corresponds to the unit-impulse response of a causal recursive system of order p or less, the a_k^o and b_k^o generated through expressions (6.48) and (6.49) will identify that recursive system perfectly. We leave a proof of this conjecture to the reader. The following example, however, illustrates this property.

EXAMPLE 6.6

Using the method of this section, find the pseudo-optimum recursive system of order $p = 1$ that approximates the ideal unit-impulse response

$$h_d(n) = 3(\tfrac{1}{2})^n u(n)$$

The optimum parameter a_1^o is specified by equation (6.49) with $p = 1$. The required coefficients are, by relationship (6.50),

$$c_{10} = \sum_{n=2}^{\infty} 3(\tfrac{1}{2})^{n-1} 3(\tfrac{1}{2})^n u(n - 1)u(n) = \tfrac{3}{2}$$

$$c_{11} = \sum_{n=2}^{\infty} 3(\tfrac{1}{2})^{n-1} 3(\tfrac{1}{2})^{n-1} u(n - 1)u(n - 1) = 3$$

thus indicating that

$$3a_1^o = -\tfrac{3}{2}$$

We therefore conclude that $a_1^o = -\frac{1}{2}$. Using expression (6.48), it further follows that $b_0^o = 3$ and $b_1^o = 0$. Thus the pseudo-optimum recursive system is governed by

$$y(n) - 0.5y(n - 1) = 3x(n)$$

It is readily shown that this recursive system's unit-impulse response is given by $h(n) = 3(\frac{1}{2})^n u(n)$, indicating a perfect match. ■

6.10 DESIGN OF PROTOTYPE LOW-PASS FILTERS

We now put to use the concepts developed in Sections 6.8 and 6.9 for synthesizing recursive filters that implement low-pass filtering operations. In particular the ideal operator is taken to have the frequency response

$$H_d(e^{j\omega}) = \begin{cases} 1 & |\omega| \le \Delta \\ 0 & \Delta \le |\omega| \le \pi \end{cases} \tag{6.53}$$

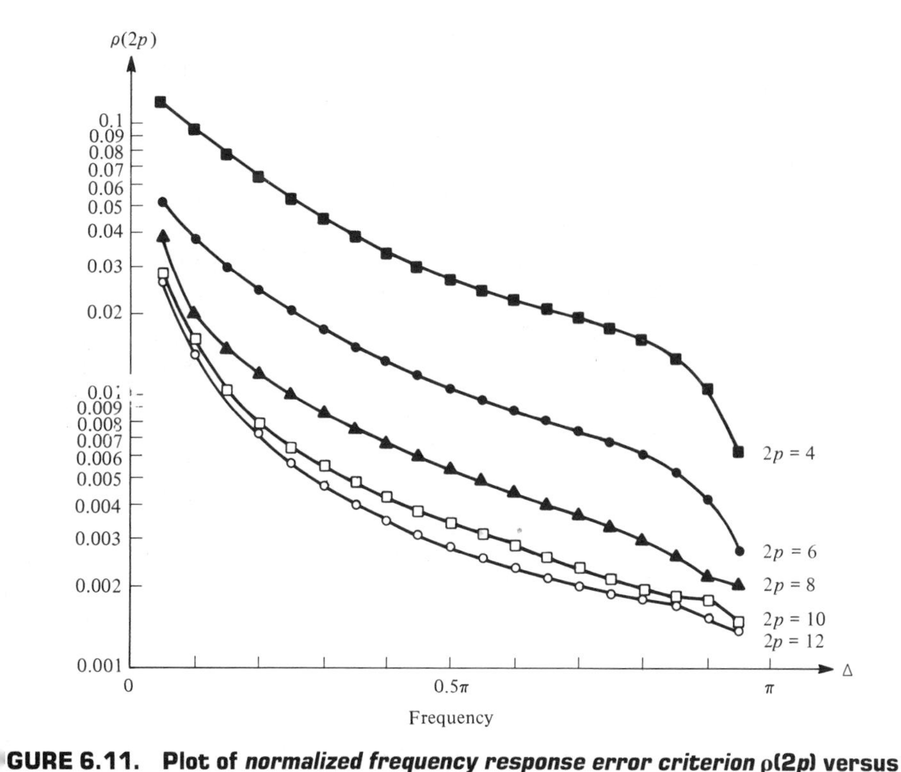

GURE 6.11. Plot of *normalized frequency response error criterion* ρ(2p) versus toff frequency for several choices of recursive filter order 2p.

and the corresponding symmetrical unit-impulse response

$$h_d(n) = \frac{\sin(\Delta n)}{\pi n} \tag{6.54}$$

For this selection the c_{mk} coefficients appearing in the system of equation (6.49) which identify the filters' a_k parameters are readily obtained using expression (6.51). This is found to result in the readily computable formula

$$c_{mk} = \frac{\cos[\Delta(m+k)]}{2\pi(m+k)} - \frac{1}{2\pi^2} \sum_{n=m+k-p}^{p} \frac{\sin[\Delta(n-m)]\sin[\Delta(n-k)]}{(n-m)(n-k)} \tag{6.55}$$

for $0 \le k \le p$ and $1 \le m \le p$. Using these entries in step 1 of Table 6.3, the a_k° and b_k° characterizing the optimal pth-order causal recursive filter component are obtained directly. An efficient order iterative procedure for obtaining these coefficients has been developed (Cadzow and Sun, 1986).

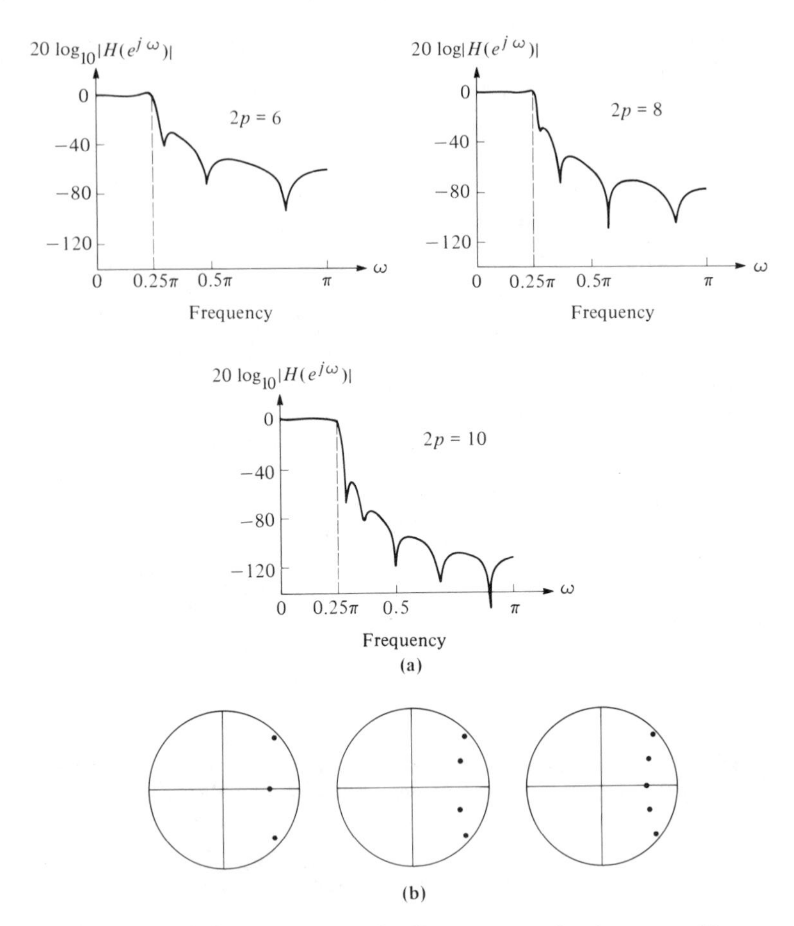

FIGURE 6.12. Frequency responses of a linear recursive low-pass filter.

To measure how closely the synthesized recursive filter depicted in Figure 6.10 approximates the dynamics of the ideal low-pass filter, the *normalized frequency response magnitude error criterion* defined by

$$\rho(p) = \frac{\int_{-\pi}^{\pi} [|H_d(e^{j\omega})| - |G_{2p}(e^{j\omega})|]^2 \, d\omega}{\int_{-\pi}^{\pi} |H_d(e^{j\omega})|^2 \, d\omega} \tag{6.50}$$

is employed, where $G_{2p}(e^{j\omega})$ is specified by relationship (6.39). This criterion takes on values exclusively in the interval [0, 1], with values close to zero indicating a good recursive filter match. To gain a feeling for the recursive filter's behavior arising from solving expressions (6.49) and (6.48), plots of the normalized frequency response magnitude match criterion versus the low-pass filter bandwidth Δ are shown in Figure 6.11 for choice of filter order $2p =$

6, 8, 10, and 12. These plots enable us to make an informed selection of the filter-order parameter $2p$ needed to achieve a prescribed match.

EXAMPLE 6.7

Design a linear recursive low-pass filter with cutoff frequency $\Delta = 0.25\pi$ radians for order selections 6, 8, and 10. To begin the design process, the coefficient expression (6.55) is evaluated for $p = 3$, 4, and 5. With these coefficients, relationships (6.49) and (6.48) are then used to obtain the required a_k^o and b_k^o parameters for the three filter-order selections $2p = 8$, 10, and 12. The resultant frequency responses are plotted in Figure 6.12a, in which the corresponding values of $\rho(2p)$ are indicated. In addition, the pole locations relative to the unit circle for the related causal filters are displayed in Figure 6.12b. Upon comparison of these magnitude results with those given for the corresponding Butterworth filters shown in Figure 6.9, it is apparent that the causal Butterworth filter has inferior discrimination characteristics. ■

6.11 SUMMARY

Some procedures for synthesizing causal nonrecursive and recursive filters to approximate the dynamics of ideal low-pass filters have been provided. Particular emphasis was given to approximating the ideal magnitude function.

SUGGESTED READINGS

The Suggested Readings of Chapter 5 provide various methods for synthesizing frequency-discrimination filters. In addition, reference to the following has been made in this chapter.

CADZOW, J. A., *Discrete-Time Systems,* Englewood Cliffs, N.J.: Prentice-Hall, Inc., 1973.

CADZOW, J. A., and Y. SUN, "Sequences with Positive-Semidefinite Fourier Transforms," *IEEE Transactions,* ASSP, Dec. 1986.

GOLD, B., and C. M. RADER, *Digital Processing of Signals,* New York: McGraw-Hill Book Company, 1969.

PAPOULIS, A., *Signal Analysis,* New York: McGraw-Hill Book Company, 1977.

PROBLEMS

6.1 Determine the weighting sequence associated with the ideal frequency response behavior

$$H_d(e^{j\omega}) = \begin{cases} 1 & \text{for } 0 \le |\omega| < \omega_1 \quad \text{and} \quad \omega_2 \le |\omega| < \omega_3 \\ 0 & \text{elsewhere for } -\pi < |\omega| < \pi \end{cases}$$

where it is assumed that $0 < \omega_1 < \omega_2 < \omega_3 < \pi$.

6.2 Determine the weighting sequence associated with bandlimited differentiator specified by

$$H_d(e^{j\omega}) = \begin{cases} j\omega & \text{for } 0 \le |\omega| < \omega_1 \\ 0 & \text{for } \omega_1 < |\omega| \le \pi \end{cases}$$

6.3 In designing a realizable filter through the operations of truncation and shifting, relationships (6.10) and (6.13) played a prominent role. Carry out the details that resulted in these expressions.

6.4 Verify the correctness of recursive expression (6.14) for computing the normalized frequency response error criterion.

6.5 Determine the order needed for a nonrecursive low-pass digital filter whose cutoff frequency is 0.30π radians so that the frequency response error criterion $\rho(q)$ is less than or equal to 0.01. Use the results shown in Figure 6.2 for selecting q. Given an expression for the weighting coefficients need in the implementing nonrecursive expression (6.6).

6.6 Develop a nonrecursive high-pass filter synthesis procedure along the lines taken in Sections 6.2 and 6.3, where the ideal frequency response is given by

$$H_{hp}(e^{j\omega}) = \begin{cases} 1 & \text{for } \omega_2 \le |\omega| \le \pi \\ 0 & \text{for } 0 \le |\omega| < \omega_2 \end{cases}$$

That is, develop expressions analogous to relationship (6.10) and (6.13).

6.7 Repeat Problem 6.6 for the bandpass filter characterized by

$$H_{bp}(e^{j\omega}) = \begin{cases} 1 & \omega_1 < |\omega| \le \omega_2 \\ 0 & \text{otherwise} \end{cases}$$

6.8 Verify that upon substituting $z = e^{j\omega}$ into expression (6.24), the Butterworth magnitude-squared frequency response (6.19) arises.

6.9 Design a Butterworth low-pass recursive filter of order $p = 4$ that has the cutoff frequency $\Delta = 0.30\pi$ radians.

6.10 Using the general procedure of Section 6.5, determine the order of the Butterworth low-pass digital filter with cutoff frequency $\Delta = 0.2\pi$ radians so that its gain for frequencies greater than 0.4π radians is no larger than 0.01.

6.11 Show that the unit-impulse response of a linear recursive system with transfer function (6.43) satisfies relationship (6.45).

6.12 Verify that the partial derivative approach taken in Section 6.9 leads to the optimum a_k parameter selection (6.49).

6.13 Consider a linear system whose generally unsymmetric unit-impulse response $\{h_d(n)\}$ has its causal and anticausal components specified by

$$h_d^c(n) = h_d(n)u(n) - 0.5h_d(0)\delta(n)$$

$$h_d^a(n) = h_d(n)u(-n) - 0.5h_d(0)\delta(n)$$

so that $h(n) = h^c(n) + h^a(n)$. Describe a procedure using the approach taken in Section 6.9 for synthesizing an approximating recursive system. *Hint:* Make a separate recursive approximation of $\{h_d^c(n)\}$ and $\{h_d^a(n)\}$.

6.14 Using the Parseval's relationship, provide a proof of relationship (6.51).

6.15 Show that if the ideal unit-impulse response $\{h_d(n)\}$ is generated by a linear recursive system of form (6.43), the a_k and b_k parameters generated through expressions (6.48) and (6.49) identify this system [i.e., $f(a_k^o, b_k^o) = 0$].

6.16 Demonstrate the conjecture made in Problem 6.15 for the specific causal system

$$y(n) - \tfrac{1}{2}y(n - 1) = 3x(n)$$

6.17 Using the unit-impulse response specified by

$$h_d(n) = 3(\tfrac{1}{4})^n u(n) - 2\delta(n)$$

determine an associated recursive model with $p = 1$ using the method of Section 6.9.

6.18 Verify expression (6.55).

CHAPTER 7

Probability Theory

7.1 INTRODUCTION

Our studies up to this point have been directed primarily toward time series that are describable by convenient mathematical formulas such as the unit-impulse, unit-step, exponential, and sinusoid signals. Although such time series do play a prominent role in signal-processing analysis, they are woefully inadequate for representing *real-world*-generated signals. Most signals encountered in practical applications contain a significant degree of *unpredictableness* (or complexity) in the sense that their future behavior cannot be predicted with *exactness* from their present and past behavior. We have only to compile a list of economic, scientific, biological, or, engineering-based time series to appreciate this fact.

It will be recalled from previous discussions that a time series may often be interpreted as being the sequentially recorded outcome of an experiment that is observed so as to reveal the quantitative nature of an underlying phenomenon. Depending on the predictiveness of a given time series, it falls into one of two categories as specified in the following definition.

Definition 7.1. A time series (or discrete-time signal) is said to be *deterministic* if its element values are known to the observer before the experiment it describes is carried out. If these elements are not known a priori, however, the time series is said to be *nondeterministic* or *random*.

Clearly, unit-impulse, unit-step, exponential, and sinusoidal time series are deterministic in behavior due to their exact predictiveness. On the other hand, a digital communication signal is properly classified as being nondeterministic before its transmission insofar as the person who ultimately receives the signal is concerned.

EXAMPLE 7.1

To illustrate the concept of predictiveness in a real-world setting, let us consider the two time series as identified by

$x_1(n)$ = sunrise time in New York City on the nth day of the year 2001, where $1 \leq n \leq 365$

$x_2(n)$ = official high temperature recorded in New York City on the nth day of the year 2001, where $1 \leq n \leq 365$

The first of these time series is deterministic, since by appealing to the geophysical laws governing the earth's and sun's position in space, we may predict with precision the sunrise times for all future dates. On the other hand, the second time series is seen to be nondeterministic because even the most gifted weather forecaster is able to provide only an informed estimate of the daily high temperatures for future days. ■

Since virtually all time series encountered in practice possess a degree of unpredictability, it is necessary that a mathematical framework be developed for their study. Clearly, this is an essential requirement if we are both to understand the salient characteristics of nondeterministic time series and have the capability of designing effective signal-processing algorithms that achieve desired objectives. Not surprisingly, probability theory plays the central role in characterizing time series that contain random (or unpredictable) behavior. In this chapter and Chapter 8 we explore those aspects of probability theory that are essential to our study of nondeterministic time series. Readers with a previous exposure to probability theory may proceed to Chapter 9.

The origins of probability theory can be traced to studies related to games of chance. These studies gave rise to the *classical* form of probability theory, where it is tacitly assumed that the underlying random experiment has a finite number of possible outcomes (e.g., N) and that each outcome has an equal probability of occurring (i.e., $1/N$). Although it is possible to modify the equally likely outcome requirement so as to treat more general situations, the classical approach to probability theory is much too restrictive for most applications. To overcome this serious shortcoming, the contemporary axiomatic approach to probability theory was introduced and developed during this century. Contemporary probability theory makes extensive use of elementary set-theory concepts. With this in mind, a brief description of those aspects of set theory needed in our study of probability theory is given next.

7.2 SET THEORY

Although the mathematical theory of sets can be quite complex in nature, we need only a few primitive concepts. To begin this development, the basic definition of a set is first offered.

Definition 7.2. A *set* is a collection of objects (or elements). Sets are typically denoted by capital letters, such as A, B, C, while the elements (or objects) of sets are designated by lowercase letters, such as x, y, z. When the element x is a member of set A, we write $x \in A$, and when the element x is not a member of set A, we write $x \notin A$. Sets are typically identified using one of three methods:

1. An explicit listing of the set's members.
2. A verbal description of the set.
3. A mathematical formula that describes the set.

Which of these representation methods is to be used in a given situation is very much dependent on the nature of the set.

EXAMPLE 7.2

The following examples illustrate the three methods of set representation.

$$A_1 = \{\text{cat, dog, car}\}$$
$$A_2 = \{\text{set of integers between 0 and 10 inclusive}\}$$
$$A_3 = \{x: 0 \leq x \leq 7\}$$

In set A_3, the right side is to be read as "the set of x that satisfy the inequality $0 \leq x \leq 7$." It is to be noted that $7 \in A_2$, and $7 \in A_3$, but $7 \notin A_1$. ■

Universal Set

In set-theory applications, the concept of the *universal set* is of fundamental importance. This set is used so extensively that the special symbol S is here reserved for it. Specifically, the universal set S is defined by

$$S = \{\text{set of all objects under consideration}\} \tag{7.1}$$

In a given context, it is generally clear what is meant by the phrase "set of all objects under consideration." Implicit in this definition, however, is the inference that all other sets to be considered must be composed exclusively of elements contained in S. Thus the descriptor "universal set" is appropriate.

It will be beneficial to provide a visual interpretation of the set-theory concepts presented to this point. In particular, let us depict the universal set as being composed of all points lying within the rectangle shown in Figure 7.1a. All

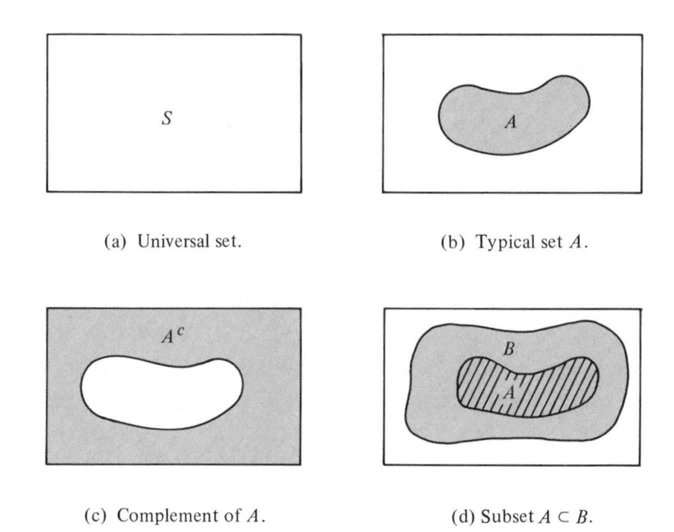

(a) Universal set.

(b) Typical set A.

(c) Complement of A.

(d) Subset $A \subset B$.

FIGURE 7.1. Visual interpretation of set-theory concepts.

other sets to be considered within this universal-set context must then be composed of a subset of these points. For example, a typical set A might be depicted as in Figure 7.1b and is interpreted as being composed of all points within and on the contour defining A (i.e., the shaded region). Moreover, all sets to be employed in a given setting must be wholly contained with this (rectangular) universal set. This visual interpretation provides a convenient means of interpreting many set-theory concepts. These visual pictures are commonly referred to as *Venn diagrams*. The reader is cautioned, however, to use this two-dimensional Venn diagram interpretation only to gain a *feel* for set-theory concepts. Venn diagrams cannot be used for proving theoretical results concerning sets.

Complement of a Set

In set-theory applications, the notion of the complement of a set arises in a natural manner. More specifically, the complement of a set A is here denoted by A^c and defined by

$$A^c = \{x \in S: x \notin A\} \tag{7.2}$$

In words, the complement set A^c is composed of all elements in S which are themselves not elements of A. Thus A^c consists of those elements in S that are left over after the elements in A have been removed. The complement of the set A may be visualized as the shaded area shown in Figure 7.1c.

To each set A, the process of determining its complement is conceptually straightforward. A certain degree of ambiguity exists, however, when applying this concept to the universal set. Since the universal set contains all elements relevant to a given application, the complement of S must contain no elements. To treat this situation, the *null set* denoted by 0 is introduced, where

$$0 = \{\text{set with no elements}\} \tag{7.3}$$

The concept of set complement can now be applied to the universal and null sets, that is,

$$S^c = 0 \qquad \text{and} \qquad 0^c = S \tag{7.4}$$

Thus the introduction of the null set yields the useful result that if A is any set in S, its complement always exist.

Subsets

The concept of subset plays a basic role in set theory as well as in probability theory. The set A is said to be a *subset* of set B if every element of A is also an element of B. We denote this subset relationship as

$$A \subset B \tag{7.5}$$

which implies that if an element $x \in A$, it must follow that $x \in B$. This subset relationship can be expressed equivalently as $B \supset A$. A visual depiction of the subset concept is given in Figure 7.1d. Once a universal set S has been agreed upon for a given application, it must follow that all sets under consideration must be subsets of S (i.e., $A \subset S$). Furthermore, for any set A, it follows trivially that $0 \subset A$, since A contains all of 0 elements (it has none).

A useful application of the subset concept involves the task of proving that two sets A and B are equal (i.e., they have the same members). The equivalence of two sets may often be disguised through their seemingly different definitions. Whatever the case, the set equality $A = B$ follows if and only if $A \subset B$ and $B \subset A$.

EXAMPLE 7.3

Let the universal set be composed of all nonnegative integers, that is,

$$S = \{0, 1, 2, 3, \ldots\}$$

The following two sets are subsets of U:

$$A = \{1, 7, 9\}$$

$$B = \{\text{set of positive odd integers}\}$$

Moreover, $A \subset B$, since the elements 1, 7, and 9 constituting A are also elements of B. ■

7.3

SET ALGEBRA

We are now in a position to introduce an algebra of sets whereby two or more sets may be combined. Central to this capability are the notions of set union and set intersection. The *union* of the sets A and B is formally denoted by $A \cup B$, where

$$A \cup B = \{x \in S: x \in A \text{ or } x \in B\} \tag{7.6}$$

Thus the union set $A \cup B$ is seen to be composed of those elements of S that are contained in either set A or set B, or possibly contained simultaneously in both. The union operation is depicted by the shaded area in Figure 7.2a. In a similar fashion, the *intersection* of the sets A and B is designated by AB and is defined by

$$AB = \{x \in S: x \in A \text{ and } x \in B\} \tag{7.7}$$

We see that the intersection set AB is composed of those elements of S that are shared in common by the sets A and B. This intersection operation is depicted in Figure 7.2b by the lined-shaded region.

EXAMPLE 7.4
Let the universal set by the set of real numbers and the sets A and B be specified by

$$A = \{-7.5, 2, \pi, -1\} \qquad \text{and} \qquad B = \{0, \pi\}$$

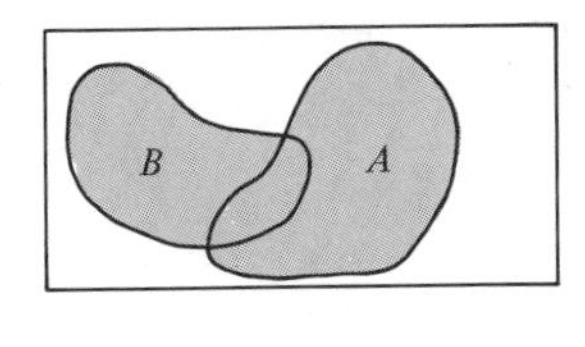

(a) $A \cup B$.

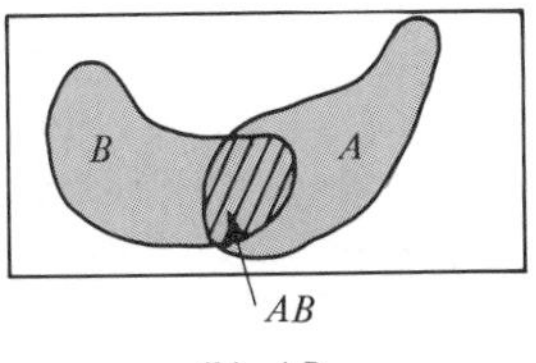

(b) AB.

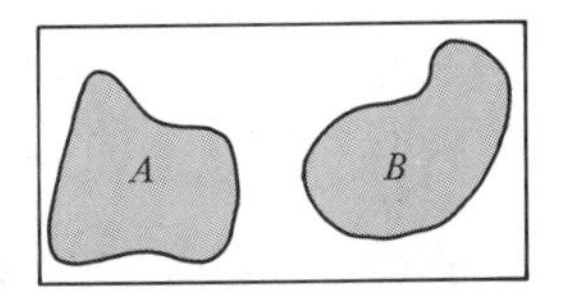

(c) $AB = 0$.

FIGURE 7.2. Concepts of (a) set union, (b) set intersection, and (c) mutually exclusive sets.

It then follows that

$$A \cup B = \{-7.5, 2, \pi, -1, 0\} \quad \text{and} \quad AB = \{\pi\}$$

Mutually Exclusive Sets

In probability theory applications, it is desirable to work with sets that contai no points in common. Two sets that possess this property are said to be mutuall exclusive. Thus the sets A and B are *mutually exclusive* if their intersection i empty, that is,

$$AB = 0 \tag{7.8}$$

Mutually exclusive events are depicted as shown in Figure 7.2c. We can exten the notion to include more than two sets. In particular, the sets A_1, A_2, A_3 . . . , are said to be *pairwise mutually exclusive* if

$$A_k A_m = 0 \qquad \text{for } k \neq m \tag{7.9}$$

This important notion will soon be put to use in a probability setting.

Set Identities

The set operations of union and intersection as defined for *two* sets may readil be extended in an obvious manner to any *finite* number of sets. For example the union of the three sets A, B, and C may be determined according to

$$A \cup B \cup C = A \cup (B \cup C) \tag{7.10}$$

where the right-hand side is observed to be equal to the union of the two set A and $B \cup C$. In a similar fashion, the intersection of these three sets may b obtained as

$$ABC = A(BC) \tag{7.11}$$

where the right side is seen to be equal to the intersection of the two sets A an BC. Clearly, we can extend this procedure in an obvious manner so as to unio (or intersect) four, five, six, and so on, sets.

With these set operation extensions, it is possible to develop a number o useful set identities. Some of the more important identities are listed in Tabl 7.1. In accordance with a comment made previously, to prove such identitie formally, one must show that the left-side set of the equality is a subset of th

TABLE 7.1 Set-Theory Identities

Identity	Law
$A \cup B = B \cup A$	Commutative laws
$AB = BA$	
$A \cup (B \cup C) = (A \cup B) \cup C$	Associative laws
$A(BC) = (AB)C$	
$A \cup (BC) = (A \cup B)(A \cup C)$	
$A(B \cup C) = (AB) \cup (AC)$	
$A0 = 0$	
$A \cup 0 = A$	
$(A \cup B)^c = A^cB^c$	
$(AB)^c = A^c \cup B^c$	
$(A^c)^c = A$	

right-side set, and, vice versa. This can be a very demanding task for some of these identities. On the other hand, a demonstration (not a proof) of their validity is readily established with the use of Venn diagrams.

7.4 NONDETERMINISTIC EXPERIMENTS

Probability theory provides a systematic basis for describing experiments which have outcomes that are uncertain in nature. More specifically, it deals with situations involving an experiment whose outcome is known only after the experiment is conducted and its results recorded (or observed). The nature of the outcome may be extremely simple, as in the case of a coin being flipped (the experiment) and the presence of a head or tail recorded. On the other hand, the experimental outcome can be quite complex, as the recording of the daily closing prices of IBM stock on the New York Stock Exchange in a given year. This process of experimentation and its associated recorded outcome are depicted in Figure 7.3, in which $\mathscr{E}$ is used to denote the underlying experiment.

Sample Space

Independent of the given experiment's complexity, it is always possible to conceptually form a set whose members are composed of all the outcomes possible

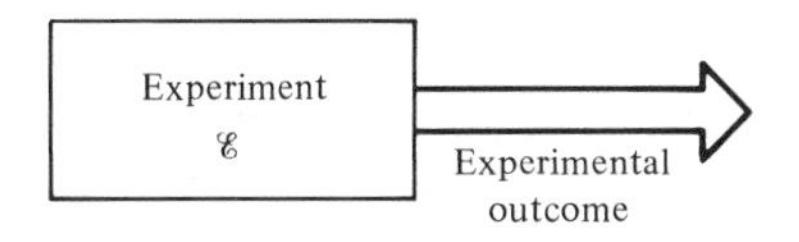

FIGURE 7.3. Concept of experiment.

from the experiment. This fundamental set is referred to as the *sample space* and is formally defined as

$$S = \{\text{set of all possible outcomes associated with a given experiment}\} \tag{7.12}$$

It is possible to list the elements (i.e., outcomes) of the sample space for relatively elementary experiments, whereas a verbal description is the best that can be done for more complex experiments.

EXAMPLE 7.5

The following two experimental situations illustrate the range of experimental complexities that may be treated by probability theory.

$$\mathscr{E}_1 = \{\text{tossing a coin and recording the face showing}\}$$

$$\mathscr{E}_2 = \{\text{recording a telephone signal over the time interval } t_0 \le t \le t_1\}$$

The sample space associated with experiment $\mathscr{E}_1$ is quite simple and is given by

$$S_1 = \{\text{h, t}\}$$

where h and t denote the experimental outcomes associated when a *head* or a *tail* is tossed, respectively. On the other hand, the sample space corresponding with experiment $\mathscr{E}_2$ contains an uncountably infinite number of possible outcomes and cannot be listed on an individual-outcome basis. ■

Events

Once an experiment has been identified and its associated sample space formed, the concept of *event* is a natural by-product. That is, upon conducting an experiment, a given event is said to have occurred if the observed experimental outcome possesses the attributes associated with that event. More formally, an *event* is composed of all members of the sample space that possess a well-defined attribute which identifies the event, that is,

$$\text{event} = \{x \in S: x \text{ possesses a given attribute}\} \tag{7.13}$$

As such, an *event* is a *subset* of the sample space. In accordance with the set-theory notions in Section 7.2, events (or sets) are denoted by capital letters, such as A, B, C.

EXAMPLE 7.6

Let the experiment consist of the tossing of two distinguishable dice, with the outcome being the number of dots showing on each face. The sample space is then composed of the 36 pairs as specified by

$$S = \{(1, 1), (1, 2), \ldots, (6, 5), (6, 6)\}$$

in which (m, n) denotes that experimental outcome where the first and second die have m and n dots showing, respectively, for $1 \leq m, n \leq 6$. Three events that could be defined on this sample space are

$$\begin{aligned} A &= \{\text{sum of dots showing is five}\} \\ &= \{(1, 4), (2, 3), (3, 2), (4, 1)\} \\ B &= \{\text{die 1 has three dots showing}\} \\ &= \{(3, 1), (3, 2), (3, 3), (3, 4), (3, 5), (3, 6)\} \\ C &= \{\text{sum of dots showing is 20}\} \\ &= 0 \end{aligned}$$

In these examples, the verbal discription gives the attribute possessed by the event, and this is followed by a list of the experimental outcomes in S that possess this attribute. ■

Although Example 7.6 is directed toward a specific application, it is clear that the notions of sample space and events can be conceptually extended in an obvious manner to describe any experiment that has random outcomes. Of particular importance from our viewpoint is an experiment in which the outcome is a random time series or a discrete-time signal. Whatever the case, however, the description of the experiment has been posed in a set-theory formulation. This is central to the modern (or axiomatic) theory of probability, which we present next.

7.5 AXIOMS OF PROBABILITY

Intuitively, the mathematics of probability is concerned with the task of assigning numbers to events, with these numbers measuring the likelihood of the events occurring on any experimental trial. Contemporary probability theory incorporates this concept in a set-theory formulation to achieve a powerful procedure for characterizing random experiments. As in many branches of mathematics, modern probability theory is based on a set of rules (axioms) that are *accepted* (without proof) as being true. Within this axiomatic framework, various theo-

S
unit mass

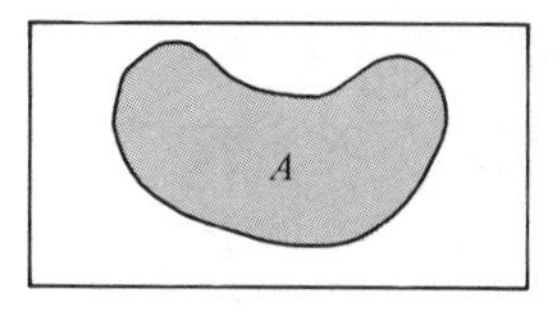

$P(A)$ = amount of mass
within shaded boundary

FIGURE 7.4. Interpretation of probability.

rems can then be developed which are essential to the useful application of probability theory. With this in mind, the following axiomatic definition of probability theory is given.

Definition 7.3. Let $\mathscr{E}$ be an experiment and S a sample space associated with that experiment. For each event (set) A contained in S, a real number $P(A)$, called the *probability of* A, is assigned such that the following properties (axioms) are satisfied:

1. $0 \leq P(A) \leq 1$
2. $P(S) = 1$
3. If A and B are mutually exclusive events, then

$$P(A \cup B) = P(A) + P(B) \tag{7.14}$$

4. if $A_1, A_2, A_3, \ldots$ are pairwise mutually exclusive events, then

$$P\left(\bigcup_{k=1}^{\infty} A_k\right) = \sum_{k=1}^{\infty} P(A_k)$$

Upon reflection, these assumed truths (or axioms) accurately reflect our preconceived notion concerning probability.

As a visual aid in understanding this set-theory approach to probability, it is beneficial to provide a mass distribution interpretation. This entails depicting the sample space as being composed of all points within the rectangular region shown in Figure 7.4a. Furthermore, let a unit of mass be distributed within this rectangle in some fashion. Then the probability that the event A occurs [i.e., $P(A)$] is equated with the amount of mass contained within the boundaries defining the set A as illustrated in Figure 7.4b. Under this setting, the four probability axioms (7.14) have a logical and consistent interpretation.

Although the axiomatic approach to probability does provide a rigorous theoretical basis for studying random phenomena, it does not address the practical issue of how values are to be assigned to the entities $P(A)$. Let the reader be comforted by the observation that this assignment can be made in an obvious manner in most applications. As such, we now direct our attention toward examining some of the more important theoretical aspects of probability theory.

A number of theorems are now given which are direct consequences of the four axioms that define probability theory. Their proofs each follow a similar line whereby an event is first appropriately decomposed into a union of mutually exclusive events and then axiom 3 is invoked. The next four theorems illustrate this widely used approach for computing probabilities.

Theorem 7.1. If 0 is the null (empty) event, then $P(0) = 0$. (7.15)

Proof. For any event A, the set identity $A = A \cup 0$ holds. Since A and 0 are mutually exclusive events, however, axiom 3 indicates that $P(A) = P(A \cup 0) = P(A) + P(0)$. It therefore follows that $P(0)$ must be equal to zero as required.

Theorem 7.2. The probabilities of event A and its complement A^c are related by

$$P(A^c) = 1 - P(A) \tag{7.16}$$

Proof. The set identity $S = A \cup A^c$ holds for any event A. Since A and A^c are mutually exclusive events, axioms 2 and 3 indicate that $1 = P(S) = P(A \cup A^c) = P(A) + P(A^c)$, which gives us the desired result.

Theorem 7.3. If A and B are any two events, then

$$P(A \cup B) = P(A) + P(B) - P(AB) \tag{7.17}$$

Proof. Let us first decompose the event $A \cup B$ into two mutually exclusive events according to the readily established set identity

$$A \cup B = A \cup (A^cB)$$

Using axiom 3, we then have

$$P(A \cup B) = P(A) + P(A^cB)$$

To complete our proof, event B is also decomposed into the two mutually exclusive events

$$B = (AB) \cup (A^cB)$$

from which axiom 3 gives

$$P(B) = P(AB) + P(A^cB)$$

Solving for $P(A^cB)$ and inserting into the expression for $P(A \cup B)$ given above yields the desired result (7.17).

Theorem 7.4. If $A \subset B$, then $P(A) \leq P(B)$.

Proof. The event B is first decomposed into the two mutually exclusive events $B = A \cup (BA^c)$. We therefore have by axiom 3 that $P(B) = P(A) + P(BA^c)$ and, since $P(BA^c) \geq 0$ from axiom 1, the theorem's result follows.

Using the axiomatic approach to probability theory as specified in Definition 7.3, it is seen that a number of theorems follow directly. Although we have examined only four of the more important of such theorems, it is to be noted that there exist numerous other theorems which are of interest in various applications. As our primary intent is that of developing only the probability framework necessary for a satisfactory study of nondeterministic time series, however, the theorems studied up to this point are sufficient.

7.6 FINITE SAMPLE SPACES

The simplest class of random experiments to study from a probabilistic viewpoint are those which have a *finite* number of possible outcomes. For such an experiment, the associate sample space must be of the form

$$S = \{a_1, a_2, \ldots, a_N\} \tag{7.18}$$

where the a_k designate the individual experimental outcomes for $1 \leq k \leq N$. A surprisingly large number of practical random experiments can be so described. We now provide a straightforward procedure for assigning a probability measure to this sample space which is consistent with the axioms of probability given in Section 7.5. Specifically, to each *elementary event* $A_k = \{a_k\}$ that describes the event in which the specific experimental outcome a_k occurs, a number $P(\{a_k\})$ is assigned such that the following two conditions are satisfied:

$$0 \leq P(\{a_k\}) \leq 1 \tag{7.19a}$$

$$\sum_{k=1}^{N} P(\{a_k\}) = 1 \tag{7.19b}$$

The number $P(A_k) = P(\{a_k\})$ is to be interpreted as being equal to *the probability that the elementary event* $\{a_k\}$ *occurs* on a given experimental trial. As suggested earlier, the basis for assigning values to the elementary entities $P(\{a_k\})$ is generally clear in a given context.

By using this notion of probability measure assignment, it is possible to consider more complex events. In particular, let the event A be composed of the m outcomes.

$$A = \{a_{i_1}, a_{i_2}, \ldots, a_{i_m}\} \tag{7.20a}$$

where the elementary outcomes $a_{i_k} \in S$. The probability that the event A occurs shall be defined by

$$P(A) = \sum_{k=1}^{m} P(\{a_{i_k}\}) \tag{7.20b}$$

That is, the probability that the compound event A occurs on a given experiment is equal to the sum of the probabilities of the individual elementary events constituting event A. It is readily shown that the method of probability assignment as given by expressions (7.19) and (7.20) is consistent with the axioms of probability theory (7.14).

EXAMPLE 7.7

Let the experiment consist of the tossing of a coin and recording whether a *head* or *tail* shows. The sample space is then $S = \{\text{h}, \text{t}\}$. If it is known that the probability that a head shows is twice as large as a tail showing, then $2P(\{t\}) = P(\{\text{h}\})$. Furthermore, since $S = \{h\} \cup \{t\}$, we have $P(\{h\}) + P(\{\text{t}\}) = 1$. We therefore conclude that

$$P(\{\text{h}\}) = \tfrac{2}{3} \qquad \text{and} \qquad P(\{\text{t}\}) = \tfrac{1}{3}$$

Equally Likely Outcomes

The classical theory of probability theory treats the finite sample space case here being considered in which the additional assumption is made that the individual outcomes composing S are *equally likely*. In particular, using the sample space specified in expression (7.18), the probability associated with the elementary events under the equally likely assumption is given by

$$P(\{a_k\}) = \frac{1}{N} \tag{7.21}$$

Moreover, if A is a more complex event that contains m elements of the sample space, then

$$P(A) = \frac{\text{number of elements in } A}{\text{number of elements in } S} = \frac{m}{N} \tag{7.22}$$

Thus, under the equally likely outcome assumption, the assignment of probability measure is equivalent to a counting scheme.

EXAMPLE 7.8

Let the experiment consist of the tossing of two distinguishable fair die as described in Example 7.6. Furthermore, let the events A and B correspond to outcomes in which the sum of the face dots showing is 7 and 11, respective that is,

$$A = \{(1, 6), (2, 5), (3, 4), (4, 3), (5, 2), (6, 1)\}$$

$$B = \{(5, 6), (6, 5)\}$$

Under the equally likely outcome assumption, which is justified by the *fa* die assertion, expression (7.22) indicates that

$$P(A) = \tfrac{6}{36} \qquad \text{and} \qquad P(B) = \tfrac{2}{36}$$

Furthermore, it is to be noted that since A and B are mutually exclusive ever (no elements in common), the probability that a 7 or an 11 sum occurs any given toss is, by axiom 3,

$$P\{A \cup B\} = P(A) + P(B) = \tfrac{8}{36}$$

This same result could be obtained by using expression (7.22) and noting th the event $A \cup B$ has eight elements.

As indicated above, probability measurement assignments under the equal likely outcome assumption are equivalent to counting the number of elemen in sets. In various applications, the concepts of *permutation* and *combinatio* play a vital role in this counting scheme. As such, a brief description of the two notions is given next.

Permutations

Suppose that there is given a set of n distinguishable elements and it is desire to find how many ways these elements can be arranged. As an example, th three elements a_1, a_2, and a_3 can be arranged in the six ways $a_1a_2a_3$, a_1a_3a $a_2a_1a_3$, $a_2a_3a_1$, $a_3a_1a_2$, and $a_3a_2a_1$, where it noted that the order of the eleme position is here taken as being important. To answer the general problem posed it is noted that the element assigned to the first position can be any one of th n elements. After this element choice has been made, however, the secon position must be filled by one of the remaining $n - 1$ elements. Similarly, afte the second element position has been assigned, the third element can be an one of the remaining $n - 2$ elements. If this assignment procedure is continue to its logical conclusion, it is apparent that the multiplicative rule

$$n! = n(n - 1)(n - 2) \cdots (3)(2)(1) \qquad (7.2$$

specifies the number of ways for arranging the n elements. The quantity $n!$ i referred to as *n factorial*.

More generally, it is desired to determine in how many ways n distinguishab elements can be arranged in groups of m, where $m \leq n$. For instance, the thre elements a_1, a_2, and a_3 can be arranged in the six groups of two a_1a_2, a_1a

a_2a_1, a_2a_3, a_3a_1, and a_3a_2. In this counting scheme, the element *order is considered important* in the sense that the entities a_1a_2 and a_2a_1 are taken as different. Upon using the multiplicative counting scheme described above, the answer is specified by the *permutation of n things taken m at a time,* as specified by

$$P_n^{(m)} = n(n-1)\cdots(n-m+1) = \frac{n!}{(n-m)!} \tag{7.24}$$

The integer m can take on the values $0, 1, 2, \ldots, n$, where the definition $0! = 1$ is invoked.

Combinations

Let us again consider the task of arranging n distinct elements into groups of m in which *order is considered unimportant*. For instance, the three elements a_1, a_2, and a_3 can be arranged into the three groups of two elements a_1a_2, a_1a_3, and a_2a_3 (note that a_1a_2 and a_2a_1 are here considered identical). The number of arrangements under this restriction is readily obtained from the permutation result above. For each specific set of m elements, there are $m!$ different arrangements of those elements. Thus the number of ways of arranging n distinct elements into groups of m is obtained by dividing expression (7.24) by this duplicating factor $m!$, that is,

$$C_n^{(m)} = \binom{n}{m} = \frac{n!}{m!(n-m)!} \tag{7.25}$$

This quantity is referred to as the *combination* of n elements *taken m at a time*. The integer m can take on the integer values $0, 1, 2, \ldots, n$.

EXAMPLE 7.9

Compute the probability of being dealt a spade flush hand (i.e., all 13 spades) in a game of bridge. For this example, the equally likely outcome sample space consists of all possible bridge hands. Since order may be considered as unimportant, the number of elements in S (i.e., number of possible bridge hands) is therefore given by

$$N = C_{52}^{(13)} = \frac{52!}{13!\ 39!}$$

On the other hand, there is only one element in S that is composed of 13 spades. Thus the probability of a spade flush must be

$$P(A) = \frac{1}{N} = \frac{13!\ 39!}{52!}$$

■

7.7 CONDITIONAL PROBABILITY

Let us now return to a more general treatment of probability theory in whic the underlying sample space is general in nature. In various applications, w seek to use readily measurable data to make inferences concerning an underlyin unmeasurable (or unobservable) entity. For example, in determining the stat of health of a patient's heart, a physician makes use of such measurable data a the patient's weight, blood pressure, family health history, electrocardiogra recordings, and so on, to provide a more informed diagnosis. Underlying thi general approach is the belief that the measured data are somehow correlate with the unmeasured entity whose state we seek to estimate. We shall now pu this useful concept into a probabilistic setting.

Let A and B be two events defined on a sample space S. Furthermore, on given trial of the underlying experiment, it is observed that the event A ha occurred. Based on this partial information, it is now desired to compute th probability that the unobserved event B has also occurred on that experimenta trial. This probability will be denoted by the symbol $P(B|A)$ and is to be rea "*the probability that event B occurs conditioned on the event A having oc curred.*" A little thought should indicate that the *uncondition probability* $P(B$ and the *conditioned probability* $P(B|A)$ are in general different. In fact, th conditioned probability is *defined* by the expression

$$P(B|A) = \frac{P(AB)}{P(A)} \quad (7.26$$

provided that $P(A) \neq 0$. This definition is now shown to conform to our pre conceived notions concerning conditional probability.

A justification for the conditional probability definition (7.26) can be mad that is based on the unit mass analogy to probability theory as depicted in Figur 7.4. Specifically, the given conditioning on event A implies that the experimenta outcome must be an element of A. As such, the event A can be interpreted a forming a reduced sample space relative to the conditional restriction here bein considered. It follows that $P(B|A)$ would correspond to that proportion of mas within boundary A that is associated with event B. This proportion is given b the ratio $P(AB)/P(A)$, which is equal to the definition of conditional probabilit specified above.

EXAMPLE 7.10

Consider a family with two children. Assume that each child is equally likel to be a girl as it is to be a boy. Under this condition, compute the probabilit that both children are girls given that at least one of the children is a girl. I this case, the prevailing sets are specified by

$$S = \{(g, g), (g, b), (b, g), (b, b)\}$$

$$A = \{\text{at least one child is a girl}\}$$
$$= \{(\text{g, g}), (\text{g, b}), (\text{b, g})\}$$
$$B = \{\text{both children are girls}\} = \{(\text{g, g})\}$$

where, for example, the entity (b, g) denotes a family whose oldest child is a boy and whose youngest child is a girl. In accordance with definition (7.26), the required conditional probability is then

$$P(B|A) = \frac{P(AB)}{P(A)} = \frac{1/4}{3/4} = \frac{1}{3}$$

where use of the facts that $AB = \{(\text{g, g})\}$ and the equally likely assumption have been made. ■

From the conditional probability definition (7.26), it is readily shown that the axioms of probability are satisfied on the reduced sample space A. Namely, if the probability that A occurs is nonzero, then

1. $0 \leq P(B|A) \leq 1$.
2. $P(A|A) = 1$.
3. If B and C are pairwise mutually exclusive events on the set A, then

$$P((B \cup C)|A) = P(B|A) + P(C|A) \tag{7.27}$$

4. If $A_1, A_2, A_3, \ldots$ are pairwise mutually exclusive events on the set A, then

$$P\left(\bigcup_{k=1}^{\infty} A_k|A\right) = \sum_{k=1}^{\infty} P(A_k|A)$$

It is to be noted that in the third axiom, the events B and C are required to be mutually exclusive relative to event A but not necessarily with respect to S. That is, any common elements shared by B and C must be contained in A^c. A similar statement holds for the fourth axiom.

7.8 PARTITIONING AND BAYES' THEOREM

The solution to many difficult probability problems can be obtained in a systematic manner by the use of sample-space *partitioning*. A partition constitutes one of the more powerful tools of probability and is now formally defined.

Definition 7.4. The events $B_1, B_2, \ldots, B_m$ constitute a *partition* of the sample space S if:

$$\text{(a) } B_iB_j = 0 \qquad \text{for } i \neq j$$

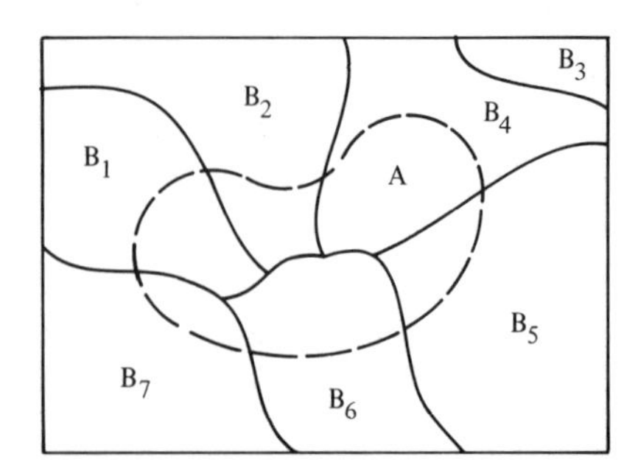

FIGURE 7.5. Partitioning of sample space.

$$\text{(b)} \quad \bigcup_{n=1}^{m} B_i = S \tag{7.28}$$

$$\text{(c)} \quad P(B_i) > 0$$

Thus a partition divides the sample space into pairwise mutually exclusive events that each have a nonzero probability of occurring. This concept is depicted in Figure 7.5.

Total Probability

To see how the partition concept is typically put to use, let it be desired to compute the probability of an event A. Due to partition requirement (7.28b), it is clear that event A can be equivalently expressed by the decomposition

$$\begin{aligned} A &= AS \\ &= AB_1 \cup AB_2 \cup \cdots \cup AB_m \end{aligned}$$

where the elements of this decomposition are depicted in Figure 7.5. Since the events AB_k constituting this decomposition are pairwise mutually exclusive, the fourth axiom of probability indicates that

$$P(A) = P(AB_1) + P(AB_2) + \cdots + P(AB_m) \tag{7.29}$$

Moreover, by employing the concept of conditional probability to each of the terms $P(AB_k)$ in this relationship, we obtain the often more useful *total probability* result

$$P(A) = P(A|B_1)P(B_1) + P(A|B_2)P(B_2) + \cdots + P(A|B_m)P(B_m) \tag{7.30}$$

The true beauty of this result lies in the fact that in many situations, the direct computation of $P(A)$ is most difficult. The process of decomposing $P(A)$ int the sum (7.30), however, can make this determination both logical and straightforward. Namely, upon making a wise choice of the partition, the individual entities $P(B_k)$ and $P(A|B_k)$ are readily evaluated.

EXAMPLE 7.11

A standard digital communication signal consists of a sequence of binary symbols (e.g., 1 and 0) that are transmitted over a channel. Due to equipment limitations and environmental effects, these symbols are occasionally misinterpreted so that a transmitted 1 (or 0) is erroneously received as a 0 (or 1). We now provide a probabilistic characterization of such a signal on an individual symbol-by-symbol basis.

A sample space associated with this experiment is specified by

$$S = \{(1, 1), (1, 0), (0, 1), (0, 0)\}$$

where, for example, the outcome (1, 0) designates the situation where a 1 is transmitted and a 0 is received (an error is made). This sample space may be partitioned using the events B_0 and B_1 as specified by

$$B_0 = \{\text{a zero symbol is transmitted}\} = \{(0, 1), (0, 0)\}$$

$$B_1 = \{\text{a one symbol is transmitted}\} = \{(1, 1), (1, 0)\}$$

Furthermore, the probabilities associated with these events are taken to be

$$P(B_0) = p \qquad \text{and} \qquad P(B_1) = 1 - p$$

where $0 \leq p \leq 1$. Conditional probabilities are used to describe the manner in which the channel effects the received binary symbol. In particular, if

$$A_0 = \{\text{a zero symbol is received}\} = \{(1, 0), (0, 0)\}$$

$$A_1 = \{\text{a one symbol is received}\} = \{(1, 1), (0, 1)\}$$

then the following conditional probabilities are taken (or assumed) to characterize the channel:

$$P(A_0|B_0) = P(A_1|B_1) = q$$

$$P(A_0|B_1) = P(A_1|B_0) = 1 - q$$

where $0 \leq q \leq 1$. That is, the correct received symbol interpretation is made with probability q, whereas an incorrect interpretation is made with probability $1 - q$. In accordance with expression (7.30) and the partition B_0, B_1, the probability that the symbol 1 has been received is given by

$$P(A_1) = P(A_1|B_0)P(B_0) + P(A_1|B_1)P(B_1)$$
$$= (1 - q)p + q(1 - p)$$

Similarly, we have

$$P(A_0) = qp + (1 - q)(1 - p)$$

We shall use these intermediate results in Example 7.12.

Bayes' Theorem

The partition concept can be used in an inverted fashion for computing t entities $P(B_k|A)$, where B_k is one of the events constituting the partition. Th is directly achieved by twice invoking the conditional probability rule so tha

$$P(B_k|A) = \frac{P(AB_k)}{P(A)} = \frac{P(A|B_k)P(B_k)}{P(A)} \qquad (7.3$$

Next, the total probability expression (7.30) is substituted for the denomina term to give

$$P(B_k|A) = \frac{P(A|B_k)P(B_k)}{\sum_{n=1}^{m} P(A|B_n)P(B_n)} \qquad 1 \le k \le m \qquad (7.3$$

This result, known as *Bayes' theorem,* plays a prominent role in many sign processing applications.

EXAMPLE 7.12

Let us consider the digital communication signal situation described in E ample 7.11. To measure the reliability to be accorded the received symb the conditional probability that a received symbol 1 was generated by a tran mitted 1 is now computed. In accordance with Bayes' theorem (7.32), have

$$P(B_1|A_1) = \frac{P(A_1|B_1)P(B_1)}{P(A_1)}$$
$$= \frac{q(1 - p)}{(1 - q)p + q(1 - p)}$$

where information provided in Example 7.11 has been used. Similarly, it c be shown that

$$P(B_1|A_0) = \frac{(1 - q)(1 - p)}{qp + (1 - q)(1 - p)}$$

If the probability of sending a zero is $p = 0.4$ and the probability of correct symbol reception is $q = 0.95$ (a reasonably reliable channel), these conditional probabilities become $P(B_1|A_1) = 0.966$ and $P(B_1|A_0) = 0.073$, indicating reasonable confidence in the validity of the symbol received. ■

7.9 INDEPENDENT EVENTS

The notion of statistically independent events has a strong connotational implication. Thus the statement that *events A and B are independent* carries the implicit meaning that knowledge that one of the events has occurred on an experimental trial provides no additional information concerning the occurrence or nonoccurrence of the other event on that trial. The essence of this meaning is captured in the following definition.

Definition 7.5. The events A and B are said to be statistically *independent* if

$$P(B|A) = P(B) \tag{7.33a}$$

or equivalently,

$$P(A|B) = P(A) \tag{7.33b}$$

In many applications, it is hypothesized that two events are independent. This assumption is normally justified by considering the nature of the underlying experiment and then making a judgment as to the lack of any meaningful interrelationship between the two events. For independent events, the conditional probability behavior leads to another equivalent definition of independence as specified by

$$P(AB) = P(A)P(B) \tag{7.34}$$

This equivalency is readily established by noting that, in general, $P(AB) = P(A|B)P(B)$, which simplifies to $P(AB) = P(A)P(B)$ if A and B are independent.

Multiple Events

The notion of statistically independent events can be extended in a logical manner. As an example, the three events A_1, A_2, and A_3 are said to be *mutually independent* if they are independent taken as (all) pairs and as a triple, that is,

$$\begin{aligned} P(A_1A_2) &= P(A_1)P(A_2) \qquad & P(A_1A_3) &= P(A_1)P(A_3) \\ P(A_2A_3) &= P(A_2)P(A_3) \qquad & P(A_1A_2A_3) &= P(A_1)P(A_2)P(A_3) \end{aligned} \tag{7.35}$$

It is important to appreciate the fact that each of these four conditions must be met for mutual independence. One may readily construct situations where three of these conditions are satisfied but the fourth is not. By continuing in this manner, the set of events $A_1, A_2, \ldots, A_m$ are said to be *mutually exclusive* if they are independent when taken as (all) pairs, triples, quadruples, . . . , and as an m-tuple. One may show that there are $2^m - m - 1$ conditions in this case.

EXAMPLE 7.13

We now consider the situation where n *independent* trials of a given experiment are made. Furthermore, the probability that a specific event A occurs on any given trial is taken to be p. Let it be desired to compute the probability that event A has occurred on precisely k of the n trials, where $0 \leq k \leq n$.

To solve this problem, let us introduce the elementary event A_m = {event A occurs on the mth trial} for $1 \leq m \leq n$. We now examine a specific situation conducive to the required objective, whereby the first k trials result in event A occurring and the last $n - k$ trials produce the event A^c. The probability that this particular experimental sequence arises is then given by

$$\begin{aligned} P(A_1 A_2 \cdots A_k A_{k+1}^c \cdots A_n^c) &= \prod_{i=1}^{k} P(A_i) \prod_{i=k+1}^{n} P(A_i^c) \\ &= p^k (1 - p)^{n-k} \end{aligned}$$

where use of the fact that the individual experimental trials are independent has been incorporated. In addition, the information that $P(A_i) = p$ and $P(A_i^c) = 1 - p$ has also been made. ■

There are other trial sequences that will result in a total of k event A occurrences (e.g., $A_1^c \cdots A_{n-k}^c A_{n-k+1} \cdots A_n$). Each of these other mutually exclusive possibilities has the same probability $p^k(1 - p)^{n-k}$ of occurring. A little thought indicates that the number of such sequences is given by $C_n^{(m)}$. We therefore conclude that

$$P(A \text{ occurs exactly } k \text{ times}) = \binom{n}{k} p^k (1 - p)^{n-k} \tag{7.36}$$

This is the well-known *Bernoulli probability* law governing independent trials of a given experiment.

7.10 SUMMARY

The axiomatic approach to probability theory has been presented in which fundamental concepts have been emphasized. This includes the notions of sample space, events, mutually exclusive events, conditional probability, partitions, and statistical independence. Although this presentation has been relatively brief, it is sufficient for the digital signal-processing applications to be subsequently examined.

SUGGESTED READINGS

There exist a number of fine textbooks on elementary probability theory that serve as good introductions to probability theory. These include the following.

BREIPOHL, A. M., *Probabilistic Systems Analysis*. New York: John Wiley & Sons, Inc., 1970.

DRAKE, A. W., *Fundamentals of Applied Probability Theory*. New York: McGraw-Hill Book Company, 1967.

HELSTROM, C. W., *Probability and Stochastic Processes for Engineers*. New York: Macmillan Publishing Company, 1984.

LARSON, H. J., and B. O. SHUBERT, *Probabilistic Models in Engineering Sciences*. New York: John Wiley & Sons, Inc., 1979.

MEYER, P. L., *Introductory Probability and Statistical Applications*. Reading, Mass.: Addison-Wesley Publishing Company, Inc., 1965.

PFEIFFER, P. E., *Concepts of Probability*. New York: McGraw-Hill Book Company, 1965.

PROBLEMS

7.1 Let the universal set consist of all integers from -5 to $+5$ inclusively. If $A = \{-2, 3, 1\}$, $B = \{3, 4\}$, and $C = \{-5, -4, 5\}$, determine the members of the following sets.

(a) AB^c **(b)** $A^c \cup B$ **(c)** $(A \cup BC)^c$
(d) $A \cup B^c \cup C$ **(e)** ABC **(f)** $(A(BC)^c)^c$

7.2 Let the universal set consist of all real numbers in the interval $-20 \leq x \leq 10$. If $A = \{x: -3 \leq x < 2\}$, $B = \{x: -2 \leq x \leq 3\}$, and $C = \{x: -5 < x \leq 6\}$, describe the members of the sets specified in Problem 7.1.

7.3 If the universal set is specified by three elements so that $S = \{a_1, a_2, a_3\}$, list the $2^3 = 8$ subsets of S. With this result in mind, show that if S contains n elements, there will exist 2^n subsets defined on S.

7.4 Let the universal set consist of all integer pairs (m, n) for which $-3 \leq m \leq 2$, $2 \leq n \leq 5$. List the elements of the following sets.

(a) $A = \{(m, n) \in S: m^2 + n^2 \leq 5\}$ **(b)** $B = \{(m, n) \in S: m \leq 2n\}$
(c) $C = \{(m, n) \in S: m^2 + n \leq 5\}$ **(d)** ABC **(e)** $A \in B^cC$

7.5 Use Venn diagrams to *illustrate* the validity of the following set identities.

(a) $A(B \cup C) = (A \cup B)(A \cup C)$ **(b)** $(A \cup B)^c = A^cB^c$
(c) $(AB)^c = A^c \cup B^c$ **(d)** $A \cup (BC) = (AB) \cup (AC)$
(e) $A \cup B = A \cup (A^cB)$ **(f)** $B = (AB) \cup (A^cB)$

7.6 Let the experiment consist of the tossing of a fair coin until a head is fi observed. Describe a universal set that describes this experiment. Does th universal set contain a finite or an infinite number of elements? List the elemen of the event A = {four or fewer tosses are needed to conclude the experiment

7.7 During an 8-hour period, a light is turned on in a room at time t_1 and subs quently turned off at time t_2 in that 8-hour period. Let t_1 and t_2 be measured hours, and the 8-hour period be taken to begin at $t = 0$. An outcome of th experiment may be described by (t_1, t_2). Describe the sample space associate with this experiment. Furthermore, describe and sketch in the (t_1, t_2) plane th events

(a) The light is off for less than 3 hours.
(b) The light is off twice as long as it is on.
(c) The light is on at time t_3 where $0 \leq t_3 \leq 8$.

7.8 Let A, B, and C be three events associated with an experiment. Using s notation, describe the following compound events.

(a) At least one of the events occur.
(b) At least two of the events occur.
(c) Exactly two of the events occur.
(d) No fewer than one of the events occur simultaneously.
(e) Event A occurs and at least one of the events B and C occur.

7.9 Let A, B, and C be three events associated with an experiment. Show that th probability that at least one of these events occurs (i.e., $A \cup B \cup C$) on a tri is given by

$$P(A \cup B \cup C) = P(A) + P(B) + P(C) - P(AB) - P(AC) - P(BC) + P(ABC)$$

7.10 Let A and B be two events associated with an experiment. Prove that

$$P(AB) \leq P(A) \leq P(A \cup B) \leq P(A) + P(B)$$

7.11 Let A and B be two events associated with an experiment in which the prob bilities $P(A)$, $P(B)$, and $P(AB)$ are known. In terms of these probabilities, com pute the probability of the following events:

(a) At least k of the events A and B occur for $k = 0, 1, 2$.
(b) Exactly k of the events A and B occur for $k = 0, 1, 2$.
(c) At least k of the events A^c and B^c occur for $k = 0, 1, 2$.

7.12 Let the three numbers 1, 2, and 3 be written down in a random order. What the probability that at least one number appears in its proper place? Generaliz this result for the case of the n numbers $1, 2, \ldots, n$.

7.13 Let an urn contain 10 balls numbered 1 to 10. If two balls are selected at random from the urn, what is the probability that the sum of their associated numbers is 7?

7.14 A hat contains n slips of papers each marked with one of the numbers 1, 2, . . . , n. Evaluate the probability that upon selecting two slips randomly from the hat, the numbers on the two slips are consecutive integers.

7.15 A binary code of length n consists of a sequence of zeros and ones. Show that there exists 2^n possible such codes.

7.16 Given the five letters a, b, c, d, and e, how many four-letter code words can be formed if:

(a) No letter may repeat?

(b) Letters may be repeated?

7.17 Using the counting scheme concept described in Section 7.6, prove the binomial theorem,

$$(x + y)^n = \sum_{m=0}^{n} \binom{n}{m} x^m y^{n-m}$$

7.18 Let an urn contain n balls of which n_1 are white and $n - n_1$ are black. If m balls are randomly selected from this urn, where $m \leq n$, prove that the probability of selecting k white and $m - k$ black balls is given by

$$\frac{\binom{n_1}{k}\binom{n - n_1}{m - k}}{\binom{n}{m}}$$

7.19 Let there be n slips of paper in which n_k of the slips have the number k written on them for $k = 1, 2, \ldots, m$. We therefore have $n_1 + n_2 + \cdots + n_m = n$. Show that the number of distinguishable arrangements of the slips (permutations) is specified by

$$\frac{n!}{n_1!\, n_2! \cdots n_m!}$$

.20 An urn contains four white and six black balls. Two balls are drawn at random from the urn and one is then examined and found to be black. What is the probability that the other ball is **(a)** white; **(b)** black?

.21 Two defective transistors are mixed with two good ones. These transistors are tested one by one until both defectives are identified. What is the probability that the last defective transistor is obtained on the **(a)** second test; **(b)** third test; **(c)** fourth test? **(d)** Add the probabilities obtained in parts (a), (b), and (c). Is this sum result surprising?

7.22 Two pairs of dice are rolled. Given that the two faces showing are different, what is the probability that: **(a)** One of the faces is a 3? **(b)** The sum of the two faces is 6?

7.23 Show that conditional probability $P(B|A)$ satisfies the axioms of conditional probability (7.27).

7.24 Events A and B associated with an experiment are independent. If the probability that A or B occurs equals 0.6, and the probability that A occurs is 0.4, find the probability that event B occurs.

7.25 Let A and B be two events associated with an experiment. Furthermore, suppose that $P(A) = 0.4$, $P(A \cup B) = 0.7$, and $P(B) = p$. For what choice of p are A and B **(a)** mutually exclusive; **(b)** statistically independent?

CHAPTER 8

Random Variables and the Expected-Value Operator

8.1 INTRODUCTION

Up to this point, our discussion of probability theory has been quite general in the sense that the underlying sample space's composition has not been restricted. The elements constituting the sample space can be numeric or nonnumeric depending on the nature of the experiment under consideration. For example, in the coin-tossing experiment, the sample space $S = \{h, t\}$ is inherently composed of nonnumeric elements. Since probability theory is a branch of mathematics, it is esthetically desirable that probability problems be posed in a strictly numerical setting. To achieve this formulation without losing the flexibility of treating nonnumeric random experiments, the concept of random variable is introduced. In this chapter we examine the concept of random variables and their statistical chracterization through the expected value operator.

8.2 UNIVARIATE RANDOM VARIABLES

In the standard description of a random experiment, the sample space lists all the possible outcomes of that experiment. It is conceptually possible to map each of these (possibly nonnumeric) outcomes into numbers using an associative procedure. This process of association constitutes the very notion of a random variable.

Definition 8.1. Let $\mathscr{E}$ be a random experiment that is described by a sample space S. A *function* X that assigns to every element $s \in S$ a number $X(s)$ is called a univariate *random variable*.

For our purposes, the number $X(s)$ is taken to be real valued. In some applications, however, it may be more natural to assign a complex value to $X(s)$. The use of the term *random variable* to describe this deterministic association rule (or function) is not literally accurate. Specifically, $X(\ )$ is a *function* whose domain is the sample space and whose range is the set of real (or complex) numbers. This misleading terminology has, however, been universally accepted.

EXAMPLE 8.1

Consider the experiment in which a normal coin is tossed and the face showing is observed. In this situation, the natural sample space is specified by $S =$ {h, t}, where h and t designate the outcomes *head* and *tail,* tossed, respectively. To represent this nonnumeric sample space in a numerical form, the following random variable selection could be used:

$$X(\mathrm{h}) = 0 \qquad X(\mathrm{t}) = 1$$

Thus the two sample spaces

$$S = \{\mathrm{h}, \mathrm{t}\} \qquad \text{and} \qquad S_x = \{0, 1\}$$

are in a sense equivalent. It should be noted that many other choices for the random variable could have been made [e.g., $X(\mathrm{h}) = -7.673$ and $X(\mathrm{t}) =$ 2.95].

Using the concept of random variable makes it always possible to provide a numerical interpretation of an experiment whose outcomes are nonnumerical. Once this random variable has been chosen, there is generated a related sample space whose elements are numerically valued, that is,

$$S_X = \{\text{set of numbers } x\text{: } x = X(s) \text{ for some element } s \in S\} \qquad (8.$$

Clearly, the composition of the elements in S_X is totally dependent on the choice of the random variable X and the original sample space.

EXAMPLE 8.2

In many applications, the experimental outcomes are either inherently numerical valued, or virtually so. An obvious selection of the random variable is then possible for such cases. This is illustrated by the experiment in which a die is tossed and the face showing is observed. A sample space describing

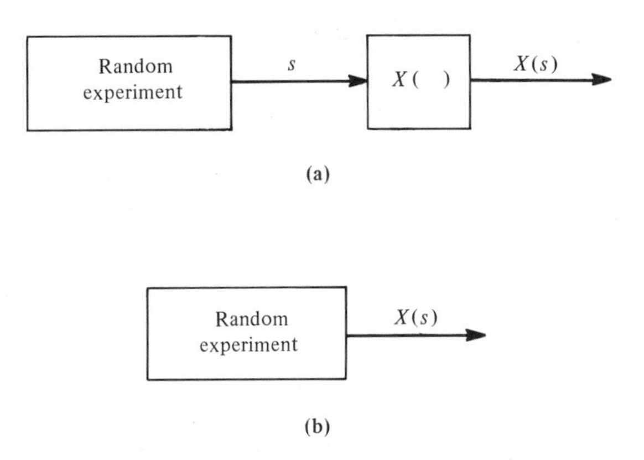

FIGURE 8.1. Concept of random variable: (a) nonnumeric experiment, (b) numeric experiment.

this experiment is $S = \{f_1, f_2, f_3, f_4, f_5, f_6\}$, where f_k denotes that outcome where the face with k dots shows. It is quite natural to select the random variable X as

$$X(f_k) = k \qquad 1 \leq k \leq 6$$

so that the related sample space is given by $S_X = \{1, 2, 3, 4, 5, 6\}$ ■

The random variable concept provides a systematic method for numericalizing random experiments. This approach may be visualized as shown in Figure 8.1a, where the experimental outcome is first observed to be s and then its associated numerical value $X(s)$ is computed. It is conceptually beneficial to envision this two-step procedure by the one-step process shown in Figure 8.1b. If the experiment produces outcomes that are inherently numerical valued (i.e., s is a number), as is usually the case in quantitative disciplines, the obvious choice $X(s) = s$ for the random variable may be made. Whatever the situation, we shall henceforth restrict our investigations to experiments that have numerically valued outcomes. We are comforted by the knowledge that no loss of generality is incurred, due to the random variable notion. Furthermore, in most real-world applications of probability theory, the underlying sample space is inherently numerically valued.

In our discussion of probability theory in Chapter 7, the concept of *event* played a central role. It will be recalled that an *event* is a *subset* of the underlying sample space. Since the sample spaces to be considered now are composed of real numbers, it follows that all events defined on this sample space must themselves be composed of real numbers (the complex-valued case is treated later). Typically, these events take on either an interval form such as

$$A = \{a < X \leq b\}$$

or a specific numerical value assignment such as

$$B = \{X = c\}$$

In these event descriptions, the event A is said to have occurred if the observed random outcome X takes on a value in the interval $(a, b]$, while event B occurs if $X = c$. Our task is to develop a systematic method for computing the probability that events such as A or B occur on any trial of the underlying random experiment.

EXAMPLE 8.3

Consider a die-tossing experiment in which the die is considered fair and the sample space is specified by $S_X = \{1, 2, 3, 4, 5, 6\}$. Under the fair-die assumption, the probability of the elementary events $A_k = \{X = k\}$ for $k = 1, 2, \ldots, 6$ is given by

$$P[X = k] = \tfrac{1}{6} \qquad \text{for } 1 \leq k \leq 6$$

■

8.3 PROBABILITY DENSITY FUNCTIONS

In accordance with the depiction of a random experiment as given in Figure 8.1, the experimental outcome may be taken to be a real number. We now provide a useful means for measuring the likelihood that an experimental outcome takes on values in any given subset of the real x axis. To effect this objective, let us conceptually distribute a unit amount of mass on the x axis in a way that reflects this likelihood. In particular, more mass is allocated to those ranges of x deemed most likely to occur, and less mass to those considered not as likely to occur. This method of unit mass distribution is governed by a mass (probability) *density function* $f_X(x)$ which by its very nature must possess the two properties

$$f_X(x) \geq 0 \qquad \text{for all } x \tag{8.2a}$$

$$\int_{-\infty}^{\infty} f_X(x)\, dx = 1 \tag{8.2b}$$

The subscript X has been appended to the density function to recognize that it characterizes the random variable X. A typical probability density function would appear as shown in Figure 8.2, and in accordance with properties (8.2) it is a nonnegative function that has an area of one.

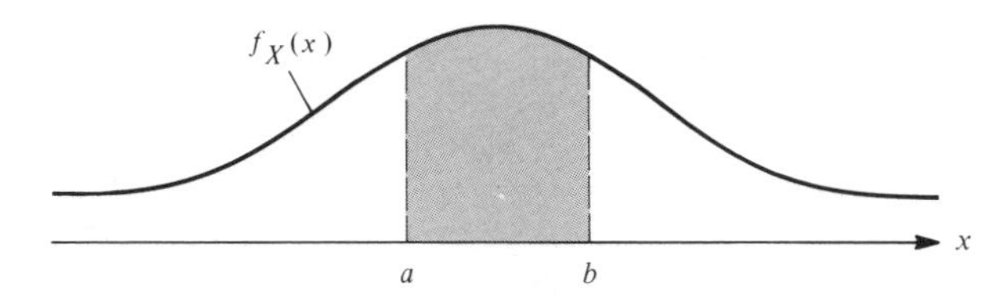

FIGURE 8.2. Probability density function.

The density function may now be used to compute systematically the probability of events defined on the sample space S_X. Of particular importance to our immediate needs is the *probability rule*

$$P[a < X \leq b] = \int_a^b f_X(x)\, dx \tag{8.3}$$

That is, the probability that the random variable X takes on a value in the interval $(a, b]$ is obtained by integrating the density function over that interval. If we use the aforementioned unit mass analogy, this integral is seen to give the amount of mass lying in that interval. This corresponds to the area of the shaded region shown in Figure 8.2.

EXAMPLE 8.4

Let the random variable X be uniformly distributed on the interval $[0, 2]$, that is,

$$f_X(x) = \begin{cases} \frac{1}{2} & 0 \leq x \leq 2 \\ 0 & \text{otherwise} \end{cases}$$

This function is seen to satisfy the prerequisite properties (8.2) and is therefore a legitimate probability density function. In accordance with relationship (8.3), the probability of the event $A = \{\frac{1}{3} < X \leq \frac{5}{3}\}$ is given by

$$P[\tfrac{1}{3} < X \leq \tfrac{5}{3}] = \int_{1/3}^{5/3} \tfrac{1}{2}\, dx = \tfrac{2}{3}$$

■

General Rule for Probability Determination

An extension of the foregoing probability determination scheme to more general events (noninterval) is possible. In particular, if A denotes a subset of the real line $-\infty < x < \infty$, then the probability that A occurs on any experimental trial is specified by the *probability rule*

$$P(A) = \int_{x \epsilon A} f_X(x)\, dx \tag{8.4}$$

That is, the entity $P(A)$ is determined by integrating the density function over all values of x contained in event A. It is seen that relationship (8.3) is a special case of this more general expression in which $A = \{a < x \leq b\}$.

8.4 PROBABILITY DISTRIBUTION FUNCTIONS

The probability density function provides a complete probabilistic description of a random variable. There exists a related function that contains the same information. The *probability distribution function,* defined formally by

$$F_X(x) = P[X \leq x] \qquad \text{for } -\infty < x < \infty \tag{8.5}$$

gives the probability that the random variable X assume a value less than or equal to the scalar x on any realization (trial) of the underlying experiment. Clearly, the distribution function constitutes a useful means of describing the behavior of a random variable. From definition (8.5) and rule (8.4), it is apparent that the probability density and distribution functions are related by

$$F_X(x) = \int_{-\infty}^{x} f_X(\alpha)\, d\alpha \tag{8.6}$$

Moreover, by taking the derivative of this integral relationship with respect to x, we have

$$f_X(x) = \frac{dF_X(x)}{dx} \tag{8.7}$$

This result follows by employing the standard definition of differentiation to the distribution function (see Problem 8.3). Thus the probability density and distribution functions are seen to be derivative–integral pairs.

The distribution function satisfies a number of useful properties that follow directly from its definition. These include

$$\begin{aligned} &\text{(a) } 0 \leq F_X(x) \leq 1 \\ &\text{(b) } F_X(x_1) \leq F_X(x_2) \qquad \text{for all } x_1 \leq x_2 \\ &\text{(c) } F_X(-\infty) = 0 \qquad \text{and} \qquad F_X(\infty) = 1 \\ &\text{(d) } P[x_1 < X \leq x_2] = F_X(x_2) - F_X(x_1) \end{aligned} \tag{8.8}$$

The distribution function is thus seen to be a real function of a real variable that exclusively takes on values in the range [0, 1]. Moreover, from property (8.8b), it is seen that the distribution function is a monotonically nondecreasing function of its argument.

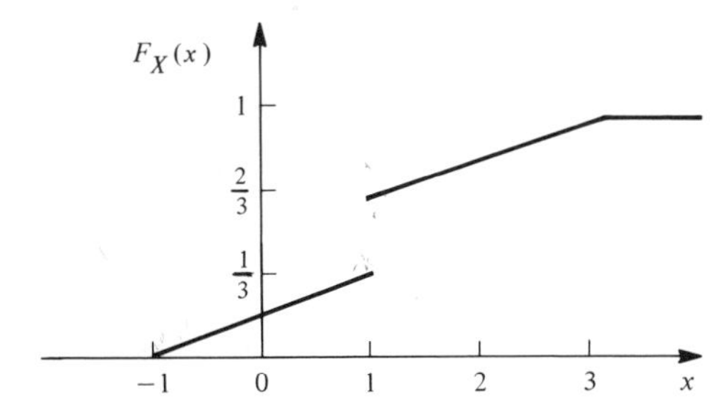

FIGURE 8.3. Distribution function for Example 8.5.

EXAMPLE 8.5

Let it be desired to determine the density function associated with the distribution function shown in Figure 8.3. Taking the derivative of this distribution function, we have

$$f_X(x) = \tfrac{1}{6}\,[u(x + 1) - u(x - 3)] + \tfrac{1}{3}\,\delta(x - 1)$$

where $\delta(x)$ and $u(x)$ denote the standard continuous unit Dirac delta and unit-step functions, respectively. It is to be noted that the Dirac function $\frac{1}{3}\delta(x - 1)$ arises due to the discontinuity in $F_X(x)$ at $x = 1$ of size $\frac{1}{3}$. ■

8.5 TYPES OF RANDOM VARIABLES

Depending on the nature of the underlying random experiment and the random variable employed, we find that the governing probability density function will exhibit three distinct types of behavior. Each behavior relates to the manner in which the unit mass is distributed on the x-axis. This mass distribution characteristic is reflected in the name given to the underlying random variable.

Continuous Random Variables

The random variable X is said to be *continuous* if the probability that it takes on any value x is always zero, that is,

$$P[X = x] = 0 \qquad \text{for all } x \in (-\infty, \infty) \tag{8.9}$$

Clearly, a random variable is continuous if and only if there is no point on the x-axis where a nonzero amount of mass is allocated. The density function and distribution function associated with a typical continuous random variable would appear as shown in Figure 8.4a. Of particular note is the fact that the distribution function associated with a continuous random variable is a continuous function

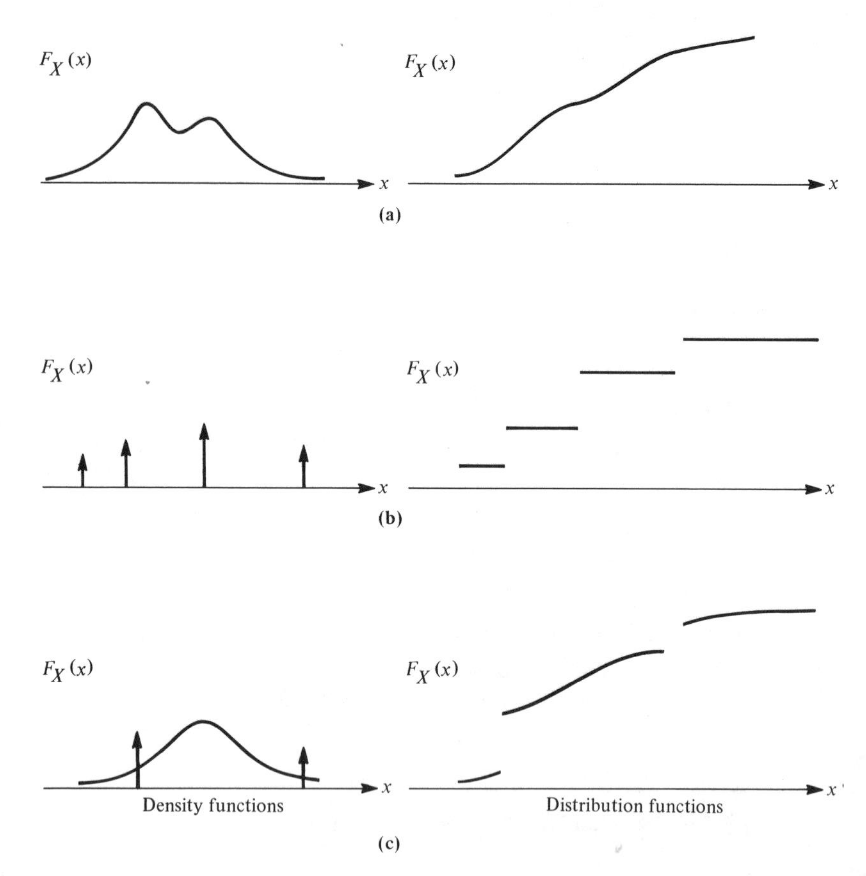

FIGURE 8.4. Three classes of random variables: (a) continuous, (b) discrete, (c) mixed.

of the variable x (i.e., it has no discontinuities). It is important to realize that although property (8.9) holds for continuous random variables, the probability that such a random variable takes on a value in a region is always nonzero, provided that $f_X(x)$ is itself nonzero over any subinterval in that region. This is made clear upon careful examination of the probability rule (8.4).

Discrete Random Variables

The random variable X is said to be *discrete* if the set of elements (i.e., numbers) in the sample space S_X is either finite or at most countably infinite.† For such cases, the associated density function will consist of a sum of weighted Dirac unit impulses, that is,

†A set is said to have a *countably infinite* number of elements if there exists a one-to-one correspondence between its elements and the nonnegative integers.

$$f_X(x) = \sum_{k=1}^{\infty} p_k \, \delta(x - x_k) \tag{8.10a}$$

The weight p_k corresponds to the probability that the random variable X takes on the value x_k (i.e., $p_k = P[X = x_k]$. These individual probabilities must therefore be nonnegative and be such that

$$\sum_{k=1}^{\infty} p_k = 1 \tag{8.10b}$$

as stipulated by axiom 2 of probability theory. The appearance of the Dirac delta functions in expression (8.10a) implies that there is p_k units of mass located at the point x_k. A visual depiction of the density and distribution functions associated with a discrete random variable is given in Figure 8.4b. The distribution function of a discrete random variable is seen to be a piecewise constant function that has discontinuities (i.e., jumps) at the points x_k.

Mixed Random Variables

It is possible to describe relevant random experiments in which the underlying sample space contains a noncountably infinite number of elements (numbers), some of which occur with nonzero probability. A random variable of this type is referred to as *mixed,* since its associated density function will contain both weighted Dirac impulses and a continuous component as well. The density and distribution functions describing a mixed random variable would appear as shown in Figure 8.4c.

Dirac Impulses and Nonzero Probability

If a nonzero amount of mass is located at a point, the density function is said to have a Dirac impulse at that point whose amplitude equals the amount of mass there concentrated. This may be readily shown by determining the probability that the random variable X takes on the specific value x_0. This probability is formally given by the limiting operation

$$\begin{aligned} P[X = x_0] &= \lim_{\epsilon \to 0^+} P[x_0 - \epsilon < X \leq x_0 + \epsilon] \\ &= \lim_{\epsilon \to 0^+} [F_X(x_0 + \epsilon) - F_X(x_0 - \epsilon)] \\ &= F_X(x_0^+) - F_X(x_0^-) \end{aligned}$$

where $F_X(x_0^+)$ and $F_X(x_0^-)$ correspond to the value of $F_X(x)$ just to the right and to the left of the point x_0, respectively. The only way for this quantity to be other than zero is for the distribution function to have a nonzero discontinuity at the point x_0. In this case the associated density function must then have a Dirac impulse of size $P[X = x_0]$ at the point x_0, while the distribution function will have a discontinuity of size $P[X = x_0]$ at the point x_0. Thus points where the density (distribution) function possesses Dirac impulses (discontinuities) are seen to correspond to values of the underlying random variable that occur with nonzero probability. Conversely, if a density (distribution) function possesses no impulses (discontinuities), the probability that the associated random variable takes on any specific value is always zero.

8.6 EXPECTED-VALUE OPERATOR AND MOMENTS

In most applications the density function associated with a random variable is not known nor is it readily estimated. Given this all-too-common situation, we still seek to characterize the random variable in some meaningful sense through the use of a more accessible probabilistic description that can be estimated from empirical experimentation. The *mean value*, (or *expected value*) defined formally by

$$\mu_X = E\{X\} = \int_{-\infty}^{\infty} x f_X(x)\, dx \tag{8.11}$$

constitutes the simplest of such measures. Using the previous interpretation of $f_X(x)$ as a mass density function, we see that the mean value gives the *center of gravity* for that mass. From a probabilistic sense, we shall see shortly that the mean value also specifies that value which the random variable X assumes *on the average* in a large number of independent trials of the random experiment.

Expected-Value Operator

In expression (8.11) the symbol E denotes the *expected-value operator*. To compute the expected value of any random variable, one literally first forms the product of that random variable in lowercase notation with the appropriate density function and then integrates that product over the interval $(-\infty, \infty)$. For example, for any function $g(X)$ of the random variable X, the expected value of the random variable $g(X)$ is given formally by

$$E\{g(X)\} = \int_{-\infty}^{\infty} g(x) f_X(x)\, dx \tag{8.12}$$

as long as this integral exists. The expected-value entity $E\{g(X)\}$ may be interpreted as representing the average value that the random variable $g(X)$ assumes on a large number of experimental trials. The expected-value operator, E, possesses two important properties which we shall use extensively henceforth. That is, the expected value of any constant α is equal to α:

$$E\{\alpha\} = \int_{-\infty}^{\infty} \alpha f_X(x)\, dx = \alpha \tag{8.13}$$

This follows directly from probability density function property (8.2b). Moreover, if the expected values of the individual random variables $g_1(X)$, $g_2(X), \ldots, g_q(X)$ each exists, the expected value of their linear combination simplifies to

$$\begin{aligned} E\left\{\sum_{k=1}^{q} \alpha_k g_k(X)\right\} &= \int_{-\infty}^{\infty}\left[\sum_{k=1}^{q} \alpha_k g_k(x)\right] f_X(x)\, dx \\ &= \sum_{k=1}^{q} \alpha_k \int_{-\infty}^{\infty} g_k(x) f_X(x)\, dx \\ &= \sum_{k=1}^{q} \alpha_k E\{g_k(X)\} \end{aligned} \tag{8.14}$$

where the α_k are arbitrary constants. The expected-value operator is said to be a *linear operator,* due to its satisfaction of this latter property. It will soon be made apparent that the expected-value operator plays an important role in linear estimation theory.

EXAMPLE 8.6

Let the random variable X be described by the density function

$$f_X(x) = \begin{cases} 1 - |x| & \text{for } -1 \le x \le 1 \\ 0 & \text{otherwise} \end{cases}$$

The mean value of this random variable, according to relationship (8.11), is given by

$$E\{X\} = \int_{-1}^{1} x(1 - |x|)\, dx = 0$$

Similarly, the expected value of the related random variable $Y = X^2$ is, by expression (8.12),

$$E\{X^2\} = \int_{-1}^{1} x^2(1 - |x|)\, dx = \tfrac{1}{6}$$

■

Moments

Motivated by the mean value definition (8.11), a further characterization of the random variable X is obtained with knowledge of its nth *moment* as defined by

$$m_n = E\{X^n\} = \int_{-\infty}^{\infty} x^n f_X(x)\, dx \tag{8.15}$$

where n is any nonnegative integer. It is seen from this definition that the zeroth moment equals 1 and the first moment corresponds to the random variable's mean value (i.e., $m_1 = \mu$). From a linear estimation viewpoint, one of the more important moments is the second or *mean-square* value.

$$m_2 = E\{X^2\} \tag{8.16}$$

The following example illustrates a typical utilization of second moments.

EXAMPLE 8.7

An important consideration in characterizing a random variable is determining the *constant value* that *best represents* the random variable (i.e., the best guess of X before its experimental realization). If m denotes a constant to be used for this purpose, a widely accepted measure of the goodness of this constant is provided by the *mean-squared-error* (MSE) *criterion* $E\{[X - m]^2\}$. If this criterion takes on a large (small) value, the constant m is said to represent the random variable poorly (well). The mean-squared-error criterion can be expressed as

$$\begin{aligned} E\{[X - m]^2\} &= E\{X^2 - 2mX + m^2\} \\ &= E\{X^2\} - 2mE\{X\} + m^2 \end{aligned}$$

where use of the expected-value operator properties (8.13) and (8.14) has been made. The optimum choice of m insofar as minimizing the MSE is readily obtained by differentiating the MSE criterion expression with respect to m and setting that result to zero, that is,

$$\frac{\partial E\{[X - m]^2\}}{\partial m} = -2E\{X\} + 2m = 0$$

The best MSE constant-value representation for the random variable X is therefore equal to the mean value

$$m^0 = E\{X\} = \mu_x$$

Moreover, the *minimum mean-squared error* (MMSE) associated with this selection is found by substitution to be

$$E\{[X - \mu_x]^2\} = E\{X^2\} - (E\{X\})^2$$

The MMSE is then seen to be explicitly dependent on the mean and mean-squared values associated with the random variable. ■

Central Moments

Motivated by the results of Example 8.7, we now consider the central moments associated with a random variable. The nth *central moment* is specified by

$$\mu_n = E\{[X - \mu_x]^n\} \tag{8.17}$$

where n is a nonnegative integer. Central moments are seen to provide a measure for how a random variable realization tends to differ from its mean value. The *second central moment* arises so frequently in applications that the special name *variance* and the symbol σ_x^2 is reserved for it. Formally, the variance of the random variable X is given by

$$\begin{aligned} \sigma_x^2 &= \text{var}\,\{X\} = E\{[X - \mu_x]^2\} \\ &= E\{X^2\} - (E\{X\})^2 \end{aligned} \tag{8.18}$$

The variance's square root, σ_x, is commonly referred to as the *standard deviation*. It was shown in Example 8.7 that the *variance measure* corresponds to the minimum MSE when the mean value was used as an a priori estimate of X. If the variance is small, this suggests that the value actually assumed by X on an experimental trial will typically be near its mean value. The variance provides a measure of how the mass associated with the density function tends to be dispersed about its center of gravity (i.e., the mean). In particular, the variance would correspond to the moment of inertia associated with the mass distribution. A large (small) value of variance suggests that this mass is widely (narrowly) dispersed about the mean value.

It is beneficial to utilize the concept of central moments to describe the results of the problem considered in Example 8.7. Since this example illustrates much of the spirit of estimation theory, it is beneficial to summarize these results in a theorem format.

Theorem 8.1. The scalar m that best represents the random variable X in the sense of minimizing the MSE criterion

$$E\{[X - m]^2\} \tag{8.19a}$$

is given by the random variable's mean value

$$m^0 = E\{X\} = \mu_x \tag{8.19b}$$

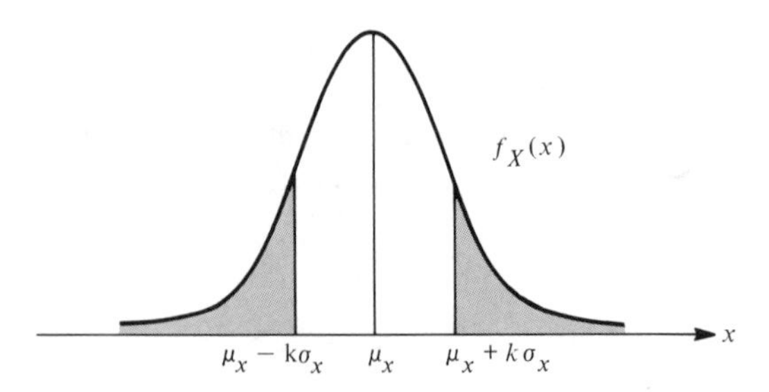

FIGURE 8.5. Chebyshev inequality.

Moreover, the associated minimum MSE corresponds to the random variable's variance, that is,

$$E\{[X - \mu_x]^2 = \sigma_x^2 \quad (8.19c)$$

Chebyshev Inequality

The Chebyshev inequality provides another mechanism for characterizing the properties of a random variable when knowledge of its associated mean and variance parameters is given but the associated density function is not known.

Theorem 8.2. Let X be a univariate random with mean μ_x and variance σ_x^2. The probability that the random variable X will differ from its mean by more than k standard deviations where $k > 0$ is bounded above by $1/k^2$, that is,

$$P[|X - \mu_x| \geq k\sigma_x] \leq \frac{1}{k^2} \quad (8.20)$$

This is the *Chebyshev inequality,* which indicates that for a fixed k, the values that X assumes on different experimental realizations tend to cluster about the mean μ_x as the variance becomes progressively smaller. These ideas are depicted in Figure 8.5, where the probability that the inequality $|X - \mu_x| \geq k\sigma$ is given by the shaded area. The Chebyshev inequality indicates that this area is always less than or equal to $1/k^2$.

8.7 COMMONLY USED DENSITY FUNCTIONS

In probabilistic studies related to nondeterministic experiments, the ability to make a convenient closed-form analysis is dependent on several factors. Of primary importance is the invoking of simple functional forms for the density functions associated with the experiment's random variables. This density function selection must be made with the dual objective of functional simplicity and

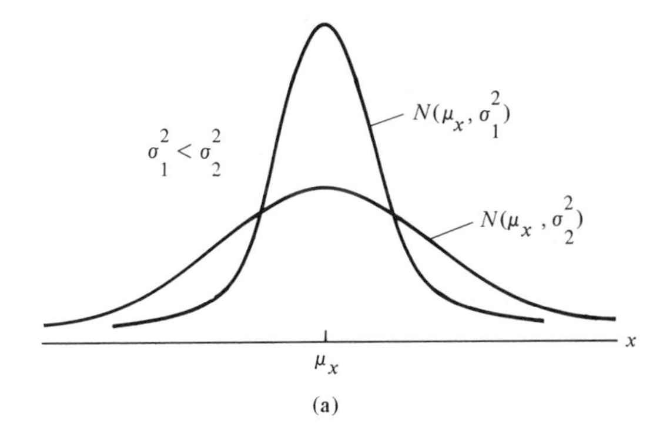

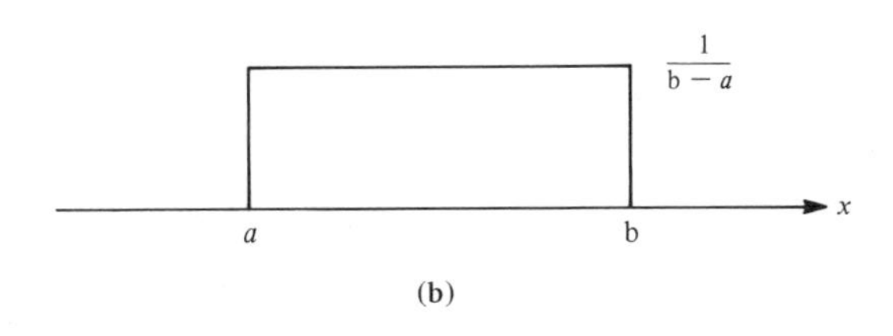

FIGURE 8.6. Density function: (a) normal; (b) uniform.

the density function's adequacy for representing the random variable's observed behavior. In this section we describe briefly some of the more commonly used density functions.

Normal Density Function

The normal (or Gaussian) random variable is the most widely employed of all random variables in applications. A normal random variable is compactly denoted by $N(\mu_x, \sigma_x)$ and has a density function specified by

$$f_X(x) = \frac{1}{\sqrt{2\pi}\,\sigma_x} e^{-(x-\mu_x)^2/2\sigma_x^2} \qquad \text{for } -\infty < x < \infty \tag{8.21}$$

in which the fixed parameters μ_x and σ_x^2 are readily shown to correspond to the random variable's *mean value* and *variance* parameters, respectively. Knowledge of these two parameters is seen to identify uniquely the normal density function. This normal density function possesses the characteristic symmetric bell shape shown in Figure 8.6a. This function is symmetric about the mean value μ_x and its breadth is seen to increase as the variance parameter σ_x^2 increases. The central limit theorem, to be discussed later, provides convincing evidence as to why the normal random variable is used so extensively in quantitatively oriented disciplines.

TABLE 8.1 Commonly Used Density Functions

Density Function	Functional Form, $f_x(x)$	Mean Value, μ_x	Variance, σ_x^2
Normal	$\frac{1}{\sqrt{2\pi}\,\sigma_x} e^{-(x-\mu_x)^2/2\sigma_x^2}$	μ_x	σ_x^2
Uniform	$\frac{1}{b-a}[u(x-a)-u(x-b)]$	$\frac{b+a}{2}$	$\frac{(b-a)^2}{12}$
Exponential	$\alpha e^{-\alpha x} u(x)$	$\frac{1}{\alpha}$	$\frac{1}{\alpha^2}$
Binomial	$\sum_{k=0}^{\infty} \binom{n}{k} p^k (1-p)^{n-k} \delta(x-k)$	np	$np(1-p)$
Poisson	$\sum_{k=0}^{\infty} \frac{e^{-\alpha}\alpha^k}{k!} \delta(x-k)$	α	α
Geometric	$\sum_{k=0}^{\infty} (1-p)^{k-1} p \delta(x-k)$	$\frac{1}{p}$	$\frac{1-p}{p^2}$

Uniform Density Function

The uniform random variable has a density function defined by

$$f_X(x) = \begin{cases} \dfrac{1}{b-a} & a \le x \le b \\ 0 & \text{otherwise} \end{cases} \tag{8.22}$$

where the real parameters a, b are such that $a < b$. A random variable so characterized is said to take on values in a uniform sense in the interval $[a, b]$, as suggested by the density function sketch shown in Figure 8.6b. This implies that the probability that X takes on a value in any subinterval of width δ contained in $[a, b]$ has the same probability of $\delta/(b - a)$ no matter where that subinterval is located. Some other more popularly used random variables and their corresponding density functions are described in Table 8.1. In this table the last three density functions are discrete in nature and are therefore described by a summation of *Dirac delta functions*.

EXAMPLE 8.8

Let us consider an experiment in which the random variable X is identified with the high temperature recorded in New York City on January 29 of a given year. Since the daily high temperature is dependent on the accumulative effect of a large number of factors (e.g., barometric pressure, wind velocity, relative humidity, solar activity, etc.), the central limit theorem to be treated in Section 9.15 suggests that the random variable X would be of a normal distribution form. In this case the mean value would be associated with the average high temperature recorded on January 29 in previous years. One disturbing feature of using a normal distribution to model X is the fact that such a distribution allows for unheard of New York City high temperatures (e.g., $X \geq 150°C$ or $X \leq -75°C$). This admitted quirk in the postulated model is mitigated by the fact that such unrealistic situations occur *on the tails* of the normal density function and therefore have very low probabilities of occurring. Using this reasoning, the invocation of a normal distribution model has become standard practice in a wide variety of applications. ■

8.8 FUNCTIONS OF UNIVARIATE RANDOM VARIABLES

In various applications the outcome of a random experiment is applied to an operator rule (or algorithm) so as to achieve some desired objective. This operation is specified mathematically by

$$Y = g(X) \tag{8.23}$$

where $g(\cdot)$ is a well-defined function. We now wish to provide a probabilistic characterization of the functionally generated random variable Y whereby knowledge of the density function on random X is presumed. In particular, a conceptually straightforward procedure for evaluating the distribution function of random variable Y is presented here. It is predicated on using the distribution function's definition,

$$\begin{aligned} F_Y(y) &= P[Y \leq y] \\ &= P[g(X) \leq y] \end{aligned} \tag{8.24}$$

where the function equality (8.23) enables us to replace Y by $g(X)$. To evaluate this latter probability, it is first necessary to find the set of x values for which $g(x) \leq y$. Once this set has been determined, the required probability (8.24) is obtained by finding the amount of mass that lies within this set. The required distribution function is specified formally by

$$F_Y(y) = \int_{g(x)\leq y} f_X(x)\, dx \tag{8.25}$$

where the integration is to be carried out over all x for which $g(x) \leq y$. This restricted range of integration is often describable by a suitable selection of the limits of integration, as the example to follow demonstrates. Once this distribution function has been obtained, the associated density function is then given by the differentiation rule

$$f_Y(y) = \frac{dF_y(y)}{dy} \tag{8.26}$$

To evaluate this derivative, it is often necessary to utilize the Leibnitz differentiating theorem.

Theorem 8.3 (Leibniz Rule of Differentiation). Let the function $f(x, t)$ be continuous and have a continuous derivative $\partial f(x, t)/\partial t$ in the xt plane over the rectangular region $a(t) \leq x \leq b(t)$, $t_1 \leq t \leq t_2$, where the functions $a(t)$ and $b(t)$ have continuous derivatives for $t_1 \leq t \leq t_2$. Then for $t_1 < t < t_2$,

$$\begin{aligned} \frac{d}{dt}\int_{a(t)}^{b(t)} f(x, t)\, dx &= f(b(t), t)\frac{db(t)}{dt} - f(a(t), t)\frac{da(t)}{dt} \\ &\quad + \int_{a(t)}^{b(t)} \frac{\partial f(x, t)}{\partial t}\, dx \end{aligned} \tag{8.27}$$

EXAMPLE 8.9

Let the random variable Y be equal to the magnitude of random variable X, that is,

$$Y = |X|$$

In accordance with expression (8.25), the distribution function of the random variable Y is given by

$$F_Y(y) = \int_{|x| \leq y} f_X(x)\, dx$$

From this integral expression, it is apparent that

$$F_Y(y) = \begin{cases} 0 & y < 0 \\ \displaystyle\int_{-y}^{y} f_X(x)\, dx & y \geq 0 \end{cases} \tag{8.28}$$

Given the density function $f_X(x)$, we may therefore straightforwardly evaluate the underlying integral. Furthermore, using the Leibniz differentiation rule, it follows that the associated density function is specified by

$$f_Y(y) = \begin{cases} 0 & y < 0 \\ f_X(y) + f_X(-y) & y \geq 0 \end{cases} \tag{8.29}$$

For example, if the density function $f_X(x)$ is uniformly distributed as

$$f_X(x) = \begin{cases} \frac{1}{3} & -1 \leq x \leq 2 \\ 0 & \text{otherwise} \end{cases}$$

it follows upon substitution into expressions (8.28) and (8.29) that

$$F_Y(y) = \begin{cases} 0 & y < 0 \\ \frac{2}{3}y & 0 \leq y \leq 1 \\ \frac{1}{3}y + \frac{1}{3} & 1 \leq y \leq 2 \end{cases}$$

and

$$f_Y(y) = \begin{cases} 0 & y < 0 \\ \frac{2}{3} & 0 < y \leq 1 \\ \frac{1}{3} & 1 < y \leq 2 \end{cases}$$

■

8.9 BIVARIATE RANDOM VARIABLES

It is possible to extend the concept of univariate random variable to describe random experiments whose outcomes are described by pairs of numbers. Such an experiment may be visualized as shown in Figure 8.7, in which the realized experimental outcome is specified by the number pair $(X_1(s), X_2(s))$. The individual entities X_1 and X_2 constituting this description are univariate random variables that are defined on the underlying sample space. These univariate random variables, when taken as the pair (X_1, X_2), are referred to as a *bivariate random variable*. The experimental outcome of a bivariate random experiment is then a 2-tuple. It will be convenient to depict individual outcomes as being points in the two-dimensional plane (x_1, x_2).

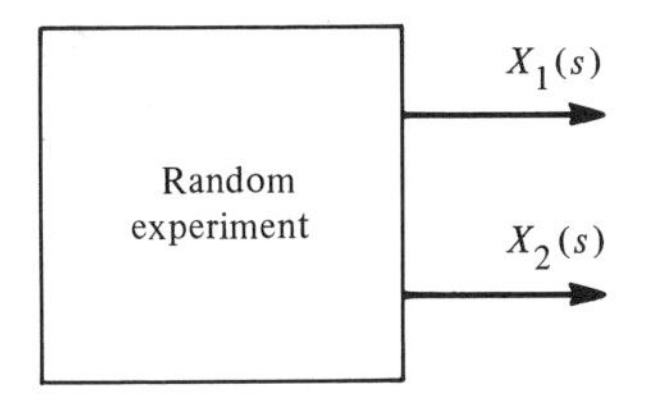

FIGURE 8.7. Random experiment characterized by a bivariate random variable.

EXAMPLE 8.10

Let the random experiment consist of a standard health examination being given to a patient by a physician. A specific selection of bivariate random variables for this experiment would be

$$X_1 = \text{blood pressure of patient}$$

$$X_2 = \text{heart rate of patient}$$ ■

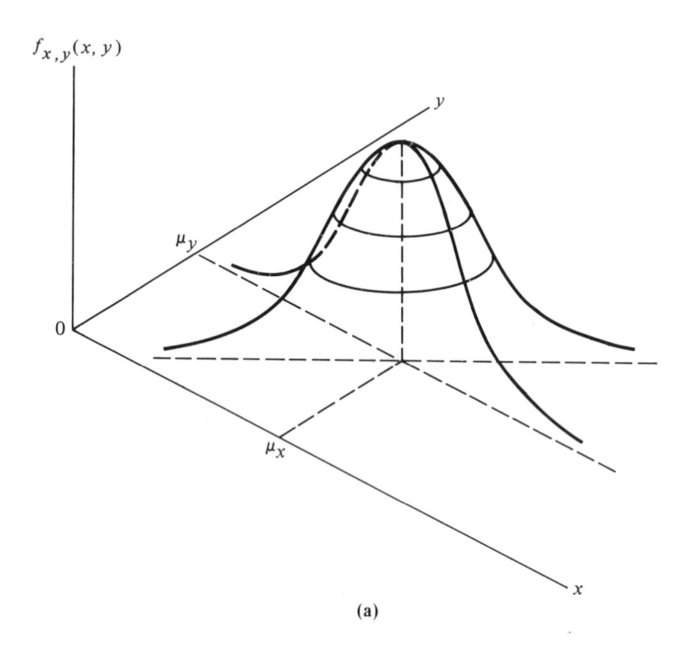

(a)

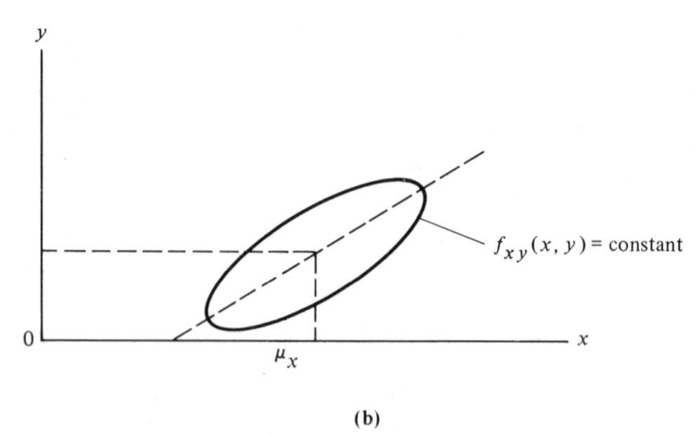

(b)

FIGURE 8.8. Typical joint density function of two Gaussian random variables and locus of the constant values of joint density functions.

Joint Density Function

The joint density function provides a means for computing the likelihood that the bivariate random variable (X_1, X_2) takes on values in any subset of the (x_1, x_2) plane. It is predicated on distributing a unit of mass in the (x_1, x_2) plane in a manner so as to reflect this likelihood. Thus more mass is distributed in regions where outcomes have a high likelihood, while less mass is allocated to regions of less likelihood. The manner in which this unit mass is distributed is described by the mass (probability) *joint density function* $f_{X_1X_2}(x_1, x_2)$. As such, the joint density function must possess the following two properties:

$$f_{X_1X_2}(x_1, x_2) \geq 0 \qquad \text{for } -\infty < x_1, x_2 < \infty \tag{8.30}$$

$$\int_{-\infty}^{\infty} \int_{-\infty}^{\infty} f_{X_1X_2}(x_1, x_2)\, dx_1\, dx_2 = 1 \tag{8.31}$$

A typical plot of the joint density function would appear as shown in Figure 8.8.

The probability that the bivariate random variable (X_1, X_2) takes on values in a given region A (an event) of the (x_1, x_2) place is formally defined by the probability rule

$$P[(X_1, X_2) \in A] = \iint_{(x_1,x_2)\in A} f_{X_1X_2}(x_1, x_2)\, dx_1\, dx_2 \tag{8.32}$$

This integral expression measures the amount of mass lying in the region A. Clearly, this relationship satisfies our previously described objective of allocating mass in a manner to reflect the likelihood behavior of the underlying random experiment.

EXAMPLE 8.11

Let the bivariate random variable (X_1, X_2) be uniformly distributed as

$$f_{X_1X_2}(x_1, x_2) = \begin{cases} 0.5 & 0 \leq x_1 \leq 2, \quad 0 \leq x_2 \leq 1 \\ 0 & \text{otherwise} \end{cases}$$

We shall now compute the probability that the sum of the random variables X_1 and X_2 is greater than or equal to 1 (i.e., $P[X_1 + X_2 \geq 1]$). Using probability expression (8.32) in conjunction with region A as depicted in Figure 8.9 [i.e., all (x_1, x_2) on and above line $x_1 + x_2 = 1$], we see that

$$P[X_1 + X_2 \geq 1] = \int_0^1 \left[\int_{1-x_2}^{2} 0.5\, dx_1 \right] dx_2 = \tfrac{3}{4}$$

■

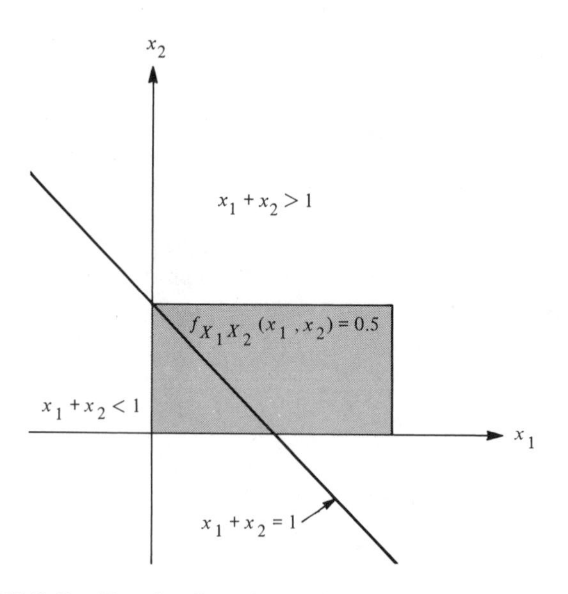

FIGURE 8.9. Density function and region *A* used in Example 8.11.

As in the univariate random variable case, there exists three classes of bivariate random variables. The bivariate random variable (X_1, X_2) is said to be *continuous* if $P[X_1 = x_1, X_2 = x_2] = 0$ for all values of x_1 and x_2. The joint density function associated with continuous bivariate random variables therefore cannot have a nonzero amount of mass located at any point in the (x_1, x_2) plane. On the other hand, the bivariate random variable (X_1, X_2) is said to be *discrete* if all the unit mass is allocated at a finite or a countably infinite set of points in the (x_1, x_2) plane. Finally, the bivariate random variable (X_1, X_2) is said to be *mixed* if the unit mass is distributed in a partially continuous and partially discrete manner.

Joint Distribution Function

As in the univariate case, the joint distribution function provides another means for describing the probability characteristics of bivariate random variables. The joint distribution is defined formally by

$$F_{X_1X_2}(x_1, x_2) = P[X_1 \leq x_1, X_2 \leq x_2] \tag{8.33}$$

and is seen to give the probability that the univariate random variables simultaneously satisfy the inequalities $X_1 \leq x_1$ and $X_2 \leq x_2$ on any given experimental trial. Clearly, knowledge of the joint distribution function for all real pairs (x_1, x_2) provides a complete probabilistic description of the bivariate random variable (X_1, X_2).

The joint probability function possesses a number of properties that are immediate consequences of its definition (8.33). Among the more important such properties are

$$\begin{aligned}
&\text{(a) } 0 \le F_{X_1X_2}(x_1, x_2) \le 1 \\
&\text{(b) } F_{X_1X_2}(x_1, x_2) \le F_{X_1X_2}(x_1 + \delta_1, x_2 + \delta_2) \quad \text{for all } \delta_1, \delta_2 \ge 0 \qquad (8.34) \\
&\text{(c) } F_{X_1X_2}(-\infty, -\infty) = 0 \quad \text{and} \quad F_{X_1X_2}(\infty, \infty) = 1
\end{aligned}$$

Thus the joint distribution takes on values exclusively in the range [0, 1] and is a monotonically nondecreasing function of its two variables.

Relationship Between Joint Density and Joint Distribution Functions

The joint density and joint distribution functions are interrelated to one another in a simple manner. Upon using the probability rule (8.32), we obtain the integral relationship

$$\begin{aligned}
F_{X_1X_2}(x_1, x_2) &= P[X_1 \le x_1, X_2 \le x_2] \\
&= \int_{-\infty}^{x_2} \int_{-\infty}^{x_1} f_{X_1X_2}(\alpha, \beta)\, d\alpha\, d\beta
\end{aligned} \tag{8.35}$$

If this expression is now differentiated with respect to the x_1 and x_2 variables using the Leibniz rule, it is found that

$$f_{X_1X_2}(x_1, x_2) = \frac{\partial F_{X_1X_2}(x_1, x_2)}{\partial x_1 \partial x_2} \tag{8.36}$$

As in the univariate case, the joint density and distribution functions are derivative–integral pairs.

8.10 MARGINAL AND JOINT DENSITY FUNCTIONS

One is always able to determine the marginal distribution functions associated with a given joint distribution function. To see how this is accomplished, we appeal to the definition of the *marginal distribution function* to obtain

$$\begin{aligned}
F_{X_1}(x_1) &= P[X_1 \le x_1] \\
&= P[X_1 \le x_1, X_2 < \infty] \\
&= F_{X_1X_2}(x_1, \infty)
\end{aligned} \tag{8.37}$$

Here we have used the fact that the events $\{X_1 \leq x_1\}$ and $\{X_1 \leq x_1, X_2 < \infty\}$ are equivalent. In a similar fashion it is found that the companion marginal distribution function is specified by

$$F_{X_2}(x_2) = F_{X_1X_2}(\infty, x_2) \tag{8.38}$$

Expressions (8.37) and (8.38) then provide the mechanism for generating the marginal distributions functions associated with a joint distribution function. From these results it is clear that the joint distribution function contains a higher level of probabilistic information concerning bivariate random variables than do the associated two marginal distribution functions. This follows since we are generally unable to reconstruct the joint distribution function associated with two given marginal distribution functions unless additional statistical information is assumed (e.g., independence of the random variables X and Y).

By using the fact that a marginal density function is the derivative of its corresponding distribution function, it is apparent from expressions (8.37) and (8.38) that

$$f_{X_1}(x_1) = \frac{\partial F_{X_1X_2}(x_1, \infty)}{\partial x_1} \tag{8.39}$$

and

$$f_{X_2}(x_2) = \frac{\partial F_{X_1X_2}(\infty, x_2)}{\partial x_2} \tag{8.40}$$

These results further illustrate the value of the joint distribution function in giving a full probabilistic characterization to bivariate random variables. It is to be noted that a related means for obtaining these marginal density functions is given by

$$\begin{aligned} f_{X_1}(x_1) &= \frac{\partial}{\partial x_1} F_{X_1X_2}(x_1, \infty) \\ &= \frac{\partial}{\partial x_1} P[X_1 \leq x_1, X_2 < \infty] \\ &= \frac{\partial}{\partial x_1} \int_{-\infty}^{x_1} \left[\int_{-\infty}^{\infty} f_{X_1X_2}(\alpha, \beta)\, d\beta \right] d\alpha \end{aligned}$$

Applying the Leibniz differentiation rule to this expression yields the desired result,

$$f_{X_1}(x_1) = \int_{-\infty}^{\infty} f_{X_1X_2}(x_1, x_2)\, dx_2 \tag{8.41}$$

In a similar fashion it is found that

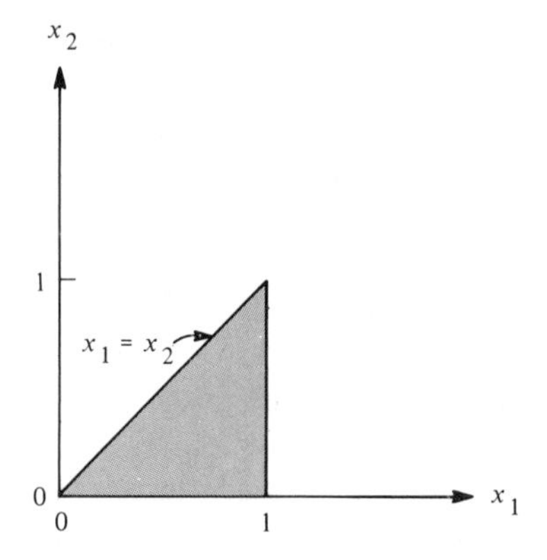

GURE 8.10. Bivariate density function for Example 8.12.

$$f_{X_2}(x_2) = \int_{-\infty}^{\infty} f_{X_1X_2}(x_1, x_2)\, dx_1 \tag{8.42}$$

EXAMPLE 8.12

Let the bivariate random variable (X_1, X_2) be uniformly distributed according to

$$f_{X_1X_2}(x_1, x_2) = \begin{cases} 2 & 0 \le x_2 \le x_1 \le 1 \\ 0 & \text{otherwise} \end{cases}$$

as depicted by the shaded region in Figure 8.10. The marginal density function $f_{X_1}(x_1)$ is by expression (8.41),

$$f_{X_1}(x_1) = \int_0^{x_1} 2\, dx_2 = 2x_1 \qquad \text{for } 0 \le x_1 \le 1$$

and is zero otherwise. Similarly,

$$f_{X_2}(x_2) = \int_{x_2}^{1} 2\, dx_1 = 2(1 - x_2) \qquad \text{for } 0 \le x_2 \le 1$$

and is zero otherwise. ■

.11 INDEPENDENT BIVARIATE RANDOM VARIABLES

For certain random experiments characterized by a bivariate random variable (X_1, X_2), it can be argued that there exists no intrinsic statistical linkage between the constituent random variables X_1 and X_2. This implies that if we have knowledge as to the value assumed by one of the random variables, that informatio conveys no additional knowledge concerning the companion random variabl Constituent random variables that satisfy this behavior are said to be (statisti cally) independent. The random variables X_1 and X_2 are *independent* if and on if their joint density function is separable as

$$f_{X_1X_2}(x_1, x_2) = f_{X_1}(x_1)f_{X_2}(x_2) \quad (8.4$$

namely, the joint density function of independent random variables is equal the product of the individual marginal density functions. An equivalent mea for characterizing independent random variables is made through their joi distribution function, which must take the form

$$F_{X_1X_2}(x_1, x_2) = F_{X_1}(x_1)F_{X_2}(x_2) \quad (8.4$$

It is a simple matter to show that expressions (8.43) and (8.44) are equivalen

EXAMPLE 8.13

Consider the random variables described in Example 8.12. Since the equali (8.43) does not hold in this case, it follows that the random variables X_1 ar X_2 are not independent. This should not be surprising, since if we are give the value assumed by X_1 (i.e., $X_1 = x_1$), this conveys the additional know edge that X_2 must take on a value between zero and x_1. On the other hand the uniformly distributed random variables considered in Example 8.11 a readily shown to be independent via relationship (8.43). This simply reflec the fact that if it is known that $X_1 = x_1$, no additional information concernir X_2 is thereby provided.

8.12 FUNCTIONS OF BIVARIATE RANDOM VARIABLES

In many applications, there is given a random variable Y which is a function the bivariate random variable (X_1, X_2), that is,

$$Y = g(X_1, X_2) \quad (8.4$$

Let it now be desired to determine the marginal density function associated wi the random variable Y, in which it is assumed that knowledge of the joint densi function $f_{X_1X_2}(x_1, x_2)$ is available. Conceptually, the solution to this problem straightforward and requires our first computing the distribution function

$$\begin{aligned} F_Y(y) &= P[Y \leq y] \\ &= P[g(X_1, X_2) \leq y] \end{aligned} \quad (8.4$$

The probability that the event $g(X_1, X_2) \leq y$ is next obtained by the two-step procedure of: (1) finding that region of the (x_1, x_2) plane for which the inequality $g(x_1, x_2) \leq y$ holds, and (2) determining the probability that the bivariate random variable (X_1, X_2) lies in this region. This systematic procedure may be given the following integral representation

$$F_Y(y) = \iint_{g(x_1,x_2)\leq y} f_{X_1X_2}(x_1, x_2)\, dx_1\, dx_2 \tag{8.47}$$

where the integration is to be taken over all (x_1, x_2) for which $g(x_1, x_2) \leq y$. This restricted range of integration is often describable by a suitable selection for the limits of integration as the example to follow demonstrates. Once this distribution function has been determined, the associated density function is then given by

$$f_Y(y) = \frac{dF_Y(y)}{dy} \tag{8.48}$$

Again, depending on the specific nature of the function $g(x_1, x_2)$ and the joint density function $f_{X_1X_2}(x_1, x_2)$, we find that this density function expression often has a convenient analytical formulation.

EXAMPLE 8.14

Let the random variable Y be related to the individual signal–noise bivariate random variables (S, N) as

$$Y = S + N$$

where S is the *signal* component and N the *noise* component. To determine the distribution function of the signal plus noise random variable Y, we note that

$$\begin{aligned} F_Y(y) &= P[Y \leq y] \\ &= P[S + N \leq y] \end{aligned}$$

The shaded region in the (s, n) plane shown in Figure 8.11 corresponds to those (s, n) for which $s + n \leq y$. The probability that the bivariate random variable (S, N) lies within this region is therefore given by

$$F_Y(y) = \int_{-\infty}^{\infty} \left[\int_{-\infty}^{y-s} f_{SN}(s, n)\, dn \right] ds$$

in which the first integral with respect to n is taken along the vertical line extending from minus infinity to the line $s + n = y$.

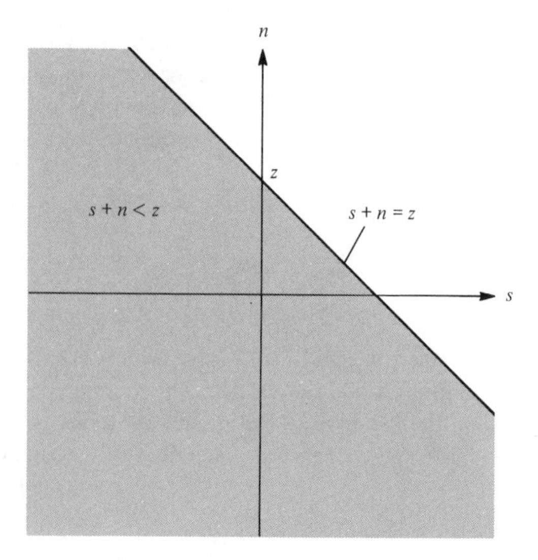

FIGURE 8.11. Region in (*s*, *n*) plane for which $s + n \leq y$.

The corresponding density function is obtained by differentiating this distribution function expression with respect to y by using the Leibniz rule. This differentiation operation yields

$$f_Y(y) = \int_{-\infty}^{\infty} f_{SN}(s, y - s)\, ds \tag{8.49}$$

Given the joint density function of the bivariate random variables (S, N), this integration operation yields the marginal density function of the signal plus noise random variable $Y = S + N$. ■

Joint Density Function

In applications described by the functional relationship (8.45), it often required to determine the joint density function associated with the bivariate random variable (X_2, Y) or possibly (X_1, Y). It is here assumed that the joint density function $f_{X_1X_2}(x_1, x_2)$ is known. If we use the same approach that led to expression (8.47), it follows that

$$\begin{aligned} F_{X_2Y}(x_2, y) &= P[X_2 \leq x_2, Y \leq y] \\ &= P[X_2 \leq x_2, g(X_1, X_2) \leq y] \\ &= \int_{-\infty}^{x_2} \left[\int_{g(x_1,\alpha)\leq y} f_{X_1X_2}(x_1, \alpha)\, dx_1 \right] d\alpha \end{aligned} \tag{8.50}$$

The corresponding joint density function is obtained by differentiating this expression with respect to x_2 and y, that is,

$$f_{X_2Y}(x_2, y) = \frac{\partial F_{X_2Y}(x_2, y)}{\partial x_2 \, \partial y} \tag{8.51}$$

For many typical applications, the inequality regions specified in integration (8.50) are manifested by suitable choices for the limits of integration.

8.13 MULTIVARIATE RANDOM VARIABLES

In this section we examine random experiments whose outcomes are described by n numbers, namely, after the experiment has been conducted, the resultant outcome is specified by the n-tuple $(X_1(s), X_2(s), \ldots, X_n(s))$. This process is depicted in Figure 8.12. The set of random variables $(X_1, X_2, \ldots, X_n)$ that characterize this random experiment are referred to as a *multivariate random variable*. We have already discussed the special cases of univariate (i.e., $n = 1$) and bivariate (i.e., $n = 2$) random variables. Our ultimate interest in multivariate random variables arises from their ability to describe finite-length segments of a random time series. In this interpretation the individual random variable X_k characterizes the kth element of the random time series.

EXAMPLE 8.15

Let the multivariate random variable $(X_1, X_2, \ldots, X_n)$ have its individual members X_k identified with the high temperature recorded in New York City on the kth day of a given year. In this case the multivariate random variable corresponds to the first n terms of a daily-high-temperature time series. Viewed in this manner, we see that the study of multivariate random variables is equivalent to the study of random time series. ■

To give a probabilistic description of multivariate random variables, we again appeal to the concept of distributing a unit amount of mass in n-space. This

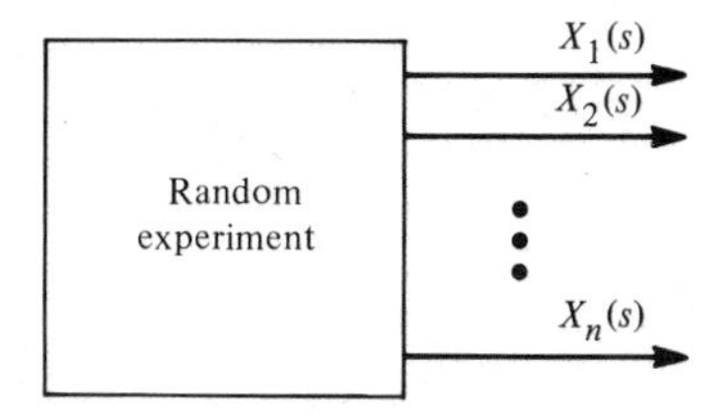

FIGURE 8.12. Random experiment characterized by a multivariate random variable.

mass is distributed in a manner which reflects the likelihood that the experimental outcomes (i.e., n-tuples) lie within various regions of n-space. The *joint density function* provides a mathematical description of how this mass is allocated. The *joint density function* thus possesses the two properties

$$f_{X_1X_2\cdots X_n}(x_1, x_2, \ldots, x_n) \geq 0 \tag{8.52}$$

$$\int_{-\infty}^{\infty}\int_{-\infty}^{\infty}\cdots\int_{-\infty}^{\infty} f_{X_1X_2\cdots X_n}(x_1, x_2, \ldots, x_n)\, dx_1 dx_2 \cdots dx_n = 1 \tag{8.53}$$

The multivariate random variable $(X_1, X_2, \ldots, X_n)$ is said to be *discrete*, *continuous*, or, *mixed* depending on whether the mass accumulates exclusively at points in n-space, is continuously distributed, or has a mixture of these distributions, respectively.

To compute the probability that an experimental outcome lies within any region A of n-space (e.g., the hypersphere $X_1^2 + X_2^2 + \cdots + X_n^2 \leq 1$), we simply evaluate the multidimensional integral

$$P[(X_1, X_2, \ldots, X_n) \in A] = \iint\cdots\int_{(x_1,\ldots,x_n)\in A} f_{X_1\cdots X_n}(x_1, \ldots, x_n)\, dx_1 \cdots dx_n \tag{8.54}$$

where the integration is to be carried out over all $(x_1, x_2, \ldots, x_n) \in A$. For many practical applications, the integration over A is manifested by appropriately fixing the limits of integration.

The most important invocation of probability relationship (8.54) is obtained for the specific selection of A given by $(X_1 \leq x_1, X_2, \leq x_2, \ldots, X_n \leq x_n)$. The probability that event A occurs is commonly referred to as the *joint distribution function* and is denoted by

$$\begin{aligned} F_{X_1X_2\cdots X_n}(x_1, x_2, \ldots, x_n) &= P[X_1 \leq x_1, X_2 \leq x_2, \ldots, X_n \leq x_n] \\ &= \int_{-\infty}^{x_n}\cdots\int_{-\infty}^{x_1} f_{X_1\cdots X_n}(\alpha_1, \ldots, \alpha_n)\, d\alpha_1 \cdots d\alpha_n \end{aligned} \tag{8.55}$$

The limits of integration here are seen to reflect the event A being considered. Upon taking the partial derivative of this result, we further obtain

$$f_{X_1\cdots X_n}(x_1, \ldots, x_n) = \frac{\partial^n}{\partial x_1 \cdots \partial x_n} Fx_1 \cdots x_n(x_1, \ldots, x_n) \tag{8.56}$$

which is in agreement with previously established results which have shown that the joint density and joint distribution functions are derivative–integral pairs.

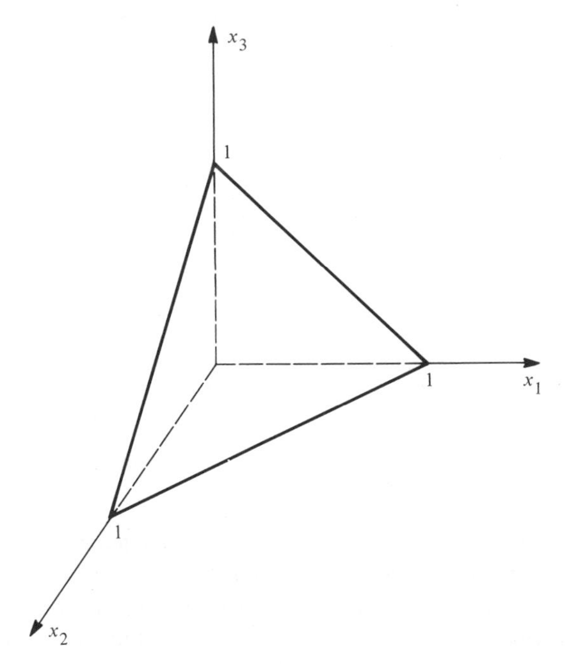

FIGURE 8.13. Region in which joint density function of Example 8.16 is defined.

The joint density and distribution functions above may be used to obtain lower-order joint density and distribution functions. In particular, from the basic definition of the joint distribution function (8.55), we find that

$$F_{X_1 \cdots X_m}(x_1, \ldots, x_m) = F_{X_1 \cdots X_n}(x_1, \ldots, x_m, +\infty, \ldots, +\infty) \quad (8.57)$$

where $m < n$. The associated marginal density function is therefore

$$f_{X_1 X_m}(x_1, \ldots, x_m) = \int_{-\infty}^{\infty} \cdots \int_{-\infty}^{\infty} f_{X_1 \cdots X_n}(x_1, \ldots, x_n)\, dx_{m+1} \cdots dx_n \quad (8.58)$$

EXAMPLE 8.16

Consider the multivariate random variable (X_1, X_2, X_3) whose joint density function is specified by

$$f_{X_1 X_2 X_3}(x_1, x_2, x_3) = \begin{cases} 6 & \text{for } x_1, x_2, x_3 \geq 0 \text{ and } 0 \leq x_1 + x_2 + x_3 \leq 1 \\ 0 & \text{otherwise} \end{cases}$$

The region of (x_1, x_2, x_3) space in which this joint density function is nonzero corresponds to the sliced cube volume depicted in Figure 8.13. The marginal density functions associated with this joint density function may be generated according to relationship (8.58). For example,

$$f_{X_1X_2}(x_1, x_2) = \int_0^{1-x_1-x_2} f_{X_1X_2X_3}(x_1, x_2, x_3)\, dx_3$$

$$= \begin{cases} 6(1 - x_1 - x_2) & \text{for } x_1, x_2 \geq 0 \text{ and } 0 \leq x_1 + x_2 \leq 1 \\ 0 & \text{otherwise} \end{cases}$$

and

$$f_{X_1}(x_1) = \int_0^{1-x_1} 6(1 - x_1 - x_2)\, dx_2$$

$$= \begin{cases} 3(1 - x_1)^2 & \text{for } 0 \leq x_1 \leq 1 \\ 0 & \text{otherwise} \end{cases}$$

■

Independent Random Variables

If the values assumed by individual members of a multivariate random variable experiment are not influenced by the values assumed by its other members, the random variables are said to be *independent*. Independent random variables have joint density and distribution functions that may be factored into the product of first-order marginal density and distribution functions, that is,

$$f_{X_1 \cdots X_n}(x_1, x_2, \ldots, x_n) = \prod_{k=1}^{n} f_{X_k}(x_k) \tag{8.59}$$

and

$$F_{X_1 \cdots X_n}(x_1, x_2, \ldots, x_n) = \prod_{k=1}^{n} F_{X_k}(x_k) \tag{8.60}$$

Expected-Value Operator

We may apply the concept of expectation to multivariate random variables. In particular, the expected value of the random variable function $g(X_1, X_2, \ldots, X_n)$ is given by

$$E\{g(X_1, \ldots, X_n)\} =$$

$$\int_{-\infty}^{\infty} \cdots \int_{-\infty}^{\infty} g(x_1, \ldots, x_n)\, f_{X_1 \cdots X_n}(x_1, \ldots, x_n)\, dx_1 \cdots dx_n \tag{8.61}$$

provided that the integral exists. The expected-value operator is also linear in this general setting, that is,

$$E\left\{\sum_{k=1}^{q} \alpha_k g_k(X_1, \ldots, X_n)\right\} = \sum_{k=1}^{q} \alpha_k E\{g_k(X_1, \ldots, X_n)\} \tag{8.62}$$

where the α_k are constants and the constituent functionally generated random variables $g_k(X_1, \ldots, X_n)$ are each assumed to have expected values. A proof of linear relationship (8.62) follows directly from expression (8.61), in which the linearity of the integral operator is employed.

8.14 SUMMARY

A random variable was shown to be a mechanism for providing numerical descriptions to experiments that have random outcomes. The ability to numericalize random experiments allows us to examine their probabilistic characterization using a common approach. Important tools in this characterization are the density and distribution functions. Knowledge of the density function enables us to compute the probability that various experimental outcomes will arise on any given trial of the experiment. Furthermore, the expected-value operator provides a convenient measure of the value that a random variable takes *on average* in a large number of trials of the random experiments. These notions will play a prominent role in our investigations of signal processing.

SUGGESTED READINGS

The Suggested Readings in Chapter 7 also provides useful descriptions of the random variable concept.

PROBLEMS

8.1 A coin is known to come up heads three times as often as tails. Let an experiment consist of the tossing of this coin four times and an associated random variable be X = {number of heads that came up in four tosses}. Determine and sketch the probability density and distribution functions of this random variable. Compute $P[1.5 \leq X \leq 2.2]$.

8.2 The fraction of oxygen in a compound is taken to be a random variable X with associated density function

$$f_X(x) = \begin{cases} \alpha x^3(2 - x) & 0 \leq x \leq 1 \\ 0 & \text{otherwise} \end{cases}$$

(a) Determine α.
(b) Evaluate $F_X(x)$.
(c) Compute $P[0.4 \leq X \leq 5]$.

8.3 Using the limit definition of the derivative of a function,

$$\frac{dh(x)}{dx} = \lim_{\Delta \to 0} \frac{h(x + \Delta) - h(x)}{\Delta}$$

prove that the derivative of distribution function (8.6) is $f_X(x)$.

8.4 A point is chosen at random on a line of length L. Let the random variable associated with this experiment be

$$X = \{\text{ratio of shortest to longest segment}\}$$

(a) Determine $F_X(x)$ and $f_X(x)$.
(b) Compute $P[0 \le X \le \frac{1}{2}]$.

8.5 Metro trains leave a certain station at 7:03, 7:18, 7:30, 7:42, and 7:58 A.M. each morning for the same destination. A rider arrives at the station at a random time between 7:00 and 7:45 A.M. and takes the first available train.
(a) Compute the probability that the rider takes the 7:30 train.
(b) Compute the probability that the rider waits less than 8 minutes before boarding the train.

8.6 A circular area with radius r meters is bombarded by mortar fire. The intensity of fire measured by hits per minute per square unit area is h_1 at the center and decreases linearly to h_2 at the perimeter of the circle. If X denotes the distance that an individual mortar hits from the center, determine the density function $f_X(x)$.

8.7 Given the following distribution functions, determine the associated density functions and ascertain whether the random variable is continuous, discrete, or of a mixed type. *Hint:* Sketch the distribution function and look for discontinuities.
(a) $F_X(x) = [1 - e^{(7-x)}]u(x - 7)$
(b) $F_X(x) = \frac{1}{5}u(x + 5) + \frac{1}{7}u(x - 3) + \frac{23}{35}u(x - 7)$
(c) $F_X(x) = \dfrac{x - 3}{x}\, u(x - 3)$
(d) $F_X(x) = \dfrac{x}{16}[u(x) - u(x - 1)] + \dfrac{3x + 2}{16}[u(x - 1) - u(x - 2)] + u(x - 2)$
where the unit-step function $u(x) = 1$ for $x \ge 0$ and is zero for $x < 0$

8.8 Determine the distribution functions associated with the following density functions.
(a) $f_X(x) = 7e^{-7x}u(x)$
(b) $f_X(x) = \frac{1}{3}\,\delta(x + 4) + \frac{1}{5}\,\delta(x + 1) + \frac{1}{3}\,\delta(x - 1) + \frac{2}{15}\,\delta(x - 7)$
(c) $f_X(x) = x[u(x) - u(x - 1)] + (1 - x)[u(x - 2) - u(x - 1)]$
(d) $f_X(x) = (1 - |x|)[u(x + 1) - u(x - 1)]$

8.9 Show that if the density function $f_X(x)$ is symmetrical about the point μ [i.e., $f_X(x + \mu) = f_x(-x + \mu)$], then $E\{X\} = \mu$.

8.10 Four minicomputer systems are selected at random from a production line in a given day for quality control testing. If properly functioning minicomputer systems are being produced with probability p and defective ones with probability $1 - p$, determine the distribution function of the random variable X where $X = \{$number of properly functioning minicomputer systems in the sample of four$\}$. Compute $E\{X\}$ and $\text{var}\{X\}$.

8.11 Let the random variable X have the density function

$$f_X(x) = \tfrac{1}{2}[u(x + 1) - u(x - 1)]$$

(a) Calculate $P[|X - \mu| \geq k\sigma]$ and sketch it as a function of k.
(b) Plot the Chebyshev bound for this random variable and compare your results with those found in part *a*.

8.12 Let the random variable X represent the signal level at the output of a video amplifier at a given time instance. Let X be exponentially distributed as

$$f_X(x) = ae^{-ax}u(x)$$

where a is a positive constant.
(a) Determine the probability that the signal amplitude exceeds $2a$ in magnitude (i.e., $|X| > 2a$).
(b) Determine the density function associated with the rectified signal $Y = |X|$.

8.13 Let the random variable X be uniformly distributed on the interval $[-1, 3]$. Determine the density functions associated with the related random variables

(a) $Y = X^2$
(b) $Z = \text{sgn } X = \begin{cases} 1 & \text{if } X \geq 0 \\ -1 & \text{if } X < 0 \end{cases}$
(c) $W = |X|$

8.14 Prove the Chebyshev inequality. *Hint:* Use the inequality

$$\sigma^2 \geq \int_{|x-\mu| \geq k\sigma} (x - \mu)^2 f_X(x)\, dx$$

and the fact that for all $|x - \mu| \geq k\sigma$ it follows that $(x - \mu)^2 \geq k^2\sigma^2$.

8.15 Let the bivariate random variable (X_1, X_2) have the joint density function

$$f_{X_1X_2}(x_1, x_2) = \begin{cases} \tfrac{9}{4}x_2 & 0 \leq x_1 \leq 1, \quad 2x_1 \leq x_2 \leq 1 \\ 2x_1x_2 & 0 \leq x_1 \leq 1, \quad 0 \leq x_2 \leq 2x_1 \end{cases}$$

Compute the marginal density functions. Are the random variables X_1 and X_2 independent?

8.16 Let the bivariate random variable (X_1, X_2) have the joint density function

$$f_{X_1X_2}(x_1, x_2) = \begin{cases} x_1^2 + \dfrac{x_1x_2}{3} & 0 < x_1 < 1, \quad 0 < x_2 < 2 \\ 0 & \text{otherwise} \end{cases}$$

(a) Determine $P[X_1 > \frac{1}{2}]$.
(b) Compute $P[X_1 + X_2 \leq 1]$.
(c) Compute $P[X_1^2 + X_2^2 \geq \frac{1}{2}]$.
(d) Determine $f_{X_1}(x_1)$ and $f_{X_2}(x_2)$. Are X_1 and X_2 independent random variables?

8.17 Let X_1 and X_2 be uniformly distributed over the square with vertices (2, 0), (1, 1), (0, 0), and (1, −1).
(a) Determine $f_{X_1}(x_1)$ and $f_{X_2}(x_2)$.
(b) Compute $P[X_1/X_2 \geq 1]$.
(c) Compute $P[X_1 + 2X_2 \leq \frac{1}{2}]$.

8.18 Let (X_1, X_2) be a bivariate random variable with joint density function

$$f_{X_1X_2}(x_1, x_2) = \begin{cases} 12x_1x_2(1 - x_2) & 0 < x_1 < 1, 0 < x_2 < 1 \\ 0 & \text{otherwise.} \end{cases}$$

(a) Determine $P[X_1 - 2X_2 > \frac{1}{4}]$ and $P[X_2/2X_1 < 1]$.
(b) Evaluate the marginal density functions.

8.19 Let the bivariate random variable (X_1, X_2) have the density function

$$f_{X_1X_2}(x_1, x_2) = \begin{cases} e^{-x_2} & 0 < x_1 < x_2 < \infty \\ 0 & \text{otherwise} \end{cases}$$

(a) Determine the marginal density functions.
(b) Compute $P[X_2 \geq 1]$ and $P[X_1 \leq 1]$.
(c) Compute $P[X_1^2 + X_2^2 \geq 1]$.

8.20 Let (X_1, X_2) be a bivariate random variable with joint density function

$$f_{X_1X_2}(x_1, x_2) = \begin{cases} \dfrac{2}{\pi} & 0 \leq x_1^2 + x_2^2 \leq 1 \quad x_2 \geq 0 \\ 0 & \text{otherwise} \end{cases}$$

(a) If $Y = X_1^2 + X_2^2$, determine $f_Y(y)$.
(b) Compute $P[Y \leq 5]$.

8.21 A switch to a control system is turned on at a random time X_1 during the day so that X_1 is uniformly distributed on (0, 24). At a later time, the switch is turned off at random time X_2 between X_1 and 24.

(a) Compute $f_{X_2}(x_2)$ and $f_{X_1X_2}(x_1, x_2)$

(b) Determine the probability that the switch is off at 5 P.M.

8.22 Let the random variable Z have density function

$$f_Z(z) = (0.2 - 0.02z)[u(z) - u(z - 10)]$$

If the random variables X_1 and X_2 correspond to the integer and fractional parts of Z, compute the joint density function $f_{X_1X_2}(x_1, x_2)$.

8.23 Let the multivariate random variable $(X_1, X_2, \ldots, X_n)$ have the joint density function

$$f_{X_1 \cdots X_n}(x_1, x_2, \ldots, x_n) = n! \qquad \text{for } x_1, x_2, \ldots, x_n \geq 0 \text{ and } 0 \leq x_1 + x_2 + \cdots + x_n \leq 1$$

and is zero otherwise. Show that the marginal density functions are given by

$$f_{X_1 \cdots X_{n-k}}(x_1, x_2, \ldots, x_{n-k}) = \frac{n!}{k!}(1 - x_1 - \cdots - x_{n-k})^k$$

for

$$x_1, x_2, \ldots, x_{n-k} \geq 0 \quad \text{and} \quad 0 \leq x_1 + x_2 + \cdots + x_{n-k} \leq 1$$

for $k = 1, 2, \ldots, n - 1$. *Hint:* Find $f_{X_1 \cdots X_{n-1}}(x_1, x_2, \ldots, x_{n-1})$ and then use an induction proof.

8.24 Let the multivariate random variable $(X_1, X_2, \ldots, X_n)$ have the joint density function

$$f_{X_1 \ldots X_5}(x_1, x_2, \ldots, x_5) = \begin{cases} x_1x_2x_3x_4e^{-(x_5x_4+x_4x_3+x_3x_2+x_2x_1+x_1)} & \text{for } x_1, x_2, \ldots, x_5 \geq 0 \\ 0 & \text{otherwise} \end{cases}$$

Determine the marginal density function $f_{X_2}(x_2)$.

CHAPTER 9

Estimation Theory: Basic Concepts

9.1 INTRODUCTION

A fundamental objective in signal processing is that of making inferences concerning experimentally obtained data. This might take such varied forms as the reconstruction of the time behavior of a signal from noise-contaminated data (filtering), the determination of whether a prescribed signal is present in a noisy environment (detection), or the forecasting of the future behavior of a signal (prediction). Independent of the specific application at hand, the ability to make informed decisions based on probabilistic considerations is central to signal-processing theory. This is in recognition of the fact that most measured data are random in nature. With this in mind, some fundamental estimation theory concepts are introduced in this chapter which are relevant to many standard signal-processing problems. Future chapters build on this foundation to establish a framework for solving a general class of signal-processing problems. Central to this development is the notion of the *correlation* existent between pairs of random variables. It will be made apparent that the correlation measure constitutes the most important probabilist characteristic of random data insofar as linear signal-processing algorithms is considered.

9.2

CORRELATION AND COVARIANCE

Our ultimate objective is that of developing signal-processing algorithms for analyzing data that have a random nature. If these algorithms are to be useful in a practical setting, it is absolutely essential that their synthesis be dependent on a minimal amount of probabilistic knowledge concerning the data being processed. This is in recognition of the fact that in most applications we know very little about the underlying data's probabilistic behavior. For instance, the joint density functions associated with random data are almost never known.

With this thought in mind, it is essential that we use signal-processing algorithms that require relatively primitive probabilistic measures. As we will see shortly, this requirement is satisfied if the algorithms being employed are linear and if a mean-squared-error criterion is used to measure performance. Under these restrictions it is found that the only prerequisite probabilistic information needed is the expected value of the product of random variable pairs. This represents a rather modest level of probabilistic information which is generally accessible or is readily estimated in most practical applications. Let us now examine some of the more salient properties of the expected value of random variable products.

The *correlation* existent between the real-valued random variables X and Y is defined to be equal to the expected value of their product, that is,

$$\begin{aligned} r_{xy} &= E\{XY\} \\ &= \int_{-\infty}^{\infty}\int_{-\infty}^{\infty} xyf_{XY}(x, y)\, dx\, dy \end{aligned} \tag{9.1}$$

Clearly, we are always able to compute this correlation scalar if the underlying joint density function is known. The converse is generally not true, however, which simply reinforces the fact that correlation represents a (significantly) lower level of probabilistic knowledge concerning the random variables X and Y. In a similar fashion, the *covariance* that exists between these real-valued bivariate random variables is defined by

$$\begin{aligned} c_{xy} &= E\{[X - \mu_x][Y - \mu_y]\} \\ &= \int_{-\infty}^{\infty}\int_{-\infty}^{\infty} (x - \mu_x)(y - \mu_y)f_{XY}(x, y)\, dx\, dy \end{aligned} \tag{9.2}$$

where μ_x and μ_y denote the mean values of the random variables X and Y, respectively. The correlation and covariance measures provide similar information concerning the random variables X and Y. This is made evident upon carrying out the multiplications inside the expected-value operator brackets defining the covariance. This is found to give

$$\begin{aligned} c_{xy} &= E\{XY - X\mu_y - \mu_x Y + \mu_x\mu_y\} \\ &= r_{xy} - \mu_x\mu_y \end{aligned} \tag{9.3}$$

where use has been made of the linearity of the expected-value operator. It is to be noted that if the mean value of at least one of the random variables X or Y is zero, then the correlation and covariance measures are identical.

EXAMPLE 9.1

Let the random variables X and Y have the joint density function

$$f_{XY}(x, y) = \begin{cases} 2 & 0 \leq y \leq x \leq 1 \\ 0 & \text{otherwise} \end{cases}$$

The mean values, correlation, and covariance measures associated with these random variables are given by

$$\mu_x = \int_0^1 \int_y^1 2x \, dx \, dy = \tfrac{2}{3} \qquad \mu_y = \int_0^1 \int_y^1 2y \, dx \, dy = \tfrac{1}{3}$$

$$r_{xy} = \int_0^1 \int_y^1 2xy \, dx \, dy = \tfrac{1}{4} \quad c_{xy} = \int_0^1 \int_y^1 2(x - \tfrac{2}{3})(y - \tfrac{1}{3}) \, dx \, dy = \tfrac{1}{36}$$

It is to be noted that $c_{xy} = r_{xy} - \mu_x\mu_y$, as required by expression (9.3). ■

Correlation Estimate

Correlation provides a measure of probabilistic association existent between the two random variables X and Y. It is possible to estimate the correlation r_{xy} empirically by conducting a large number of trials of the underlying experiment. This will result in the recorded outcomes (x_k, y_k) for $k = 1, 2, \ldots, N$. Since the correlation is equal to the expected value of the random variable product XY, an intuitively appealing estimate (informed guess) of the correlation would take the form

$$\hat{r}_{xy} = \frac{1}{N} \sum_{k=1}^{N} x_k y_k \tag{9.4}$$

The caret (^) is used to indicate that $\hat{r}_{xy}$ is an estimate of the underlying correlation r_{xy}. Under certain restrictions, it can be shown that this estimate approaches the desired value r_{xy} as the number of experimental trials N becomes unbounded.

It is convenient to depict the general nature of the correlation by means of a scatter diagram. A *scatter diagram* is simply a plot of the recorded outcomes (x_k, y_k) for $1 \leq k \leq N$ made in the (x, y) plane. Typical scatter diagrams might

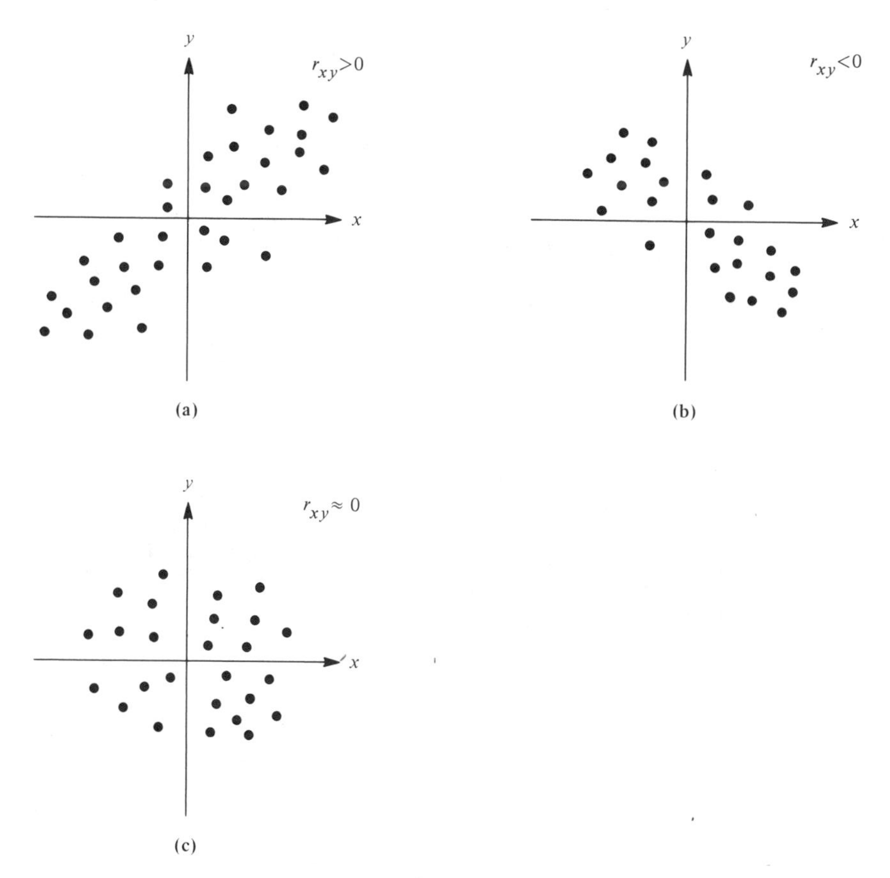

FIGURE 9.1. Scatter diagrams.

appear as shown in Figure 9.1. If these plotted outcomes show a strong tendency to lie in either the first or third quadrants, as illustrated in Figure 9.1a, this suggests a positive-valued correlation. Similarly, a negative-valued correlation is indicated whenever a large majority of the points lie within the second and fourth quadrants, as shown in Figure 9.1b. If neither of these tendencies is evident, as depicted in Figure 9.1c, the underlying correlation will tend to be near zero.

9.3 BOUNDS ON THE CORRELATION AND COVARIANCE

In our analysis of random data, it is useful to provide a bounding on the expected value of random variable products. The Schwarz inequality provides a particularly convenient means for achieving this objective.

Theorem 9.1 (Schwarz Inequality). Let X and Y be real-valued random variables with finite second moments so that $E\{X^2\} < \infty$ and

$E\{Y^2\} < \infty$. It then follows that the associated correlation and covariance measures are bounded according to

$$r_{xy}^2 = [E\{XY\}]^2 \leq E\{X^2\}E\{Y^2\} \tag{9.5}$$

and

$$c_{xy}^2 = [E\{[X - \mu_x][Y - \mu_y]\}]^2 \leq \sigma_x^2\sigma_y^2 \tag{9.6}$$

where μ_x, μ_y, σ_x^2, and σ_y^2 correspond to the mean values and variances of the X and Y random variables. Furthermore, equality holds in these relationships if and only if X and Y are linearly related as

$$Y = aX \tag{9.7}$$

where a is a scalar.

This theorem is straightforwardly proved by considering the nonnegative-valued random variable $[Y - aX]^2$, where a is an arbitrary real-valued scalar. Upon taking the expected values of this nonnegative real random variable, we have

$$\begin{aligned} 0 &\leq E\{[Y - aX]^2\} \\ &= E\{Y^2\} - 2aE\{XY\} + a^2E\{X^2\} \end{aligned}$$

The right-hand side of this inequality is always nonnegative for all choices of the scalar a. It achieves its minimum value for the selection $a^o = E\{XY\}/E\{X^2\}$, as is readily verified by using standard calculus methods. Inserting this minimizing choice for a into the inequality above results in

$$0 \leq E\{Y^2\} - \frac{E^2\{XY\}}{E\{X^2\}}$$

from which inequality (9.5) follows. It is a simple matter to show that equality holds if and only if X and Y are linearly related as in expression (9.7). To prove the second inequality (9.6), we simply consider the random variables $W = X - \mu_x$ and $Z = Y - \mu_y$ and employ inequality (9.5) to put a bound on $E\{WZ\}$.

Correlation Coefficient

The covariance c_{xy} is defined to be equal to the expected value of the product of the random variables $X - \mu_x$ and $Y - \mu_y$. The value assumed by this measure is seen to be dependent on the magnitudes that the individual random variables X and Y themselves assume. If we are to use the covariance measure to make meaningful inferences concerning any mutual probabilistic association that exists between random variable pairs, however, it is essential that this magnitude dependency be removed. This normalization is readily achieved by introducing the related *correlation coefficient* as defined by

$$\rho_{xy} = \frac{c_{xy}}{\sigma_x \sigma_y} \tag{9.8}$$

where σ_x^2 and σ_y^2 denote the variances associated with the random variables X and Y, respectively. The Schwarz inequality (9.6) then yields the following characterization of the correlation coefficient.

Theorem 9.2. Let the real-valued random variables X and Y possess finite second-order moments. It then follows that their associated correlation coefficient satisfies

$$-1 \leq \rho_{xy} \leq 1 \tag{9.9}$$

Moreover, the conditions $\rho_{xy} = \pm 1$ occur with probability 1 provided that X and Y are linearly related as $Y = aX + b$. Furthermore, the correlation coefficient of the linearly related random variables $W = aX + b$ and $Z = cY + d$ is specified by $\rho_{wz} = \rho_{xy}$ for all choice of the constants a, b, c, and d.

From this theorem it is seen that the correlation coefficient takes on values exclusively in the interval $[-1, 1]$. If it should assume a value close to ± 1, there is a strong presumption of a near linear relationship existent between the random variables X and Y. On the other hand, a value near zero would suggest the lack of any meaningful linear relation. Thus the behavior of the correlation coefficient can serve as a useful tool insofar as justifying the use of linear models. With these thoughts in mind, we examine the important topic of linear models.

9.4 LINEARLY RELATED BIVARIATE RANDOM VARIABLES

In signal-processing applications, one of the primary objectives is that of extracting useful information from empirically obtained data. To gain a feeling for how this is accomplished, let us consider an experiment that has an associated real-valued random variable pair (X, Y). Owing to the nature of this experiment, however, we are able to observe only the outcome of the univariate random variable X, but not its companion univariate random variable Y. This situation is depicted in Figure 9.2, where the values assumed by the random variables are $X = x$ and $Y = y$. The problem to be addressed is that of using the observation $X = x$ to make an *informed guess* as to the value that was assumed by Y the unobserved companion random variable (i.e., y). From a practical viewpoint, it is desirable that a solution to this problem entail only a minimal

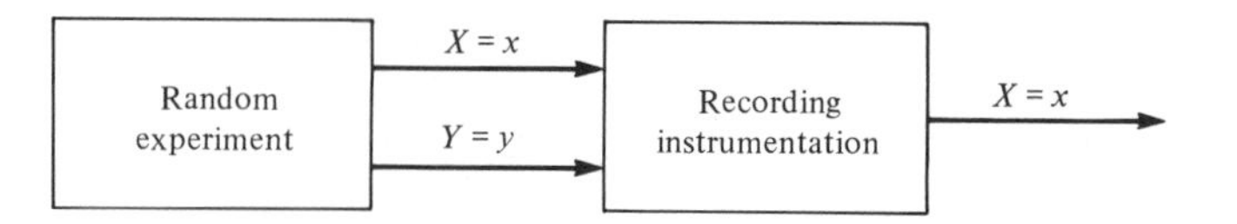

FIGURE 9.2. Random experiment under consideration.

amount of probabilistic preknowledge concerning the bivariate random variables.

With regard to the latter point, a considerable simplification arises in the artificial situation in which the bivariate random variables are perfectly linearly related as

$$Y = aX + b \tag{9.10}$$

where the real-valued parameters a and b are taken as known. Of necessity, any experimental outcome arising in such a situation must lie on the line depicted in Figure 9.3. Clearly, given the experimental realization $X = x$, it follows directly that the unobserved companion random variable Y must take on the value $y = ax + b$. Thus, when two bivariate random variables are related in a known linear fashion, knowledge of the value assumed by either component is sufficient to determine the value assumed by the other component. The true significance of a linear model, however, is manifested in the rather modest statistical information needed in its formulation. This is important when considering the relevant case in which X and Y are not linearly related (see Section 9.5).

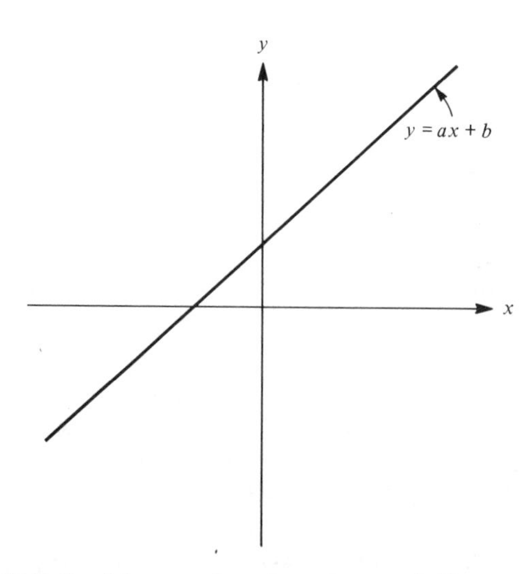

FIGURE 9.3. Linear related random variables.

Scatter Diagrams and Linear Models

Although the imposition of a linear model offers desirable benefits from an analysis viewpoint, the question as to the adequacy of such a model must be addressed. To test empirically for the appropriateness of a linear model, we might conduct a large number of experimental trials which result in the observed outcome pairs (x_k, y_k) for $k = 1, 2, \ldots, N$. To test for linearity, these N outcomes can be plotted in the (x, y) plane to form the associated *scatter diagram*. If these outcomes all lie on one line, it is strongly suggested that the random variables X and Y are perfectly linearly related, as in expression (9.10). In most applications, however, this ideal behavior does not follow. More typically, the N points will either be closely grouped about some line, as shown in Figure 9.4a and b, or be widely dispersed with no apparent linear trend, as shown in Figure 9.4c. Depending on the nature of the scatter diagram, we are able to judge the prudence of invoking a linear model approximation.

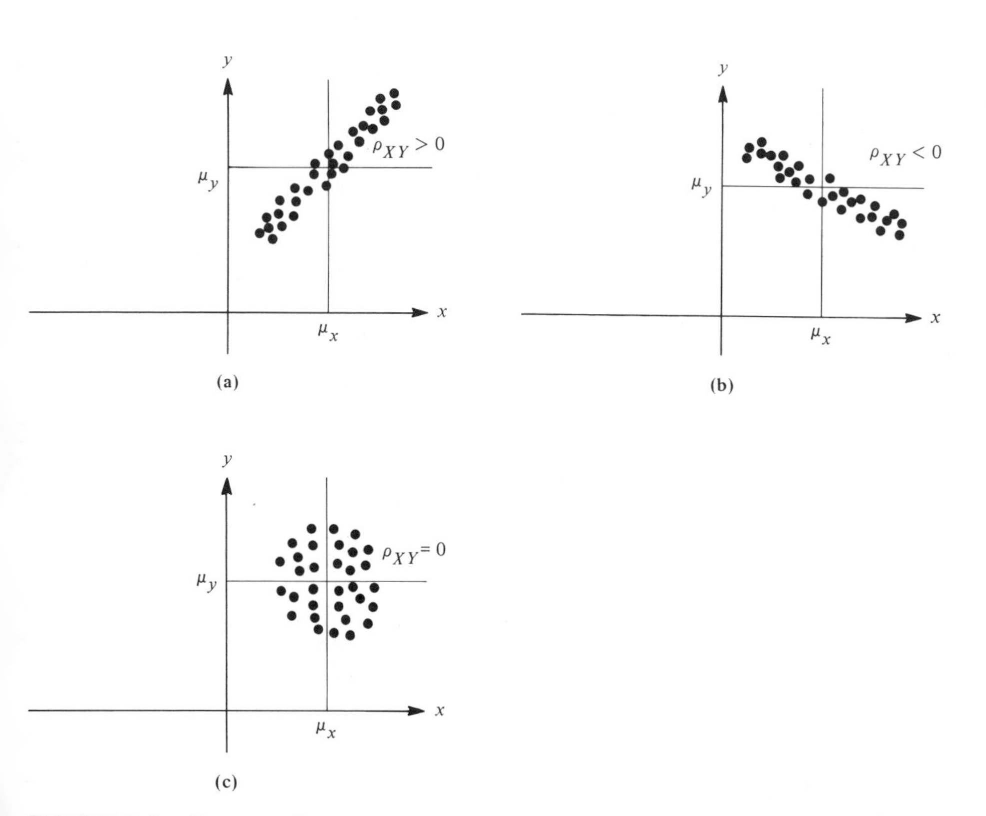

FIGURE 9.4. Scatter diagrams corresponding to a random variable pair for which a linear fit is appropriate, as in parts (a) and (b), or inappropriate, as in part (c).

9.5 OPTIMAL LINEAR ESTIMATOR: BIVARIATE RANDOM VARIABLES

It often happens that a linear model provides a reasonably adequate representation for a set of real-valued bivariate random variables. Even though (X, Y) may not be perfectly linearly related, there may be much to be gained in terms of modeling simplicity by approximating Y through the linear relationship

$$\hat{Y} = aX + b \tag{9.11}$$

In this approximation it is desired to select the parameters a and b characterizing this relationship so that the *linear estimator* $\hat{Y}$ best represents Y in some meaningful sense. This will entail analyzing the associated *error random variable* $Y - \hat{Y}$. To effect a tractable analysis, it is beneficial to use the standard *mean-squared-error* (MSE) *criterion,*

$$E\{[Y - \hat{Y}]^2\} = E\{[Y - aX - b]^2\} \tag{9.12}$$

to provide a basis for judging the goodness of the linear representation above. If this nonnegative criterion can be made suitably close to zero for appropriate choices of the parameters a and b, the linear estimator $\hat{Y}$ is said to be a good representation for Y. If this is not the case, however, the choice of a linear model can be of questionable value.

Upon carrying out the squaring required in expression (9.12) and using the expected-value operator's linearity, the MSE criterion is equivalently expressed as

$$E\{[Y - \hat{Y}]^2\} = E\{Y^2\} + a^2E\{X^2) + b^2 - 2aE\{XY\} - 2bE\{Y\} + 2abE\{X\} \tag{9.13}$$

Our objective is then that of selecting values for the parameters a and b which characterize the postulated linear relationship (9.11) so as to render this MSE criterion a minimum. This particular parameter selection is said to give rise to the *optimum linear estimator* of the random variable Y. To obtain these optimum parameters, we simply set to zero the partial derivatives of the MSE criterion with respect to the a and b parameters. Upon solving the resultant set of two equations in the two unknowns a^o and b^o, the optimum parameters are readily found and give rise to the following theorem.

Theorem 9.3. Let X and Y be real-valued random variables that describe a random experiment. The optimum linear estimator

$$\hat{Y} = a^oX + b^o \tag{9.14}$$

of the random variable Y in the sense of minimizing the MSE criterion $E\{[Y - \hat{Y}]^2\}$ has its parameters specified by

$$a^o = \frac{c_{xy}}{\sigma_x^2} \quad \text{and} \quad b^o = \mu_y - a^o\mu_x \tag{9.15}$$

where $\mu_x = E\{X\}$, $\mu_y = E\{Y\}$, $\sigma_x^2 = \text{var}\{X\}$, $\sigma_y^2 = \text{var}\{Y\}$, and $c_{xy} = E\{[X - \mu_x][Y - \mu_y]\}$. Moreover, the corresponding minimum MSE is given by

$$\begin{aligned} E\{[Y - a^oX - b^o]^2\} &= \sigma_y^2 - \frac{c_{xy}^2}{\sigma_x^2} \\ &= (1 - \rho_{xy}^2)\sigma_y^2 \end{aligned} \tag{9.16}$$

The details of proof for this theorem are straightforward and are left to the reader as an exercise. As suggested earlier, the optimal linear estimator is seen to be completely characterized by the mean values, variances, and covariances of the two random variables being modeled. Although we have considered here only linear models between *two* random variables, a similar statement will be shown to hold for linear models of multivariate random variables.

The results of Theorem 9.3 provide useful insights into some important aspects of estimation theory. In particular, it has been argued previously that the mean value μ_y provides a reasonably *good* estimate (an informed guess) of the value that the random variable Y will assume on any realization (trial) of the underlying random experiment. The MSE associated with this estimate was found to equal the variance of Y (i.e., σ_y^2). Once the experiment has been conducted, however, and the value assumed by the companion bivariate random variable X observed (i.e., $X = x$), the optimal MSE estimate of the unobserved random variable Y as given by

$$\hat{y} = a^o x + b^o \tag{9.17}$$

is a better estimate than the unconditioned estimate μ_y. This follows directly from expression (9.16), where the MSE error associated with this new estimate was found to be $(1 - \rho_{xy}^2)\sigma_y^2$. Since the correlation coefficient lies in the interval $-1 \leq \rho_{xy} \leq 1$, this MSE error is smaller that the MSE error, σ_y^2, which is associated with the mean value estimate μ_y, provided that $\rho_{xy} \neq 0$. Moreover, if $\rho_{xy} = \pm 1$, the MSE error associated with the new liner estimate is identically zero. This simply reflects the point made previously that $\rho_{xy} = \pm 1$ if X and Y are perfectly linearly related.

It is to be noted that relationship (9.11) is referred to as a linear *estimator* while relationship (9.17) is called a linear *estimate*. The terminology estimator and estimate will hereafter be used to distinguish between the random variable description (before the observations are made) of an estimation scheme and its numerical realization (after the observations are made), respectively.

EXAMPLE 9.2

Let the random variables X and Y have the joint density function

$$f_{XY}(x, y) = \begin{cases} 2 & 0 \leq y \leq x \leq 1 \\ 0 & \text{otherwise} \end{cases}$$

Determine the optimal linear estimate of Y based on X. From this density function, it is clear that the random variables X and Y are not linearly related (i.e., the unit mass is not distributed along a line). Nonetheless, it is often desired to find the best approximate linear model. Using the results obtained in Example 9.1, it follows that the required optimal linear estimator has its parameters specified by relationship (9.15), that is,

$$a^{\text{o}} = 0.5 \qquad \text{and} \qquad b^{\text{o}} = 0$$

Thus the optimal linear estimator of random variable Y is given by

$$\hat{Y} = 0.5X$$

and the corresponding minimum MSE (9.16) is found to be

$$E\{[Y - \tfrac{1}{2}X]^2\} = \tfrac{1}{24}$$

The optimal linear estimate $\hat{y}_1 = \frac{1}{2}x$ and the standard mean value estimate $\hat{y}_2 = u_y = \frac{1}{3}$ are depicted in Figure 9.5. It is noted that for these *uniformly*

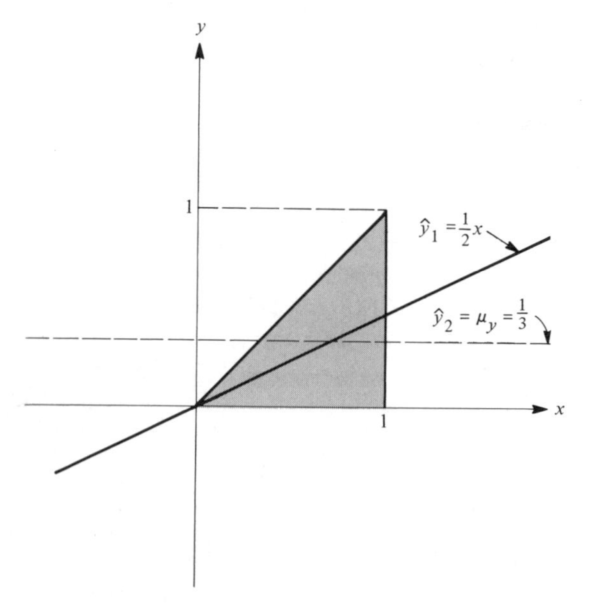

FIGURE 9.5. Optimal linear estimate and mean value estimate for Example 9.2.

distributed random variables, the optimal linear estimate corresponds to the center of gravity of the mass lying along the vertical line through point x. This very property ensures that the MSE is minimum. ■

9.6 UNCORRELATED AND ORTHOGONAL RANDOM VARIABLES

As we shall shortly see, Theorem 9.3 indicates that the use of a linear estimate for random variable pairs is questionable when the correlation coefficient is at or near zero. To emphasize the importance attached to this condition, the notion of uncorrelated random variables is now introduced.

Definition 9.1. The real-valued random variables X and Y are said to be *uncorrelated* if their covariance is zero, that is,

$$c_{xy} = E\{[X - \mu_x][Y - \mu_y]\} = 0 \tag{9.18}$$

Furthermore, these random variables are said to be *orthogonal* if their correlation is zero, that is,

$$r_{xy} = E\{XY\} = 0 \tag{9.19}$$

If X and Y are uncorrelated random variables, Theorem 9.3 shows that the optimal linear estimator of Y based on X is specified by $Y = \mu_y$. In other words, knowledge of the value assumed by X conveys no additional information concerning Y insofar as linear relationships are concerned. This is a most remarkable but simple result and its implications should be carefully digested and understood by the reader.

If the random variables X and Y are statistically independent, it follows that $c_{xy} = E\{X - \mu_x\}E\{Y - \mu_y\} = 0$. Thus independent random variable are *always uncorrelated*. This is not surprising, since by the very notion of independence, it is meant that X and Y share no common probabilistic association (linear or nonlinear). It is therefore natural that such random variables must be uncorrelated. One is cautioned, however, not to presume that uncorrelated random variables must always be independent. This is certainly not the case, as is readily demonstrated by a simple counterexample (e.g., see Problem 9.8). Independence implies that no associational probabilistic relationship whatsoever exists between X and Y, while uncorrelatedness implies only that no *linear* associational relationship exists.

9.7 OPTIMAL ESTIMATOR: BIVARIATE RANDOM VARIABLES

Although linear estimators are useful in many applications, they may prove unsatisfactory in certain situations (e.g., uncorrelated random variables). In this section we generalize the concepts made up to this point so as to consider optimal *nonlinear* estimators. As before, we consider an experiment whose outcome is described by the real-valued bivariate random variable (X, Y). Moreover, it is assumed that after a realization of the underlying experiment is made, we are able to observe the value assumed by X (i.e., x) but that the value of Y remains unobserved. Since these two random variables are interrelated in a probabilistic manner through their joint density function, this suggests that knowledge of the observation $X = x$ may convey useful statistical information concerning its unobserved companion random variable Y. We hope that this new information will enable us to make a more informed estimate of the value assumed by Y.

With these thoughts in mind, we now seek a functional relationship for Y in terms of X that takes into account the underlying joint density function. In a formal sense, we seek an estimator of Y based on X that takes the functional form

$$\hat{Y} = \phi(X) \tag{9.20}$$

where $\phi(\cdot)$ represents some yet to be determined *estimator function*. The terminology estimator is here being used to emphasize the fact that the function $\phi(X)$ represents an estimate of Y conditioned on the knowledge of X. Our objective is to find that specific estimator function which best represents the unobserved random variable Y. In accordance with previous reasoning, the mean-squared-error criterion

$$E\{[Y - \phi(X)]^2\} = \int_{-\infty}^{\infty} \int_{-\infty}^{\infty} [y - \phi(x)]^2 f_{XY}(x, y)\, dx\, dy \tag{9.21}$$

is used to measure the goodness of this representation. The smaller this MSE criterion is made, the better the estimate $\hat{Y} = \phi(X)$ is said to represent Y. In contrast with Section 9.5, where the estimator function was restricted to be linear, we here place no such restriction on the function $\phi(x)$. As such, this more general estimator function typically provides better estimates than does its restricted linear counterpart.

The task at hand then is to find that estimator function ϕ which solves the minimization problem

$$\min_{\phi} E\{[Y - \phi(X)]^2\}$$

where the minimization is to be taken over all possible selections of candidate estimator functions ϕ. To determine that optimal estimator function ϕ^{o} which solves this minimization problem, we make use of a standard approach taken in the *calculus of variations*. This entails expressing a candidate estimator function in the form

$$\phi(x) = \phi^o(x) + \epsilon\eta(x) \tag{9.22}$$

where $\phi^o(x)$ is the optimum estimator function that is being sought, $\eta(x)$ is a *perturbation function,* and ϵ is an arbitrary real-valued scalar. All candidate estimator functions can be decomposed in this manner. By fixing the perturbation function $\eta(x)$ and letting ϵ vary, a class of single-parameter estimator functions is thereby constructed. If the candidate estimator function (9.22) is substituted into the MSE expression (9.21), we obtain

$$E\{[Y - \phi(X)]^2\} = \int_{-\infty}^{\infty}\int_{-\infty}^{\infty} [y - \phi^o(x) - \epsilon\eta(x)]^2 f_{XY}(x, y)\, dx\, dy$$

Upon examination of the right side of this relationship, it is seen that once the integration is carried out, this MSE criterion is a quadratic function of the scalar ϵ. As the scalar ϵ is varied between plus and minus infinity, this MSE takes on its absolute minimum value at $\epsilon = 0$, due to the hypothesized optimality of ϕ^o. It therefore follows that upon differentiating this MSE relationship with respect to ϵ and evaluating that derivative at $\epsilon = 0$, the resultant derivative must equal zero. The required derivative is obtained by interchanging the order of the integration and differentiation operations to give

$$\frac{\partial E\{[Y - \phi(X)]^2\}}{\partial\epsilon} = -2\int_{-\infty}^{\infty}\int_{-\infty}^{\infty} [y - \phi^o(x) - \epsilon\eta(x)]\eta(x) f_{XY}(x, y)\, dx\, dy$$

Evaluating this derivative at $\epsilon = 0$ and setting the result to zero as described above gives, after rearrangement of terms,

$$\int_{-\infty}^{\infty}\left[\int_{-\infty}^{\infty} [y - \phi^o(x)] f_{XY}(x, y)\, dy\right]\eta(x)\, dx = 0 \tag{9.23}$$

This homogeneous integral relationship represents the necessary condition that the optimal estimator function must satisfy to minimize the MSE criterion. Since this relation must hold for all choices of the perturbation function $\eta(x)$, it follows that the term in brackets must be zero for all values of x, that is,

$$\int_{-\infty}^{\infty} [y - \phi^o(x)] f_{XY}(x, y)\, dy = 0 \tag{9.24}$$

If this were not the case, the necessary condition for optimality would be violated by selecting $\eta(x)$ equal to the function of x contained within the braces in expression (9.23). Thus expression (9.24) constitutes a necessary condition for an optimal estimator selection. In carrying out the indicated integration in expression (9.24), it is first noted that the term $\phi^o(x)$ is not a function of the integration variable y and may be therefore taken outside the integral. Upon using this fact and the marginal density function identity (8.41), the closed-form expression for the optimal estimator function is found to be

$$\phi^o(x) = \frac{1}{f_X(x)} \int_{-\infty}^{\infty} y f_{XY}(x, y)\, dy \qquad (9.2$$

where it is tacitly assumed that $f_X(x) \neq 0$. We may arbitrarily set $\phi^o(x) =$ for those values of x in which $f_X(x) = 0$.

Relationship (9.25) constitutes a straightforward procedure for determinir that value of y [i.e., $\hat{y} = \phi^o(x)$] which best represents the random variable Y the minimum MSE sense given that the random variable X has been observe to equal x. The optimal estimator function (9.25) is seen to require knowledg of the underlying joint density function. Unfortunately, this knowledge is ty ically unavailable in most applications. Finally, the associated optimal cond tional estimator $\phi^o(X)$ is simply obtained by replacing x by X everywhere appears in expression $\phi^o(x)$.

EXAMPLE 9.3

Let us consider the bivariate random variable (X, Y), which is distribute inside the shaded region shown in Figure 9.6 according to

$$f_{XY}(x, y) = \tfrac{15}{4} x(1 - y^2) \qquad 0 \leq y \leq x \leq 1$$

To determine the optimal estimator function associated with this joint densit function, we employ expression (9.25). This entails first carrying out th intermediate evaluations

$$f_X(x) = \int_0^x f_{XY}(x, y)\, dy = \begin{cases} \tfrac{5}{4}x^2(3 - x^2) & 0 \leq x \leq 1 \\ 0 & \text{otherwise} \end{cases}$$

and

$$\int_0^x y f_{XY}(x, y)\, dy = \begin{cases} \tfrac{15}{16}x^3(2 - x^2) & 0 \leq x \leq 1 \\ 0 & \text{otherwise} \end{cases}$$

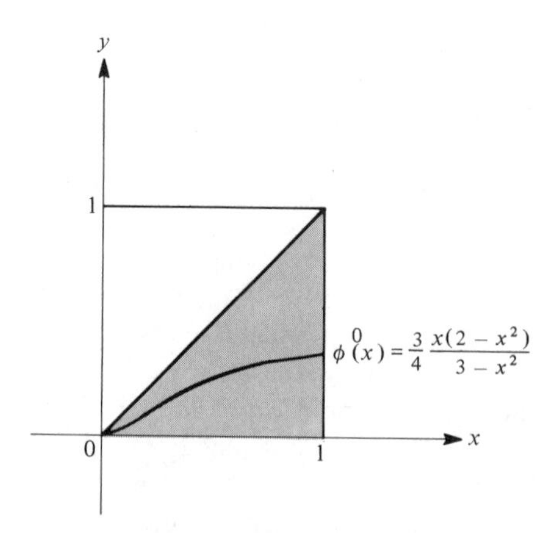

FIGURE 9.6. Conditional expectation function.

Thus the optimal estimator function becomes

$$\phi^o(x) = \begin{cases} \dfrac{3}{4}\dfrac{x(2 - x^2)}{3 - x^2} & 0 \le x \le 1 \\ 0 & \text{otherwise} \end{cases}$$

which corresponds to the curve within the shaded region shown in Figure 9.6. Interchanging the roles X and Y, it can be shown that the optimal estimator of X, given that $Y = y$, is

$$\phi^o(y) = \frac{2}{3}\frac{1 + y + y^2}{1 + y}$$

■

9.8 CONDITIONAL DENSITY FUNCTION

We shall now give an interpretation of the optimal estimator function just described which provides a useful insight into estimation theory. This entails first taking the term $1/f_X(x)$ inside the integral of expression (9.25) and then defining the so-called *conditional density function*

$$f_Y(y|x) = \frac{f_{XY}(x, y)}{f_X(x)} \tag{9.26}$$

This conditional density function is here taken to be dependent on the variable y, with x being considered as a fixed parameter (i.e., the observation $X = x$). We have here used the symbol $f_Y(y|x)$ to distinguish the conditioned density function from its unconditioned counterpart $f_Y(y)$. The behavior of the conditional density function as a function of y is seen to correspond to the intersection of the surface $f_{XY}(x, y)$ with the plane parallel to the y axis in three-dimensional space as depicted in Figure 9.7. The conditional density function possesses all the features normally associated with a standard density function in that it is nonnegative and has unit area, that is,

$$f_Y(y|x) \ge 0 \qquad \int_{-\infty}^{\infty} f_Y(y|x)\, dy = 1 \tag{9.27}$$

It is to be noted that division of the joint density function $f_{XY}(x, y)$ by the marginal density function $f_X(x)$ provides the required normalization needed so that $f_Y(y|x)$ has unity area. This probabilistic interpretation can be formally described by considering the so-called conditional distribution function, which has as its derivative the conditional density function (9.26).† The reader is

†The conditional distribution function (conditioned on $X = x$) is defined to be

$$F_Y(y|x) = \lim_{\Delta \to 0} \frac{P[x \le X \le x + \Delta,\, Y \le y]}{P[x \le X \le x + \Delta]}$$

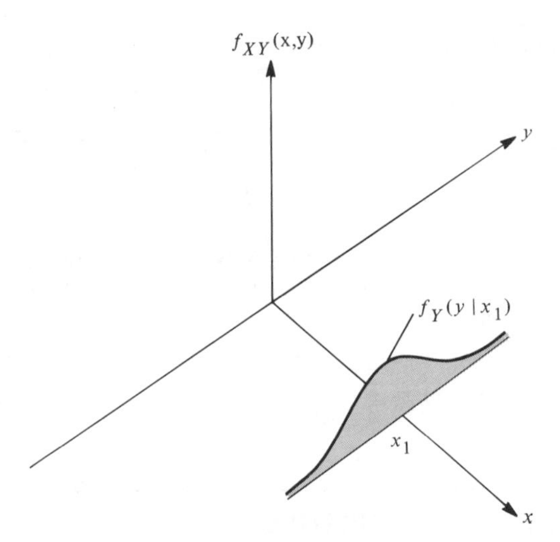

FIGURE 9.7. Conditional density function $f_y(y \mid x_1)$ depicted as a scalar version of the intersection of the joint density function with the plane $x = x_1$.

encouraged to consult the probability literature for this alternative derivation.

From the remarks above, it is apparent that the conditional density function (9.26) represents the marginal density function description of the random variable Y conditioned on the observation that $X = x$. It therefore follows that conditional probabilities such as

$$P[y_1 < Y \leq y_2 | X = x] = \int_{y_1}^{y_2} f_Y(y|x)\, dy \tag{9.28}$$

may be readily computed once knowledge of the conditional density function is available. As a matter of fact, virtually all of the characteristics associated with standard random variables carry over to conditioned random variables (e.g., means, variances, moments, expectations, etc.).

In the development of conditional experiments above, it should be apparent that the roles played by the random variables X and Y can be interchanged. Thus the conditional density function

$$f_X(x|y) = \frac{f_{XY}(x, y)}{f_Y(y)} \tag{9.29}$$

would characterize the random variable X in which the observation (realization) $Y = y$ is available. A simple manipulation of this expression with relationship (9.26) is found to result in the *Bayes* relationship

$$f_Y(y|x) = \frac{f_X(x|y) f_Y(y)}{f_X(x)} \tag{9.30}$$

This Bayes conditional density function relationship plays an important role in estimation theory.

9.9

OPTIMAL ESTIMATOR AND CONDITIONAL EXPECTATION

Using the interpretation of the conditional density function given above, we obtain the expected value of the random variable Y conditioned on the observation $X = x$:

$$E\{Y|X = x\} = \int_{-\infty}^{\infty} y f_Y(y|x)\, dy \tag{9.31}$$

This *conditional expectation* gives the average value that the random variable Y takes on in a large number of experiments in which $X = x$. If this conditional expectation is compared with the optimal conditional estimation function (9.25), it is found that they are identical. With this in mind, we now summarize the results pertaining to the important optimal estimation problem considered in Section 9.7.

Theorem 9.4. Let (X, Y) be a real-valued bivariate random variable associated with an experiment for which the value realized by the random variable X is observed to be x. The optimal estimator function $\phi^o(X)$ that best represents the unobserved random variable member Y in the sense of minimizing the MSE criterion

$$E\{[Y - \phi(X)]^2\} = \int_{-\infty}^{\infty} [y - \phi(x)]^2 f_{XY}(x, y)\, dx\, dy \tag{9.32}$$

is specified by the conditional expected value

$$\begin{aligned} \phi^o(x) &= E\{Y|X = x\} \\ &= \int_{-\infty}^{\infty} y f_Y(y|x)\, dy \end{aligned} \tag{9.33}$$

Moreover, the minimal error is *orthogonal* to the optimal estimator function, that is,

$$E\{[Y - \phi^o(X)]\phi^o(X)\} = 0 \tag{9.34}$$

and the corresponding minimum MSE value is therefore given by

$$E\{[Y - \phi^o(X)]^2\} = E\{Y^2\} - E\{\phi^o(X)^2\} \tag{9.35}$$

Optimal Conditioned MSE

Once the experiment that gives rise to the bivariate random variable (X, Y) h[...] been conducted and the realization $X = x$ is made, it is desirable to measu[...] how well the estimate $\hat{y} = \phi^o(x)$ approximates the unobserved value assume[...] by the random variable Y. The optimal *conditioned MSE* as specified by

$$E\{[Y - \phi^o(x)]^2\} = \int_{-\infty}^{\infty} [y - \phi^o(x)]^2 f_Y(y|x)\, dy \qquad (9.3\ldots$$

provides a worthwhile measure of the goodness of approximation. This cond[...] tioned MSE is seen to be a function of the observation value x. The smaller th[...] condition MSE value is for a given observation x, the more confidence we ha[...] in the corresponding estimate $\hat{y} = \phi^o(x)$. Similarly, less credibility is given [...] the estimate $\phi^o(x)$ for those observations x which render this conditioned MS[...] a relatively large value. A plot of the optimal conditioned MSE (9.36) versus [...] provides a useful means for displaying this goodness of approximation.

Linearity of Conditional Expectation Operator

The conditional density function may be used for computing the condition[...] expectation of general functions of bivariate random variables. For exampl[...] the conditional expectation of the bivariate function $g(X, Y)$ is formally give[...] by

$$E\{g(X, Y)|X = x\} = \int_{-\infty}^{\infty} g(x, y) f_Y(y|x) \qquad (9.3\ldots$$

provided that this integral exists. It is to be noted that the conditional MSE us[...] in expression (9.36) corresponds to the particular function choice $g(X, Y)$ [...] $[Y - \phi^o(x)]^2$. The conditional expectation operator is *linear,* since it is readi[...] shown that

$$E\left\{\left[\alpha_0 + \sum_{k=1}^{q} \alpha_k g_k(X, Y)\right]\middle| X = x\right\} = \alpha_0 + \sum_{k=1}^{q} \alpha_k E\{g_k(X, Y)|X = x\} \qquad (9.3\ldots$$

where the α_k are scalars and it is assumed that the conditional expectation f[...] each of the functions $g_k(X, Y)$ exists.

9.10 BIVARIATE GAUSSIAN RANDOM VARIABLES

When analyzing random data, it is often assumed that the constituent rando[...] variables are *Gaussianly* (normally) *distributed*. The central limit theorem di[...]

cussed in Section 9.15 provides a theoretical basis for justifying a Gaussian random variable description. This being the case, we now briefly examine the salient features of such bivariate random variables. The real-valued bivariate random variable (X, Y) is said to be Gaussianly distributed with parameters μ_x, μ_y, σ_x^2, σ_y^2, and ρ_{xy} if its joint density function is of the form

$$f_{XY}(x, y) = \frac{1}{2\pi\sigma_x\sigma_y\sqrt{1 - \rho_{xy}^2}} e^{-a(x,y)} \tag{9.39a}$$

in which the quadratic exponent term is given by

$$a(x, y) = \frac{1}{2(1 - \rho_{xy}^2)}\left[\frac{(x - \mu)^2}{\sigma_x^2} - 2\rho_{xy}\frac{(x - \mu_x)(y - \mu_y)}{\sigma_x\sigma_y} + \frac{(y - \mu_y)^2}{\sigma_y^2}\right] \quad \text{for } -\infty < x, y < \infty \tag{9.39b}$$

The bivariate normal density function possesses a number of interesting properties, among which are that the associated marginal density functions as specified by relationships (8.41) and (8.42) are given by

$$f_X(x) = \frac{1}{\sqrt{2\pi\sigma_x^2}} e^{-(x-\mu_x)^2/2\sigma_x^2} \qquad -\infty < x < \infty$$

$$f_Y(y) = \frac{1}{\sqrt{2\pi\sigma_y^2}} e^{-(y-\mu_y)^2/2\sigma_y^2} \qquad -\infty < y < \infty$$

Moreover, the covariance between the random variables X and Y is readily shown to be

$$c_{xy} = \rho_{xy}\sigma_x\sigma_y$$

in which ρ_{xy} designates the correlation coefficient associated with the random variables X and Y. This correlation coefficient must satisfy $-1 \leq \rho_{xy} \leq 1$. Thus a bivariate Gaussian density function is seen to be completely described by the means, variances, and the correlation coefficient characterizing the constituent individual random variables X and Y.

When the correlation coefficient $\rho_{xy} = 0$, the joint density function (9.39) is factorable as

$$f_{XY}(x, y) = f_X(x)f_Y(y) \tag{9.40}$$

This implies that *uncorrelated* ($\rho_{xy} = 0$) bivariate Gaussian random variables are also *statistically independent* random variables. It should be noted that for non-Gaussian bivariate random variables, we cannot in general assert that uncorrelatedness implies independence. On the other hand, independence alwa implies uncorrelatedness.

Optimal Estimator: Bivariate Normal Random Variables

In determining the optimal estimator for the random variable Y conditioned the observation that $X = x$, knowledge of the associated conditional densi function $f_Y(y|x)$ is required. When the bivariate random variable (X, Y) is Gau sianly distributed (9.39), it can be shown that the required conditional densi function is of the normal density form

$$f_Y(y|x) = \frac{1}{\sqrt{2\pi\sigma^2}} e^{-(y-\mu)^2/2\sigma^2} \qquad (9.41$$

where

$$\mu = \mu_y + \frac{\rho_{xy}\sigma_y}{\sigma_x}(x - \mu_x) \qquad (9.41$$

$$\sigma^2 = \sigma_y^2(1 - \rho_{xy}^2) \qquad (9.41$$

In accordance with relationship (9.33), it follows that for bivariate Gaussi random variables, the *optimal MSE estimate function* has the simple linear for

$$\begin{aligned} \hat{y}^o = \phi^o(x) &= E\{Y|X = x\} \qquad (9.4 \\ &= \mu_y + \frac{\rho_{xy}\sigma_y}{\sigma_x}(x - \mu_x) \end{aligned}$$

The attractiveness of this optimal estimator function arises from the simplici of its linear functional form (i.e., it depends on only five parameters) and t computational ease with which the estimate is computed (i.e., one multiplicati and one addition). With these thoughts in mind, it is not surprising that line estimators are so extensively employed. That is, when the underlying bivaria random variables are Gaussian, as is often the case, the *optimal estimator linear*. In such situations, we simultaneously obtain a simplicity in impleme tation with best MSE performance. This is indeed a very significant observatio

9.11 OPTIMAL LINEAR ESTIMATOR: MULTIVARIATE RANDOM VARIABLES

Although the bivariate random variable estimate problem is of academic intere it does not represent the estimation problems often encountered in practice. Mc typically, there is available a larger set of data upon which to base an estimate of an unobserved random variable. This being the case, let us now consider a random experiment that is characterized by the $n + 1$ real-valued random variables $(X_1, X_2, \ldots, X_n, Y)$. After conducting the experiment; the values assumed by the first n of these random variables are observed and recorded, that is,

$$X_k = x_k \qquad 1 \le k \le n \tag{9.43}$$

The task at hand is to use these observations to make an estimate (i.e., an informed guess) of the value assumed by the unobserved random variable Y. To simplify the resultant analysis and to ease the probabilistic information needed to synthesize the associated estimator, the estimate of the unobserved random variable Y shall be restricted here to have the linear form

$$\hat{y} = \sum_{k=1}^{n} a_k x_k + b \tag{9.44}$$

where the a_k and b are real-valued parameters.

Our objective is to select the parameters governing this linear estimate rule so that the computed estimate $\hat{y}$ provides a good approximation of the value assumed by Y. The effectiveness of this estimate may be measured by determining how well the related estimator (a random variable)

$$\hat{Y} = \sum_{k=1}^{n} a_k X_k + b \tag{9.45}$$

represents the random variable Y. As was the case for bivariate random variables, the standard mean-squared-error criterion $E\{[Y - \hat{Y}]^2\}$ will be used to gauge the adequacy of approximation. Upon substituting the linear form (9.45) into this MSE criterion, it is readily shown that

$$\begin{aligned} E\{[Y - \hat{Y}]^2\} &= E\{Y^2\} + b^2 + \sum_{k=1}^{n} \sum_{m=1}^{n} a_k a_m E\{X_k X_m\} \\ &\quad + 2b \sum_{k=1}^{n} a_k E\{X_k\} - 2bE\{Y\} - 2 \sum_{k=1}^{n} a_k E\{X_k Y\} \end{aligned} \tag{9.46}$$

where use of the expected value operator's linearity has been employed.

Given values for the parameters $a_1, \ldots, a_n, b$, the MSE criterion (9.46) provides a widely accepted measure for determining the goodness of approximation of Y by the linear form $\hat{Y}$. The optimal choice of estimator parameters corresponds to that selection which minimizes this MSE criterion. The optimum estimator parameters are formally obtained, using standard calculus techniques, by setting the partial derivatives of this MSE criterion with respect to these parameters each equal to zero. Upon carrying out this procedure, it is found that the following theorem characterizes the optimal linear estimator.

Theorem 9.5. Let $(X_1, X_2, \ldots, X_n, Y)$ be a multivariate real-valued random variable that describe a random experiment. The optimum linear estimator

$$\hat{Y}^o = \sum_{k=1}^{n} a_k^o X_k + b^o \tag{9.47}$$

of the random variable Y in the sense of minimizing the MSE criterion $E\{[Y - \hat{Y}]^2\}$ is obtained by solving the n linear system of equations

$$\sum_{k=1}^{n} c_{xx}(m, k) a_k^o = c_{xy}(m) \qquad 1 \leq m \leq n \tag{9.48}$$

for the n optimum parameters $a_1^o, a_2^o, \ldots, a_n^o$, and then setting

$$b^o = E\{Y\} - \sum_{k=1}^{n} E\{X_k\} a_k^o \tag{9.49}$$

The coefficients constituting the linear system of equations (9.48) are given by the covariance entities

$$c_{xy}(m) = E\{X_m Y\} - E\{X_m\}E\{Y\} \qquad 1 \leq m \leq n$$

$$c_{xx}(m, k) = E\{X_m X_k\} - E\{X_m\}E\{X_k\} \qquad 1 \leq k, m \leq n$$

The minimum MSE criterion value corresponding to this optimal parameter selection is specified by

$$E\{[Y - \hat{Y}^o]^2\} = \sigma_y^2 - \sum_{k=1}^{n} c_{xy}(k) a_k^o \tag{9.50}$$

The details of proof are straightforward and left as an exercise to the reader. As a final note, it is seen from the b^o selection (9.49) that the expected value of the error $Y - \hat{Y}^o$ is zero, as should be anticipated.

EXAMPLE 9.4

Let the multivariate random variable (X_1, X_2, Y) have the joint density function

$$f_{X_1 X_2 Y}(x_1, x_2, y) = \begin{cases} 6 & x_1, x_2, y \geq 0 \text{ and } 0 \leq x_1 + x_2 + y \leq 1 \\ 0 & \text{otherwise} \end{cases}$$

This joint density function was studied in Example 8.16, where it was found that the following marginal density functions hold:

$$f_{X_1X_2}(x_1, x_2) = \begin{cases} 6(1 - x_1 - x_2) & x_1, x_2 \geq 0 \text{ and } 0 \leq x_1 + x_2 \leq 1 \\ 0 & \text{otherwise} \end{cases}$$

$$f_{X_kY}(x_k, y) = \begin{cases} 6(1 - x_k - y) & x_k, y \geq 0 \text{ and } 0 \leq x_k + y \leq 1 \\ 0 & \text{otherwise} \end{cases}$$

$$f_{X_k}(x_k) = \begin{cases} 3(1 - x_k)^2 & 0 \leq x_k \leq 1 \\ 0 & \text{otherwise} \end{cases}$$

$$f_Y(y) = \begin{cases} 3(1 - y)^2 & 0 \leq y \leq 1 \\ 0 & \text{otherwise} \end{cases}$$

where $k = 1$ and 2 for the two middle density functions. The entities required to evaluate expressions (9.48) and (9.49) are readily shown to be

$$E\{X_1\} = E\{X_2\} = E\{Y\} = \tfrac{1}{4}$$

$$c_{xx}(1, 1) = c_{xx}(2, 2) = \tfrac{3}{80}$$

$$c_{xx}(1, 2) = c_{xx}(2, 1) = c_{xy}(1) = c_{xy}(2) = -\tfrac{1}{80}$$

$$\sigma_y^2 = \tfrac{3}{80}$$

Thus the required set of equations (9.48) for the optimal a_k^o coefficients are

$$\tfrac{3}{80}\, a_1^o - \tfrac{1}{80}\, a_2^o = -\tfrac{1}{80}$$

$$-\tfrac{1}{80}\, a_1^o + \tfrac{3}{80}\, a_2^o = -\tfrac{1}{80}$$

It is readily shown that the solution to this system of equations is specified by

$$a_1^o = a_2^o = -\tfrac{1}{2}$$

The b^o parameter is, by relationship (9.49),

$$b^o = \tfrac{1}{4} - \tfrac{1}{4}(-\tfrac{1}{2}) - \tfrac{1}{4}(-\tfrac{1}{2}) = \tfrac{1}{2}$$

We therefore find that the optimal linear estimator of Y is governed by

$$\hat{Y} = \tfrac{1}{2} - \tfrac{1}{2}X_1 - \tfrac{1}{2}X_2$$

while the corresponding minimum mean-squared error is given by expression (9.50), that is,

$$E\{[Y - Y^o]^2\} = \tfrac{3}{80} - \tfrac{1}{160} - \tfrac{1}{160} = \tfrac{1}{40} \qquad \blacksquare$$

9.12 CORRELATION AND COVARIANCE MATRICES

The system of equations (9.48) governing the optimal selection of the line estimator a_k parameters is seen to be completely characterized by the covarianc of the constituent random variables $(X_1, X_2, \ldots, X_n, Y)$ identifying the u derlying experiment. To provide another interpretation as to why this is so, is beneficial to provide an algebraic setting to the estimation problem. This begun by first introducing the $n \times 1$ *random vector*

$$\mathbf{X} = [X_1, X_2, \ldots, X_n]' \quad (9.5$$

whose elements correspond to the individual random variables X_k whose valu are observed in the random experiment. The prime (′) superscript here us denotes the operation of *vector transposition* whereby a row vector is chang into its related column vector, and vice versa. In addition, the boldface symb is used to denote the fact that **X** is a random vector and not a random variab (scalar).

Mean Vector

The expected value (or mean) of the random vector **X** is defined to be equal that $n \times 1$ vector whose components are equal to $\mu_k = E\{X_k\}$ for $1 \leq k \leq$ This mean vector therefore has the structure

$$\begin{aligned} \boldsymbol{\mu}_{\mathbf{x}} &= E\{\mathbf{X}\} \\ &= [\mu_1, \mu_2, \ldots, \mu_n]' \end{aligned} \quad (9.5$$

Unlike the random vector, this mean vector is composed of the determinist scalars μ_k and is therefore itself not random. The mean vector can be interprete as the average value assumed by the random vector **X** on a large number independent experimental trials.

Correlation Matrix

The $n \times n$ *correlation matrix* associated with the random vector (9.51) formally defined by

$$R_{xx} = E\{\mathbf{XX}'\} \quad (9.5$$

The $n \times n$ random matrix $\mathbf{XX}'$ appearing inside the expected-value-operatc brackets has as its (k, m)th component the random variable product X_kX_m fc

$1 \leq k, m \leq n$. Taking the expected value of this component as indicated in definition (9.53) gives the (k, m) element of the correlation matrix, that is,

$$r_{xx}(k, m) = E\{X_k X_m\} \qquad 1 \leq k, m \leq n \tag{9.54}$$

These correlation entities then form the elements of the correlation matrix R_{xx}. This being the case, it is seen that the $n \times n$ correlation matrix (9.53) has the structure

$$R_{xx} = \begin{bmatrix} r_{xx}(1, 1) & r_{xx}(1, 2) & \cdots & r_{xx}(1, n) \\ r_{xx}(2, 1) & r_{xx}(2, 2) & \cdots & r_{xx}(2, n) \\ \cdot & \cdot & & \cdot \\ \cdot & \cdot & & \cdot \\ \cdot & \cdot & & \cdot \\ r_{xx}(n, 1) & r_{xx}(n, 2) & \cdots & r_{xx}(n, n) \end{bmatrix} \tag{9.55}$$

From expression (9.54) it is seen that the elements of the correlation matrix satisfy $r_{xx}(k, m) = r_{xx}(m, k)$ for $1 \leq k, m \leq n$. This indicates that the correlation matrix is symmetric so that $R'_{xx} = R_{xx}$. Furthermore, using the readily established identity $0 \leq E\{[\mathbf{a}'\mathbf{X}]^2\} = \mathbf{a}' R_{xx}\mathbf{a}$, where $\mathbf{a}$ is an arbitrary $n \times 1$ real vector, it follows that the correlation matrix is positive semidefinite.† In signal-processing applications, extensive use is made of the symmetry and positive semidefiniteness of the correlation matrix. A full development of these properties and their use is made in the companion text.

Covariance Matrix

The $n \times n$ *covariance matrix* associated with the random vector (9.51) is specified by

$$C_{xx} = E\{[\mathbf{X} - \boldsymbol{\mu}_\mathbf{x}][\mathbf{X} - \boldsymbol{\mu}_\mathbf{x}]'\} \tag{9.56}$$

in which the (k, m)th element of the random matrix $[\mathbf{X} - \boldsymbol{\mu}_\mathbf{x}][\mathbf{X} - \boldsymbol{\mu}_\mathbf{x}]'$ is equal to $[X_k - \mu_k][X_m - \mu_m]$ for $1 \leq k, m \leq n$. The expected value of this term is seen to correspond to the covariance existent between the random variable pairs (X_k, X_m), that is,

$$c_{xx}(k, m) = E\{[X_k - \mu_k][X_m - \mu_m]\} \qquad 1 \leq k, m \leq n \tag{9.57}$$

As in the correlation case, these covariance terms constitute the elements of the covariance matrix

†A real-valued $n \times n$ matrix P is said to be positive semidefinite if $\mathbf{a}'P\mathbf{a} \geq 0$ holds for all $n \times 1$ real-valued vectors $\mathbf{a}$.

$$C_{xx} = \begin{bmatrix} c_{xx}(1, 1) & c_{xx}(1, 2) & \cdots & c_{xx}(1, n) \\ c_{xx}(2, 1) & c_{xx}(2, 2) & \cdots & c_{xx}(2, n) \\ \vdots & \vdots & & \vdots \\ c_{xx}(n, 1) & c_{xx}(n, 2) & \cdots & c_{xx}(n, n) \end{bmatrix} \tag{9.58}$$

It is a simple matter to show that the covariance matrix is also symmetric and positive semidefinite.

The covariance and correlation matrices associated with the random variables $(X_1, X_2, \ldots, X_n)$ are each seen to be symmetric and positive semidefinite. Furthermore, they are closely related, as is made evident upon carrying out the *outer products* within the expected-value-operator brackets in expression (9.56). This yields

$$\begin{aligned} C_{xx} &= E\{\mathbf{X}\mathbf{X}' - \mathbf{X}\boldsymbol{\mu}_\mathbf{x}' - \boldsymbol{\mu}_\mathbf{x}\mathbf{X}' + \boldsymbol{\mu}_\mathbf{x}\boldsymbol{\mu}_\mathbf{x}'\} \\ &= R_{xx} - \boldsymbol{\mu}_\mathbf{x}\boldsymbol{\mu}_\mathbf{x}' \end{aligned} \tag{9.59}$$

where use of the expected value's linearity has been made. The following theorem points up the importance of the correlation and covariance matrices.

Theorem 9.6. Let $(X_1, X_2, \ldots, X_n, Y)$ be a multivariate real-valued random variable that describes a random experiment. The linearly related random variable

$$\begin{aligned} \hat{Y} &= \sum_{k=1}^{n} a_k X_k + b \\ &= \mathbf{a}'\mathbf{X} + b \end{aligned} \tag{9.60}$$

where $\mathbf{a}' = [a_1, a_2, \ldots, a_n]$ has its mean value and variance given by

$$\begin{aligned} \mu_{\hat{y}} &= \sum_{k=1}^{n} a_k E\{X_k\} + b \\ &= \mathbf{a}'\boldsymbol{\mu}_\mathbf{x} + b \end{aligned} \tag{9.61}$$

and

$$\sigma_{\hat{y}}^2 = \mathbf{a}'C_{xx}\mathbf{a} \tag{9.62}$$

in which C_{xx} is the $n \times n$ covariance matrix (9.58). If $\hat{Y}$ as specified by expression (9.60) constitutes a linear estimator of Y, the optimal linear estimator is one that minimizes the MSE criterion $E\{[Y - \hat{Y}]^2\}$. The parameters $a_1^o, a_2^o, \ldots, a_n^o, b^o$, which govern an optimal estimator, satisfy the consistent system of linear equations

$$C_{xx}\mathbf{a}^o = \mathbf{c}_{xy} \tag{9.63}$$

where the $n \times 1$ covariance vector $\mathbf{c}_{xy}$ has as its components $E\{X_kY\} - E\{X_k\}E\{Y\}$ for $1 \le k \le n$. The optimal parameter b^o is then given by

$$\begin{aligned} b^o &= \mu_y - \sum_{k=1}^{n} E\{X_k\}a_k^o \\ &= \mu_y - \boldsymbol{\mu}_x'\mathbf{a}^o \end{aligned} \tag{9.64}$$

while the resultant MSE is specified by

$$\begin{aligned} E\{[Y - \hat{Y}^o]^2\} &= \sigma_y^2 - \sum_{k=1}^{n} c_{xy}(k)a_k^o \\ &= \sigma_y^2 - \mathbf{c}_{xy}'\mathbf{a}^o \end{aligned} \tag{9.65}$$

A proof that expressions (9.61) and (9.62) give the required mean value and variance, respectively, is straightforward and therefore not given. The optimality conditions (9.63) to (9.65) are simply reexpressions of the results of Theorem 9.5.

EXAMPLE 9.5

Consider the multivariate random variable (X_1, X_2, Y) described in Example 9.4. The required mean vector, covariance vector, correlation matrix, and covariance matrix associated with this multivariate random variable are then specified by

$$\boldsymbol{\mu}_x = \begin{bmatrix} \frac{1}{4} \\ \frac{1}{4} \end{bmatrix} \qquad R_{xx} = \begin{bmatrix} \frac{1}{10} & \frac{1}{20} \\ \frac{1}{20} & \frac{1}{10} \end{bmatrix}$$

$$\mathbf{c}_{xy} = \begin{bmatrix} -\frac{1}{80} \\ -\frac{1}{80} \end{bmatrix} \qquad C_{xx} = \begin{bmatrix} \frac{3}{80} & -\frac{1}{80} \\ -\frac{1}{80} & \frac{3}{80} \end{bmatrix}$$

It is readily shown that the optimal a_k^o parameters arising from relationship (9.63) are $a_1^o = a_2^o = -\frac{1}{2}$, while expressions (9.64) and (9.65) give $b_0 = \frac{1}{2}$ and $E\{[Y - Y^o]^2\} = \frac{1}{40}$, respectively. As anticipated, these results are in agreement with those found in Example 9.4. ■

9.13 OPTIMAL ESTIMATOR: MULTIVARIATE RANDOM VARIABLES

Although linear estimators do perform admirably in many applications, it may be advisable to consider more general estimators in certain situations. With this in mind, let us explore the case in which the random experiment is characterized by the $n + 1$ real-valued random variables $(X_1, X_2, \ldots, X_n, Y)$. Furthermore, after conducting the experiment, the values assumed by the first n of these random variables are observed and recorded as

$$X_k = x_k \qquad 1 \leq k \leq n \tag{9.66}$$

We now wish to use these observations to make an informed estimate of the unobserved random variable Y. This estimate will take on the general functional form

$$\hat{y} = \phi(x_1, x_2, \ldots, x_n) \tag{9.67}$$

in which ϕ is a generally nonlinear function of the n real-valued parameters x_k for $1 \leq k \leq n$. The task at hand is to select the function ϕ so that the estimate (9.67) best approximates the value assumed by Y (i.e., y).

To measure the goodness-of-estimation approximation, the MSE criterion will again be employed. In particular, the accuracy with which the random variable estimator

$$\hat{Y} = \phi(X_1, X_2, \ldots, X_n) \tag{9.68}$$

approximates the random variable Y is measured by the MSE criterion

$$E\{[Y - \phi(X_1, X_2, \ldots, X_n)]^2\} \tag{9.69}$$

We now wish to select the estimator function ϕ so as to render this criterion a minimum value. Using an adaption of the calculus-of-variations approach taken in Section 9.7, we readily obtain the following theorem characterizing the optimal estimator.

Theorem 9.7. Let $(X_1, X_2, \ldots, X_n, Y)$ be real-valued random variables associated with a random experiment in which the observations $X_k = x_k$ for $1 \leq k \leq n$ are made. It then follows that the optimum estimate $\phi^o(x_1, x_2, \ldots, x_n)$ which best represents the unobserved random variable realization $Y = y$ in the sense of minimizing the MSE criterion

$$E\{[Y - \phi(X_1, X_2, \ldots, X_n)]^2\} \tag{9.70}$$

is given by the conditional expectation

$$\begin{aligned} \phi^o(x_1, x_2, \ldots, x_n) &= \int_{-\infty}^{\infty} y f_Y(y|x_1, \ldots, x_n)\, dy \\ &= E\{Y|X_k = x_k \text{ for } 1 \leq k \leq n\} \end{aligned} \tag{9.71}$$

where the conditional density function is specified by

$$f_Y(y|x_1, \ldots, x_n) = \frac{f_{X_1 \cdots X_n Y}(x_1, \ldots, x_n, y)}{f_{X_1 \cdots X_n}(x_1, \ldots, x_n)}$$

Moreover, the minimal error is orthogonal to the optimal estimator function, that is,

$$E\{[Y - \phi^o(X_1, X_2, \ldots, X_n)]\phi^o(X_1, X_2, \ldots, X_n)\} = 0 \quad (9.72)$$

and the corresponding miminum MSE value is therefore given by

$$E\{[Y - \phi^o(X_1, X_2, \ldots, X_n)]^2\} = E\{Y^2\} - E\{\phi^o(X_1, X_2, \ldots, X_n)^2\} \quad (9.73)$$

EXAMPLE 9.6

Consider the multivariate random variable (X_1, X_2, Y) described in Example 9.4. To determine the optimal estimator of Y conditioned on $X_1 = x_1$ and $X_2 = x_2$, we first find the conditional density function. This yields

$$f_Y(y|x_1, x_2) = \begin{cases} \dfrac{1}{1 - x_1 - x_2} & 0 \le y \le 1 - x_1 - x_2 \\ 0 & \text{otherwise} \end{cases}$$

The optimal estimate function according to expression (9.71) is given by

$$\phi^o(x_1, x_2) = \int_0^{1-x_1-x_2} y \frac{1}{1 - x_1 - x_2} dy$$

$$= \frac{1 - x_1 - x_2}{2}$$

which is seen to be linear. Thus, for this problem, the optimal estimate happens to be linear. Furthermore, the conditioned minimum mean-squared error is, by expression (9.73),

$$E\{[Y - \mathbf{Y}^o]^2 | X_1 = x_1, X_2 = x_2\}$$

$$= \int_0^{1-x_1-x_2} \left(y - \frac{1 - x_1 - x_2}{2}\right)^2 \frac{1}{1 - x_1 - x_2} dy$$

$$= \tfrac{1}{12}(1 - x_1 - x_2)^2$$

■

9.14 MULTIVARIATE GAUSSIAN RANDOM VARIABLES

As suggested earlier, when analyzing random data it is frequently assumed that the underlying random variables are *Gaussianly* distributed. In this section we provide a brief exposure to Gaussian distributed multivariate random variables. The covariance matrix introduced in Section 9.12 plays an important role in this characterization. The real-valued multivariate random variable $(X_1, X_2, \ldots, X_n)$ is said to be *Gaussianly distributed* if its joint density function is of the form

$$f_{X_1 \cdots X_n}(x_1, \ldots, x_n) = \frac{1}{\sqrt{(2\pi)^n \det(C_{xx})}} e^{-(1/2)(\mathbf{x}-\boldsymbol{\mu}_x)' C_{xx}^{-1}(\mathbf{x}-\boldsymbol{\mu}_x)} \tag{9.74}$$

where $\mathbf{x}$ and $\boldsymbol{\mu}_x$ are each $n \times 1$ vectors with components x_k and $E\{X_k\}$, respectively, for $1 \leq k \leq n$ and C_{xx} is the associated $n \times n$ covariance matrix (9.58). The symbol $\det(C_{xx})$ denotes the determinant of the covariance matrix. Because of its special form, a multivariate Gaussian density function possesses a number of salient properties, which are given in the following theorem.

Theorem 9.8. Let $\mathbf{X}$ be a $n \times 1$ Gaussianly distributed random vector as specified by the joint density function (9.74). It then follows that:

(a) All the marginal density functions associated with $\mathbf{X}$ are Gaussian.
(b) The conditional density function of each component X_k given its $n - 1$ other component values is Gaussian.
(c) The conditional expectation of the component X_k given its $n - 1$ other component values is linear, that is,

$$E\{X_k | X_i = x_i \text{ for } 1 \leq i \leq n, i \neq k\} = a_0 + \sum_{\substack{i=1 \\ i \neq k}}^{n} a_i x_i$$

(d) If each of the random variables X_k are pairwise uncorrelated (i.e., $c_{x_i x_j} = 0$ for $i \neq j$), they are also independent.

Optimal Estimator: Multivariate Gaussian Case

From the viewpoint of estimation theory, Theorem 9.8(c) is of particular relevance. To see why this is so, let us consider the set of Gaussian random variables $(X_1, \ldots, X_n, Y)$ in which values for the first n components are given (i.e., $X_k = x_k$ for $1 \leq k \leq n$). We now seek to form an estimate for the value realized by the unobserved random variable component Y that is based on the realizations $x_1, x_2, \ldots, x_n$. If a MSE criterion is chosen for measuring goodness of estimation, Theorem 9.7 indicates that the optimal estimate is the conditional expectation and is generally nonlinear. It follows from Theorem 9.8, however, that this optimal estimate for the Gaussian case is of the linear form

$$\hat{y} = \phi(x_1, \ldots, x_n) = \sum_{k=1}^{n} a_k x_k + b \tag{9.75}$$

The optimal values for the coefficients in this expression may then be obtained by a straightforward application of Theorem 9.6.

9.15 CENTRAL LIMIT THEOREM

By this time the reader may be somewhat puzzled by our disposition toward studying random variables that are Gaussianly distributed. This inclination is clearly influenced by the rich analytical dividends that accrue when invoking a Gaussian distribution assumption. Foremost in this regard is the knowledge that the optimal estimate associated with multivariate Gaussian random variables has a linear form. The research literature concerning linear estimators and Gaussianly distributed random time series is indeed both rich and extensive.

Although the reasoning above might provide an abstract justification for studying Gaussianly distributed random variables, the typical engineer and scientist is concerned with more practical matters. Fortunately, the invocation of a Gaussian distribution assumption can be justified in a surprisingly large number of applications. The primary reason for this is due to the following theorem.

Theorem 9.9 (Central Limit Theorem). Let $X_1, X_2, \ldots, X_n$ be a set of independent random variables with $E\{X_k\} = \mu_k$, var $\{X_k\} = \sigma_k^2$ for $k = 1, 2, \ldots, n$. Furthermore, let

$$X = X_1 + X_2 + \cdots + X_n$$

Then, under very general conditions and appropriately *large* values of n, the associated univariate random variable

$$Z_n = \frac{X - \sum_{k=1}^{n} \mu_k}{\sum_{k=1}^{n} \sigma_k^2} \tag{9.76}$$

is approximately Gaussianly distributed with mean zero and variance one [i.e., $N(0, 1)$].

The implications of this theorem are quite clear. If the experimental outcome associated with a random phenomenon can be considered as the accumulative sum of a large number of random effects, the outcome tends to be Gaussianly distributed. As we shall see in Chapter 10, widely accepted time series models fit this description very nicely. Thus the assumption of Gaussianly distributed random variables is attractive from a practical as well as a theoretical viewpoint.

EXAMPLE 9.7

One of the more interesting applications of the central limit theorem is found in random number generation. Suppose that it is desired to approximate a Gaussian-distributed random number generator $N(0, 1)$. It is assumed that there is available a random number generator that provides independent samples of a uniformly distributed random variable which takes on values in the interval (0, 1). To approximate the required Gaussianly distributed random variable, a commonly accepted procedure is first to obtain 12 independent samples from the available uniform random number generator, that is, x_1, $x_2, \ldots, x_{12}$. Next, the required approximately Gaussian distributed sample is computed according to

$$z = \sum_{k=1}^{12} (x_k - 0.5)$$

The reason why this represents an acceptably good approximation of a Gaussian-distributed random variable follows from studying the related random variable,

$$Z = \sum_{k=1}^{12} (X_k - 0.5) \tag{9.77}$$

This random variable is readily shown to have zero mean and unity variance. Moreover, the central limit theorem indicates that Z will be approximately $N(0, 1)$. This systematic procedure for generating Gaussian samples is widely employed. ■

9.16 SUMMARY

In this chapter, various concepts from probability theory that are essential to the study of random time series were presented. Since one of our main objectives in time series analysis is that of filtering, smoothing and prediction, the concept of minimum mean-squared-error estimation was of fundamental importance. In particular, the major development of this chapter concerned the multivariate random variable $(X_1, X_2, \ldots, X_n, Y)$ and how we provide an optimal estimator of the random variable Y from realizations of its other n constituent members (i.e., $X_k = x_k$ for $1 \le k \le n$). This estimator took the functional form

$$\hat{Y} = \phi(X_1, X_2, \ldots, X_n) \tag{9.78}$$

where ϕ was referred to as the estimator function. We then sought an estimator function for which the estimator $\hat{Y}$ best approximates Y.

If the mean-squared-error (MSE) criterion

$$E\{[Y - \hat{Y}]^2\} \tag{9.79}$$

is selected as the mechanism for measuring goodness of estimate, it was shown that the conditional expectation

$$\hat{y}^{o} = E\{Y|X_k = x_k \text{ for } 1 \leq k \leq n\} \tag{9.80}$$

provided the minimum mean-squared-error (MMSE) estimation. This conditional expectation is (conceptually) straightforwardly computed by use of the conditional density function associated with this problem. In general, this conditional expectation is a nonlinear function of the realizations $(x_1, x_2, \ldots, x_n)$.

To make the synthesis of estimators tractable (conditional density functions are usually not known), it was found that the linear estimator structure

$$\hat{Y} = \sum_{k=1}^{n} a_k X_k + b \tag{9.81}$$

had particularly attractive computational and statistical features. These arise from the fact that the parameters associated with the minimum MSE linear estimator require only mean and correlation information. Specifically, the minimum MSE linear estimator coefficients are obtained by solving the linear system of equations

$$C_{xx}\mathbf{a}^{o} = \mathbf{c}_{xy} \tag{9.82}$$

where C_{xx} is the $n \times n$ covariance matrix associated with the multivariate random variables $(X_1, X_2, \ldots, X_n)$, $\mathbf{a}^{o} = [a_1^{o}, a_2^{o}, \ldots, a_n^{o}]'$ is the optimum linear estimator coefficient vector, and $\mathbf{c}_{xy} = [c_{x_1y}, \ldots, c_{x_ny}]'$ is the $n \times 1$ covariance vector. Moreover, the minimum MSE for this coefficient selection is

$$E\{[Y - \hat{Y}]^2\} = \sigma_y^2 - \mathbf{c}_{xy}'\mathbf{a}^{o} \tag{9.83}$$

Although the optimal linear estimator (9.81) will generally have an estimation performance that is inferior to the optimal conditional expectation estimator (9.80), it has been used extensively in numerous applications. This is due to the relative modest statistical information needed in its generation (i.e., means and covariances) and the fact that in many applications the estimation performance achieved with the optimal linear estimator is satisfactory for the purposes at hand. Moreover, when the multivariate random variables are normally distributed, the optimal estimator (9.80) is found to be linear. With these thoughts in mind, we shall henceforth concentrate our efforts primarily on linear estimator structures.

SELECTED READINGS

See Suggested Readings in Chapter 7.

PROBLEMS

9.1 Let the bivariate random variable (X, Y) be distributed as

$$f_{XY}(x, y) = \begin{cases} \dfrac{x(x - y)}{8} & 0 \leq x \leq 2, \quad -x \leq y \leq x \\ 0 & \text{otherwise} \end{cases}$$

Compute the associated mean values μ_x and μ_y, variances σ_x^2 and σ_y^2, the correlation and covariance measures, and the correlation coefficient.

9.2 Let the joint density function of the bivariate random variable (X, Y) be specified by

$$f_{XY}(x, y) = \begin{cases} e^{-x} & y > 0, \quad x > y \\ 0 & \text{otherwise} \end{cases}$$

Evaluate the associated mean values μ_x and μ_y, variances σ_x^2 and σ_y^2, the correlation and covariance measures, and the correlation coefficient.

9.3 Let the bivariate random variable (X, Y) be governed by the joint density function

$$f_{XY}(x, y) = \begin{cases} x^2 + \frac{1}{3}xy & 0 \leq x \leq 1, \quad 0 \leq y \leq 2 \\ 0 & \text{otherwise} \end{cases}$$

Determine the related mean values μ_x and μ_y, variances σ_x^2 and σ_y^2, the correlation and covariance measures, and the correlation coefficient.

9.4 Let the discrete bivariate random variable (X, Y) be distributed as

$$f_{XY}(x, y) = \tfrac{1}{8}\,\delta(x, y) + \tfrac{1}{4}\,\delta(x - 1, y - 2) + \tfrac{1}{16}\,\delta(x - 2, y - 2) + \tfrac{9}{16}\,\delta(x + 1, y)$$

Evaluate the corresponding mean values μ_x and μ_y, variances σ_x^2 and σ_y^2, the correlation and covariance measures, and the correlation coefficient.

9.5 Show that the Schwarz inequalities (9.5) and (9.6) are satisfied for the bivariate random variables specified in **(a)** Problem 9.1; **(b)** Problem 9.2; **(c)** Problem 9.3; **(d)** Problem 9.4.

9.6 Determine the optimal linear estimate of Y based on X in the MSE sense for the bivariate random variables described in **(a)** Problem 9.1; **(b)** Problem 9.2; **(c)** Problem 9.3; **(d)** Problem 9.4.

9.7 Let the bivariate random variable (X, Y) be uniformly distributed over the square with vertices $(1, 1)$, $(1, -1)$, $(-1, -1)$, and $(-1, 1)$. Show that the random variables X and Y are independent and uncorrelated.

9.8 Let the random variable Y be specified by $Y = X^2$, in which X is uniformly distributed on the interval $[-1, 1]$. Show that these random variables are uncorrelated although they are clearly dependent.

9.9 Let the bivariate random variables (X, Y) be distributed as

$$f_{XY}(x, y) = \alpha^2 e^{-\alpha x} \qquad x > 0, \quad 0 < y < x$$

Are X and Y uncorrelated or independent?

9.10 Let the random variable $\hat{Y}^o = a^o X + b^o$ be the optimal linear estimator of the random variable Y in the MSE sense. Prove the following.
(a) $E\{\hat{Y}^o\} = E\{Y\}$.
(b) The minimal error random variable $Y - \hat{Y}^o$ and the optimal linear estimator $\hat{Y}^o$ are orthogonal, that is, $E\{[Y - \hat{Y}^o]\hat{Y}^o\} = 0$.

9.11 Let X, Y, and Z be pairwise uncorrelated random variables with variances σ_x^2, σ_y^2, and σ_y^2, respectively. If $U = X + Y$ and $V = Y + Z$, evaluate the correlation coefficient ρ_{uv}.

9.12 Let the conditional expectation of the bivariate random variable (X, Y) be linear, that is,

$$E\{X|Y = y\} = \alpha y + \beta \qquad \text{and} \qquad E\{Y|X = x\} = \gamma x + \delta$$

Determine ρ_{xy}, $E\{X\}$, and $E\{Y\}$.

9.13 Let the bivariate random variable (X, Y) have the joint density function

$$f_{XY}(x, y) = \begin{cases} \alpha[1 - |x|] & -1 \le x \le 1, \quad 0 \le y \le 1 \\ 0 & \text{otherwise} \end{cases}$$

(a) Determine the appropriate value for the scalar α.
(b) Compute the correlation coefficient ρ_{xy}.
(c) Determine the conditional density functions $f_X(x|y)$ and $f_Y(y|x)$.
(d) Evaluate the optimal estimate functions $\phi^o(x) = E\{Y|X = x\}$ and $\phi^o(y) = E\{X|Y = y\}$.

9.14 Compute the conditional density functions and the optimal estimator of Y based on X for the bivariate random variables described in **(a)** Problem 9.1; **(b)** Problem 9.2; **(c)** Problem 9.3; **(d)** Problem 9.4.

9.15 Consider the bivariate random variable (X, Y) that is uniformly distributed in the square with vertices $(1, 0)$, $(0, 1)$, $(-1, 0)$, and $(0, -1)$.
(a) Show that these random variables are uncorrelated but not independent.
(b) Determine the optimal estimate function $\phi^o(x) = E\{Y|X = x\}$.

9.16 Let (X, Y) be a bivariate Gaussian random variable with means μ_x and μ_y, variances σ_x^2 and σ_y^2, and correlation coefficient ρ_{xy}. Prove that the conditional density function $f_Y(y|x)$ is Gaussian. Moreover, find the mean and variance of this conditional density function.

9.17 Carry out the details of proof for Theorem 9.5.

9.18 Verify that the mean value and variance of the random variable (9.60) are given by expressions (9.61) and (9.62), respectively.

9.19 Let $\hat{X}$ be a linear estimator of the mean value of random variable X, that is,

$$\hat{X} = a_1 X_1 + a_2 X_2 + \cdots + a_N X_N$$

where X_k are uncorrelated samples of X. Show that:
(a) This linear estimator is unbiased (i.e., $E\{\hat{X}\} = E\{X\}$) if and only if $\sum_{k=1}^{N} a_k = 1$.
(b) The *minimum variance* unbiased linear estimate of the mean as defined by

$$\min_{\sum_{k=1}^{N} a_{k=1}} E\{|\hat{X} - E\{\hat{X}\}|^2\}$$

is the sampled mean (i.e., $a_k^o = 1/N$ for $1 \le k \le N$). *Hint:* Use the Lagrange multiplier method.

9.20 Carry out the details of Example 9.4.

9.21 Let (X_1, X_2, X_3) be a multivariate random vector whose mean vector and covariance matrix are given by

$$u_x = \begin{bmatrix} -1 \\ 3 \\ 2 \end{bmatrix} \qquad C_{xx} = \begin{bmatrix} \frac{1}{2} & \frac{1}{4} & \frac{1}{3} \\ \frac{1}{4} & 2 & \frac{2}{3} \\ \frac{1}{3} & \frac{2}{3} & 1 \end{bmatrix}$$

Determine the linear MMSE estimator of X_3 based on X_1 and X_2.

9.22 Let $\mathbf{X} = [X_1, X_2, \ldots, X_n]'$ be a $n \times 1$ real random vector whose $n \times 1$ mean vector is denoted by $\mathbf{u}_x$ and whose $n \times n$ covariance matrix is designated by C_{xx}. If the $m \times 1$ real random vector $\mathbf{Y}$ is related to $\mathbf{X}$ according to

$$\mathbf{Y} = A\mathbf{X} + \mathbf{b}$$

where $\mathbf{b}$ is a given real $m \times 1$ vector and A is a given real $m \times n$ matrix, determine the mean vector and covariance matrix associated with the random vector $\mathbf{Y}$.

9.23 Let (X_1, X_2, X_3) be a multivariate random vector whose mean and covariance matrix are given by

$$\boldsymbol{\mu}_x = \begin{bmatrix} \frac{1}{2} \\ \frac{1}{2} \\ \frac{1}{2} \end{bmatrix} \qquad C_{xx} = \begin{bmatrix} \frac{3}{4} & \frac{1}{4} & 0 \\ \frac{1}{4} & \frac{3}{4} & \frac{1}{4} \\ 0 & \frac{1}{4} & \frac{3}{4} \end{bmatrix}$$

Determine the linear MMSE estimator of X_3 based on X_1 and X_2.

9.24 Let the density function of the random variable X have the exponential form

$$f_x(x|\theta) = \tfrac{1}{2}e^{-|x-\theta|} \qquad -\infty < x < \infty$$

where θ is an unknown parameter. Furthermore, let $X_1, X_2, \ldots, X_n$ constitute n independent samples of this random variable. Compute $E\{X\}$ and var $\{X\}$, where $X = X_1 + X_2 + \cdots + X_n$.

9.25 Show that the random variable specified in equation (9.77) has a zero mean and a variance of unity.

9.26 Sketch a proof for the central limit theorem. Use the Chebyshev inequality in your proof.

CHAPTER 10

Random Signals

10.1

INTRODUCTION

In analyzing data in the form of a time series (i.e., a sequence of numbers), the elements constituting the time series are typically made known only after an underlying experiment has been conducted and its outcome observed. Such time series are said to be random in nature due to our inability to predict precisely their behavior before the experiment has been conducted. By using the concept of random variables introduced in Chapter 8, it is possible to give a more mathematically based definition of a random time series.

Definition 10.1. A *random time series* is composed of a sequence of univariate random variables

$$\{\mathbf{x}(n)\} = \ldots, \mathbf{x}(-1), \mathbf{x}(0), \mathbf{x}(1), \mathbf{x}(2), \ldots$$

in which the ordering parameter n is arbitrarily referred to as *time*.

It is to be noted that the lowercase boldface symbol [e.g., $\mathbf{x}(n)$] is here used to denote a random variable instead of the more traditional uppercase notation [e.g., $X(n)$] employed in Chapters 8 and 9. This notation is adopted because in system theory studies, uppercase symbols are invariably used to denote transforms [e.g., $X(z)$, $X(e^{j\omega})$].

Once the experiment that characterizes a random time series has been conducted, the values assumed by the constituent random variables are observed, that is,

$$\mathbf{x}(n) = x(n) \qquad \text{for } n = 0, \pm 1, \pm 2, \ldots$$

The observed *data* sequence $\{x(n)\}$ then represents a *realization* of the underlying random time series $\{\mathbf{x}(n)\}$. Our ultimate objective is to manipulate this data set, using appropriate signal-processing algorithms, so as to extract desired information contained within the data. Intuitively, it is apparent that an appropriate choice of the algorithm to be employed should be linked in some manner to the probabilistic description of the underlying random time series. The most complete description of a random time series requires knowledge of all the joint density functions associated with its constituent random variable members. As might be expected, this level of information is almost never available in any meaningful application. As suggested from the results of Chapter 9, however, there exist many relevant applications in which correlation information is all that is needed.

In much of what follows, we are concerned with signal-processing algorithms that are linear in structure, and a mean-square-error criterion is employed to measure signal-processing performance. Under such conditions it is found that the resultant optimal linear processor is exclusively dependent on the mean and correlation behavior of the time series being processed. Fortunately, this relatively low-level probabilistic information is often either available or is readily estimated from the empirically derived data being analyzed. With this in mind, we now seek to characterize the mean value and covariance behavior of various classes of random time series. This characterization, in turn, will enable us to provide a meaningful study of the important signal-processing applications of *filtering*, *smoothing*, and *prediction*.

10.2 COMPLEX-VALUED RANDOM TIME SERIES

To ensure that our treatment of random time series is general enough so as to accommodate applications found in such important disciplines as communications, sonar, and radar, it is necessary to expand our treatment to complex-valued random time series of the form

$$\mathbf{x}(n) = \mathbf{x}_1(n) + j\mathbf{x}_2(n) \tag{10.1}$$

where $j = \sqrt{-1}$. The constituent real-valued random time series $\{\mathbf{x}_1(n)\}$ and $\{\mathbf{x}_2(n)\}$ are commonly referred to as the *real* and *imaginary components*, respectively.† It is to be noted that by adopting this more general representation,

†In communication theory applications, the *real* and *imaginary* time series components correspond to the so-called *in-phase* and *quadrature components* of a modulated signal.

we are still able to study real-valued time series simply by setting the imaginary component $\mathbf{x}_2(n)$ equal to zero.

Mean Value Sequence

In accordance with previously made observations, the mean value and correlation behavior of the foregoing complex-valued time series are of immediate interest. The mean value sequence associated with time series (10.1) is readily obtained by taking the expected value of its general nth element, which gives

$$\begin{aligned} \mu_{\mathbf{x}}(n) &= E\{\mathbf{x}(n)\} \\ &= E\{\mathbf{x}_1(n)\} + jE\{\mathbf{x}_2(n)\} \\ &= \mu_{\mathbf{x}_1}(n) + j\mu_{\mathbf{x}_2}(n) \end{aligned} \tag{10.2}$$

in which the real and imaginary time series components have the associated mean value sequences $\mu_{\mathbf{x}_1}(n) = E\{\mathbf{x}_1(n)\}$ and $\mu_{\mathbf{x}_2}(n) = E\{\mathbf{x}_2(n)\}$, respectively. In arriving at this result, we have used the linearity of the expected value operator. As expression (10.2) explicitly indicates, the mean value sequence is generally dependent on the time index n. This reflects the fact that in many time series, there is an underlying trend in the data that is manifested in a time-changing mean value behavior.

Autocorrelation Sequence

The correlation behavior of the complex-valued time series (10.1) is formally defined by the expression

$$r_{\mathbf{xx}}(n_1, n_2) = E\{\mathbf{x}(n_1)\overline{\mathbf{x}}(n_2)\} \tag{10.3}$$

where an overbar is used to denote the operation of complex conjugation. It is necessary to include this complex conjugation operation because of the generally complex-valued nature of the underlying time series. Since the univariate random variables $\mathbf{x}(n_1)$ and $\mathbf{x}(n_2)$ appearing in this correlation relationship are associated with the same time series, we shall hereafter refer to the entity $r_{\mathbf{xx}}(n_1, n_2)$ as the *autocorrelation sequence*. This two-dimensional sequence is deterministic in nature and depends on the two integer variables n_1 and n_2. Upon substituting the time series representation (10.1) into this autocorrelation expression and then regrouping terms, it is readily shown that

$$r_{\mathbf{xx}}(n_1, n_2) = r_{\mathbf{x}_1\mathbf{x}_1}(n_1, n_2) + r_{\mathbf{x}_2\mathbf{x}_2}(n_1, n_2) - jr_{\mathbf{x}_1\mathbf{x}_2}(n_1, n_2) + jr_{\mathbf{x}_2\mathbf{x}_1}(n_1, n_2) \tag{10.4}$$

7 $3n_1$

The constituent real-valued autocorrelation and cross-correlation terms that appear on the right side of this expression are given by

$$r_{\mathbf{x}_1\mathbf{x}_1}(n_1, n_2) = E\{\mathbf{x}_1(n_1)\mathbf{x}_1(n_2)\} \tag{10.5a}$$

$$r_{\mathbf{x}_2\mathbf{x}_2}(n_1, n_2) = E\{\mathbf{x}_2(n_1)\mathbf{x}_2(n_2)\} \tag{10.5b}$$

$$r_{\mathbf{x}_1\mathbf{x}_2}(n_1, n_2) = E\{\mathbf{x}_1(n_1)\mathbf{x}_2(n_2)\} \tag{10.5c}$$

$$r_{\mathbf{x}_2\mathbf{x}_1}(n_1, n_2) = E\{\mathbf{x}_2(n_1)\mathbf{x}_1(n_2)\} \tag{10.5d}$$

It is important to reemphasize the point that these correlation entities are generally dependent on the specific integer values assumed by each of the time indices n_1 and n_2. However, when the real and imaginary components of the time series possess a wide-sense stationary type of time invariance to be described shortly, it is only necessary to know the time separation $n_1 - n_2$, not the individual time instants. This greatly simplifies the resultant correlation description.

EXAMPLE 10.1

Let the real-valued random time series $\{\mathbf{x}(n)\}$ be specified by

$$\mathbf{x}(n) = 2n\mathbf{v}(n) - 5$$

for all integers values of n, where the $\mathbf{v}(n)$ are independent samples of a univariate random variable that is uniformly distributed on the interval $[-1, 4]$. The mean value of the nth element of this time series is given by

$$\begin{aligned}\mu_{\mathbf{x}}(n) &= E\{\mathbf{x}(n)\} = 2nE\{\mathbf{v}(n)\} - 5 \\ &= 3n - 5\end{aligned}$$

where use has been made of the fact that $E\{\mathbf{v}(n)\} = 1.5$. In a similar fashion, the autocorrelation is specified by

$$\begin{aligned}r_{\mathbf{xx}}(n_1, n_2) &= E\{\mathbf{x}(n_1)\mathbf{x}(n_2)\} \\ &= E\,\{[2n_1\mathbf{v}(n_1) - 5][2n_2\mathbf{v}(n_2) - 5]\} \\ &= 4n_1n_2E\{\mathbf{v}(n_1)\mathbf{v}(n_2)\} - 10n_1E\{\mathbf{v}(n_1)\} - 10n_2E\{\mathbf{v}(n_2)\} + 25\end{aligned}$$

Due to the assumption of the $\mathbf{v}(n)$ random variable independency, it is readily shown that $E\{\mathbf{v}(n_1)\mathbf{v}(n_2)\}$ equals $\frac{13}{3}$ if $n_1 = n_2$ and $\frac{9}{4}$ if $n_1 \neq n_2$ [or, more compactly, $\frac{9}{4} + \frac{25}{12}\delta(n_1 - n_2)$]. We therefore have

$$r_{\mathbf{xx}}(n_1, n_2) = [9 + \tfrac{25}{3}\,\delta(n_1 - n_2)]n_1n_2 - 15(n_1 + n_2) + 25 \quad \blacksquare$$

EXAMPLE 10.2

Let the complex-valued random time series $\{\mathbf{x}(n)\}$ be governed by the relationship

$$\mathbf{x}(n) = 5\mathbf{v}(n) + j4\mathbf{y}(n)$$

where $\mathbf{v}(n)$ and $\mathbf{y}(n)$ are independent samples of two zero-mean real-valued univariate Gaussian random variables with variances $\sigma_{\mathbf{v}}^2$ and $\sigma_{\mathbf{y}}^2$, respectively, and correlation coefficient $\rho_{\mathbf{vy}}$. Upon applying the definitions given above, it is found that the mean of $\mathbf{x}(n)$ is zero, that is,

$$\mu_{\mathbf{x}}(n) = 0$$

and the autocorrelation is specified by

$$\begin{aligned} r_{\mathbf{xx}}(n_1, n_2) &= E\{[5\mathbf{v}(n_1) + j4\mathbf{y}(n_1)][5\mathbf{v}(n_2) - j4\mathbf{y}(n_2)]\} \\ &= [25\sigma_{\mathbf{v}}^2 + 16\sigma_{\mathbf{y}}^2]\delta(n_1 - n_2) \end{aligned}$$

10.3 STATIONARITY IN THE MEAN, AUTOCORRELATION, AND AUTOCOVARIANCE

Upon examining the behavior of a time series realization (i.e., the observed data), it often happens that the randomness characterizing the time series seems to remain the same as time evolves. Namely, if one extracts several subintervals of sufficient length from a given time series realization, each of the subintervals would display a similarity in time behavior. If this is the case, the underlying time series is said to possess a *stationarity* behavior. An examination of the two time series realizations shown in Figure 10.1 would suggest that one possesses this stationary feature whereas the other does not. We provide next a more quantified measure of exactly what is meant by the term *stationarity*. This involves the introduction of increasingly more restrictive forms of stationarity.

Stationary in the Mean

A random time series is said to be *stationary in the mean* if its individual elements take on values that fluctuate about the same fixed level. In such cases, the time series is said to have a level trend that remains invariant as time evolves. The following definition provides a formal description of this property.

Definition 10.2. The random time series $\{\mathbf{x}(n)\}$ is said to be *stationary in the mean* if the expected value of all its elements are identical, that is,

$$E\{\mathbf{x}(n)\} = \mu_1 + j\mu_2 \qquad \text{for all } n \tag{10.6}$$

(a) Stationary.

(b) Not stationary.

FIGURE 10.1. Example of stationary and nonstationary time series.

where $\mu_1 = E\{\mathbf{x}_1(n)\}$ and $\mu_2 = E\{\mathbf{x}_2(n)\}$ are the constant mean values associated with the real and imaginary time series components.

The time series considered in Example 10.2 is seen to be stationary in the mean, whereas that in Example 10.1 is not. Stationary in the mean describes the lowest level of time series statistical stationarity. From a linear filtering viewpoint, however, it is necessary to invoke the next higher level of stationarity, namely, correlation stationarity.

Stationary in the Autocorrelation and Autovariance

In accordance with expression (10.3), the autocorrelation description is seen to be dependent on the two time instants n_1 and n_2 at which the time series elements in the expectation $E\{\mathbf{x}(n_1)\overline{\mathbf{x}}(n_2)\}$ are being evaluated. If the time behavior of a time series is such that this expected value depends only on the time differences $n_1 - n_2$ and not on the specific time instants n_1 and n_2, then the time series is said to be stationary in the autocorrelation. In particular, the time series $\{\mathbf{x}(n)\}$ is said to be *stationary in the autocorrelation* if the equality

$$\begin{aligned} r_{\mathbf{xx}}(n_1, n_2) &= E\{\mathbf{x}(n_1)\overline{\mathbf{x}}(n_2)\} \\ &= E\{\mathbf{x}(n_1 + m)\overline{\mathbf{x}}(n_2 + m)\} \end{aligned}$$

holds for all selections of the integers m, n_1, and n_2. In a similar fashion, the time series $\{\mathbf{x}(n)\}$ is said to be *stationary in the autocovariance* if the equality

$$\begin{aligned} c_{xx}(n_1, n_2) &= E\{[\mathbf{x}(n_1) - \mu_{\mathbf{x}}(n_1)][\overline{\mathbf{x}}(n_2) - \overline{\mu}_{\mathbf{x}}(n_2)]\} \\ &= E\{[\mathbf{x}(n_1 + m) - \mu_{\mathbf{x}}(n_1 + m)][\overline{\mathbf{x}}(n_2 + m) - \overline{\mu}_{\mathbf{x}}(n_2 + m)]\} \end{aligned}$$

holds for all selections of the integers m, n_1, and n_2. For such time series, the autocorrelation and autocovariance are seen to be dependent only on the time argument difference (or lag) $n_1 - n_2$. We now formalize this most important concept.

Definition 10.3. The time series $\{\mathbf{x}(n)\}$ is said to be *stationary in the autocorrelation* if its autocorrelation sequence is dependent only on the *lag variable* $n = n_1 - n_2$ in the sense that

$$\begin{aligned} r_{\mathbf{xx}}(n) &= E\{\mathbf{x}(k)\overline{\mathbf{x}}(k - n)\} \\ &= E\{\mathbf{x}(k + n)\overline{\mathbf{x}}(k)\} \end{aligned} \tag{10.7a}$$

holds for all integer values of k and n. Furthermore, if the time series $\{x(n)\}$ is stationary in the mean and autocorrelation, it is also *stationary in the autocovariance,* since

$$\begin{aligned} c_{xx}(n) &= E\{[\mathbf{x}(k) - \mu_{\mathbf{x}}][\overline{\mathbf{x}}(k - n) - \overline{\mu}_{\mathbf{x}}]\} \\ &= r_{\mathbf{xx}}(n) - |\mu_{\mathbf{x}}|^2 \end{aligned} \tag{10.7b}$$

holds for all integer selections of k and n.

As examples, the time series studied in Example 10.2 satisfies properties (10.7) and is therefore stationary in the autocorrelation and autocovariance; the time series examined in Example 10.1 does not. In this chapter we direct our attention primarily to the autocorrelation sequence and its associated stationarity behavior. As expression (10.7b) indicates, however, autocovariance stationary follows directly when the time series is also stationary in the mean.

Upon substituting the general representation (10.1) for a complex-valued time series into relationship (10.7a), it is seen that stationarity in the autocorrelation requires that

$$r_{\mathbf{xx}}(n) = r_{\mathbf{x}_1\mathbf{x}_1}(n) + r_{\mathbf{x}_2\mathbf{x}_2}(n) - jr_{\mathbf{x}_1\mathbf{x}_2}(n) + jr_{\mathbf{x}_2\mathbf{x}_1}(n)$$

Thus the real and imaginary components of such a time series must themselves be stationary in the autocorrelation. Clearly, the assumption of *autocorrelation stationarity* dramatically reduces the statistical information needed in characterizing an autocorrelation sequence. Namely, an autocorrelation sequence that is generally a function of two integer time variables then simplifies to a function of one-integer variable. It is indeed fortunate that many practical time series possess (or approximately possess) this stationarity property. For those that do not, it is often possible to decompose the time series into shorter subsequences over which the *stationary in the autocorrelation* assumption is reasonably valid. We may then analyze these subsequences using stationary modeling assumptions to effect a particular objective.

EXAMPLE 10.3

In this example we consider a system's oriented problem related to random time series. Let us consider the causal linear system as governed by the first-order recursive equation

$$y(n) - ay(n - 1) = x(n)$$

in which a is a real-valued parameter. Furthermore, let this system be driven by a random amplitude impulse type of input, as specified by

$$\mathbf{x}(n) = \mathbf{v}\delta(n)$$

In this excitation expression, $\mathbf{v}$ is a real-valued univariate random variable with mean $\mu_{\mathbf{v}}$ and variances $\sigma_{\mathbf{v}}^2$. The system's response to this random excitation is readily found to be

$$\mathbf{y}(n) = \mathbf{v}(a)^n u(n)$$

and is seen to consist of a sequence of exponentially weighted real-valued random variables. Upon taking the expected value of this expression, the mean value of the response time series is found to be

$$\mu_{\mathbf{y}}(n) = \mu_{\mathbf{v}}(a)^n u(n)$$

The response's autocorrelation sequence is formally specified by expression (10.3), that is,

$$r_{\mathbf{yy}}(n_1, n_2) = E\{[\mathbf{v}(a)^{n_1}u(n_1)][\mathbf{v}(a)^{n_2}u(n_2)]\}$$

If we use the fact that $u(n_1)$, $u(n_2)$, and a are constants, this expectation is found to simplify to

$$r_{\mathbf{yy}}(n_1, n_2) = (\sigma_{\mathbf{v}}^2 + \mu_{\mathbf{v}}^2)(a)^{n_1+n_2}u(n_1)u(n_2)$$

It is seen from these results that the mean value is an exponential function of time n, while the autocorrelation sequence does not depend exclusively on the lag variable $n = n_1 - n_2$. Thus the time series $\{\mathbf{x}(n)\}$ is neither stationary in the mean nor in the autocorrelation. ■

10.4 WIDE-SENSE STATIONARITY

In what follows we are concerned primarily with the class of random time series that are stationary in both the mean and the autocorrelation. A random time series that possesses both of these two properties in addition to having fini power is said to be *wide-sense stationary*. This concept is of fundamental ir portance to signal-processing theory and is therefore formally defined.

Definition 10.4. The random time series $\{\mathbf{x}(n)\}$ is said to be *wide-sen stationary* if it is stationary in the mean and autocorrelation and possesses finite variance, that is,

$$\text{(a) } E\{\mathbf{x}(n)\} = \mu_{\mathbf{x}} \qquad \text{for all } n \qquad (10.8$$

$$\text{(b) } r_{\mathbf{xx}}(n) = E\{\mathbf{x}(k)\bar{\mathbf{x}}(k - n)\} \qquad \text{for all } k \text{ and } n \qquad (10.8$$

$$\text{(c) } \sigma_{\mathbf{x}}^2 = r_{\mathbf{xx}}(0) - |\mu_x|^2 \qquad \text{is finite} \qquad (10.8$$

As we demonstrate shortly, the synthesis of optimal linear filters, smoother and predictors becomes a relatively simple matter in the case of wide-sens stationary random time series. This feature, in conjunction with the observatio that many time series encountered in practice are approximately wide-sens stationary, provides the motivation for our study of this important class of tim series.

The autocorrelation sequence associated with a wide-sense stationary tim series possesses a number of salient properties that follow from the definin equations (10.8). Among the more important properties, we have

$$\text{(a) } |r_{\mathbf{xx}}(n)| \leq r_{\mathbf{xx}}(0) \qquad (10.9a$$

$$\text{(b) } r_{\mathbf{xx}}(-n) = \bar{r}_{\mathbf{xx}}(n) \qquad (10.9b$$

$$\text{(c) } r_{\mathbf{xx}}(-n) = r_{\mathbf{xx}}(n) \qquad \text{if } \{\mathbf{x}(n)\} \text{ is real valued} \qquad (10.9c$$

The first property stipulates that the zero-lag autocorrelation element is large than or equal to any of the other autocorrelation lag's magnitudes. To establis this property, one simply uses the Schwarz inequality (9.5) with $X = \mathbf{x}(k)$ an $Y = \bar{\mathbf{x}}(k - n)$. In the second property, the autocorrelation sequence is seen t be a complex conjugate symmetric function of the lag variable n. This propert follows from

$$\begin{aligned} r_{\mathbf{xx}}(-n) &= E\{\mathbf{x}(k)\bar{\mathbf{x}}(k + n)\} \\ &= E\{\overline{\mathbf{x}(k + n)\bar{\mathbf{x}}(k)}\} \\ &= \bar{r}_{\mathbf{xx}}(n) \end{aligned}$$

where the operations of complex conjugation and expectation have been inter changed in going from the second line to the third line. The third property is direct consequence of the second property because for real-valued random tim series, the autocorrelation sequence is always real.

Normalized Autocorrelation Sequence

It is often beneficial to *normalize* the autocorrelation sequence by dividing each of its elements by $r_{xx}(0)$. The resulting sequence is referred to as the *normalized autocorrelation sequence,* that is,

$$\rho_{xx}(n) = \frac{r_{xx}(n)}{r_{xx}(0)} \tag{10.10}$$

This sequence possesses the following properties:

(a) $\rho_{xx}(0) = 1$

(b) $|\rho_{xx}(n)| \leq 1$

(c) $\rho_{xx}(-n) = \bar{\rho}_{xx}(n)$

which are simply the normalized versions of the autocorrelation sequence properties (10.9).

Interpretation of the Autocorrelation Sequence

The autocorrelation sequence gives a convenient procedure for measuring the statistical relationship existent between the time series element $x(k)$ and its n-lag companion element $x(k - n)$. If there is a strong similarity between these two elements, $r_{xx}(n)$ tends to be relatively large in magnitude. On the other hand, dissimilar behavior is manifested in $r_{xx}(n)$ being close to zero. An alternative interpretation of the autocorrelation sequence is provided by considering the time series $\{x(k)\}$ and its n shifted version $\{x(k - n)\}$, as shown in Figure 10.2. If these two time series are very similar (or dissimilar) functions of time,

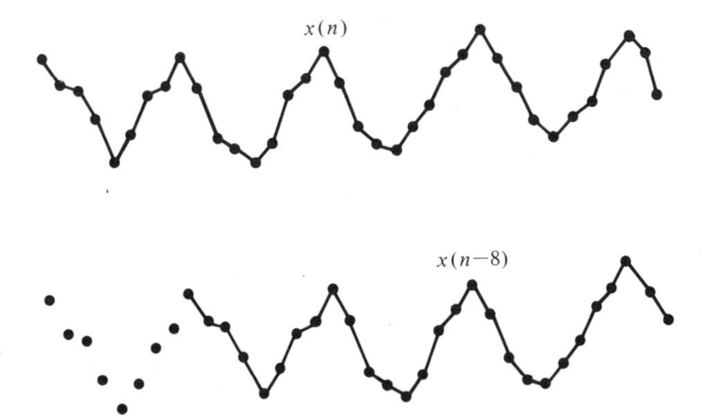

FIGURE 10.2. Time series $\{x(n)\}$ and its right-shifted version $\{x(n-8)\}$, suggesting that $r_{xx}(8) \approx r_{xx}(0)$.

it is found that the autocorrelation element $r_{xx}(n)$ tends to be relatively large (o small) in magnitude.

To illustrate the correlated structure of a random time series, it is prudent t sketch the autocorrelation or normalized autocorrelation sequences versus th lag argument n. In most applications it is found that the autocorrelation sequenc tends to decay to zero as the lag variable n grows. This simply reflects th normal situation in which the time series elements $x(k)$ and $x(k - n)$ tend to b uncorrelated [i.e., $r_{xx}(n) \approx 0$] for sufficiently large values of the lag paramete (time separation). A typical autocorrelation lag plot would appear as shown i Figure 10.3. It should be noted, however, that there exist meaningful time serie for which $r_{xx}(n)$ does not decay to zero for large lags. This is demonstrated i Section 10.8 for the important class of random sinusoids in additive white-nois time series.

10.5 HIGHER-ORDER STATIONARITY

From the developments in Section 10.4 it is clear that the concept of stationarity implies some form of time invariance. Thus stationarity in the mean character izes a time series whose individual elements all possess the same mean value Wide-sense stationarity time series have a pairwise correlation behavior that i dependent only on the time separation of the members constituting the pair Other forms of stationarity are related to the density functions characterizing a time series. As an example, a time series is said to be *stationary of order 1* i the marginal density functions characterizing its constituent elements are the same, that is,

$$f^{(x)}_{\mathbf{x}(n_1)} = f^{(x)}_{\mathbf{x}(n_2)} \tag{10.11}$$

holds for all integers n_1 and n_2. Clearly, a time series that has first-order sta tionarity is also stationary in the mean, but the converse need not hold.

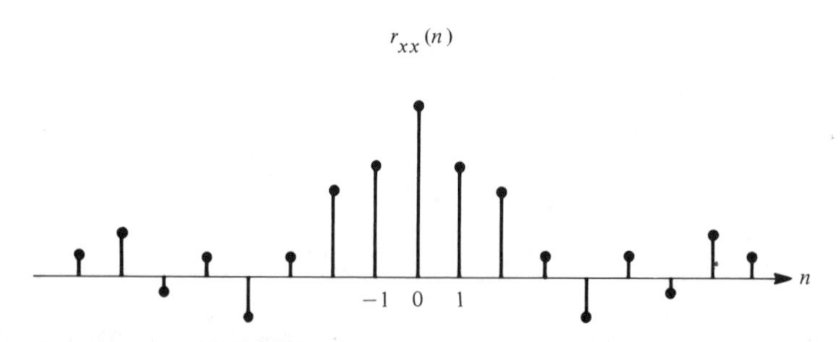

FIGURE 10.3. Typical autocorrelation sequence.

In a similar fashion, a time series is said to be *stationary of order 2* if the joint density function property

$$f_{\mathbf{x}(n_1)\mathbf{x}(n_2)}(x_1, x_2) = f_{\mathbf{x}(n_1+m)\mathbf{x}(n_2+m)}(x_1, x_2) \tag{10.12}$$

holds for all choices of the integers m, n_1, and n_2. In this case the joint density functions characterizing element pairs separated by the same time increment $n_1 - n_2$ are identical. It is readily shown that if a time series is second-order stationary, it is also wide-sense stationary; however, the converse need not follow. Thus wide-sense stationarity is a weaker probabilistic condition than is stationary of order 2. Continuing on in this manner, we next define higher orders of stationarity. For instance, a kth-order-stationary time series would have joint density functions that satisfy

$$f_{\mathbf{x}(n_1)\cdots\mathbf{x}(n_k)}(x_1, \ldots, x_k) = f_{\mathbf{x}(n_1+m)\cdots\mathbf{x}(n_k+m)}(x_1, \ldots, x_k) \tag{10.13}$$

which holds for all choices of the integers $m, n_1, \ldots, n_k$.

As suggested in Section 10.4, when using a time-invariant linear signal processor in combination with a mean-squared-error criterion, the statistical information required is completely contained in the mean value and autocorrelation sequences of a time series. With this in mind, it is apparent that the assumption of wide-sense stationarity imposes the weakest condition needed for synthesizing optimum time-invariant linear mean-squared-error signal processors. It is because of this feature that much of the signal-processing literature is concerned with linear operators and wide-sense stationary time series.

0.6

POWER IN A RANDOM TIME SERIES

When analyzing signal-processing algorithms, it is often necessary to compute the power associated with a random time series. For any finite-length random time series as represented by

$$\mathbf{x}(n_1), \mathbf{x}(n_1 + 1), \ldots, \mathbf{x}(n_2)$$

this computation is straightforward. It simply entails using the standard definition of *average power estimate* as given by

$$\frac{1}{n_2 - n_1 + 1} \sum_{n=n_1}^{n_2} |\mathbf{x}(n)|^2$$

in which the integers n_1 and n_2 identify the time instants at which the time series observations begin and end, respectively. Since the summand terms $|\mathbf{x}(n)|^2$ are random variables, it follows immediately that this power measure is itself random variable. This reflects the fact that in any realization of the finite-lengt time series specified above, the computed power may take on a range of value: The *average* power associated with the time series is then obtained by takin the expected value of the average-power estimate random variable above. Th is found to yield

$$P_{\mathbf{x}} = \frac{1}{n_2 - n_1 + 1} \sum_{n=n_1}^{n_2} E\{|\mathbf{x}(n)|^2\} \tag{10.1}$$

Thus the required power measure is seen to be equal to the average of the mea squared values [i.e., the $E\{|\mathbf{x}(n)|^2\}$ terms] of the time series.

When the time series being analyzed possesses infinite length, its associate power level can be measured by considering the limit expression

$$P_{\mathbf{x}} = E\left\{ \lim_{N \to +\infty} \frac{1}{2N+1} \sum_{n=-N}^{N} |\mathbf{x}(n)|^2 \right\}$$

In evaluating the expected value of this infinite summation, a subtle assumptic is now made, namely, the order of the expectation and summation are inte changed to obtain the desired result,

$$P_{\mathbf{x}} = \lim_{N \to +\infty} \frac{1}{2N+1} \sum_{n=-N}^{N} E\{|\mathbf{x}(n)|^2\} \tag{10.1}$$

Although this operator interchange is always valid for the case in which only finite sum of random variable is involved, it is possible to construct pathologic examples for which this is not true in the infinite sum of random variables ca here being considered. Fortunately, in virtually all practical applications, it possible to interchange the order of the expectation and infinite summatic operators. We leave the uninteresting pathological cases to the more abstract inclined.

If the underlying random time series is composed of elements that have co stant mean-squared-value behavior, a considerable simplification in the pow measure arises. For such situations, the power measure simplifies to $P_{\mathbf{x}}$ $E\{|\mathbf{x}(n)|^2\}$ for both the finite- and infinite-length cases. We have therefore esta lished the following important result.

Theorem 10.1. The power associated with a time series $\{\mathbf{x}(n)\}$ that has constant mean-squared value $E\{|\mathbf{x}(n)|^2\}$ for all n is given by

$$P_{\mathbf{x}} = E\{|\mathbf{x}(n)|^2\} \tag{10.1}$$

It is to be noted that a time series will always possess a constant mean-squar value if it satisfies the stronger conditions of being either stationary of order or wide-sense stationary. For wide-sense stationary time series, the correspon ing power is seen to correspond to the autocorrelation's zero-lag term $r_{xx}(0)$, that is,

$$\begin{aligned} P_{\mathbf{x}} &= r_{\mathbf{xx}}(0) \\ &= \sigma_{\mathbf{x}}^2 + |\mu_{\mathbf{x}}|^2 \end{aligned} \tag{10.17}$$

where $\mu_{\mathbf{x}}$ and $\sigma_{\mathbf{x}}^2$ denote the mean value and variance, respectively, associated with the random variable $\mathbf{x}(n)$.

10.7 WHITE NOISE

When seeking to measure the salient features of a linear system, its response to the deterministic unit impulse and sinusoidal excitations played an important role. Since the unit-impulse input has a Fourier transform that is a constant (i.e., 1), this input excites all *spectral modes* of a linear system. As such, the unit-impulse response is of particular value when identifying an unknown linear operator. On the other hand, a real sinusoidal input has a Fourier transform that is zero everywhere except for two Dirac delta unit impulses. A sinusoidal input then excites only discrete spectral modes, and as a consequence, the corresponding response provides a well-defined characterization of the linear operator's dynamics only at these two spectral frequencies.

When analyzing wide-sense stationary time series, it is logical to ask the question as to whether there exist prototype random time series that have properties analogous to those of the deterministic unit impulse and sinusoidal time series. The answer to this insightful question is in the affirmative and the analogous random time series possess autocorrelation sequences which are equal to the deterministic unit-impulse and cosine sequences. We explore these prototype random time series with the foreknowledge that they play a fundamental role in our study of wide-sense stationary processes.

With these thoughts in mind, let us now characterize the essential features of a wide-sense stationary time series $\{\mathbf{w}(n)\}$ whose mean value is zero and whose autocorrelation sequence is equal to the weighted unit-impulse sequence

$$r_{\mathbf{ww}}(n) = E\{\mathbf{w}(k)\overline{\mathbf{w}}(k - n)\} = \sigma_{\mathbf{w}}^2\, \delta(n) \tag{10.18}$$

We refer to any zero-mean random time series that possesses this impulse-type autocorrelation sequence behavior as being *white noise*. This terminology arises from the fact that this particular autocorrelation sequence has as its Fourier transform the constant $\sigma_{\mathbf{w}}^2$. As we will see shortly, this implies that the underlying random time series contains all spectral (i.e., frequency) components with an equal power level. This property, however, is precisely that which is possessed by *white light*. Hence the terminology white noise has a well-based foundation in physics. Furthermore, we shall hereafter reserve the descriptive notation $\{\mathbf{w}(n)\}$ to designate a white-noise time series.

Real-Valued White Noise

It is prudent to consider separately the cases of real-valued and complex-valued white noise. Upon examination of the defining property of white noise (10.18), a straightforward procedure for generating a zero-mean real-valued white noise process is now specified.

Definition 10.5. The real-valued wide-sense stationary random time series $\{\mathbf{w}(n)\}$ is said to be *white noise* if the following two conditions hold:

(a) Its individual real-valued random variable elements $\mathbf{w}(n)$ each have zero mean and variance $\sigma_{\mathbf{w}}^2$.

(b) Its elements are pairwise uncorrelated, so that $E\{\mathbf{w}(n_1)\mathbf{w}(n_2)\} = 0$ for $n_1 \neq n_2$.

Under these conditions, the autocorrelation sequence associated with the white-noise time series satisfies the whiteness condition $r_{\mathbf{ww}}(n) = \sigma_{\mathbf{w}}^2\delta(n)$.

To generate a real-valued white-noise time series, we thus need access to a *random number* generator which sequentially computes zero-mean uncorrelated samples of a random variable.† Frequently, these random variable samples satisfy the stronger condition of being independent. It is to be noted that the particular distribution which governs the white-noise time series elements is not specified. If these elements are Gaussianly distributed, the term *Gaussian white noise* is often used in identifying the corresponding time series. A similar terminology would be used for describing other distribution types, such as uniform white noise, exponential white noise, and Poisson white noise.

Complex-Valued White Noise

When characterizing a complex-valued random white-noise time series, it is necessary to characterize statistically the relationships that exist between its real and imaginary random time series elements,

$$\mathbf{w}(n) = \mathbf{w}_1(n) + j\mathbf{w}_2(n)$$

so that the fundamental property (10.18) is satisfied. The zero mean value requirement is seen to imply that $\mu_{\mathbf{w}} = 0$, indicating that the real and imaginary random time series $\{\mathbf{w}_1(n)\}$ and $\{\mathbf{w}_2(n)\}$ must each have zero means. Substituting the above expression for $\mathbf{w}(n)$ into relationship (10.3), the corresponding autocorrelation sequence is specified by

†From a strict interpretation of Definition 10.5, the white-noise random variables $\mathbf{w}(n)$ need not be governed by the same distribution. They need only have zero means, the same variance, and must be pairwise uncorrelated. Furthermore, the zero-mean condition in part (a) can be dropped in which case the resultant autocorrelation becomes $r_{\mathbf{ww}}(n) = (\sigma_{\mathbf{w}}^2 + \mu_{\mathbf{x}}^2)\delta(n)$, where $\mu_x = E\{\mathbf{x}(n)\}$.

$$r_{\mathbf{ww}}(n) = E\{[\mathbf{w}_1(n_1) + j\mathbf{w}_2(n_1)][\mathbf{w}_1(n_2) - j\mathbf{w}_2(n_2)]\}$$

We now seek to impose probabilistic properties on the random variable elements $\mathbf{w}_1(n)$ and $\mathbf{w}_2(n)$ so that this expectation is equal to the unit impulse behavior (10.18) required of white noise. The following definition provides the conditions.

Definition 10.6. The complex-valued wide-sense stationary time series $\{\mathbf{w}(n)\} = \{\mathbf{w}_1(n)\} + j\{\mathbf{w}_2(n)\}$ is said to be *white noise* if the following conditions hold:

(a) The real-valued random variables $\mathbf{w}_1(n)$ and $\mathbf{w}_2(n)$ each have zero means and constant finite variances $\sigma^2_{\mathbf{w}_1}$ and $\sigma^2_{\mathbf{w}_2}$, respectively, for all values of n.

(b) The real-valued random variables $\mathbf{w}_1(n)$ and $\mathbf{w}_2(n)$ are each white noise sequences so that

$$E\{\mathbf{w}_1(m)\mathbf{w}_1(n)\} = \sigma^2_{\mathbf{w}_1}\delta(n - m)$$

$$E\{\mathbf{w}_2(m)\mathbf{w}_2(n)\} = \sigma^2_{\mathbf{w}_2}\delta(n - m)$$

(c) The real and imaginary elements are orthogonal:

$$E\{\mathbf{w}_1(m)\mathbf{w}_2(n)\} = 0 \qquad \text{for all } m, n$$

Under these conditions, the mean value and autocorrelation sequences associated with the white-noise time series are specified by

$$\mu_{\mathbf{w}}(n) = 0 \qquad \text{and} \qquad r_{\mathbf{ww}}(n) = (\sigma^2_{\mathbf{w}_1} + \sigma^2_{\mathbf{w}_2})\,\delta(n)$$

In this model of complex-valued white noise, the real and imaginary components of the time series are each seen to be real-valued white noise that are mutually uncorrelated.

10.8 RANDOM SINUSOID TIME SERIES

In various communication applications, the need to study random sinusoidal time series arises in a natural manner. For example, pulsed sinusoidal bursts are often employed in digital communication systems for transmitting information. With this example providing motivation, we now characterize the *random sinusoid* time series specified by

$$\mathbf{x}(n) = a\cos(\omega_0 n + \boldsymbol{\theta}) \qquad -\infty < n < \infty \tag{10.19}$$

where a and ω_0 are real-valued parameters identifying the amplitude and frequency of the sinusoid, respectively, and $\boldsymbol{\theta}$ is a random phase variable that is uniformly distributed in $(-\pi, \pi]$. Thus a random sinusoid time series is seen to have a deterministic sinusoid waveform with a random phase displacement. The mean value of a general term of this time series is zero since the expected-value expression

$$
\begin{aligned}
E\{\mathbf{x}(n)\} &= \int_{-\pi}^{\pi} f_\theta(\theta) x(n)\, d\theta \\
&= \int_{-\pi}^{\pi} \frac{1}{2\pi} a \cos(\omega_0 n + \theta)\, d\theta = 0
\end{aligned}
$$

is seen to be proportional to the area of the cosine function over one period of θ. The associated autocorrelation sequence is next obtained in accordance with

$$
\begin{aligned}
r_{\mathbf{xx}}(n_1, n_2) &= E\{\mathbf{x}(n_1)\bar{x}(n_2)\} \\
&= \int_{-\pi}^{\pi} \frac{a^2}{2\pi} \cos(\omega_0 n_1 + \theta) \cos(\omega_0 n_2 + \theta)\, d\theta \\
&= \frac{a^2}{4\pi} \int_{-\pi}^{\pi} (\cos[\omega_0(n_1 + n_2) + 2\theta] + \cos[\omega_0(n_1 - n_2)]\, d\theta \\
&= \frac{a^2}{2} \cos[\omega_0(n_1 - n_2)]
\end{aligned}
\tag{10.20}
$$

Since the random sinusoidal time series has a constant mean value (i.e., zero) and an autocorrelation sequence that depends only on the time difference $n_1 - n_2$ and not on the specific values of n_1 and n_2, it follows that it is a wide-sense stationary process. With this result in hand, we now offer the following more general result, whose proof is obtained in a similar fashion.

Theorem 10.2. The real-valued random sinusoidal time series as specified by

$$
\mathbf{x}(n) = \sum_{k=1}^{p} a_k \cos(\omega_k n + \boldsymbol{\theta}_k) \qquad -\infty < n < \infty \tag{10.21}
$$

is wide-sense stationary, where a_k and ω_k are real-valued amplitude and frequency parameters, respectively, and the $\boldsymbol{\theta}_k$ are independent uniformly distributed random variables on $(-\pi, \pi]$. The associated mean value and autocorrelation sequences are specified by

$$
E\{\mathbf{x}(n)\} = 0 \qquad -\infty < n < \infty \tag{10.22}
$$

$$
r_{\mathbf{xx}}(n) = \frac{1}{2} \sum_{k=1}^{p} a_k^2 \cos(\omega_k n) \qquad -\infty < n < \infty \tag{10.23}
$$

In applications entailing complex-valued sinusoidal time series, a simple extension of the foregoing development provides the following useful result.

Theorem 10.3. The complex-valued random sinusoidal time series as specified by

$$\mathbf{x}(n) = \sum_{k=1}^{p} a_k e^{j(\omega_k n + \boldsymbol{\theta}_k)} \tag{10.24}$$

is wide-sense stationary, where a_k are generally complex-valued amplitudes, ω_k are the corresponding real-valued frequency parameters, and the $\boldsymbol{\theta}_k$ are independent random variables uniformly distributed on the interval $(-\pi, \pi]$. This time series has its mean and autocorrelation sequences specified by

$$E\{\mathbf{x}(n)\} = 0 \qquad -\infty < n < \infty \tag{10.25}$$

$$r_{\mathbf{xx}}(n) = \sum_{k=1}^{p} |a_k|^2 e^{j\omega_k n} \qquad -\infty < n < \infty \tag{10.26}$$

0.9 RANDOM SINUSOIDS IN ADDITIVE WHITE NOISE

A surprising large number of relevant applications are concerned with the detection of random sinusoids embedded in additive white noise. As an example, in adaptive radar array processing, it is desired to detect the presence of sinusoids and then estimate their associated amplitudes and frequencies.

Real-Valued Random Sinusoids in Additive White Noise

The real-valued random signal model

$$\mathbf{x}(n) = \mathbf{w}(n) + \sum_{k=1}^{p} a_k \cos(\omega_k n + \boldsymbol{\theta}_k) \tag{10.27}$$

describes a real-valued time series in which $\{\mathbf{w}(n)\}$ designates an additive zero-mean real-valued white noise of variance σ_w^2 and the sinusoid phase angles $\boldsymbol{\theta}_k$ are again taken to be independent random variables uniformly distributed on $(-\pi, \pi]$. It is typically assumed that the white-noise components $\mathbf{w}(n)$ and the random sinusoidal terms are pairwise uncorrelated. Under these conditions, the time series (10.27) mean value sequence is specified by

$$\mu_{\mathbf{x}}(n) = 0 \qquad -\infty < n < \infty \tag{10.28}$$

Similarly, the associated autocorrelation sequence is given by

$$r_{\mathbf{x}}(n_1, n_2) = E\{\mathbf{x}(n_1)\overline{\mathbf{x}}(n_2)\}$$
$$= \sigma_w^2\delta(n_1 - n_2) + \frac{1}{2}\sum_{k=1}^{p} a_k^2 \cos\left[\omega_k(n_1 - n_2)\right] \quad (10.29)$$

In arriving at this result, we have used the results of Theorem 10.2 and the assumption that the white-noise and sinusoidal components are uncorrelated. Since the mean value behavior is constant (i.e., zero) and the autocorrelation sequence depends only on the time difference $n_1 - n_2$, it follows that random time series (10.27) under the given assumptions is wide-sense stationary. A more compact representation for its autocorrelation sequence that reflects the wide-sense stationarity property is therefore given by

$$r_{xx}(n) = \sigma_w^2\delta(n) + \frac{1}{2}\sum_{k=1}^{p} a_k^2 \cos(\omega_k n) \qquad -\infty < n < \infty \quad (10.30)$$

Complex-Valued Random Sinusoids in Additive White Noise

The results above may be directly extended to the complex-valued random time series case. In particular, consider the random time series as governed by

$$\mathbf{x}(n) = \mathbf{w}(n) + \sum_{k=1}^{p} a_k e^{j(\omega_k n + \theta_k)} \quad (10.31)$$

in which $\mathbf{w}(n)$ is zero-mean complex-valued additive white noise with variance $\sigma_{\mathbf{w}}^2$ and the $\boldsymbol{\theta}_k$ are independent random variables uniformly distributed on $(-\pi, \pi]$. Under the assumption that the white-noise and sinusoidal components are pairwise uncorrelated, it is straightforwardly shown that this time series is wide-sense stationary with zero mean value

$$\mu_{\mathbf{x}}(n) = 0 \quad (10.32)$$

and autocorrelation sequence

$$r_{\mathbf{xx}}(n) = \sigma_{\mathbf{w}}^2\delta(n) + \sum_{k=1}^{p} |a_k|^2 e^{j\omega_k n} \quad (10.33)$$

We shall use the results captured in expressions (10.30) and (10.33) in our future developments related to optimal linear estimators.

10.10 SPECTRAL DENSITY FUNCTION

In a large number of analytically based studies, it is beneficial to investigate the statistical behavior of a random wide-sense stationary time series in the transform domain. To begin this development, let us consider the z-transform of the deterministic autocorrelation sequence as defined by

$$R_{\mathbf{xx}}(z) = \sum_{n=-\infty}^{\infty} r_{\mathbf{xx}}(n)z^{-n} \tag{10.34}$$

The transform $R_{\mathbf{xx}}(z)$ is referred to as the *complex spectral density function*. In a similar manner, the Fourier transform of the autocorrelation sequence is a real-valued-positive semidefinite function and is called the *spectral density function;* it is given by

$$S_{\mathbf{xx}}(\omega) = \sum_{n=-\infty}^{\infty} r_{\mathbf{xx}}(n)e^{-j\omega n} \tag{10.35}$$

It will be shown shortly that this real-valued nonnegative spectral density function $S_{\mathbf{xx}}(\omega)$ characterizes the manner in which power is distributed in the random time series $\{\mathbf{x}(n)\}$ as a function of frequency. As indicated in Chapter 3, the spectral density function is obtained by evaluating the complex spectral density function on the unit circle [i.e., $S_{xx}(\omega) = R_{xx}(e^{j\omega})$]. It is for this reason that $R_{xx}(z)$ is referred to as the complex spectral density function. In what follows it is notationally preferable to use the simplified description $S_{\mathbf{xx}}(\omega)$ to denote the spectral density function rather than the more cumbersome equivalent $R_{\mathbf{xx}}(e^{j\omega})$. It should be emphasized that the z-transform entity (10.34) is not really a density function, since it does not generally possess the required properties of a density function (e.g., being real and nonnegative valued).

The spectral density function and the autocorrelation sequence are seen to be Fourier transform pairs, as is evident from expression (10.35). Thus to obtain the autocorrelation sequence associated with a given spectral density function, we simply evaluate the inverse Fourier transform integral

$$r_{\mathbf{xx}}(n) = \frac{1}{2\pi} \int_{-\pi}^{\pi} S_{\mathbf{xx}}(\omega)e^{j\omega n}\, d\omega \tag{10.36}$$

In certain applications, the power spectral density function is more useful than the equivalent autocorrelation sequence when depicting the salient characteristics of a random time series (e.g., random sinusoidal time series). The converse can also be true as is the case for white noise.

The behavior of the spectral density function $S_{\mathbf{xx}}(\omega)$ as a function of the frequency parameter ω provides a meaningful characterization of the underlying random time series. It is found that if a time series is rich in certain frequencies, its associated spectral density tends to be peaked over those same frequencies, and vice versa. Thus the spectral density function provides a good signature for the spectral content of a wide-sense stationary time series.

EXAMPLE 10.4

Consider the real-valued wide-sense stationary random time series $\{\mathbf{x}(n)\}$ whose autocorrelation sequence is specified by

$$r_{\mathbf{xx}}(n) = \mu_{\mathbf{x}}^2 + \sigma_{\mathbf{x}}^2 a^{|n|}$$

in which $\mu_{\mathbf{x}} = E\{x(n)\}$, $\sigma_{\mathbf{x}}^2$ denotes time series variance, and the real parameter a has a magnitude of less than 1. The associated spectral density function is then given by

$$\begin{aligned} S_{\mathbf{xx}}(\omega) &= \sum_{n=-\infty}^{\infty} (\mu_{\mathbf{x}}^2 + \sigma_{\mathbf{x}}^2 a^{|n|})e^{-j\omega n} \\ &= 2\pi\mu_{\mathbf{x}}^2\delta(\omega) + \frac{(1 - a^2)\sigma_{\mathbf{x}}^2}{1 + a^2 - 2a\cos\omega} \end{aligned}$$

The expected-value term $\mu_{\mathbf{x}}^2$ appearing in $r_{\mathbf{xx}}(n)$ is seen to produce an impulse at zero frequency, while the exponential term is low (high) frequency in nature if the parameter a is positive (negative). ■

Some of the more commonly appearing autocorrelation sequences and their associated spectral density functions are given in Table 10.1 and plotted in Figure 10.4. It is worthwhile to comment briefly on the spectral behavior of the various entries there listed. The first entry typically corresponds to the nonzero constant mean component of a wide-sense stationary time. It is seen to give rise

TABLE 10.1 Commonly Occurring Spectral Density Functions

Time Series, $\{\mathbf{x}(n)\}$	Autocorrelation, $r_{\mathbf{xx}}(n)$	Spectral Density Function, $S_{\mathbf{xx}}(\omega)$ $-\pi < \omega \le \pi$
Constant	$\mu_{\mathbf{x}}^2$	$2\pi\mu_{\mathbf{x}}^2\delta(\omega)$
White noise	$\sigma_{\mathbf{x}}^2\delta(n)$	$\sigma_{\mathbf{x}}^2$
Bandlimited white noise, $\lvert a\rvert < 1$	$\sigma_{\mathbf{x}}^2 \dfrac{\sin(\omega_1 n)}{\pi n}$	$\sigma_{\mathbf{x}}^2[u(\omega + \omega_1) - u(\omega - \omega_1)]$
First-order exponential	$\sigma_{\mathbf{x}}^2 a^{\lvert n\rvert}$	$\dfrac{(1 - a^2)\sigma_{\mathbf{x}}^2}{1 + a^2 - 2a\cos\omega}$
Complex sinusoids	$\sum_{k=1}^{p} \lvert a_k\rvert^2 e^{\omega_k n}$	$2\pi \sum_{k=1}^{p} \lvert a_k\rvert^2 \delta(\omega - \omega_1)$
Real sinusoids	$\sum_{k=1}^{p} a_k^2 \cos(\omega_k n)$	$\pi \sum_{k=1}^{p} a_k^2[\delta(\omega + \omega_k) + \delta(\omega - \omega_k)]$

to a Dirac impulse spectral density function located at zero frequency. Thus the appearance of peaky behavior at $\omega = 0$ in $S_{xx}(\omega)$ is indicative of a nonzero mean in the associated time series. On the other hand, the second and third entries suggest that a level component in the spectral density function is seen to correspond to either a white-noise or a bandlimited-white-noise term in the

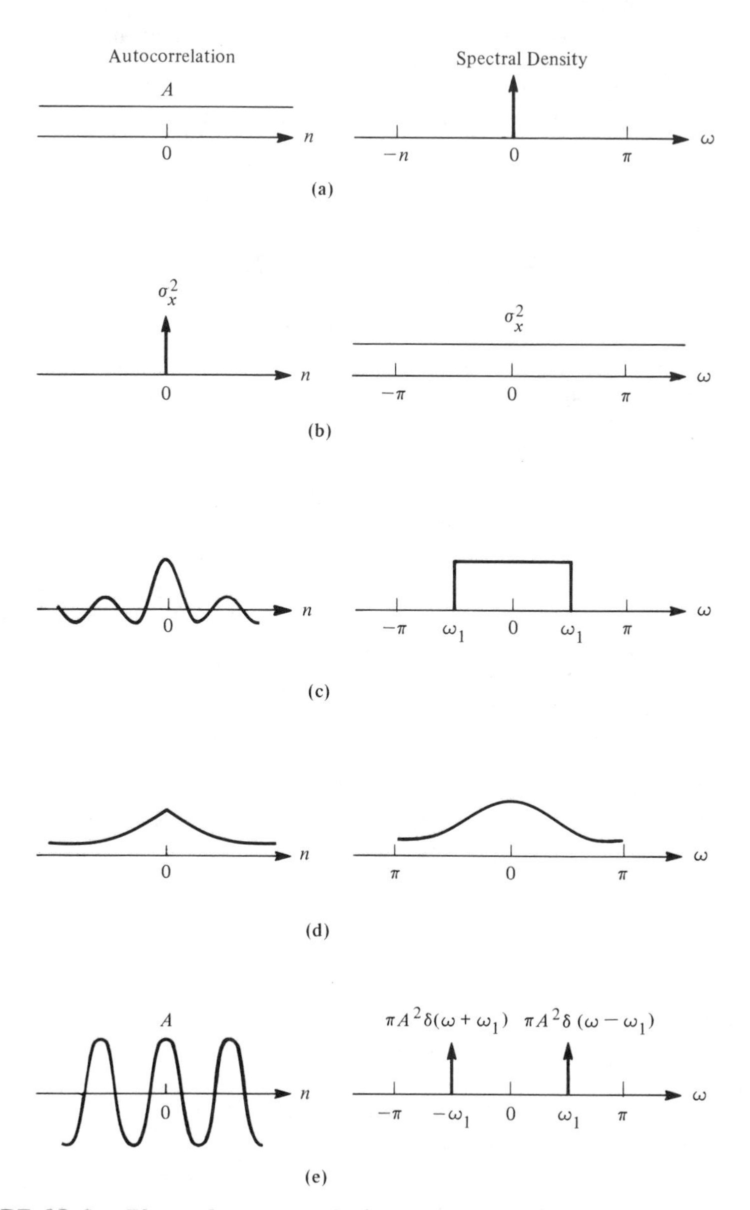

IGURE 10.4. Plots of autocorrelation and spectral density pairs: (a) constant;) white noise; (c) bandlimited white noise; (d) first-order exponential; (e) sinusoidal.

time series. The first-order exponential time series is seen to produce a spectral density function that has a relatively broad peak about $\omega = 0$ (π) if the damping parameter a is positive (negative). Finally, random sinusoidal-type time series are seen to produce Dirac impulses located at the sinusoidal frequencies. Peaky behavior in the spectral density function is therefore seen to be associated with a random sinusoidal (or narrow band) component in the time series. As a final comment, information regarding the sign of the random sinusoid amplitude a_k is lost in the autocorrelation and spectral density function due to their appearance in squared form (a_k^2). Thus phase information is seen to be lost when looking at second-order statistics.

10.11 PROPERTIES OF THE SPECTRAL DENSITY FUNCTION

As its name suggests, the *spectral density function* has properties normally associated with a density function in that it characterizes the power content of wide-sense stationary time series as a function of frequency. Some of these more important properties are captured in Theorem 10.4. The periodicity property (10.38) follows directly from the periodic property associated with all Fourier transforms. On the other hand, the symmetrical property (10.40) follows from the autocorrelation sequence condition $r_{xx}(n) = r_{xx}(-n)$ satisfied by all real-valued time series. For complex-valued time series, the autocorrelation property $r_{xx}(n) = \bar{r}_{xx}(-n)$ ensures the real valuedness of the spectral density function, but it is no longer a symmetric function of frequency. To establish a proof of properties (10.37) and (10.39), we need the developments made in Chapter 11. The importance of the spectral density function concept cannot be overemphasized. It is found to be of exceptional use in theoretical studies as well as in practical considerations.

Theorem 10.4. Let the wide-sense stationary time series $\{\mathbf{x}(n)\}$ have the power spectral density function $S_{\mathbf{xx}}(\omega)$. It follows that this function is real nonnegative,

$$S_{\mathbf{xx}}(\omega) \geq 0 \tag{10.37}$$

as well as being periodic with period 2π, so that

$$S_{\mathbf{xx}}(\omega + 2\pi) = S_{\mathbf{xx}}(\omega) \tag{10.38}$$

Furthermore, the power in the wide-sense stationary time series $\{\mathbf{x}(n)\}$ in the frequency interval $\omega_1 \leq \omega \leq \omega_2$ is given by

$$P_{\mathbf{x}}(\omega_1, \omega_2) = \int_{\omega_1}^{\omega_2} S_{\mathbf{xx}}(\omega)\, d\omega \tag{10.39}$$

Finally, if the time series $\{\mathbf{x}(n)\}$ is *real* valued, the spectral density function is a symmetric function, that is,

$$S_{\mathbf{xx}}(\omega) = S_{\mathbf{xx}}(-\omega) \tag{10.40}$$

10.12 CROSS-CORRELATION AND CROSS-SPECTRAL DENSITY FUNCTIONS

In describing the autocorrelation sequence associated with a complex-valued random time series, it was necessary to introduce the cross-covariance relationship between the real and imaginary components of the time series. The concept of crosscorrelation has an even more general setting when characterizing the statistical relationship that exists between any two random time series. In particular, the cross-correlation sequence associated with the two wide-sense stationary time series $\{\mathbf{x}(n)\}$ and $\{\mathbf{y}(n)\}$ is defined to be

$$\begin{aligned} r_{\mathbf{xy}}(n) &= E\{\mathbf{x}(k+n)\overline{\mathbf{y}}(k)\} \\ &= E\{\mathbf{x}(k)\overline{\mathbf{y}}(k-n)\} \end{aligned} \tag{10.41}$$

This cross-correlation sequence is seen to possess the property

$$r_{\mathbf{yx}}(n) = \bar{r}_{\mathbf{xy}}(-n) \tag{10.42}$$

where the order of the subscripts is to be noted. Upon taking the Fourier transform of the cross-correlation sequence (10.41), we obtain the *cross-spectral density function*

$$S_{\mathbf{xy}}(\omega) = \sum_{n=-\infty}^{\infty} r_{\mathbf{xy}}(n)e^{-j\omega n} \tag{10.43}$$

This function is of particular value when studying the statistical relationship between a linear operator's excitation and corresponding response time series. The cross-correlation sequence associated with a given cross-spectral density function is obtained by evaluating the inverse Fourier transform integral

$$r_{\mathbf{xy}}(n) = \frac{1}{2\pi}\int_{-\infty}^{\infty} S_{\mathbf{xy}}(\omega)e^{j\omega n}\,d\omega \tag{10.44}$$

Thus the cross-correlation sequence and cross-spectral density function are Fourier transform pairs.

10.13 REMOVAL OF MEAN VALUE

It is common practice to remove the mean value (or an estimate of such) from data before applying a signal processing algorithm to the data. The effects of this preprocessing on the autocorrelation and spectral density functions are easily established. In particular, let $\{\mathbf{x}(n)\}$ be a wide-sense stationary random time series whose mean value is $\mu_{\mathbf{x}}$. Let us now consider the auxiliary time series as specified by

$$\tilde{\mathbf{x}}(n) = \mathbf{x}(n) - \mu \tag{10.45}$$

in which μ is a constant. If the constant μ equals $\mu_{\mathbf{x}}$, as is normally the objective, the time series $\{\tilde{\mathbf{x}}(n)\}$ has zero mean. Whatever the case, this auxiliary time series is simply a level-shifted version of the original time series and its mean value is specified by $\mu_{\tilde{\mathbf{x}}} = \mu_{\mathbf{x}} - \mu$.

Upon using the formal definition (10.7), it follows directly that the autocorrelation sequence associated with the auxiliary time series (10.45) is given by

$$r_{\tilde{\mathbf{x}}\tilde{\mathbf{x}}}(n) = r_{\mathbf{xx}}(n) - \mu_{\mathbf{x}}\bar{\mu} - \mu\bar{\mu}_{\mathbf{x}} + \mu\bar{\mu} \tag{10.46}$$

The autocorrelations of the two time series therefore differ by a constant. Upon taking the Fourier transform of expression (10.46), it is found that the spectral density functions of the two time series are related as

$$S_{\tilde{\mathbf{x}}\tilde{\mathbf{x}}}(\omega) = S_{\mathbf{xx}}(\omega) - 2\pi(\mu_x\bar{\mu}_{\mathbf{x}} + \mu\bar{\mu}_{\mathbf{x}} - \mu\bar{\mu})\delta(\omega) \tag{10.47}$$

Thus the spectral density functions differ by a Dirac impulse of strength $-2\pi(\mu_x\mu + \mu\bar{\mu} - \mu\bar{\mu})$ located at $\omega = 0$ and are otherwise identical.

Special Case $\mu = \mu_x$

As suggested above, the constant μ is typically set equal to the mean value $\mu_{\mathbf{x}}$ of the original time series. In this case the autocorrelation and spectral density function associated with the auxiliary time series (10.45) simplify to

$$r_{\tilde{x}\tilde{x}}(n) = r_{xx}(n) - |\mu_{\mathbf{x}}|^2 \tag{10.48}$$

and

$$S_{\tilde{\mathbf{x}}\tilde{\mathbf{x}}}(\omega) = S_{\mathbf{xx}}(\omega) - 2\pi|\mu_x|^2\delta(\omega) \tag{10.49}$$

It is to be noted that for the case $\mu = \mu_x$ now being considered, the autocorrelation $r_{\tilde{\mathbf{x}}\tilde{\mathbf{x}}}(n)$ corresponds to the autocovariance associated with the original time series $\{\mathbf{x}(n)\}$. The effect of removing the mean value from original time series is manifested in the auxiliary spectral density function $S_{\tilde{\mathbf{x}}\tilde{\mathbf{x}}}(\omega)$ not having a Dirac impulse behavior at $\omega = 0$. This observation is a direct consequence of the fact that the original spectral density function $S_{\mathbf{xx}}(\omega)$ has a Dirac impulse of strength $2\pi|\mu_{\mathbf{x}}|^2$ at $\omega = 0$.

10.14 SUMMARY

In this chapter the mean value and correlation characterization of random time series has been examined, with the ultimate objective of synthesizing linear systems (or signal processors) that minimize a mean-squared-error criterion. It was found that the assumption of wide-sense stationarity provides a meaningful simplification to the structure of the autocorrelation sequence. Fortunately, many random time series encountered in practical applications are either wide-sense stationary in behavior, or they may be decomposed into sub-time series that are individually wide-sense stationary in behavior.

SUGGESTED READINGS

FREEMAN, H., *Discrete-Time Systems*. New York: John Wiley & Sons, Inc., 1965.

LARSON, H. J., and B. O. SHUBERT, *Probabilistic Models in Engineering Sciences*, Vols. I and II. New York: John Wiley & Sons, 1979.

O'FLYNN, M., *Probabilities and Random Variables, and Random Processes*. New York: Harper & Row, Publishers, Inc., 1982.

OPPENHEIM, A. V., and R. W. SCHAFER, *Digital Signal Processing*. Englewood Cliffs, N.J.: Prentice-Hall, Inc., 1975.

PRIESTLY, M. B., *Spectral Analysis and Time Series*, Vols. I and II. London: Academic Press, Inc. (London) Ltd., 1981.

SCHWARTZ, M., and L. SHAW, *Signal Processing*. New York: McGraw-Hill Book Company, 1975.

TRETTER, S. A., *Discrete-Time Signal Processing*. New York: John Wiley & Sons, Inc., 1976.

PROBLEMS

10.1 Consider the real-valued random time series specified by

$$\mathbf{x}(n) = 2\mathbf{v}(n) - 3$$

where $\{\mathbf{v}(n)\}$ is a sequence of uncorrelated random variables each of which has mean $\mu_{\mathbf{v}}$ and variance $\sigma_{\mathbf{v}}^2$. Determine the mean value and autocorrelation sequence for $\{\mathbf{x}(n)\}$. Comment on the stationarity of this time series.

10.2 Determine the mean value and autocorrelation sequence associated with the time series

$$\mathbf{x}(n) = 3\mathbf{v}(n) - 5\mathbf{v}(n-1) + 4$$

where $\{\mathbf{v}(n)\}$ is a sequence of uncorrelated random variables each with mean $\mu_{\mathbf{v}}$ and variance $\sigma_{\mathbf{v}}^2$. Is this time series stationary in the mean and autocorrelation?

10.3 Consider the time series specified by

$$\mathbf{x}(n) = -7n^2\mathbf{v}(n) + 3n - 5$$

where $\{\mathbf{v}(n)\}$ is a sequence of uncorrelated random variables each of which has mean value $\mu_{\mathbf{v}}$ and variance $\sigma_{\mathbf{v}}^2$. Compute the mean value and autocorrelation sequence associated with this time series and make comments relative to its stationarity.

10.4 Let $\{\mathbf{x}(n)\}$ be a random time series whose individual elements are correlated random variables. Determine the quantities $E\{\mathbf{x}(n)\}$, var $\{\mathbf{x}(n)\}$, and $r_{\mathbf{xx}}(n, n+1)$ when

(a) $$f_{\mathbf{x}(n)\mathbf{x}(n+1)}(x_1, x_2) = \tfrac{1}{3}\,\delta(x_1+1)\,\delta(x_2+1) + \tfrac{1}{6}\,\delta(x_1+1)\,\delta(x_2-1) + \tfrac{1}{6}\,\delta(x_1-1)\,\delta(x_2+1) + \tfrac{1}{3}\,\delta(x_1-1)\,\delta(x_2-1)$$

(b) $$f_{\mathbf{x}(n)\mathbf{x}(n+1)}(x_1, x_2) = \begin{cases} 2 & 0 \le x_1 \le x_2 \le 1 \\ 0 & \text{otherwise} \end{cases}$$

10.5 Let $\{\mathbf{x}(n)\}$ and $\{\mathbf{y}(n)\}$ be two random time series which are uncorrelated so that cov $[\mathbf{x}(n), \mathbf{y}(m)] = 0$ for all integers m and n. If a third time series is constructed according to

$$\mathbf{z}(n) = \alpha\mathbf{x}(n) + \beta\mathbf{y}(n)$$

where α and β are scalars, determine $E\{\mathbf{z}(n)\}$ and $r_{\mathbf{zz}}(m, n)$ in terms of $E\{\mathbf{x}(n)\}$, $E\{\mathbf{y}(n)\}$, $r_{\mathbf{xx}}(m, n)$, and $r_{\mathbf{yy}}(m, n)$.

10.6 Generalize the results of Example 10.3, in which the parameter a is complex and the excitation is given by

$$\mathbf{x}(n) = (\mathbf{v} + j\mathbf{y})\delta(n)$$

where $\mathbf{v}$ and $\mathbf{y}$ are real-valued random variables with means $\mu_{\mathbf{v}}$ and $\mu_{\mathbf{y}}$ and variances $\sigma_{\mathbf{v}}^2$ and $\sigma_{\mathbf{y}}^2$, respectively, and covariance $\sigma_{\mathbf{v}}\sigma_{\mathbf{y}}\rho_{\mathbf{vy}}$. In particular, show that

$$\mu_{\mathbf{y}}(n) = (\mu_{\mathbf{v}} + j\mu_{\mathbf{y}})(a)^n u(n)$$

$$r_{\mathbf{yy}}(n) = (\sigma_{\mathbf{v}}^2 + \sigma_{\mathbf{y}}^2 + |\mu_{\mathbf{v}}|^2 + |\mu_{\mathbf{y}}|^2)(a)^{n_1}(\bar{a})^{n_2}u(n_1)u(n_2)$$

10.7 Compute the power associated with the time series considered in Problems **(a)** 10.1; **(b)** 10.2; **(c)** 10.3.

10.8 Let the random time series $\{\mathbf{x}(n)\}$ have the autocorrelation $r_{\mathbf{xx}}(n) = \sigma_x^2(a)^{|n|}$, where $|a| < 1$. Compute the average power associated with this time series.

10.9 Show that if the wide-sense stationary random time series $\{\mathbf{x}(n)\}$ is composed of uncorrelated random variables, it is a random nonzero mean, white-noise time series with autocorrelation sequence $(\sigma_{\mathbf{x}}^2 + |\mu_{\mathbf{x}}|^2)\delta(n)$, where $\mu_{\mathbf{x}} = E\{\mathbf{x}(n)\}$ and $\sigma_{\mathbf{x}}^2 = \text{var}\{x(n)\}$.

10.10 Verify the correctness of the autocorrelation sequences associated with random sinusoids as specified in **(a)** equation (10.23); **(b)** equation (10.26).

10.11 For the random sinusoids in white-noise time series (10.27), verify expressions (10.28) and (10.30).

10.12 Determine the power spectral densities functions associated with the following autocorrelation sequences.
(a) $r_{\mathbf{xx}}(n) = 3\delta(n) + 2\cos(0.5\pi n)$
(b) $r_{\mathbf{xx}}(n) = -4\delta(n+1) + 7\delta(n) - 4\delta(n-1)$
(c) $r_{\mathbf{xx}}(n) = b_1\,\delta(n+1) + b_0\,\delta(n) + b_1\,\delta(n-1)$
where $b_0 \geq |b_1|$. Use a direct evaluation of expression (10.35) in your solution.

10.13 Verify the entries of Table 10.1.

10.14 Consider the time series generated by

$$\mathbf{x}(n) = 2\mathbf{w}(n) - 7$$

where $\{\mathbf{w}(n)\}$ is a sequence of uncorrelated random variables with zero mean and variance 5 (i.e., white noise). Show that

$$S_{xx}(\omega) = 20 + 98\pi\delta(\omega)$$

CHAPTER 11

Linear Systems and Random Signals

11.1

INTRODUCTION

Although the study of random signals is important in its own right, the true beauty of signal processing is revealed upon characterizing the effect that linear operations have on such signals. To be specific, let us consider the situation depicted in Figure 11.1, in which a random time signal excites a time-invariant linear system. Fundamental questions immediately suggest themselves concerning the statistical nature of the response signal thereby generated. In particular, how are the excitation signal's probabilistic measures, such as its mean, variance, and autocorrelation, manifested in the response signal? For example, one intuitively expects that the response's autocorrelation is somehow linked to the excitation's autocorrelation and the linear operator's dynamics.

In this chapter we answer fundamental issues of this nature. Our ultimate goal is to be able to use these probabilistic interrelationships to synthesize linear operators that achieve prescribed signal-processing objectives. With this capability it is possible to realize such fundamental signal operations as filtering, estimation, and prediction. It is shown that a rather thorough theoretical analysis is made possible when the random signals employed are wide-sense stationary and the linear operator is time invariant. In applications where signal stationarity

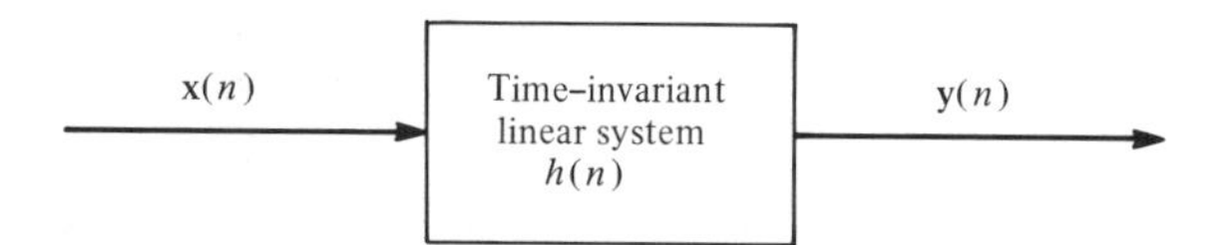

FIGURE 11.1. Response of a time-invariant linear operator to a random excitation.

is not globally true, however, this analysis may still be used by employing data windowing to isolate time intervals over which signal stationarity is reasonably satisfied.

11.2 LINEAR SYSTEM'S RESPONSE TO RANDOM EXCITATIONS

Signal processing constitutes one of the more important applications of linear system theory. In this section we examine some of the more basic probabilistic properties of the response associated with the general stable time-invariant linear system

$$\mathbf{y}(n) = \sum_{k=-\infty}^{\infty} h(k)\mathbf{x}(n - k) \tag{11.1}$$

when it is excited by a random input. For purposes of generality, the excitation can be complex valued. We have here not restricted the system to be causal, although this condition is readily effected upon setting the unit-impulse response elements $h(n) = 0$ for negative values of n. Whatever the case, the random excitation time series $\{\mathbf{x}(n)\}$ is seen to give rise to the response time series $\{\mathbf{y}(n)\}$, which is itself random. This cause-and-effect relationship is depicted in Figure 11.1. The response random variable $\mathbf{y}(n)$ is seen to be a linear combination of the excitation random variable elements. As such, it is apparent that the statistics of the response will somehow be interrelated with those of the excitation through the linear system's weighting (or unit-impulse response) elements $h(n)$. We shall now obtain the mean value, autocorrelation, and cross-correlation sequences characterizing the random response for general random excitations. When the excitation happens to be wide-sense stationary, it is shown that the response is also wide-sense stationary.

Response's Mean Value Sequence

The mean value behavior of the response time series is obtained directly by taking the expected value of relationship (11.1). Using the linearity of the expected value operator, we find that

$$\begin{aligned}\mu_{\mathbf{y}}(n) &= E\{\mathbf{y}(n)\} = E\left\{\sum_{k=-\infty}^{\infty} h(k)\mathbf{x}(n-k)\right\} \\ &= \sum_{k=-\infty}^{\infty} h(k)\mu_{\mathbf{x}}(n-k)\end{aligned} \tag{11.2}$$

where $\mu_{\mathbf{x}}(n) = E\{\mathbf{x}(n)\}$. Thus the response's mean value sequence is seen to be a convolution of the linear system's weighting sequence with the excitation's mean value sequence. In arriving at this result, the operations of summation and expectation have been interchanged. This is valid, provided that the system being analyzed is stable.

Response's Autocorrelation and Autocovariance Sequences

The autocorrelation sequence associated with the response time series is obtained in a similar fashion by using the fundamental definition

$$r_{\mathbf{yy}}(n_1, n_2) = E\{\mathbf{y}(n_1)\overline{\mathbf{y}}(n_2)\}$$

Upon substituting expression (11.1) into this relationship and using the expected value operator's linearity, we obtain

$$\begin{aligned}r_{\mathbf{yy}}(n_1, n_2) &= E\left\{\left[\sum_{k=-\infty}^{\infty} h(k)\mathbf{x}(n_1-k)\right]\left[\sum_{m=-\infty}^{\infty} \overline{h}(m)\overline{\mathbf{x}}(n_2-m)\right]\right\} \\ &= \sum_{k=-\infty}^{\infty}\sum_{m=-\infty}^{\infty} h(k)\overline{h}(m)r_{\mathbf{xx}}(n_1-k, n_2-m)\end{aligned} \tag{11.3}$$

where the order of the expectation and summation operators have been interchanged. If the excitation time series autocorrelation behavior and the linear system's unit-impulse response are known, evaluation of this double summation enables us to systematically determine the response's autocorrelation sequence.

In a similar fashion, the response's autocovariance sequence is obtained by appealing to the basic definition

$$\begin{aligned}c_{\mathbf{yy}}(n_1, n_2) &= E\{[\mathbf{y}(n_1) - \mu_{\mathbf{y}}(n_1)][\overline{\mathbf{y}}(n_2) - \overline{\mu}_{\mathbf{y}}(n_2)]\} \\ &= r_{\mathbf{yy}}(n_1, n_2) - \mu_{\mathbf{y}}(n_1)\overline{\mu}_{\mathbf{y}}(n_2)\end{aligned} \tag{11.4}$$

This autocovariance sequence expression is seen to depend on the underlying mean value and autocorrelation sequence. Thus upon substitution of relationships (11.2) and (11.3), it is found that

$$c_{\mathbf{yy}}(n_1, n_2) = \sum_{k=-\infty}^{\infty}\sum_{m=-\infty}^{\infty} h(k)\overline{h}(m)c_{\mathbf{xx}}(n_1-k, n_2-m) \tag{11.5}$$

Comparing this autocovariance expression with the corresponding autocorrelation expression (11.3), we see that they are functionally equivalent. In particular, the impact that the linear system's unit-impulse response has on these entities is made apparent through these convolution summations.

Cross-Correlation and Cross-Covariance of Excitation and Response

When analyzing the performance properties of linear signal processors, the cross-correlation sequence relating the excitation and response time series inevitably arises. This cross-correlation is formally given by

$$r_{\mathbf{yx}}(n_1, n_2) = E\{\mathbf{y}(n_1)\bar{\mathbf{x}}(n_2)\}$$

After substitution of relationship (11.1) into this expression and then using the expected value operator's linearity, we have

$$\begin{aligned} r_{\mathbf{yx}}(n_1, n_2) &= E\left\{\left[\sum_{k=-\infty}^{\infty} h(k)\mathbf{x}(n_1 - k)\right]\bar{\mathbf{x}}(n_2)\right\} \\ &= \sum_{k=-\infty}^{\infty} h(k)r_{\mathbf{xx}}(n_1 - k, n_2) \end{aligned} \tag{11.6}$$

As in the response autocorrelation expression, we are able to compute the cross-correlation if complete knowledge of $h(n)$ and $r_{\mathbf{xx}}(n_1, n_2)$ is available. In a similar fashion, the cross-covariance between the excitation and response time series is found to be

$$\begin{aligned} c_{\mathbf{yx}}(n_1, n_2) &= E\{[\mathbf{y}(n_1) - \mu_{\mathbf{y}}(n_1)][\mathbf{x}(n_2) - \bar{\mu}_{\mathbf{x}}(n_2)]\} \\ &= \sum_{k=-\infty}^{\infty} h(k)c_{\mathbf{xx}}(n_1 - k, n_2) \end{aligned} \tag{11.7}$$

Again, the similarity between the cross-correlation and cross-covariance expressions is manifested through the same convolution summation format.

11.3 LINEAR SYSTEM'S RESPONSE TO WIDE-SENSE STATIONARY EXCITATIONS

When a stable time-invariant linear system is excited by a wide-sense stationary time series, the corresponding response is itself wide-sense stationary. This follows directly from the fact that the excitation's mean value and autocorrelation sequences are of the form

$$\mu_{\mathbf{x}}(n) = \mu_{\mathbf{x}} \qquad \text{and} \qquad r_{\mathbf{xx}}(n_1, n_2) = r_{\mathbf{xx}}(n_1 - n_2)$$

When this mean value sequence behavior is substituted into relationship (11.2), the response time series' mean value sequence is seen to be equal to the constant:

$$\mu_{\mathbf{y}}(n) = \mu_{\mathbf{y}} = \mu_{\mathbf{x}} \sum_{k=-\infty}^{\infty} h(k)$$

Similarly, the response's autocorrelation sequence becomes

$$r_{\mathbf{yy}}(n_1, n_2) = \sum_{k=-\infty}^{\infty} \sum_{m=-\infty}^{\infty} h(k)\bar{h}(m) r_{\mathbf{xx}}(n_1 - n_2 + m - k)$$

This autocorrelation sequence is seen to be a function of the time difference $n_1 - n_2$, and not on the specific individual time indices n_1 and n_2. Thus the response autocorrelation sequence can be equivalently described by the lag variable expression

$$r_{\mathbf{yy}}(n) = \sum_{k=-\infty}^{\infty} \sum_{m=-\infty}^{\infty} h(k)\bar{h}(m) r_{\mathbf{xx}}(n + m - k)$$

These relationships for mean value and autocorrelation, however, constitute the properties that describe a wide-sense stationary time series. To emphasize the importance that is attached to this result, the following theorem is offered.

Theorem 11.1. Let the stable time-invariant linear system characterized by

$$y(n) = \sum_{k=-\infty}^{\infty} h(k)x(n - k) \tag{11.8}$$

be excited by a wide-sense stationary time series $\{\mathbf{x}(n)\}$ whose mean value and autocorrelation sequences are given by $\mu_{\mathbf{x}}$ and $r_{\mathbf{xx}}(n)$, respectively. The corresponding response sequence $\{\mathbf{y}(n)\}$ is also wide-sense stationary with constant mean value

$$\mu_{\mathbf{y}}(n) = \mu_{\mathbf{y}} = \mu_{\mathbf{x}} \sum_{k=-\infty}^{\infty} h(k) \tag{11.9}$$

and associated autocorrelation sequence

$$r_{\mathbf{yy}}(n) = \sum_{k=-\infty}^{\infty} \sum_{m=-\infty}^{\infty} h(k)\bar{h}(m) r_{\mathbf{xx}}(n + m - k) \tag{11.10}$$

Moreover, the cross-correlation between the response and excitation time series is specified by the convolution summation

$$r_{\mathbf{yx}}(n) = \sum_{k=-\infty}^{\infty} h(k) r_{\mathbf{xx}}(n - k) \tag{11.11}$$

From these results it is clear that the wide-sense stationarity of a time series is not altered by a time-invariant linear operation. This property can be of value when making an analysis of a linear digital signal processor's effect on wide-sense stationary data. Much of our future development employs this result directly.

EXAMPLE 11.1

Consider the stable causal linear system

$$y(n) + ay(n - 1) = bx(n)$$

in which a and b are real numbers, with a having a magnitude less than 1. Let this system be excited by a wide-sense stationary white-noise process $\mathbf{x}(n) = \mathbf{w}(n)$ with autocorrelation sequence

$$r_{\mathbf{ww}}(n) = \sigma_{\mathbf{w}}^2 \delta(n)$$

It is now desired to determine the corresponding response's autocorrelation sequence and the cross-correlation sequence existent between the white-noise excitation and response time series. Using the results of Theorem 11.1 and the fact that the weighting sequence of this system is given by $h(n) = b(-a)^n u(n)$, we have

$$r_{\mathbf{yy}}(n) = b^2 \sigma_{\mathbf{w}}^2 \sum_{k=0}^{\infty} \sum_{m=0}^{\infty} (-a)^k (-a)^m \delta(n + m - k)$$

For nonnegative values of n, this double summation simplifies to

$$\begin{aligned} r_{\mathbf{yy}}(n) &= b\sigma_{\mathbf{w}}^2 \sum_{m=0}^{\infty} (-a)^m (-a)^{n+m} \\ &= \frac{b\sigma_{\mathbf{w}}^2}{1 - a^2} (-a)^n \qquad \text{for } n \geq 0 \end{aligned}$$

Since the autocorrelation sequence for a real-valued time series is a symmetric function of n, we therefore have established that

$$r_{\mathbf{yy}}(n) = \frac{b\sigma_{\mathbf{w}}^2}{1 - a^2} (-a)^{|n|} \qquad \text{for all } n \tag{11.12}$$

In a similar fashion, it is found that

$$\begin{aligned} r_{\mathbf{yx}}(n) &= b\sigma_{\mathbf{w}}^2 \sum_{k=0}^{\infty} (-a)^k \delta(n-k) \\ &= b\sigma_{\mathbf{w}}^2 (-a)^n u(n) \end{aligned} \tag{11.13}$$

■

11.4 TRANSIENT RESPONSE BEHAVIOR

In any realistic signal-processing application, the time series data to be processed is not observable over the infinite past. Typically, the data observations will begin at a finite time instant, which without loss of generality is here taken to correspond to $n = 0$. More specifically, the random causal time series to be processed has the mathematical form

$$\mathbf{x}(n) = \mathbf{x}_1(n)u(n) \tag{11.14}$$

where $\{\mathbf{x}_1(n)\}$ corresponds to a wide-sense stationary random time series that is observable only for nonnegative time. The autocorrelation sequence associated with this one-sided random time series is formally given by

$$\begin{aligned} r_{\mathbf{xx}}(n_1, n_2) &= E\{\mathbf{x}(n_1)\overline{\mathbf{x}}(n_2)\} \\ &= E\{\mathbf{x}_1(n_1)u(n_1)\overline{\mathbf{x}}_1(n_2)u(n_2)\} \\ &= r_{\mathbf{x}_1\mathbf{x}_1}(n_1 - n_2)u(n_1)u(n_2) \end{aligned} \tag{11.15}$$

where use of the fact the random time series $\{\mathbf{x}_1(n)\}$ is wide-sense stationary has been made. It is to be noted that the one-sided random time series (11.14) is not a wide-sense stationary time series, since its autocorrelation depends on the specific values of n_1 and n_2, not on their difference $n_1 - n_2$ [e.g., in general $r_x(4, 5) \neq r_x(-1, 0) = 0$].

If the one-sided random time series (11.14) is applied to a time-invariant linear system whose weighting sequence is given by $\{h(n)\}$, expression (11.3) indicates that the corresponding response's autocorrelation sequence is specified by

$$\begin{aligned} r_{\mathbf{yy}}(n_1, n_2) & \\ &= \sum_{k=-\infty}^{\infty} \sum_{m=-\infty}^{\infty} h(k)\overline{h}(m) r_{\mathbf{x}_1\mathbf{x}_1}(n_1 - n_2 + m - k)u(n_1 - k)u(n_2 - m) \\ &= \sum_{k=-\infty}^{n_1} \sum_{m=-\infty}^{n_2} h(k)\overline{h}(m) r_{\mathbf{x}_1\mathbf{x}_1}(n_1 - n_2 + m - k) \end{aligned} \tag{11.16}$$

This response autocorrelation sequence is in general dependent on the specific values assumed by the time indices n_1 and n_2 because of their appearance in the upper limits of the two summations. Thus the response time series is not wide-sense stationary in behavior. This simply reflects the fact that a linear system's response to a nonstationary excitation must itself be nonstationary.

Although the response time series $\mathbf{y}(n)$ is nonstationary, it becomes increasingly more stationary in behavior for large positive values of n so long as the linear system is stable. This behavior is a direct consequence of the approximation

$$\sum_{k=-\infty}^{n_1} \sum_{m=-\infty}^{n_2} h(k)\bar{h}(m) r_{\mathbf{x}_1\mathbf{x}_1}(n_1 - n_2 + m - k) \approx \sum_{k=-\infty}^{\infty} \sum_{m=-\infty}^{\infty} h(k)\bar{h}(m) r_{\mathbf{x}_1\mathbf{x}_1}(n_1 - n_2 + m - k)$$

which becomes more exact as the upper limit indices n_1 and n_2 each take on increasing large positive values. This is, of course, contingent on the stability of the linear system, which in turn implies that $h(n)$ approaches zero for large positive n.

In summary, if a stable time-invariant linear system is excited by the one-sided stationary-type random time series (11.14), the corresponding response becomes increasingly stationary in behavior for large positive values of time. This condition suggests that the stationary behavior prevails once the transient effects caused by the initial excitation of the stable linear system have decayed away to zero. After a sufficiently long period, the steady-state behavior of the response will dominate and the response thereafter takes on an increasingly stationary behavior. The required time period over which the transient effect is significant is directly proportional to the time required for $h(n)$ to decay away sufficiently close to zero.

EXAMPLE 11.2

Consider the stable causal linear system characterized by

$$y(n) + ay(n - 1) = bx(n)$$

in which a is a real paramter whose magnitude is less than 1. Let this system be excited by the real-valued *one-sided white-noise* time series

$$\mathbf{x}(n) = \mathbf{w}(n)u(n) \tag{11.17}$$

where $\{\mathbf{w}(n)\}$ is a sequence of uncorrelated random variables each with zero mean and variance $\sigma_{\mathbf{w}}^2$ (i.e., white noise). This excitation is often referred to as being ''pseudo white'' because it is deterministic in behavior for negative time [i.e., $\mathbf{x}(n) = 0$ for $n < 0$] and white in nature for nonnegative time. To compute the mean value, autocorrelation, and cross-correlation sequences of this system's response, it is necessary first to determine the excitation's mean value and autocorrelation sequences. Since the random variables $\mathbf{w}(n)$ have zero mean value, it therefore follows that

$$\mu_{\mathbf{x}}(n) = 0 \qquad \text{for all } n$$

Using this expression and the fact that the linear system's unit impulse response is

$$h(n) = b(-a)^n u(n)$$

we compute the response's mean value using expression (11.2), that is,

$$\mu_{\mathbf{y}}(n) = 0$$

The excitation signal's autocorrelation sequence is given by expression (11.15), which becomes

$$\begin{aligned} r_{\mathbf{xx}}(n_1, n_2) &= E\{[\mathbf{w}(n_1)u(n_1)][\mathbf{w}(n_2)u(n_2)]\} \\ &= \sigma_{\mathbf{w}}^2\delta(n_1 - n_2)u(n_1)u(n_2) \end{aligned}$$

The response's autocorrelation sequence is formally specified by relationship (11.16). Since the excitation $x(n)$ is identically zero for negative n, it must follow that

$$r_{\mathbf{yy}}(n_1, n_2) = 0 \qquad \text{for either } n_1 < 0 \text{ or } n_2 < 0$$

When the indices n_1 and n_2 are both nonnegative, the response autocorrelation expression becomes

$$r_{\mathbf{yy}}(n_1, n_2) = b^2\sigma_{\mathbf{w}}^2 \sum_{k=0}^{n_1} \sum_{m=0}^{n_2} (-a)^{k+m}\delta(n_1 - n_2 + m - k) \qquad \text{for } n_1, n_2 \geq 0$$

It is readily shown that this double summation can equivalently be expressed as

$$r_{\mathbf{yy}}(n_1, n_2) = \begin{cases} b^2\sigma_{\mathbf{w}}^2(-a)^{n_2-n_1}\dfrac{1 - a^{2(n_1+1)}}{1 - a^2} & \text{for } n_2 \geq n_1 \\[2ex] b^2\sigma_{\mathbf{w}}^2(-a)^{n_1-n_2}\dfrac{1 - a^{2(n_2+1)}}{1 - a^2} & \text{for } n_1 \geq n_2 \end{cases}$$

In a similar fashion, the cross-correlation is found to simplify to

$$r_{\mathbf{yx}}(n_1, n_2) = \begin{cases} b\sigma_{\mathbf{w}}^2(-a)^{n_1-n_2} & \text{for } n_1 \geq n_2 \geq 0 \\ 0 & \text{otherwise} \end{cases}$$

Upon careful examination of the response's mean value, autocorrelation, and cross-correlation sequences here obtained, it is found that for sufficiently large positive values of n_1 and n_2, they agree with the wide-sense stationary results obtained in Example 11.1. This simply reflects the fact that the nonstationary response time series here determined becomes increasingly wide-sense stationary in behavior as the transient response decays away after application of the one-sided random input at $n = 0$. ■

11.5 NOISE POWER REDUCTION RATIO

An important function of linear systems in signal-processing applications resides in their ability to reduce the effects of corruptive additive noise. To illustrate this point, let the time series being processed have the additive form

$$\mathbf{x}(n) = \mathbf{s}(n) + \mathbf{w}(n)$$

in which $\mathbf{s}(n)$ denotes a useful *signal* component and $\mathbf{w}(n)$ an undesired additive-*white-noise* component. If this time series is used as the input to a stable time-invariant linear system with weighting sequence $\{h(n)\}$, the resultant response may be decomposed into its signal and white-noise-generated components according to

$$\begin{aligned}\mathbf{y}(n) &= \sum_{k=-\infty}^{\infty} h(k)\mathbf{s}(n-k) + \sum_{k=-\infty}^{\infty} h(k)\mathbf{w}(n-k) \\ &= \mathbf{y}_s(n) + \mathbf{y}_w(n)\end{aligned} \tag{11.18}$$

In the standard filtering application, it is desired to select (or synthesize) the linear system's weighting sequence so that the system passes undistorted the signal component while filtering out the noise component, that is,

$$\mathbf{y}_s(n) = \mathbf{s}(n) \qquad \text{and} \qquad \mathbf{y}_w(n) = 0 \tag{11.19}$$

In most applications, this ideal behavior can be realized only approximately. Thus a more practical goal is that of transmitting the signal in a relatively undistorted fashion while reducing the noise component's power to a lower level. For wide-sense stationary noise, it was shown in Section 10.8 that the noise power level is measured by $E\{\mathbf{y}_w^2(n)\}$. It is this power measure that is to be reduced by a proper choice of the linear system's weighting sequence.

We shall now examine a time-invariant linear system's ability to reduce noise power transfer when the noise is zero mean and white in nature. In particular, it is assumed that the autocorrelation sequence associated with the additive-noise time series $\{\mathbf{w}(n)\}$ is given by

$$\begin{aligned}r_{\mathbf{ww}}(n) &= E\{|\mathbf{w}(n)|^2\}\delta(n) \\ &= \sigma_{\mathbf{w}}^2\delta(n)\end{aligned}$$

where it is recalled that the power in the excitation noise is given by $\sigma_{\mathbf{w}}^2$. In accordance with expression (11.10), the autocorrelation sequence of the corresponding response to this white-noise excitation is given by

$$\begin{aligned} r_{\mathbf{y_w y_w}}(n) &= \sum_{k=-\infty}^{\infty} \sum_{m=-\infty}^{\infty} h(k)\bar{h}(m) r_{\mathbf{ww}}(n + m - k) \\ &= E\{|\mathbf{w}(n)|^2\} \sum_{k=-\infty}^{\infty} h(k)\bar{h}(k - n) \end{aligned}$$

The power in this noise component is then given by the value of the autocorrelation sequence $r_{\mathbf{y_w y_w}}(n)$ at lag $n = 0$, that is,

$$E\{|\mathbf{y_w}(n)|^2\} = E\{|\mathbf{w}(n)|^2\} \sum_{k=-\infty}^{\infty} |h(k)|^2$$

Thus a measure of a linear system's ability to reduce additive white noise is provided by the *noise power reduction ratio,* defined by

$$\rho = \frac{E\{|\mathbf{y_w}(n)|^2\}}{E\{|\mathbf{w}(n)|^2\}} = \sum_{k=-\infty}^{\infty} |h(k)|^2 \tag{11.20}$$

The noise power reduction ratio is seen to be equal to the ratio of the response's noise power to the excitation's noise power. As expression (11.20) indicates, a linear system's ability to filter out white noise is directly linked to that system's unit-impulse response.

It is recalled that the overall desired signal-processing objective is to select the linear system's weighting sequence so that the dual requirements (11.19) are satisfied. The first requirement [i.e., $\mathbf{y}_s(n) = \mathbf{s}(n)$] often manifests itself as a constraint on the system weighting sequence. In such cases, we then seek that weighting sequence which satisfies that constraint and minimizes the noise power reduction ratio (11.20). The next example illustrates this filter design philosophy.

EXAMPLE 11.3

Let us consider the general stable causal time-invariant linear system characterized by

$$y(n) = \sum_{k=0}^{\infty} h(k)x(n - k)$$

In various applications it is required that the steady-state response of this system to the unit-step input (the signal) be characterized by $y(n) = 1$ for large positive values of n. This behavior is seen to require that the linear system's weighting sequence satisfy

$$\sum_{k=0}^{\infty} h(k) = 1$$

Thus, in seeking a linear system of a given structure (e.g., nonrecursive or recursive) that has a small noise power reduction ratio, this unit-step-response constraint is often invoked. ■

11.6 SPECTRAL CHARACTERIZATION OF LINEAR SYSTEMS

It was shown in Section 11.3 that when a stable time-invariant linear system is excited by a wide-sense stationary time series, the response is also wide-sense stationary. In particular, if the excitation time series has an autocorrelation sequence $r_{\mathbf{xx}}(n)$, the response's autocorrelation sequence is specified by

$$r_{\mathbf{yy}}(n) = \sum_{k=-\infty}^{\infty} \sum_{m=-\infty}^{\infty} h(k)\bar{h}(m) r_{\mathbf{xx}}(n + m - k)$$

where $\{h(n)\}$ denotes the weighting sequence characterizing the linear system. To determine the linear system's effect on the spectral content of the excitation, we shall now evaluate the spectral density function associated with $r_{\mathbf{yy}}(n)$. This spectral density is formally specified by

$$\begin{aligned} S_{\mathbf{yy}}(\omega) &= \sum_{n=-\infty}^{\infty} r_{\mathbf{yy}}(n) e^{-j\omega n} \\ &= \sum_{n=-\infty}^{\infty} \left[\sum_{k=-\infty}^{\infty} \sum_{m=-\infty}^{\infty} h(k)\bar{h}(m) r_{\mathbf{xx}}(n + m - k) \right] e^{-j\omega n} \end{aligned}$$

Interchanging the order of summation and then making the change of summation variables of $p = n + m - k$ for n, we have

$$\begin{aligned} S_{\mathbf{yy}}(\omega) &= \sum_{k=-\infty}^{\infty} h(k) e^{-j\omega k} \sum_{m=-\infty}^{\infty} \bar{h}(m) e^{j\omega m} \sum_{p=-\infty}^{\infty} r_{\mathbf{xx}}(p) e^{-j\omega p} \\ &= H(e^{j\omega}) \bar{H}(e^{j\omega}) S_{\mathbf{xx}}(\omega) \\ &= |H(e^{j\omega})|^2 S_{\mathbf{xx}}(\omega) \end{aligned}$$

where $H(e^{j\omega})$ denotes the frequency response associated with the linear system. We have thereby proven the following useful characterization of linear systems.

Theorem 11.2. Let the stable time-invariant linear system governed by

$$y(n) = \sum_{k=-\infty}^{\infty} h(k)x(n - k) \quad (11.21)$$

be excited by a wide-sense stationary time series whose spectral density function is given by $S_{\mathbf{xx}}(\omega)$. The corresponding wide-sense stationary response then has the spectral density function

$$S_{\mathbf{yy}}(\omega) = |H(e^{j\omega})|^2 S_{\mathbf{xx}}(\omega) \quad (11.22)$$

where $H(e^{j\omega})$ is the frequency response of the linear system as given by

$$H(e^{j\omega}) = \sum_{n=-\infty}^{\infty} h(n)e^{-j\omega n} \quad (11.23)$$

Furthermore, the cross-spectral density function between the response and excitation time series is specified by

$$S_{\mathbf{yx}}(\omega) = H(e^{j\omega})S_{\mathbf{xx}}(\omega) \quad (11.24)$$

When analyzing the relationship between the wide-sense stationary excitation and response associated with a stable time-invariant linear operation, Theorem 11.2 provides a convenient tool. In particular, relationship (11.22) suggests that one may spectrally shape the response in a desired fashion by a proper choice of the system frequency response. The experienced user is able to conceptualize in the frequency domain and thereby make meaningful signal-processing design decisions. In more theoretically based studies, however, it is often more convenient to use the z-transform approach in obtaining the required time series characterization. This becomes apparent in the developments to be made in Chapter 12 and in the companion volume. With this thought in mind, the following analogy to Theorem 11.2 is offered. Its proof is straightforward and is left as an exercise.

Theorem 11.3. Let the stable time-invariant linear system governed by

$$y(n) = \sum_{k=-\infty}^{\infty} h(k)x(n - k) \quad (11.25)$$

be excited by a wide-sense stationary time series whose complex spectral density function is given by $R_{xx}(z) = Z[r_{xx}(n)]$. The corresponding wide-sense stationary response then has the complex spectral density function

$$R_{\mathbf{yy}}(z) = H(z)\overline{H}(\overline{z}^{-1})R_{\mathbf{xx}}(z) \quad (11.26)$$

where $H(z)$ is the transfer function of the linear system as given by

$$H(z) = \sum_{n=-\infty}^{\infty} h(n)z^{-n} \quad (11.27)$$

and $\overline{H}(\overline{z}^{-1})$ is the transfer function of its conjugated time transposed image as specified by

$$\overline{H}(\overline{z}^{-1}) = \sum_{n=-\infty}^{\infty} \overline{h}(-n)z^{-n} \tag{11.28}$$

For real-valued $\{h(n)\}$, it is readily shown that $\overline{H}(\overline{z}^{-1}) = H(z^{-1})$. Furthermore, the complex cross-spectral density function between the response and excitation time series is specified by

$$R_{\mathbf{yx}}(z) = H(z)R_{\mathbf{xx}}(z) \tag{11.29}$$

As suggested in Theorem 11.3, the z-transform provides an exceptionally powerful analytical tool for studying wide-sense stationary time series and their processing by time-invariant linear systems. We shall use the z-transform extensively in our study of linear signal processors as typified by filters, smoothers, and prediction. To illustrate the power of the z-transform, we offer the following simple example.

EXAMPLE 11.4

Let us solve the problem treated in Example 11.1 using the z-transform approach. In this example, the causal linear system being analyzed is governed by

$$y(n) + ay(n-1) = bx(n)$$

in which a and b are real parameters where $|a| < 1$. Moreover, let this system be excited by a real-valued wide-sense stationary white-noise process whose autocorrelation sequence is specified by

$$r_{\mathbf{xx}}(n) = \sigma_{\mathbf{w}}^2\delta(n)$$

We shall now use the results of Theorem 11.3 to determine the response's autocorrelation and the excitation–response crosscorrelation sequences. To begin, we use the fact that this system's transfer function and the complex spectral density function associated with the excitation are given by

$$H(z) = \frac{bz}{z+a} \qquad \text{and} \qquad R_{\mathbf{xx}}(z) = \sigma_{\mathbf{w}}^2$$

Thus, according to expression (11.26), we have the response's complex spectral density function

$$R_{\mathbf{yy}}(z) = \frac{bz}{z+a}\left(\frac{bz^{-1}}{z^{-1}+a}\right)\sigma_{\mathbf{w}}^2 \qquad \text{for } |a| < |z| < |a|^{-1}$$

To obtain the corresponding response's autocorrelation sequence, we now determine the inverse z-transform of this function. Using the partial-fraction expansion method, we have

$$R_{\mathbf{yy}}(z) = \frac{b^2\sigma_{\mathbf{w}}^2}{1 - a^2}\left(\frac{z}{z + a} + \frac{1}{1 + az} - 1\right)$$

Upon taking the inverse z-transform, the required response autocorrelation is found to be

$$\begin{aligned} r_{\mathbf{yy}}(n) &= \frac{b^2\sigma_{\mathbf{w}}^2}{1 - a^2}\left[(-a)^n u(n) + (-a)^{-n} u(-n) - \delta(n)\right] \\ &= \frac{b^2\sigma_{\mathbf{w}}^2}{1 - a^2}(-a)^{|n|} \end{aligned}$$

In a similar fashion, from expression (11.29) it is found that

$$R_{\mathbf{yx}}(z) = \left(\frac{bz}{z + a}\right)\sigma_{\mathbf{w}}^2 \quad \text{so that} \quad r_{\mathbf{yx}}(n) = \sigma_{\mathbf{w}}^2 b(-a)^n u(n)$$

These results are in agreement with those obtained in Example 11.1 using the much more cumbersome time-domain approach. ■

11.7 FREQUENCY-DISCRIMINATION FILTERING

The frequency-discrimination capabilities possessed by a linear system constitute one of its most important functions. To illustrate how this capability may be put to use, let us again consider the classical problem in which the random time series being processed is of the additive form

$$\mathbf{x}(n) = \mathbf{s}(n) + \boldsymbol{\eta}(n) \tag{11.30}$$

In this expression, the random *signal* component $\mathbf{s}(n)$ represents useful information whose time behavior is obscured by the presence of the corruptive additive-*noise* component $\boldsymbol{\eta}(n)$ (which need not be white in nature). Our objective is to obtain a better estimate of the signal's time behavior by processing the given data (11.30) by means of a linear time-invariant operation as depicted in Figure 11.2. Idealistically, the system's response would be given by

$$\mathbf{y}(n) = \mathbf{s}(n)$$

indicating that the linear system has *filtered out* the corruptive additive-noise component while allowing the signal component to pass through undistorted. Although this ideal objective is almost never achievable, we now determine conditions under which it may be closely approximated.

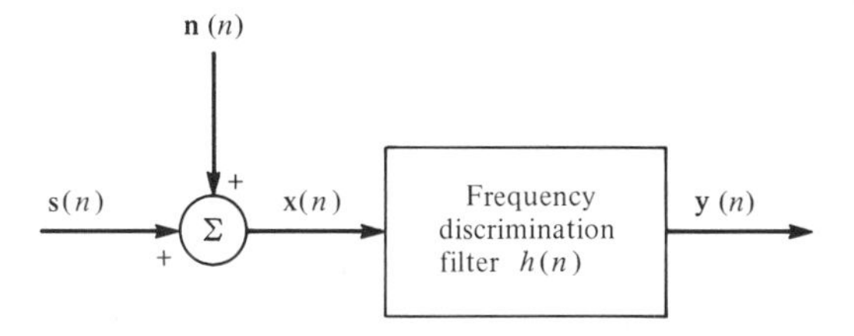

FIGURE 11.2. Frequency-discrimination filtering.

It is first assumed that the individual random time series (11.30) being processed are wide-sense stationary and that the signal and noise components are zero mean and uncorrelated. Under these restrictions, the autocorrelation sequence associated with the excitation is given by

$$r_{\mathbf{xx}}(n) = r_{\mathbf{ss}}(n) + r_{\eta\eta}(n)$$

where $r_{\mathbf{ss}}(n)$ and $r_{\eta\eta}(n)$ denote the autocorrelation sequences of the signal and noise components, respectively. It then follows that the spectral density function of the time series $\{x(n)\}$ is specified by

$$S_{\mathbf{xx}}(\omega) = S_{\mathbf{ss}}(\omega) + S_{\eta\eta}(\omega)$$

Thus, when the signal and corruptive noise components are uncorrelated, the wide-sense stationary time series being processed has a spectral density function that is separable into the sum of the signal and noise spectral density functions. We now use this structural form for evolving a simple procedure for obtaining a better estimate of the signal's behavior than is available in the original given time series (11.30).

If the time series (11.30) is applied to a time-invariant linear system whose frequency response is $H(e^{j\omega})$, Theorem 11.2 indicates that the corresponding system response $\{y(n)\}$ will have the spectral density function

$$\begin{aligned} S_{\mathbf{yy}}(\omega) &= |H(e^{j\omega})|^2 S_{\mathbf{xx}}(\omega) \\ &= |H(e^{j\omega})|^2 [S_{\mathbf{ss}}(\omega) + S_{\eta\eta}(\omega)] \end{aligned}$$

In order that the ideal objective $\mathbf{y}(n) = \mathbf{s}(n)$ be met, it is clear that the system's frequency response $H(e^{j\omega})$ must be selected so that

$$|H(e^{j\omega})|^2 S_{\mathbf{ss}}(\omega) = S_{\mathbf{ss}}(\omega) \tag{11.31}$$

and

$$|H(e^{j\omega})|^2 S_{\eta\eta}(\omega) = 0 \tag{11.32}$$

A little thought will convince you that this dual condition can be satisfied if and

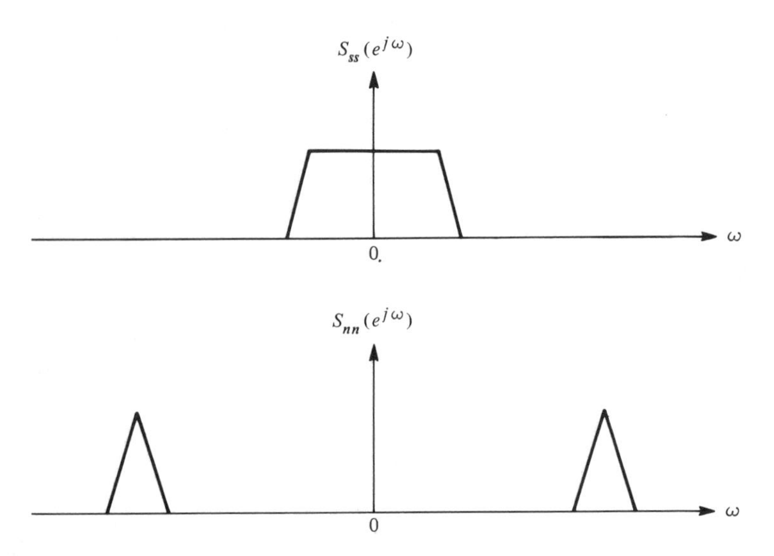

FIGURE 11.3. Disjoint signal and noise spectral density functions.

only if the signal and noise spectral density functions are *pairwise disjoint,* th is,

$$S_{ss}(\omega)S_{\eta\eta}(\omega) = 0 \qquad \text{for all } \omega \tag{11.3}$$

In other words, spectral disjointness requires that there exist no frequencies f which the signal and noise spectral density functions are simultaneously nonzer This spectral disjoint condition is depicted in Figure 11.3.

Filter Frequency Response

If the disjoint condition (11.33) is met, it is possible to synthesize a line system that perfectly filters out (i.e., removes) the additive-noise compone while passing the signal component. That is, the linear system with the frequenc response characteristic

$$H(e^{j\omega}) = \begin{cases} 1 & \text{all } \omega \text{ for which } S_{ss}(\omega) > 0 \\ 0 & \text{all } \omega \text{ for which } S_{ww}(\omega) > 0 \\ \text{arbitrary} & \text{all other frequencies} \end{cases} \tag{11.3}$$

achieves the additive-noise removal. The response spectral density function a sociated with this *frequency discrimination filter* is then given by

$$S_{yy}(\omega) = S_{ss}(\omega)$$

Since the response and signal spectral density functions are equal, we might tempted erroneously to conclude that the original objective of having $\mathbf{y}(n) = \mathbf{s}(n)$ has been met. As the next example illustrates, this need not be true. Due to the specific selection (11.34), however, it does follow that $\mathbf{y}_s(n) = \mathbf{s}(n)$ and therefore that $\mathbf{y}(n) = \mathbf{s}(n)$, as desired.

EXAMPLE 11.5

Let the wide-sense stationary time series $\{\mathbf{x}(n)\}$ be applied to the causal stable linear system characterized by

$$y(n) - ay(n-1) = -ax(n) + x(n-1)$$

in which a is a real-valued parameter whose magnitude is less than 1. It is readily established that the frequency response of this system satisfies

$$|H(e^{j\omega})| = 1 \qquad \text{for all } \omega$$

As such, the system is commonly referred to as an *all-pass system* (e.g., see Section 5.10). The response spectral density function is, in accordance with Theorem 11.2, then given by

$$S_{\mathbf{yy}}(\omega) = S_{\mathbf{xx}}(\omega)$$

Thus although the two time series $\{\mathbf{x}(n)\}$ and $\{\mathbf{y}(n)\}$ are nontrivially different, they are seen to possess the same spectral density function. ■

In virtually all practical applications, the signal and noise spectral density functions are not perfectly disjoint in the sense of expression (11.33). It is still possible, however, to suitably adapt the foregoing concepts to achieve an overall improved signal enhancement. That is, the linear system's frequency response should be selected to be approximately zero for frequencies where the noise's spectral content dominates the signal's and simultaneously the response's signal component, given by

$$\mathbf{y}_s(n) = \sum_{k=-\infty}^{\infty} h(k)\mathbf{s}(n-k)$$

should be a reasonable facsimile of the signal $\mathbf{s}(n)$. To achieve this objective, it is necessary to take into account the specific characteristics of the signal component. Since this is very dependent on the particular application being considered, the system (or filter) designer must exercise a degree of ingenuity in this matter.

In summary, when the spectral components of the signal and additive noise are disjoint, one may perfectly remove the additive-noise component by applying the noise-contaminated data to an appropriately chosen frequency discrimination filter. Although this noise-removal procedure is theoretically straightforward, its implementation in practical situations is difficult at best. Furthermore, in vi tually all practical applications, the following two realistic conditions almo always prevail:

1. The signal and noise spectral density functions are not spectrally disjoint.
2. Only a finite length of realized data [i.e., $x(n)$ for $1 \leq n \leq L$] is ever availab for signal processing.

In this realistic situation, it is not possible to remove the additive noise perfect by a linear filtering operation. Nonetheless, we can still invoke the spirit frequency-discrimination filtering to remove, at least partially, the contaminatin effects of the additive noise. This entails using a finite-length filtering operatic whose associated frequency response is such as to pass the significant sign spectral components while significantly reducing that portion of the spectr noise component that lies outside this frequency region. Even though the r sultant response signal will still contain the effects of the additive noise, significant enhancement in signal recognition is often achievable. A systemat procedure for effecting this more practical signal processing objective is give in Chapter 12.

11.8 PROPERTIES OF THE SPECTRAL DENSITY FUNCTION

The concept of spectral density function was introduced in Chapter 10. It w there postulated as being a function that measures the spectral power of a wid sense stationary time series. With the developments of Section 11.6, we a now in a position to justify this interpretation. In particular, suppose that it we desired to measure the power contained in the time series $\{\mathbf{x}(n)\}$ over the fr quency interval $[\omega_1, \omega_2]$. To achieve this power measure, we could in princip apply this time series to an ideal bandpass filter whose frequency respon satisfies

$$H_i(e^{j\omega}) = \begin{cases} 1 & \omega_1 \leq \omega \leq \omega_2 \\ 0 & \text{all other } \omega \in (-\pi, \pi] \end{cases} \quad (11.3$$

The power in the filter response signal $\{\mathbf{y}(n)\}$ would then provide the requir spectral power information. This ideal signal processing is depicted in Figu 11.4. Using the results of Section 11.6, we find that the ideal bandpass filter response has as its associated spectral density function

$$S_{\mathbf{yy}}(\omega) = |H_i(e^{j\omega})|^2 S_{\mathbf{xx}}(\omega) \quad (11.3$$

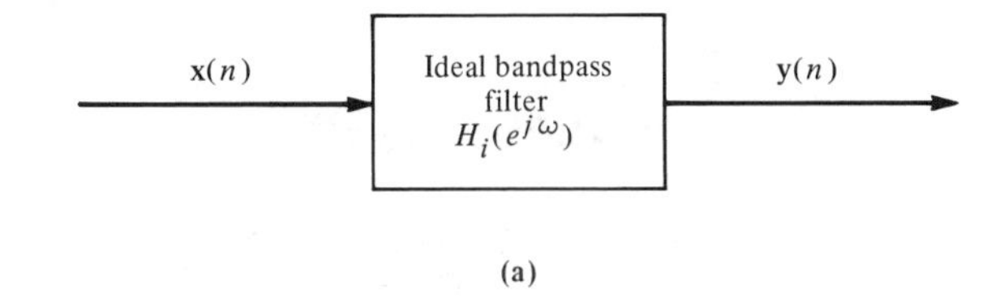

(a)

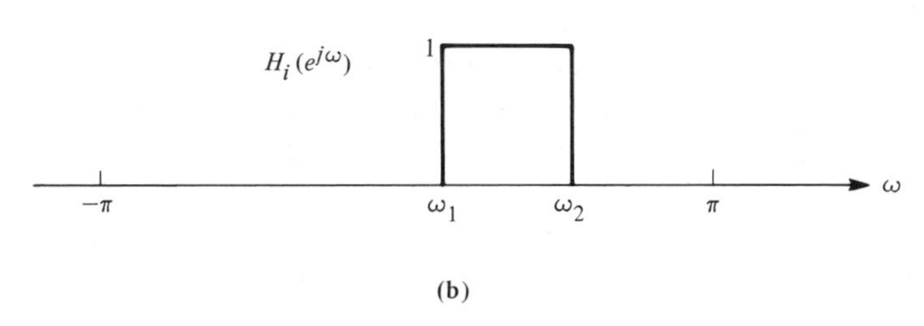

(b)

IGURE 11.4. Ideal bandpass filtering.

Moreover, the power contained in the wide-sense stationary response is formally specified by

$$E\{|\mathbf{y}(n)|^2\} = r_{\mathbf{yy}}(0) = \frac{1}{2\pi}\int_{-\infty}^{\infty} S_{yy}(\omega)\, d\omega$$
$$= \frac{1}{2\pi}\int_{\omega_1}^{\omega_2} S_{\mathbf{xx}}(\omega)\, d\omega$$

where use of relationship (11.36) and the ideal bandpass frequency response (11.35) behavior has been made. We have therefore established the fact that the power measure

$$P_{\mathbf{x}}(\omega_1, \omega_2) = \frac{1}{2\pi}\int_{\omega_1}^{\omega_2} S_{\mathbf{xx}}(\omega)\, d\omega \tag{11.37}$$

gives the expected value of the power contained in the time series $\{\mathbf{x}(n)\}$ over the frequency interval $[\omega_1, \omega_2]$. Since the power measure can never be negative, this further implies that the spectral density function itself can never be negative, that is,

$$S_{xx}(\omega) \geq 0 \qquad \text{for all } \omega \tag{11.38}$$

Properties (11.37) and (11.38), which characterize the spectral density function $S_{\mathbf{xx}}(\omega)$, are those normally associated with a density function. Thus the terminology spectral density function aptly describes the entity $S_{xx}(\omega)$.

11.9

A CLASSICAL SIGNAL-PROCESSING PROBLEM

We shall now use the concepts developed in this chapter to synthesize an optimal linear signal processor. Although our attention is focused on a common problem that arises in several disciplines, the philosophy of approach here taken is directly extendable to other signal-processing problems. In our development, many of the design compromises that need be made in any practical application are illustrated. Moreover, several useful analytical *tricks of the trade* are applied in this investigation. With these objectives in mind, let us now define the problem to be considered.

Signal-Processing Problem

Let it be desired to determine the value of a real-valued parameter s which characterizes some phenomenon. To achieve this objective, a sequential set of measurements of this parameter are made. Due to a variety of factors, each measurement is in error. This error is representable as an additive noise, so that the sequence of measurements has the form

$$\mathbf{x}(n) = s + w(n) \qquad \text{for } n \geq 0 \tag{11.39}$$

where the measurements are taken to begin at $n = 0$.† This one-sided data format reflects the condition normally met in practical applications, whereby measurements are not available from the infinite past but begin at some finite time. The real-valued additive noise terms $\mathbf{w}(n)$ are each assumed to be statistically independent from measurement to measurement and to have a mean value of zero and a variance of $\sigma_{\mathbf{w}}^2$. Under this assumption, the contaminating additive noise can be viewed as being a one-sided white-noise process. Our objective, then, is to use the data measurements (11.39) in an effective manner so as to obtain a quality estimate of the parameter s.

To effect the estimate of parameter s, let us apply a realization of the measurement time series (11.39) to a *stable causal time-invariant linear system* whose

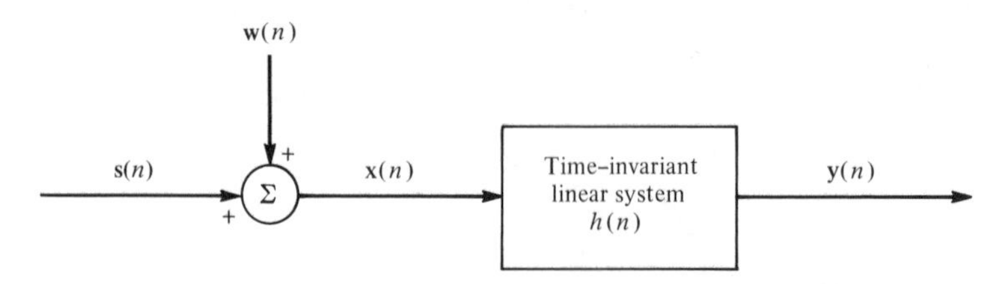

FIGURE 11.5. Filtering operation used to estimate parameter *s*.

†For example, $x(n)$ might correspond to the nth noise contaminated measurement of the moon's gravity parameter s taken from an orbiting satellite.

real-valued weighting sequence is denoted by $\{h(n)\}$. We shall refer to this system as a *filter,* since its primary function is to filter out the additive-noise component while passing undistorted the signal component [i.e., $su(n)$]. This signal-processing operation is conveniently depicted as shown in Figure 11.5. Under the stated conditions on the excitation and filter structure, the corresponding filter's response may be decomposed as

$$\begin{aligned}\mathbf{y}(n) &= \sum_{k=0}^{\infty} h(k)\mathbf{x}(n - k) \\ &= s \sum_{k=0}^{n} h(k) + \sum_{k=0}^{n} h(k)\mathbf{w}(n - k) \\ &= y_s(n) + \mathbf{y}_w(n) \qquad \text{for } n \geq 0\end{aligned} \tag{11.40}$$

in which the first and second summation terms correspond to the *signal-* and *noise-*induced response components, respectively. This ability to express the filter response in this additive manner results from the assumptions that: (1) the excitation noise is additive, and (2) the filter is linear. Our objective now is to select the linear filter's weighting sequence so that the signal component approaches the value s for large n and simultaneously, the noise component takes on appropriately small-magnitude values. If this objective is realized, then for appropriately large n the overall response $\mathbf{y}(n)$ will consist of small fluctuations about the constant level s.

With these signal-processing objectives in mind, it is useful to examine the mean value and variance behavior of the filter response $\mathbf{y}(n)$. The response's mean value is seen to be

$$\begin{aligned}\mu_{\mathbf{y}}(n) &= E\{\mathbf{y}(n)\} \\ &= s\left[\sum_{k=0}^{n} h(k)\right]u(n)\end{aligned} \tag{11.41}$$

where use of the zero-mean condition imposed on the additive noise has been made in arriving at this result. The unit-step multiplier term $u(n)$ here appears in recognition that the response is identically zero for negative time due to the one-sided excitation (11.39). Similarly, the response's variance is given by

$$\begin{aligned}\sigma_{\mathbf{y}}^2(n) &= E\{[\mathbf{y}(n) - \mu_{\mathbf{y}}(n)]^2\} \\ &= \sigma_{\mathbf{w}}^2\left[\sum_{k=0}^{n} h^2(k)\right]u(n)\end{aligned} \tag{11.42}$$

where the whiteness assumption on the additive noise has been employed. If the linear filter is to achieve its ideal goal of estimating the parameter s and removing the additive-noise effects, it is apparent that the filter's weightin sequence $\{h(n)\}$ must be selected so that the two conditions

$$\mu_y(n) = s \qquad \text{and} \qquad \sigma_y^2(n) = 0 \tag{11.4}$$

are met simultaneously for reasonably moderate positive values of n. A littl thought should convince the reader that the satisfaction of these two conditior would imply that the filter response is given by $\mathbf{y}(n) = s$ for suitably larg values of n. Unfortunately, simultaneous satisfaction of these two conditions not possible. Nonetheless, we still seek choices for the filter's weighting se quence so that the first condition is satisfied while the second condition is ap proximately met. In Sections 11.10 to 11.12 we give three different approache to the filter synthesis problem, which entail different structures for the line filter.

11.10 NONRECURSIVE FILTER SOLUTION TO THE CLASSICAL PROBLEM

In our first approach, we constrain the linear filter to have the nonrecursiv structure

$$\mathbf{y}(n) = \sum_{k=0}^{q} h(k)\mathbf{x}(n - k) \tag{11.4}$$

The task at hand is then to select this nonrecursive filter's order q and associate weighting coefficients $h(0), h(1), \ldots, h(q)$ so that the resultant filter respons to the excitation (11.39) provides a suitably accurate estimate of the paramet s. Upon examination of this nonrecursive expression, it is apparent that for time $n \geq q$, each of the excitation elements $\mathbf{x}(n - k)$ is generally nonzero. Thus th so-called *steady-state* response of this nonrecursive filter is given by

$$\mathbf{y}(n) = s\sum_{k=0}^{q} h(k) + \sum_{k=0}^{q} h(k)\mathbf{w}(n - k) \qquad \text{for } n \geq q$$

During the time interval $0 \leq n < q$, the nonrecursive filter response is in transient phase, since some of the excitation elements in expression (11.44) a zero.

For this nonrecursive filter, the filter's steady-state mean value and varianc are found to equal the two constant values

$$\mu_y(n) = s\sum_{k=0}^{q} h(k) \qquad \text{for } n \geq q \tag{11.4}$$

$$\sigma_{\mathbf{y}}^2(n) = \sigma_{\mathbf{w}}^2 \sum_{k=0}^{q} h^2(k) \qquad \text{for } n \geq q \tag{11.4}$$

as is evident from the application of expressions (11.41) and (11.42), respectively, to the nonrecursive filter here being considered. Since we wish the filter's response to provide an unbiased estimate of the unknown parameter s, it is seen from expression (11.45) that the nonrecursive filter's weighting coefficients must be selected so as to satisfy the constraint

$$\sum_{k=0}^{q} h(k) = 1 \tag{11.47}$$

It is apparent that there exist an infinite number of filter weighting coefficient choices that satisfy this condition. From this infinity of possibilities, however, a desirable selection would be the one that minimizes the response's variance (11.46). Such a selection will give rise to a filter response whose steady-state behavior (i.e., $n \geq q$) will fluctuate about the constant value s, and this fluctuation will have the smallest power (i.e., variance) possible. With these thoughts in mind, it follows that the corresponding *optimal unbiased-minimum variance nonrecursive filter* is obtained by solving the following constrained minimization problem:

$$\min_{\sum_{k=0}^{q} h(k)=1} \sum_{k=0}^{q} h^2(k)$$

The solution to this constrained problem is readily obtained by introducing the auxiliary Lagrange function

$$g(h) = \sum_{k=0}^{q} h^2(k) + \lambda\left[1 - \sum_{k=0}^{q} h(k)\right]$$

in which λ is a scalar-valued Lagrange multiplier. If we take the partial of this auxiliary function with respect to the nth weighting coefficient element $h(n)$ and setting that derivative to zero, the necessary conditions for a minimum are found to be

$$\frac{\partial g(h)}{\partial h(n)} = 2h^{o}(n) - \lambda = 0 \qquad \text{for } 0 \leq n \leq q$$

From this expression it is seen that the optimal filter coefficients must be all equal to the constant λ. From the constraint requirement (11.47), this constant must be

$$h^{o}(n) = \frac{1}{q+1} \qquad \text{for } 0 \leq n \leq q \tag{11.48}$$

This optimal nonrecursive filter is seen to correspond to the classical *unbiased mean value estimator,* which has been extensively examined in the statistical literature. The corresponding steady-state response variance (11.46) for this optimal filter coefficient selection is seen to be

$$\sigma_{\mathbf{y}}^2(n) = \frac{1}{q+1}\sigma_{\mathbf{w}}^2 \qquad \text{for } n \geq q \tag{11.49}$$

The optimal nonrecursive filter thereby reduces the additive-noise power by a factor $q + 1$ while passing in an undistorted manner the constant parameter s.

It is now beneficial to provide a probabilistic description of the optimal nonrecursive filter's steady-state response as specified by

$$\mathbf{y}(n) = s + \frac{\mathbf{w}(n) + \mathbf{w}(n-1) + \cdots + \mathbf{w}(n-q)}{q+1} \tag{11.50}$$

where $n \geq q$. For moderately large values of filter order q, the central limit theorem indicates that these response random variables are approximately Gaussianly distributed with mean value s and variance $\sigma_{\mathbf{w}}^2/(q+1)$. It then follows that the probability that $\mathbf{y}(n)$ deviates from its mean value s by less than the scalar ϵ is approximately given by the normal probability law

$$P[|\mathbf{y}(n) - s| < \epsilon] = \frac{\sqrt{q+1}}{\sqrt{2\pi}\sigma_{\mathbf{w}}} \int_{-\epsilon}^{\epsilon} e^{-(q+1)x^2/2\sigma_{\mathbf{w}}^2}\, dx \tag{11.51}$$

This expression can be used to fix a value for the filter order q whereby the filter response stays within a prescribed range of its mean value s with a given probability. Moreover, by selecting q large enough, we can make this probability be as close to 1 as possible.

The optimal nonrecursive filter's response elements possess a degree of correlation due to the filtering operation; namely, the following response covariance behavior is readily established:

$$E\{[\mathbf{y}(n_1) - s][\mathbf{y}(n_2) - s]\} = \begin{cases} \dfrac{\sigma_{\mathbf{w}}^2}{(q+1)^2}(q + 1 - |n_2 - n_1|) & \text{for } |n_2 - n_1| \leq q \\ 0 & \text{otherwise} \end{cases} \tag{11.52}$$

where $n_1, n_2 \geq q$. Thus the response elements $\mathbf{y}(n_1)$ and $\mathbf{y}(n_2)$ have a correlation behavior that decreases to zero in a linear fashion as a function of time separation $|n_2 - n_1|$. Moreover, the filter response elements become uncorrelated for time separators larger than the filter order q. This correlation behavior arises from the fact that when the time separation is less than $q + 1$ and $n_1 \leq n_2$, the two response elements $\mathbf{y}(n_1)$ and $\mathbf{y}(n_2)$ share the common excitation white-noise term

$$\frac{\mathbf{w}(n_1) + \mathbf{w}(n_1 + 1) + \cdots + \mathbf{w}(n_2 - q)}{q + 1}$$

The expected value of this quantity squared yields the covariance expression above.

A few words are now appropriate concerning the selection of the filter-order parameter q. Conceptually, we could reduce the steady-state variance (11.49) to any desirable level by selecting q to be sufficiently large. Depending on the particular application being considered, however, we find that a large choice for q may give rise to two potentially serious shortcomings: (1) Since the time to reach steady-state filtering behavior is equal to q, the corresponding long transient time might not be acceptable in a given application, and (2) in those cases in which the parameter s being estimated is a "slowly" changing function of time, the filter length parameter q should be selected small enough so that the change in s over any $q + 1$ length interval is acceptably small. In either of these situations, a large choice for q would be unacceptable in order to achieve a large noise-reduction capability.

11.11 RECURSIVE FILTER SOLUTION TO THE CLASSICAL PROBLEM

In the second approach to the parameter estimation problem given Section 11.9, we shall use a simple causal recursive first-order filter as governed by

$$\mathbf{y}(n) + a\mathbf{y}(n - 1) = b\mathbf{x}(n) \tag{11.53}$$

The unit-impulse response of this linear system is readily found to be

$$h(n) = b(-a)^n u(n) \tag{11.54}$$

When the given excitation (11.39) is applied to this recursive filter, the resultant response is readily shown to have its mean value and variance given by

$$\mu_{\mathbf{y}}(n) = \frac{b[1 - (-a)^{n+1}]}{1 + a} su(n) \tag{11.55}$$

$$\sigma_{\mathbf{y}}^2(n) = \frac{b^2[1 - (a^2)^{n+1}]}{1 - a^2} \sigma_{\mathbf{w}}^2 u(n) \tag{11.56}$$

These results are obtained by using the specific unit-impulse response (11.54) in the general linear system expressions (11.41) and (11.42), respectively. For these mean value and variance relationships to be well behaved, it is apparent that the filter parameter a must be required to have a magnitude less than 1. We shall henceforth impose this filter stability requirement.

If the recursive filter's steady-state response (i.e., large positive n) is to provide an unbiased estimate of parameter s, it is clear from expression (11.55) that this entails that the filter parameter b be selected so that

$$b = 1 + a \tag{11.57}$$

For this asymptotic unbiased selection of b it follows that for sufficiently large positive values of n (needed to ensure that $(a)^n \to 0$), the filter's mean value and variance simplify to

$$\mu_y(n) = s \tag{11.58}$$

for suitably large positive n

$$\sigma_y^2(n) = \frac{1 + a}{1 - a}\sigma_w^2 \tag{11.59}$$

Our task is to then select the filter parameter a with a number of considerations in mind: (1) to have stable filter dynamics so that $|a| < 1$, (2) to have an acceptable small response variance, and (3) to have a reasonably fast responding filter. Upon examination of expression (11.59), it is seen that the variance of the response can be reduced to any desired small level by selecting a to be suitably close to -1. Unfortunately, such a selection leads to a very sluggish (i.e., long time constant) filter. Thus the objectives of reduced response variance and fast response times are incompatible. This incompatibility is depicted in Figure 11.6, where a plot of the filter's noise power reduction ratio versus the

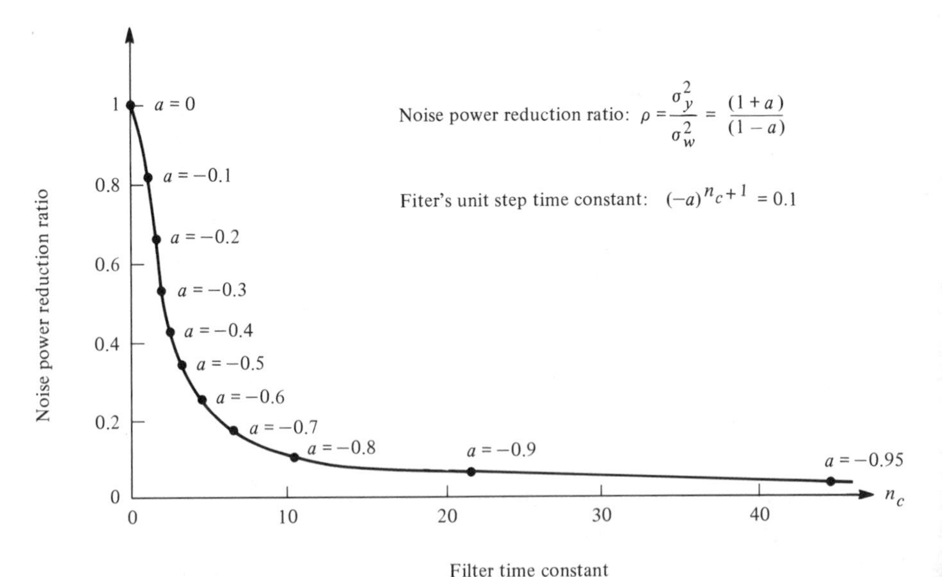

FIGURE 11.6. Optimal recursive filter's noise power reduction ratio versus unit-step time-constant plot.

filter time constant is shown. From the expression (11.59), the filter's noise power reduction ratio is seen to be given by

$$\rho = \frac{\sigma_{\mathbf{y}}^2}{\sigma_{\mathbf{w}}^2} = \frac{1 + a}{1 - a} \tag{11.60}$$

The filter time constant is here taken to correspond to the number of iterations required so that the filter's unit-step response is within 10% of its steady-state value [i.e., $(-a)^{n+1} = 0.1$].

With the thoughts above in mind, let us now specify a choice for the filter parameter a, which keeps the response variance at an acceptable small value while maintaining a degree of filtering speed. In effect, we then require that the recursive filter's noise power reduction factor (11.60) be selected to achieve a prescribed value ρ (e.g., $\rho = 0.05$). Solving expression (11.60) for a in terms of ρ, we obtain the desired filter coefficients for an unbiased response:

$$a = \frac{\rho - 1}{1 + \rho} \qquad \text{and} \qquad b = 1 + a = \frac{2\rho}{1 + \rho}$$

Thus the first-order recursive filter as characterized by

$$y(n) = \frac{2\rho}{1 + \rho} x(n) + \frac{1 - \rho}{1 + \rho} y(n - 1) \tag{11.61}$$

will have a steady-state response to the excitation (11.39) that has a mean value s and a variance of $\rho\sigma_w^2$, where $0 < \rho < 1$. If we use arguments similar to those taken for the optimum nonrecursive filter, the steady-state response elements $\mathbf{y}(n)$ will be approximately Gaussianly distributed with mean value s and variance $\rho\sigma_w^2$. Moreover, the autocovariance behavior of these steady-state response elements are readily shown to possess the wide-sense covariance behavior

$$c_{\mathbf{yy}}(n_1, n_2) = \rho(-a)^{|n_1 - n_2|} \tag{11.62}$$

11.12 OPTIMAL LINEAR FILTER SOLUTION TO THE CLASSICAL PROBLEM

In the first two approaches taken for estimating the parameter s, we constrained the linear filter to be either of a fixed nonrecursive or a fixed recursive structure. By so restricting the filter structure, it is indeed possible that the resultant filtering performance might be inferior to that possessed by an unconstrained linear filter structure. With this possibility in mind, let us now consider the unconstrained causal time-invariant filter as represented by expression (11.40). To measure the degree to which the filter response's signal component approaches the desired value s, let us introduce the *error signal*

$$\begin{aligned} e(n) &= s - y_s(n) \\ &= s\left[1 - \sum_{k=0}^{n} h(k)\right] \qquad \text{for } n \geq 0 \end{aligned}$$

Our design objective is then to select the linear filter's weighting coefficients so as to render this error signal a minimum in some sense while maintaining an acceptable noise power reduction performance. The solution to the following constrained minimization problem meets this objective.

$$\min_{\sum_{k=0}^{\infty} h^2(k) = \rho} \sum_{n=0}^{\infty} \left[1 - \sum_{k=0}^{n} h(k)\right]^2 \tag{11.63}$$

The squared-error criterion here being minimized in effect causes the error signal to be driven to zero while the constraint requires that the minimization be made over all causal, time-invariant linear filters whose noise power reduction ratio (11.20) equals a prescribed value ρ.

To obtain the solution to this constrained minimization problem, we shall again appeal to a Lagrangian solution procedure. Namely, consider the auxiliary Lagrange function

$$g(h) = \sum_{n=0}^{\infty} \left[1 - \sum_{k=0}^{n} h(k)\right]^2 + \lambda\left[\sum_{k=0}^{\infty} h^2(k) - \rho\right]$$

where λ is a scalar-valued Lagrange multiplier. To obtain the optimum filter weight coefficients, we set to zero the partial derivatives of this auxiliary functional with respect to the elements $h(m)$:

$$\frac{\partial g(h)}{\partial h(m)} = -2 \sum_{n=m}^{\infty} \left[1 - \sum_{k=0}^{n} h(k)\right] + 2\lambda h(m) = 0$$

for $0 \leq m < +\infty$. Thus the optimum filter weight coefficients are obtained by solving the infinite set of equations

$$\lambda h(m) = \sum_{n=m}^{\infty} \left[1 - \sum_{k=0}^{n} h(k)\right] \qquad \text{for } 0 \leq m < \infty$$

A method for analytically obtaining this solution is to replace m by $m + 1$ in this expression and then form the difference $\lambda h(m) - \lambda h(m + 1)$, which gives

$$\lambda h(m) - \lambda h(m+1) = 1 - \sum_{k=0}^{m} h(k) \qquad 0 \le m < \infty$$

Next, we take the z-transform of each side of this difference equation to obtain

$$\lambda H(z) - \lambda z[H(z) - h(0)] = \frac{z}{z-1} - \frac{z}{z-1} H(z)$$

Solving this expression for the optimal filter transfer function then gives

$$H(z) = \frac{h(0)z\{z - [1 + \lambda h(0)]/\lambda h(0)\}}{z^2 - [(2\lambda + 1)/\lambda]z + 1}$$

This filter transfer function has at least one of its poles on or outside the unit circle because the product of its poles is seen to be 1. Unless this pole is canceled by the filter zero located at $z_1 = [1 + \lambda h(0)]/\lambda h(0)$, the resultant optimal filter is unstable and therefore has a noise power reduction ratio of $+\infty$, not ρ, as required. Thus the parameters λ and $h(0)$ must be interrelated so as to create this pole–zero cancellation. It is readily shown that if

$$\lambda = \frac{1 - h(0)}{h^2(0)}$$

the required pole–zero cancellation follows and the filter transfer function simplifies to

$$H(z) = \frac{zh(0)}{z - [1 - h(0)]} \tag{11.64}$$

The corresponding optimal filter's weighting coefficient is then given by

$$h(n) = h(0)[1 - h(0)]^n u(n)$$

Finally, the scalar $h(0)$ must be selected so that the noise power reduction ratio constraint imposed in the miminization problem (11.63) is satisfied. This is readily shown to require that

$$h(0) = \frac{2\rho}{1 + \rho} \tag{11.65}$$

Upon substituting this condition into relationship (11.64), the desired optimum linear filter transfer function is found to be

$$H^o(z) = \frac{2\rho}{1 + \rho}\left[\frac{z}{z - [(1 - \rho)/(1 + \rho)]}\right]$$

The difference equation that corresponds to this optimal transfer function is then

$$y(n) = \frac{2\rho}{1 + \rho}x(n) + \frac{1 - \rho}{1 + \rho}y(n - 1) \tag{11.66}$$

This filtering expression is identical to that given by relationship (11.61) as found when using the approach taken in Section 11.11. It is important to appreciate the fact that in arriving at the optimum filter (11.66), however, we did not require the filter to have any prespecified structure as was done in Section 11.11. It just happens that for the additive white-noise problem being considered here, the optimum causal time-invariant linear filter turns out to have a first-order recursive structure. If the additive noise were correlated, these two approaches would in general yield different results.

In Sections 11.10 to 11.12 we have examined three approaches to synthesizing a linear filter to be used in estimating the value of a constant parameter from measurements that were contaminated by additive white noise. The resultant optimal filters that arose from these investigations were of the nonrecursive structure

$$y(n) = \frac{1}{q + 1}\sum_{k=0}^{q} x(n - k) \tag{11.67}$$

and the recursive structure

$$y(n) = \frac{2\rho}{1 + \rho}x(n) + \frac{1 - \rho}{1 + \rho}y(n - 1) \tag{11.68}$$

Some salient characteristics that depict the dynamical qualities of these two filter structures are given in Table 11.1. The noise power reduction factor measures the ability of the filter to reduce the contaminating effects of the additive white noise. Similarly, the number of input element values that must be stored per response element determination provides a measure of the signal-processing hardware complexity. The smaller this number is, the more efficient the algorithm is said to be. Clearly, the recursive filter enjoys a decided advantage for the normal situations in which $q >> 1$. Finally, the unit-step response provides an indication of how rapidly the filter reaches its steady-state behavior. The quicker a system reaches steady state, the sooner reliable estimates of the parameter s may be obtained from the filter response. It is to be noted that although the recursive filter never quite reaches steady state, its unit-step response gets very close to 1 as $(-a)^n$ approaches zero.

TABLE 11.1 Salient Characteristics of Optimum Linear Recursive and Nonrecursive Filters

	Optimum Nonrecursive Filter	Optimum Recursive Filter
Noise power reduction, $\sigma_{\mathrm{y}}^2/\sigma_{\mathrm{w}}^2$	$\dfrac{1}{q+1}$	ρ
Storage requirement per response element	$q + 1$	2
Unit-step response	$\begin{cases} \dfrac{n+1}{q+1} & \text{for } 0 \le n < q \\ 1 & \text{for } n \ge q \end{cases}$	$\left[1 - \left(\dfrac{\rho - 1}{\rho + 1}\right)^{n+1}\right] u(n)$

EXAMPLE 11.6

Consider the situation in which it is desired that the noise power reduction ratio be equal to $\frac{1}{9}$. For the optimal linear nonrecursive and recursive filters, the required parameter selections from Table 11.1 are

$$q = 8 \qquad \text{and} \qquad \rho = \tfrac{1}{9}$$

A plot of the unit-step responses of the optimum linear nonrecursive and recursive filters is shown in Figure 11.7. Steady-state behavior for the nonrecursive filter is seen to commence at $n = 9$, while at $n = 17$ the recursive filter is just approaching its steady-state behavior, since $y(17) = 0.9775$. ■

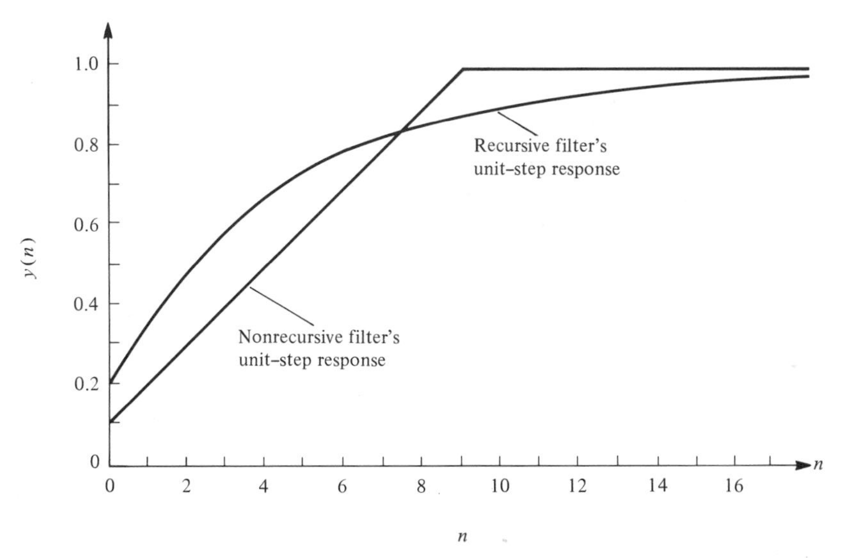

FIGURE 11.7. Unit-step responses for the optimal nonrecursive ($q = 8$) and optimal recursive ($p = \frac{1}{9}$) filters, each of which have the same noise power reduction ratio of $\frac{1}{9}$.

11.13

ERGODIC TIME SERIES

Although the assumption of wide-sense stationarity provides a welcome simplification to the statistical characterization of random time series, there still exist practical difficulties in taking this approach. Foremost is the determination of the mean values and correlation information that characterize the random time series being analyzed. Generally, these statistical measures are not known a priori and must be therefore estimated from the experimentally obtained data being analyzed. To illustrate this point, let it be assumed that a partial realization of the time series $\{\mathbf{x}(n)\}$ has been made and results in the time series data

$$x(-N), x(-N+1), \ldots, x(N) \tag{11.69}$$

in which no loss of generality is incurred in selecting the symmetric observation interval to be $-N \leq n \leq N$. We refer to this data set as a partial realization since only a finite number (i.e., $2N + 1$) of data have been collected. A little thought will convince the reader that in any practical experiment, it is possible to collect only a finite amount of data.

If the random time series $\{\mathbf{x}(n)\}$ from which the empirical data (11.69) arose was wide-sense stationary, there exist intuitively appealing methods for estimating the underlying mean value and autocorrelation behavior of the time series. For example, the standard unbiased estimate

$$\overline{\mu}_{\mathbf{x}} = \frac{1}{2N+1} \sum_{n=-N}^{N} x(n) \tag{11.70}$$

would serve as a logical candidate for estimating the mean value parameter $\mu_{\mathbf{x}}$. It is possible, however, to hypothesize wide-sense stationary time series for which this entity would provide a poor estimate of the mean no matter how large the data length N is made. With this possibility in mind, we shall henceforth be concerned primarily with better-behaved time series, which are known as ergodic in the mean.

Definition 11.1. The stationary in the mean time series $\{\mathbf{x}(n)\}$ is said to be *ergodic in the mean* if the limit

$$\lim_{N\to+\infty} \frac{1}{2N+1} \left[\sum_{n=-N}^{N} x(n) \right] = E\{\mathbf{x}(n)\} = \mu_{\mathbf{x}} \tag{11.71}$$

is satisfied with probability 1.

In essence, the concept of ergodicity in the mean allows us to interchange the time-averaging operator and the ensemble expectation operator $E\{\mathbf{x}(n)\}$. It is to be noted from this definition that for a time series to be *ergodic in the mean,* it must of necessity be *stationary* in the mean.

Continuing in this manner, we can extend the concepts of ergodicity in an obvious manner when computing autocovariance and cross-covariance measures. For instance, a logical method for estimating the autocorrelation sequence associated with a wide-sense stationary time series is given by the standard estimate

$$\hat{r}_{\mathbf{xx}}(n) = \frac{1}{2N+1}\left[\sum_{k=-N}^{N} x(k)\bar{x}(k-n)\right] \tag{11.72}$$

where the lag argument n is normally restricted to be suitably smaller than N in magnitude. In a smaller fashion, the cross-correlation between two wide-sense stationary time series may be approximated by using the estimate

$$\hat{r}_{\mathbf{xy}}(n) = \frac{1}{2N+1}\sum_{k=-N}^{N} x(k)\bar{y}(k-n) \tag{11.73}$$

As in the mean value case, these estimates are of good behavior on only a small but important subclass of wide-sense stationary time series to be now defined.

Definition 11.2. The wide-sense stationary time series $\{\mathbf{x}(n)\}$ is said to be *ergodic in the autocorrelation* if the limit operation

$$\begin{aligned}\lim_{N\to+\infty} \frac{1}{2N+1}\left[\sum_{k=-N}^{N} x(k)\bar{x}(k-n)\right] &= E\{\mathbf{x}(k)\mathbf{x}(k-n)\} \\ &= r_{\mathbf{xx}}(n)\end{aligned} \tag{11.74}$$

is satisfied with probability 1. Similarly, the two wide-sense stationary time series $\{\mathbf{x}(n)\}$ and $\{\mathbf{y}(n)\}$ are said to be *ergodic in the cross-correlation* if the limit operation

$$\begin{aligned}\lim_{N\to+\infty} \frac{1}{2N+1}\left[\sum_{k=-N}^{N} x(k)\bar{y}(k-n)\right] &= E\{\mathbf{x}(k)\bar{\mathbf{y}}(k-n)\} \\ &= r_{\mathbf{xy}}(n)\end{aligned} \tag{11.75}$$

is satisfied with probability 1.

11.14 SUMMARY

One of the more important developments in this chapter was the determination of how a random time series autocorrelation behavior is changed upon passage through a stable time-invariant linear system. This characterization, in turn, suggested the possibility of using a linear system to effect various signal-processing objectives, such as frequency discrimination, filtering, smoothing, and prediction. When the random time series being processed (i.e., the excitation) is wide-sense stationary, it was further shown that the associated response of a stable time-invariant linear system is also wide-sense stationary. This condition then leads to a remarkable simplification in the signal-processing characterization of time-invariant linear operators. We shall exploit this development in future chapters.

Utilization of the Fourier and z-transforms led to convenient and valuable methods for analyzing the salient features of wide-sense stationary time series. In particular, the spectral density function was shown to provide a direct procedure for measuring the spectral content of such random time series. One of the more attractive features found in using a transform approach was that of relating the spectral density functions associated with a time-invariant linear operators excitation and response time series. The response's spectral density function was shown to be equal to the excitation's spectral density function multiplied by the squared magnitude of the linear system's frequency response. This simple relationship illustrates, in a rather dramatic fashion, the exciting possibilities of using a linear system for spectral filtering.

SUGGESTED READINGS

The Suggested Readings in Chapter 10 cover the material treated in this chapter.

PROBLEMS

11.1 Let the causal linear system

$$y(n) = x(n) - \tfrac{1}{2}x(n - 1) + \tfrac{1}{16}x(n - 2)$$

be excited by a real-valued zero-mean white-noise excitation whose autocorrelation sequence is specified by $r_{\mathbf{xx}}(n) = \sigma_{\mathbf{w}}^2\delta(n)$. Determine $r_{\mathbf{yy}}(n)$ and $r_{\mathbf{xy}}(n)$.

11.2 Carry out the details that led to the response autocorrelation and covariance and cross-correlation expressions (11.3), (11.5), and (11.6), respectively.

11.3 Let the causal linear system characterized by

$$y(n) - \tfrac{1}{2}y(n - 1) = x(n)$$

be excited by the real-valued one-sided white-noise time series

$$\mathbf{x}(n) = \mathbf{w}(n)u(n)$$

where the $\mathbf{w}(n)$ are uncorrelated random variables each with zero mean and variance $\sigma_{\mathbf{w}}^2$. Compute **(a)** $\mu_{\mathbf{y}}(n)$; **(b)** $r_{\mathbf{yy}}(n_1, n_2)$; **(c)** $r_{\mathbf{yx}}(n_1, n_2)$. Characterize the behavior of these entities for large positive values of n, n_1, and n_2.

11.4 Let the linear system characterized by

$$y(n) - \tfrac{3}{4}y(n-1) + \tfrac{1}{8}y(n-2) = \tfrac{1}{4}x(n-1)$$

be excited by the real-valued one-sided white-noise time series

$$\mathbf{x}(n) = \mathbf{w}(n)u(n)$$

where the $\mathbf{w}(n)$ are uncorrelated random variables with zero mean $\mu_{\mathbf{w}} = 0$ and variance $\sigma_{\mathbf{w}}^2$. Compute **(a)** $\mu_{\mathbf{y}}(n)$; **(b)** $r_{\mathbf{yy}}(n_1, n_2)$; **(c)** $r_{\mathbf{yx}}(n_1, n_2)$. Characterize the behavior of these entities for large positive values of n, n_1, and n_2.

11.5 Let the random time series $\mathbf{x}(n)$ be the response of the linear system

$$\mathbf{x}(n) - \tfrac{1}{2}\mathbf{x}(n-1) = \mathbf{v}\delta(n) + \mathbf{w}\delta(n-1)$$

where $\mathbf{v}$ and $\mathbf{w}$ are real-valued random variables with means $\mu_{\mathbf{v}} = 1$, $\mu_{\mathbf{w}} = 2$, variances $\sigma_{\mathbf{v}}^2 = 3$ and $\sigma_{\mathbf{w}}^2 = 1$, and correlation coefficient $\rho_{\mathbf{vw}} = \frac{1}{4}$. Compute **(a)** $\mu_{\mathbf{x}}(n)$ and **(b)** $r_{\mathbf{xx}}(n_1, n_2)$. Characterize the behavior of these entities for large positive values of n, n_1, and n_2.

11.6 Consider the linear system characterized by

$$y(n) = x(n) + x(n-1) + \tfrac{1}{4}x(n-2)$$

Determine the response's mean value sequence, its autocorrelation sequence, and the cross-correlation between the excitation and response when:

(a) The excitation is real-valued white noise [i.e., $\mathbf{x}(n) = \mathbf{w}(n)$] with variance $\sigma_{\mathbf{w}}^2 = 2$.

(b) The excitation is real-valued one-sided white noise [i.e., $\mathbf{x}(n) = \mathbf{w}(n)u(n)$] with variance $\sigma_{\mathbf{w}}^2 = 2$.

(c) The response characteristics found in parts (a) and (b) will be wide-sense stationary and nonstationary, respectively. Compare the mean values $\mu_{\mathbf{y}}$ and $\mu_{\mathbf{y}}(n)$ found in parts (a) and (b) for suitably large positive values of n. Make a similar comparison for the autocorrelation $r_{\mathbf{yy}}(n)$ and $r_{\mathbf{yy}}(n_1, n_2)$ for suitably large positive values of n_1 and n_2.

11.7 Evaluate the noise power reduction ratio for the following causal systems.

(a) $y(n) - \frac{1}{2}y(n-1) = x(n) + 3x(n-1)$

(b) $y(n) = x(n) + 3x(n-1) - 5x(n-2)$

(c) $y(n) = x(n) + x(n-1) + 0.75y(n-1) - \frac{1}{8}y(n-2)$

11.8 Consider the time-invariant linear system characterized by

$$y(n) + ay(n - 1) = bx(n)$$

Find values for the real parameters a and b so that the noise power reduction ratio $\Sigma_{k=0}^{\infty} |h(k)|^2$ is minimized subject to the constraint that the frequency response at $\omega = \pi/2$ has a magnitude of $\frac{1}{2}$.

11.9 Let the linear discrete system governed by

$$y(n) - 1.3y(n - 1) + 0.36y(n - 2) = x(n)$$

be excited by a real-valued zero-mean, unit-variance white-noise time series

(a) Determine $\mu_{\mathbf{y}}(n)$ and $r_{\mathbf{yy}}(n)$.
(b) Determine the spectral density functions $S_{\mathbf{yy}}(\omega)$ and $S_{\mathbf{yx}}(\omega)$.
(c) Determine the noise power reduction ratio.
(d) Make a plot of $S_{\mathbf{xx}}(\omega)$, $|H(e^{j\omega})|$, $S_{\mathbf{yy}}(\omega)$, and $S_{\mathbf{yx}}(\omega)$.

11.10 Determine the power spectral density for the response of the causal systems
(a) $y(n) = x(n) - x(n - 1) + \frac{1}{3}y(n - 1)$
(b) $y(n) = b_0x(n) + b_1x(n - 1) + b_2x(n - 2)$
when the real-valued excitation is white in nature with autocorrelation $r_{\mathbf{xx}}(n) = \sigma_{\mathbf{w}}^2\delta(n)$. Furthermore, obtain the cross-spectral density functions $S_{\mathbf{xy}}(\omega)$.

11.11 Select values for the b_0, b_1, and b_2 parameters of the system

$$y(n) = b_0x(n) + b_1x(n - 1) + b_2x(n - 2)$$

so that its noise power reduction ratio is minimized subject to the constraint that $b_0 + b_1 + b_2 = 1$. This constraint ensures that the steady-state unit-step response is $y(n) = 1$ for $n \geq 2$.

11.12 Consider the linear system characterized by

$$y(n) = \sum_{k=0}^{q} b_kx(n - k)$$

which is excited by the wide-sense stationary time series

$$\mathbf{x}(n) = s + \mathbf{v}(n)$$

In this excitation, s is an unknown constant and the autocorrelation sequence characterizing the zero-mean noise $\{\mathbf{v}(n)\}$ is given by

$$r_{\mathbf{vv}}(n) = \tfrac{1}{2}\delta(n - 1) + \delta(n) + \tfrac{1}{2}\delta(n + 1)$$

It is desired to select the filter's coefficients b_k so that its response $\{\mathbf{y}(n)\}$ provides an estimate for the unknown parameter s.

(a) In order that $E\{\mathbf{y}(n)\} = s$, show that the filter parameters must satisfy $b_0 + b_1 + \cdots b_q = 1$.

(b) For the class of filters that satisfy the constraint imposed in part (a), find one that minimizes the variance of the response.

(c) Plot the response's variance as a function of filter length q.

11.13 Select values for the a and b parameters of the stable causal system

$$y(n) = bx(n) - ay(n-1)$$

so that its steady-state unit-step response approximately satisfies $y(n) = 1$ for large positive n, and its noise power reduction ratio is $\rho = 0.25$.

11.14 Consider the simple first-order linear system

$$y(n) = b_0 x(n) + b_1 x(n-1)$$

(a) Select the coefficients b_0 and b_1 so that when this system is excited by a real zero-mean unit-variance white-noise process, the resultant response will have the autocorrelation behavior

$$r_{\mathbf{yy}}(n) = -\tfrac{1}{4}\delta(n+1) + \tfrac{5}{4}\delta(n) - \tfrac{1}{4}\delta(n-1)$$

(b) Is the selection of the b_0 and b_1 coefficients for achieving this autocorrelation behavior unique? If not, give another choice.

11.15 Let a wide-sense stationary time series $\{\mathbf{y}(n)\}$ have an autocorrelation sequence which is identically zero for all lags $|n| \geq q + 1$, while its remaining lag values

$$r_{\mathbf{yy}}(-q), \quad r_{\mathbf{yy}}(-q+1), \quad \cdots, \quad r_{\mathbf{yy}}(q)$$

are otherwise known. It is desired to realize a time series that possesses this autocorrelation behavior. With this goal in mind, consider the linear system

$$y(n) = \sum_{k=0}^{q} b_k x(n-k)$$

in which the excitation $\{\mathbf{x}(n)\}$ is taken to be a white-noise process. Describe a procedure for selecting the filter b_k coefficients so that the response's autocorrelation sequence is in agreement with the values above. *Hint:* Use the z-transform interpretation of the spectral density function and equate coefficients of equal powers of z.

11.16 Prove the complex spectral density function relationships (11.26) and (11.29).

11.17 Let the causal linear system

$$y(n) = x(n) + 0.75y(n - 1) - \tfrac{1}{8}y(n - 2)$$

be excited by a real-valued wide-sense stationary signal whose autocorrelation is specified by

$$r_{\mathbf{xx}}(n) = 3(0.5)^{|n|}$$

Using the z-transform approach, determine $r_{\mathbf{yy}}(n)$ and $r_{\mathbf{xy}}(n)$.

11.18 Let the linear system governed by

$$y(n) - \tfrac{1}{2}y(n - 1) = x(n) + x(n - 1)$$

be excited by a wide-sense stationary time series with zero mean and autocorrelation sequence

$$r_{\mathbf{xx}}(n) = (\tfrac{1}{4})^{|n|}$$

Compute the following.
(a) $S_{\mathbf{yy}}(\omega)$ and $S_{\mathbf{yx}}(\omega)$
(b) $r_{\mathbf{yy}}(n)$ and $r_{\mathbf{yx}}(n)$
(c) $\sigma_{\mathbf{y}}^2$

11.19 The linear system as characterized by

$$y(n) + ay(n - 1) = (1 + a)x(n)$$

is often employed as a low-pass digital filter by selecting $a \approx -1$. Determine the response's autocorrelation sequence when the excitation is bandlimited white noise with spectral density function

$$S_{\mathbf{xx}}(\omega) = \begin{cases} \sigma^2 & |\omega| \leq \omega_c \\ 0 & \omega_c < |\omega| \leq \pi \end{cases}$$

11.20 Verify the autocovariance expression (11.52) for the optimal nonrecursive filter.

11.21 Verify the mean value and variance expressions (11.55) and (11.56), respectively, for the first-order recursive system (11.53).

11.22 Prove that the selection $\lambda = [1 - h(0)]/h^2(0)$ yields the desired pole–zero cancellation leading to transfer function (11.64). Moreover, show that linear recursive system (11.66) has its noise power reduction ratio equal to the parameter ρ.

11.23 Establish the validity of the entries in Table 11.1.

CHAPTER 12

Optimum MSE Filtering, Estimation, and Prediction: Finite Memory

12.1 INTRODUCTION

As demonstrated in Chapter 11, a signal-processing problem of considerable interdisciplinary interest concerns the task of extracting useful information from measured data. In many applications, an information-bearing signal may be modified by distortion of some form and additive noise may further obscure its observation. As examples, consider the following:

1. A received digital telephone message that is transmitted over an imperfect communication channel.
2. Radar return reflections from an airborne object obtained in an electrically unstable atmospheric environment.
3. An electrocardiogram recording of a fetal heartbeat that is masked by the mother's heartbeat.

In each of these situations, it is desired to use the noise-contaminated measurement data to recover an underlying information-bearing signal. This task is conceptually accomplished by applying the data to a linear filter which has been appropriately designed so that the filter's response best approximates the desired information signal. To simplify the presentation, the signals are initially considered to be real valued. The filter synthesis results obtained, however, are readily extended to the complex-valued signal case as is subsequently shown.

In this chapter we develop a systematic procedure for synthesizing linear filters to accomplish the signal recovery operation described above. It is here assumed that all signals considered are either wide-sense stationary or are approximately wide-sense stationary over the time interval in which the data measurements are made. Fortunately, this assumption is valid in many relevant applications and its invocation enables us to give a relatively straightforward means for synthesizing the required linear filter. We shall adopt the mean-square-error criterion for determining the optimum linear filter. It will be shown that this criterion results in our having to solve a linear system of equations for the optimum filter's weighting coefficients.

In all practical applications, there is available only a finite amount of data to be processed. Due to this finite data length, we initially restrict our investigation to linear filters that have finite memory. The fundamental synthesis problem then involves assigning values to the filter weighting coefficients (finite in number) so that a prescribed filter objective is achieved. Upon completion of the finite memory linear filter study, we consider infinite memory linear filters in Chapter 13. Infinite memory filters are useful in those situations where sufficient data are available (i.e., long data sequences). The resultant infinite memory linear filter synthesis is found to entail a spectral factorization of an associated spectral density function.

12.2 CLASSICAL LINEAR FILTERING: REAL-VALUED DATA

To begin our study, we consider the classical filtering problem. In particular, let it be desired to extract an information-bearing signal from measurements that have been contaminated by additive noise, that is,

$$\mathbf{x}(n) = \mathbf{s}(n) + \boldsymbol{\eta}(n) \tag{12.1}$$

In this expression, $\{\mathbf{s}(n)\}$ represents the underlying real-valued *information signal* and $\{\boldsymbol{\eta}(n)\}$ is a corruptive real-valued *additive noise* that obscures our observation of the information signal. To extract the desired signal from these noise-contaminated measurements, we apply a realization of the measurement signal (12.1) to the real causal linear filter as governed by

$$y(n) = \sum_{k=0}^{p} h(k)x(n - k) \tag{12.2}$$

This pth-order nonrecursive filter is said to have *finite memory,* since its response is seen to be exclusively dependent on a linear combination of the present and the p most recent excitation elements. This finite memory filter may be implemented by means of the tapped-delay-line configuration shown in Figure 12.1a,

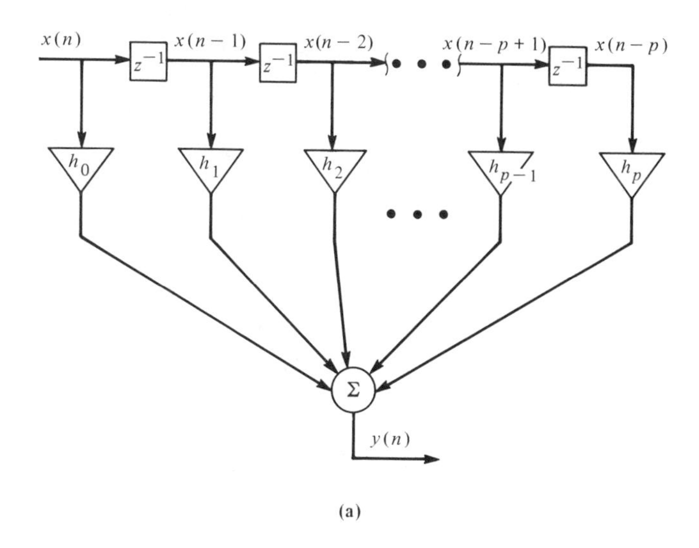

(a)

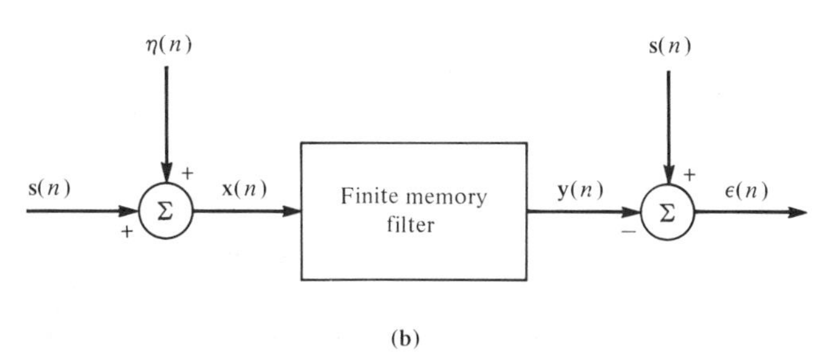

(b)

FIGURE 12.1. Finite memory filtering: (a) tapped-delay-line implementation of filter; (b) system configuration.

although other forms of realization (e.g., a lattice structure) are often preferred. The filter synthesis problem is concerned with selecting this filter's weighting coefficients $h(k)$ so that the response $\{y(n)\}$ best approximates the desired signal $\{s(n)\}$ in some sense. A brief description of the filter's effectiveness relative to these filter coefficients is now presented.

To measure the degree to which the filter's response differs from the desired underlying information signal, the *filter error* signal

$$\boldsymbol{\epsilon}(n) = \mathbf{s}(n) - \mathbf{y}(n) \tag{12.3}$$

is introduced. This filter error signal is conceptually generated in accordance with the configuration shown in Figure 12.1b. Our ultimate objective is to select the filter weighting coefficients so that this filter error signal is made reasonably close to zero. The *mean-squared-error* (MSE) criterion as defined by

$$E\{\boldsymbol{\epsilon}^2(n)\} = E\{[\mathbf{s}(n) - \mathbf{y}(n)]^2\} \tag{12.4}$$

provides a useful means for measuring how successfully this objective is achieved. This criterion is seen to be a function of the filter's weighting coefficients through its dependency on the filter response signal $\mathbf{y}(n)$.

When the desired signal $\{\mathbf{s}(n)\}$ and the additive noise $\{\boldsymbol{\eta}(n)\}$ are wide-sense stationary, the MSE criterion has a convenient closed-form expression in terms of the filters real-valued weighting coefficients. Specifically, upon substituting the response of filter (12.2) to the excitation $\mathbf{x}(n)$, it is found that

$$\begin{aligned} E\{\boldsymbol{\epsilon}^2(n)\} &= E\left\{\left[\mathbf{s}(n) - \sum_{k=0}^{p} h(k)\mathbf{x}(n-k)\right]\left[\mathbf{s}(n) - \sum_{m=0}^{p} h(m)\mathbf{x}(n-m)\right]\right\} \\ &= r_{\mathbf{ss}}(0) - 2\sum_{k=0}^{p} r_{\mathbf{sx}}(k)h(k) + \sum_{k=0}^{p}\sum_{m=0}^{p} r_{\mathbf{xx}}(m-k)h(k)h(m) \end{aligned} \quad (12.5)$$

where

$$r_{\mathbf{sx}}(n) = E\{\mathbf{s}(n+k)\mathbf{x}(k)\}$$

$$r_{\mathbf{ss}}(0) = E\{\mathbf{s}^2(n)\}$$

and

$$r_{\mathbf{xx}}(n) = E\{\mathbf{x}(n+k)\mathbf{x}(k)\}$$

The MSE criterion is seen to be a quadratic function of the filter's weighting coefficients. As we shall show, it is this quadratic behavior that gives rise to a feasible procedure for selecting the optimum filter's weighting coefficients.

12.3 OPTIMUM FILTER: THE WIENER–HOPF EQUATIONS

An optimum MSE filter is defined to be one that renders the MSE criterion (12.5) a minimum. By using standard calculus concepts, a necessary condition for a minimizing weighting coefficient selection is obtained by simply setting to zero the partial derivatives of this MSE criterion with respect to each of the filter's real-valued weighting coefficients. These partial derivatives are seen to be given by

$$\frac{\partial E\{\boldsymbol{\epsilon}^2(n)\}}{\partial h(m)} = -2r_{\mathbf{sx}}(m) + 2\sum_{k=0}^{p} r_{\mathbf{xx}}(m-k)h(k) \qquad \text{for } 0 \le m \le p$$

Thus the optimum filter's weighting coefficients must satisfy the following consistent system of linear equations:

$$\sum_{k=0}^{p} r_{\mathbf{xx}}(n-k)h^{o}(k) = r_{\mathbf{sx}}(n) \qquad \text{for } 0 \le n \le p \quad (12.6)$$

These relationships, commonly referred to as *Wiener–Hopf equations,* provide the mechanism for synthesizing the optimum filter. In particular, using any of a variety of procedures (e.g., Gaussian elimination), this system of equations is solved for the required optimum filter weighting coefficients.

The minimum MSE value associated with the foregoing optimum weighting coefficient selection is next obtained by substituting Wiener–Hopf expression (12.6) into the MSE criterion (12.5) to obtain

$$\begin{aligned} E\{\boldsymbol{\epsilon}^{o}(n)^2\} &= r_{\mathbf{ss}}(0) - \sum_{k=0}^{p}\sum_{m=0}^{p} r_{\mathbf{xx}}(m-k)h^{o}(k)h^{o}(m) \\ &= r_{\mathbf{ss}}(0) - \sum_{k=0}^{p} r_{\mathbf{sx}}(k)h^{o}(k) \end{aligned} \tag{12.7}$$

Any other filter weighting coefficient selection not satisfying the Wiener–Hopf equations will result in a larger value for the MSE criterion. Thus relationships (12.6) and (12.7) characterize the optimal linear filter in which the signal-processing objective is that of transmitting the information signal in an undistorted manner while rejecting the additive noise component.

Required Correlation Entries

To carry out the filter synthesis as described by the Wiener–Hopf equations (12.6), it is necessary to obtain values for the correlation elements that appear in these equations. These correlation terms are formally given by

$$\begin{aligned} r_{\mathbf{xx}}(n) &= E\{\mathbf{x}(n+k)\mathbf{x}(k)\} \\ &= E\{[\mathbf{s}(n+k) + \boldsymbol{\eta}(n+k)][\mathbf{s}(k) + \boldsymbol{\eta}(k)]\} \\ &= r_{\mathbf{ss}}(n) + r_{\mathbf{s}\boldsymbol{\eta}}(n) + r_{\mathbf{s}\boldsymbol{\eta}}(-n) + r_{\boldsymbol{\eta\eta}}(n) \end{aligned} \tag{12.8}$$

and

$$\begin{aligned} r_{\mathbf{sx}}(n) &= E\{\mathbf{s}(n+k)[\mathbf{s}(k) + \boldsymbol{\eta}(k)]\} \\ &= r_{\mathbf{ss}}(n) + r_{\mathbf{s}\boldsymbol{\eta}}(n) \end{aligned} \tag{12.9}$$

It often happens that the information signal and the additive noise are orthogonal so that $r_{s\eta}(n) = 0$. In this commonly occurring situation, the correlation elements take on the simplified form

$$r_{\mathbf{xx}}(n) = r_{\mathbf{ss}}(n) + r_{\boldsymbol{\eta\eta}}(n) \qquad \text{and} \qquad r_{\mathbf{sx}}(n) = r_{\mathbf{ss}}(n) \tag{12.10}$$

Whatever the case, upon examination of the Wiener–Hopf equations (12.6), it is found that values for these two correlation sequences are required only on the interval $0 \le n \le p$ to determine the optimum set of filter weighting coefficients.

EXAMPLE 12.1
Let the information-bearing real random signal $\{\mathbf{s}(n)\}$ have autocorrelation $r_{\mathbf{ss}}(n) = 4(0.5)^{|n|}$ and the orthogonal real additive white noise have variance 2 so that $r_{\eta\eta}(n) = 2\delta(n)$. It is desired to synthesize an optimum causal MSE filter of order $p = 2$. Due to the orthogonal noise assumption, the required correlation lags are specified by (12.10) so that $r_{\mathbf{xx}}(n) = 4(0.5)^{|n|} + 2\delta(n)$ and $r_{\mathbf{sx}}(n) = 4(0.5)^{|n|}$. Upon substituting these lags into the Wiener–Hopf equations (12.6) with $p = 2$, we have

$$\begin{aligned} 6h^o(0) + 2h^o(1) + h^o(2) &= 4 \\ 2h^o(0) + 6h^o(1) + 2h^o(2) &= 2 \\ h^o(0) + 2h^o(1) + 6h^o(2) &= 1 \end{aligned}$$

It is readily shown that the solution to these equations is

$$h^o(0) = \tfrac{53}{85}, \qquad h^o(1) = \tfrac{10}{85}, \qquad h^o(2) = \tfrac{2}{85}$$

thereby indicating that the optimum MSE filter of order 2 is characterized by

$$y(n) = \tfrac{53}{85}x(n) + \tfrac{10}{85}x(n-1) + \tfrac{2}{85}x(n-2)$$

Moreover, the minimum MSE error (12.7) associated with this optimum filter is

$$E\{(\boldsymbol{\epsilon}^{o^2}(n)\} = 4 - (\tfrac{53}{85})4 - (\tfrac{10}{85})2 - (\tfrac{2}{85}) = \tfrac{106}{85}$$

12.4 ORTHOGONALITY PROPERTY OF OPTIMUM FILTER

The optimum linear filter as governed by the Wiener–Hopf equations (12.6) possesses a useful statistical property in that the associated minimum MSE signal $\{\boldsymbol{\epsilon}^o(n)\}$ and the data being filtered $\{\mathbf{x}(n)\}$ are *orthogonal* in a restricted sense. That is, the following fundamental *orthogonal* condition holds:

$$r_{\boldsymbol{\epsilon}^o\mathbf{x}}(m) = E\{\boldsymbol{\epsilon}^o(n)\mathbf{x}(n-m)\} = 0 \qquad \text{for } 0 \le m \le p \tag{12.11}$$

where the interval of orthogonality $0 \le m \le p$ is seen to correspond exactly with the time interval over which the filter's weighting coefficients are nonzero. A proof of this very important property follows directly by recalling that the optimum weighting coefficients were obtained by setting the partial of the MSE criterion (12.5) with respect to $h(m)$ equal to zero, that is,

$$\frac{\partial E\{\boldsymbol{\epsilon}^2(n)\}}{\partial h(m)} = 2E\left\{\boldsymbol{\epsilon}^o(n)\,\frac{\partial \boldsymbol{\epsilon}^o(n)}{\partial h(m)}\right\} = 0$$

where the operations of expectation and differentiation have been interchanged. From relationships (12.3) and (12.2), we see that $\partial\boldsymbol{\epsilon}(n)/\partial h(m) = -\partial\mathbf{y}(n)/\partial h(m) = -\mathbf{x}(n - m)$, which when inserted into this equation yields the orthogonality condition (12.11).

It is recalled from Chapter 8 that orthogonality and uncorrelatedness are equivalent notions provided that at least one of the time series $\{\boldsymbol{\epsilon}^o(n)\}$ or $\{\mathbf{x}(n)\}$ has zero mean. With this restriction in mind, we may then interpret relationship (12.11) as indicating that the optimum linear filter operation has extracted all the correlated information concerning $\mathbf{s}(n)$ that is contained in the random variables $\mathbf{x}(n), \mathbf{x}(n - 1), \ldots, \mathbf{x}(n - p)$. Moreover, it is precisely this property that distinguishes an optimal filter from its suboptimal counterparts.

In various signal-processing applications, an often unstated requirement is that of generating orthogonal (or uncorrelated) signals. As shown above, the optimal MSE filter provides a mechanism for effecting this objective. It achieves this goal only partially, however, since for $\{\mathbf{x}(n)\}$ to be totally orthogonal with $\{\boldsymbol{\epsilon}^o(n)\}$ requires that $r_{\boldsymbol{\epsilon}^o\mathbf{x}}(n)$ equal zero for all integer values of n. If we require a completely uncorrelated condition, it is generally necessary to use an optimal linear filter whose weighting sequences are nonzero from minus to plus infinity. Clearly, this requirement is incompatible with the processing of data that has finite length.

12.5 GENERAL OPTIMUM MSE LINEAR FILTERING

In Sections 12.3 and 12.4 we examined the classical filtering problem, which was concerned with extracting a signal from an additive-noise-corrupted version of that signal. This problem is a special case of a more general problem to be now considered. The data to be processed are taken to be a finite length realization of the real-valued wide-sense stationary random time series:

$$\{\mathbf{x}(n)\} \tag{12.12}$$

We now wish to process a realization of these random elements with the finite memory real filter

$$y(n) = \sum_{k=0}^{p} h(k)x(n - k) \tag{12.13}$$

so that the response $\{\mathbf{y}(n)\}$ best approximates a prescribed *desired real-valued signal* $\{\mathbf{d}(n)\}$. Conceptually, this desired signal is somehow embedded in the data (12.12) being processed. The selection of the desired signal is very dependent on the particular application under consideration. As a specific example, the classical filtering problem considered in Section 12.2 corresponded to the selection $\mathbf{d}(n) = \mathbf{s}(n)$, where the measurement data are taken to be $\mathbf{x}(n) = \mathbf{s}(n) + \boldsymbol{\eta}(n)$. Whatever the case, the MSE criterion, specified by

$$E\{\boldsymbol{\epsilon}^2(n)\} = E\{[\mathbf{d}(n) - \mathbf{y}(n)]^2\} \tag{12.14}$$

is used to measure the degree to which $\mathbf{y}(n)$ approximates $\mathbf{d}(n)$. This criterion is seen to be a function of the filter's weighting coefficients $h(k)$ due to their presence in the response term $\mathbf{y}(n)$. A visual depiction of the general filtering problem is given in Figure 12.2. The filter synthesis problem is then concerned with selecting the filter's weighting coefficients so as to render this MSE criterion a minimum.

As just indicated, this general filtering problem is seen to encompass the classical filtering problem studied in Sections 12.2 and 12.3 as a special case. By leaving the data measurments $\{\mathbf{x}(n)\}$ and the desired signal $\{\mathbf{d}(n)\}$ initially unspecified, however, it is possible to characterize a broader class of signal-processing problems in the guise of a single general MSE filtering problem. We therefore avoid the unnecessary duplicative effort needed to study separately each problem belonging to this class. How one adapts the results of this general filter synthesis approach to specific choices for the desired signal $\{\mathbf{d}(n)\}$ is illustrated in Section 12.6. With this thought in mind, we begin the analysis of the postulated general filtering problem.

Under the assumption that the signals $\{\mathbf{d}(n)\}$ and $\{\mathbf{x}(n)\}$ are wide-sense stationary, the value assumed by the MSE criterion (12.14) for a general finite memory linear filter of form (12.13) is given by

$$\begin{aligned} E\{\boldsymbol{\epsilon}^2(n)\} &= E\left\{\left[\mathbf{d}(n) - \sum_{k=0}^{p} h(k)\mathbf{x}(n-k)\right]\left[\mathbf{d}(n) - \sum_{m=0}^{p} h(m)\mathbf{x}(n-m)\right]\right\} \\ &= r_{\mathbf{dd}}(0) - 2\sum_{k=0}^{p} r_{\mathbf{dx}}(k)h(k) + \sum_{k=0}^{p}\sum_{m=0}^{p} r_{\mathbf{xx}}(m-k)h(k)h(m) \end{aligned} \tag{12.15}$$

where

$$r_{\mathbf{dd}}(0) = E\{\mathbf{d}^2(n)\}, \quad r_{\mathbf{dx}}(n) = E\{\mathbf{d}(n+k)\mathbf{x}(k)\}, \quad r_{\mathbf{xx}}(n) = E\{\mathbf{x}(n+k)\mathbf{x}(k)\}$$

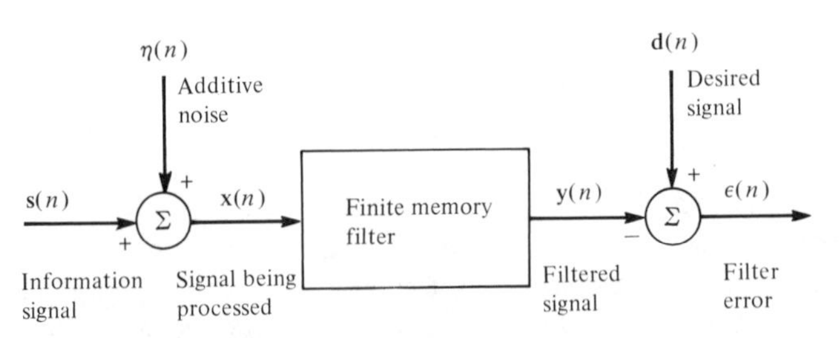

FIGURE 12.2. General linear filtering problem.

The optimum selection of the filter's weighting coefficients are obtained by setting to zero the partial derivatives of the MSE criterion with respect to the weighting coefficients. These partial derivatives are specified by

$$\frac{\partial E\{\epsilon^2(n)\}}{\partial h(m)} = -2r_{\mathbf{dx}}(m) + 2\sum_{k=0}^{p} r_{\mathbf{xx}}(m-k)h(k) \qquad \text{for } 0 \le m \le p$$

Thus the optimum filter weighting coefficients must satisfy the following consistent system of Wiener–Hopf equations:

$$\sum_{k=0}^{p} r_{\mathbf{xx}}(n-k)h^{o}(k) = r_{\mathbf{dx}}(n) \qquad \text{for } 0 \le n \le p \tag{12.16}$$

Furthermore, the minimum MSE value associated with this optimum coefficient selection is obtained by substituting relationship (12.16) into expression (12.15), which gives

$$\begin{aligned} E\{\boldsymbol{\epsilon}^{o^2}(n)\} &= r_{\mathbf{dd}}(0) - \sum_{k=0}^{p}\sum_{m=0}^{p} r_{\mathbf{xx}}(m-k)h^{o}(k)h^{o}(m) \\ &= r_{\mathbf{dd}}(0) - \sum_{k=0}^{p} r_{\mathbf{dx}}(k)h^{o}(k) \end{aligned} \tag{12.17}$$

We conclude that the optimum general linear filter is characterized by relationships (12.16) and (12.17).

Orthogonality Property

As in all optimum linear mean square error filtering problems associated with wide-sense stationary processes, the optimum error signal is partially orthogonal to the observed signal in the sense that

$$E\{\boldsymbol{\epsilon}^{o}(n)\mathbf{x}(n-m)\} = 0 \qquad \text{for } 0 \le m \le p \tag{12.18}$$

A proof of this result is obtained directly by noting that the partial derivative $\partial\{\boldsymbol{\epsilon}^2(n)\}/\partial h(m) = -2E\{\boldsymbol{\epsilon}(n)\mathbf{x}(n-m)\}$ equals zero for an optimal filter coefficient selection with $0 \le m \le p$. This property is noteworthy in that it indicates that the optimal filter estimate $\mathbf{y}(n)$ extracts all the correlated information relative to $\mathbf{d}(n)$ that is contained in the data set $\mathbf{x}(n)$, $\mathbf{x}(n-1)$, . . . , $\mathbf{x}(n-p)$. The reader is encouraged to appreciate the significance of this important property.

In summary, an optimum set of weighting coefficients for the general filtering problem considered in this section is obtained by solving the consistent system of Wiener–Hopf equations (12.16). The coefficients of this system of linear equations are given by the correlation lag elements $r_{\mathbf{xx}}(n)$ and $r_{\mathbf{dx}}(n)$ for $0 \le n \le p$. How one determines these lag elements depends critically on the nature of the observed signal $\{\mathbf{x}(n)\}$ and the desired signal $\{\mathbf{d}(n)\}$. An important class of signal-processing problems that falls into the general filtering problem is briefly examined in Section 12.6.

Complex-Valued Data

Our discussions up to this point have been exclusively involved with the processing of real-valued data. As suggested earlier, however, a number of relevant applications are concerned with complex-valued data. Fortunately, it is possible to straightforwardly extend the concepts used in the real data case to arrive at optimum filters for complex data signal processing. A brief outline is now given for making this adaptation. The filtering operation under consideration is again given by the nonrecursive relationship

$$y(n) = \sum_{k=0}^{p} h(k)x(n-k) \tag{12.19a}$$

in which the $h(k)$ filter weighting coefficients are now allowed to take on complex values. These coefficients may be expressed in their rectangular form

$$h(k) = \alpha_k + j\beta_k \tag{12.19b}$$

where the real parameters α_k and β_k are the real and imaginary components of $h(k)$, respectively. The task at hand is to then select the filter weighting coefficients so that the filter's response $\{\mathbf{y}(n)\}$ best approximates a desired signal $\{\mathbf{d}(n)\}$ in the least squared error magnitude sense as measured by

$$E\{|\boldsymbol{\epsilon}(n)|^2\} = E\{\boldsymbol{\epsilon}(n)\bar{\boldsymbol{\epsilon}}(n)\} \tag{12.19c}$$

where $\bar{\boldsymbol{\epsilon}}(n)$ denotes the complex conjugate of the filter error signal as specified by

$$\boldsymbol{\epsilon}(n) = \mathbf{d}(n) - \mathbf{y}(n)$$

It is necessary to use the magnitude of the error squared due to the complex valued behavior of the error signal. In the analysis to be given now, the signals $\{\mathbf{x}(n)\}$ and $\{\mathbf{d}(n)\}$ are each taken to be complex-valued wide-sense stationary random time series. Upon substitution of expression (12.19a) into this MSE criterion, it is readily shown that

$$E\{|\boldsymbol{\epsilon}(n)|^2\} = r_{\mathbf{dd}}(0) - \sum_{k=0}^{p} [h(k)\bar{r}_{\mathbf{dx}}(k) + \bar{h}(k)r_{\mathbf{dx}}(k)] + \sum_{k=0}^{p}\sum_{m=0}^{p} h(k)\bar{h}(m)r_{\mathbf{xx}}(m-k) \tag{12.19d}$$

It is next noted that this MSE criterion is a function of the real parameters α_k and β_k as is made clear upon substituting expression (12.19b) into criterion (12.19d). The optimum filter weighting coefficients are defined to be that particular selection which renders this MSE criterion its minimum value. By applying standard calculus techniques, an optimal choice for the weighting coefficients is obtained by simply setting to zero the partial derivatives $\partial E\{|\boldsymbol{\epsilon}(n)|^2\}/\partial\alpha_k$ and $\partial E\{|\boldsymbol{\epsilon}(n)|^2\}/\partial\beta_k$ for $0 \leq k \leq p$ This procedure is found to give rise to the following system of linear Wiener–Hopf equations

$$\sum_{k=0}^{p} r_{\mathbf{xx}}(n - k)h^{o}(k) = r_{\mathbf{dx}}(n) \qquad \text{for } 0 \leq n \leq p \qquad (12.19e)$$

which characterize the optimal filter weighting coefficients. It is interesting that these same sets of equations identify the optimal filter parameters in the real-valued data case. It is important to note, however, that in the complex data case here being considered, the optimum weighting coefficients will in general be complex-valued. This is due to the fact that the correlation lags $r_{\mathbf{xx}}(n)$ and $r_{\mathbf{dx}}(n)$ will themselves be complex valued for complex-valued time series.

The value assumed by the MSE criterion is determined by substituting relationship (12.19e) into expression (12.19d). This is found to result in

$$E\{|\boldsymbol{\epsilon}^{o}(n)|^2\} = r_{\mathbf{dd}}(0) - \sum_{k=0}^{p} r_{\mathbf{dx}}(k)\bar{h}^{o}(k) \qquad (12.19f)$$

Furthermore, the optimal filter coefficients are found to result in the following orthogonality property

$$E\{\boldsymbol{\epsilon}^{o}(n)\bar{\mathbf{x}}(n - m)\} = 0 \qquad \text{for } 0 \leq m \leq p \qquad (12.19g)$$

Not surprisingly, these results simplify to those obtained earlier for the analogous real-valued random data case.

12.6

OPTIMUM FILTERING, SMOOTHING, AND PREDICTION

To illustrate how the developments in Section 12.5 can be used in practical applications, let us examine the special case in which the *desired signal* is of the form

$$\mathbf{d}(n) = \mathbf{s}(n + m) \qquad (12.20a)$$

where m is a fixed integer. The function of the optimum linear filter is to operate on the additive-noise contaminated observed signal

$$\mathbf{x}(n) = \mathbf{s}(n) + \boldsymbol{\eta}(n) \tag{12.20b}$$

so as to filter out the noise term while producing a response signal $\mathbf{y}(n)$ that approximates $\mathbf{s}(n + m)$. The nature of the optimum filter is linked inherently to selection of the integer m. For $m = 0$, we have the classical *filter* problem considered in Section 12.2, in which the filter is to remove the additive noise term while passing undistorted the information signal $\{\mathbf{s}(n)\}$. On the other hand, if m is a negative integer, the resultant optimum *smoother* filter is to produce an "m-delayed" version of the information signal. Similarly, when m is positive, the associated optimum *predictor* filter is required to produce an m-step prediction of the information signal. Table 12.1 summarizes these various choices for m.

TABLE 12.1 Optimum Filter Associated with the Desired Signal Selection $\mathbf{d}(n) = \mathbf{s}(n + m)$

Index	Optimum Filter's Function
$m = 0$	Filtering
$m \leq -1$	Smoothing
$m \geq 1$	Prediction

For the optimum filter, smoother, and predictor, the associated correlation lag entries $r_{\mathbf{xx}}(n)$ appearing in the Wiener–Hopf equations (12.16) or (12.19e) are formally specified by

$$\begin{aligned} r_{\mathbf{xx}}(n) &= E\{[\mathbf{s}(n + k) + \boldsymbol{\eta}(n + k)][\bar{\mathbf{s}}(k) + \bar{\boldsymbol{\eta}}(k)]\} \\ &= r_{\mathbf{ss}}(n) + r_{\mathbf{s}\boldsymbol{\eta}}(n) + \bar{r}_{\mathbf{s}\boldsymbol{\eta}}(-n) + r_{\boldsymbol{\eta}\boldsymbol{\eta}}(n) \end{aligned} \tag{12.21}$$

where the signals involved are taken to be complex-valued wide-sense stationary. If they are real valued, then all the correlation sequences involved are also real valued. This is the primary feature that distinguishes real- and complex-valued time series. The required cross-correlation lag elements $r_{\mathbf{dx}}(n)$ are explicitly dependent on the choice of m, as is made evident by noting that

$$\begin{aligned} r_{\mathbf{dx}}(n) &= E\{\mathbf{d}(n + k)\bar{\mathbf{x}}(k)\} \\ &= E\{[\mathbf{s}(n + k + m)][\bar{\mathbf{s}}(k) + \bar{\boldsymbol{\eta}}(k)]\} \\ &= r_{\mathbf{ss}}(n + m) + r_{\mathbf{s}\boldsymbol{\eta}}(n + m) \qquad 0 \leq n \leq p \end{aligned} \tag{12.22}$$

Clearly, the optimum filter's weighting coefficients arising from solution of the Wiener–Hopf equations (12.16) or (12.19e) are dependent on m through these cross-correlation lag elements.

By taking the general filtering approach as represented by the Wiener–Hopf equations (12.16) or (12.19e), we are able to study the signal-processing operations of filtering, smoothing, and prediction in a single problem format. This simply entails making an obvious choice for the observed signal $\{\mathbf{x}(n)\}$ and the desired signal $\{\mathbf{d}(n)\}$, which in turn is manifested in the nature of the autocorrelation lags $r_{\mathbf{xx}}(n)$ and cross-correlation lags $r_{\mathbf{dx}}(n)$.

EXAMPLE 12.2

Let us consider the real-valued information signal and additive noise described in Example 12.1. Design the optimum linear signal processor of order $p = 2$ associated with the observed signal $\mathbf{x}(n) = \mathbf{s}(n) + \boldsymbol{\eta}(n)$ and the desired signal selections (a) $\mathbf{d}(n) = \mathbf{s}(n)$, (b) $\mathbf{d}(n) = \mathbf{s}(n - 1)$, and (c) $\mathbf{d}(n) = \mathbf{s}(n + 1)$. For this application, we have $r_{\mathbf{xx}}(n) = 4(0.5)^{|n|} + 2\delta(n)$, while expression (12.22) indicates that

$$r_{\mathbf{dx}}(n) = r_{\mathbf{ss}}(n + m) = 4(0.5)^{|n+m|}$$

where $m = 0, -1$, and 1 for cases (a), (b), and (c), respectively. In accordance with relationship (12.16), the optimum signal processor is obtained by solving the following system of three linear equations in the three unknown optimum weighting coefficients:

$$6h^{o}(0) + 2h^{o}(1) + h^{o}(2) = 4(0.5)^{|m|}$$

$$2h^{o}(0) + 6h^{o}(1) + 2h^{o}(2) = 4(0.5)^{|1+m|}$$

$$h^{o}(0) + 2h^{o}(1) + 6h^{o}(2) = 4(0.5)^{|2+m|}$$

Using any standard linear equation solving routine, we can readily show that the optimum weighting coefficients and associated minimum MSE error are as listed below for $m = 0, -1$, and 1. It is apparent that the filter coefficients vary considerably as a function of m. As might be anticipated, smoothing ($m = -1$) provides the smallest MSE, followed by filtering ($m = 0$) and then prediction ($m = 1$).

m	$h^{o}(0)$	$h^{o}(1)$	$h^{o}(2)$	$E\{\epsilon^{o}(n)^2\}$
0	53/85	10/85	2/85	106/85
−1	2/17	10/17	2/17	20/17
1	53/170	10/170	2/170	563/170

■

12.7 ITERATIVE SOLUTION OF THE WIENER–HOPF EQUATIONS

As shown in Section 12.5, the optimum MSE filter of order p is obtained by solving the system of linear Wiener–Hopf equations (12.16) or (12.19c). If standard procedures such as Gaussian elimination are employed, this solution will take on the order of p^3 multiplication and addition operations. Fortunately, due to the special form of the Wiener–Hopf equations, it is possible to obtain the required solution in a far more efficient manner. For both the real- and complex-valued data cases, this approach is found to entail sequentially solving the Wiener–Hopf equations

$$\sum_{k=0}^{m} r_{\mathbf{xx}}(n - k)h_m^{\text{o}}(k) = r_{\mathbf{dx}}(n) \qquad \text{for } 0 \leq n \leq m \tag{12.23}$$

as the filter-order parameter m is sequenced through the values $1, 2, \ldots, p$. In this solution process, the subscript m has been appended to the optimal weighting coefficients $h_m^{\text{o}}(k)$ so as to recognize explicitly their dependency on the filter order. To effect this iterative solution, it will also be necessary to have the solution [i.e., the $g_m(k)$ for $0 \leq k \leq m$] to the following auxiliary system of linear equations:

$$\sum_{k=0}^{m} r_{\mathbf{xx}}(n - k)g_m(k) = \begin{cases} 0 & \text{for } 0 \leq n \leq m - 1 \\ 1 & \text{for } n = m \end{cases} \tag{12.24}$$

The subscript m has also been appended to the solution elements $g_m(k)$ to denote their dependency on the filter order.

The task at hand is then to use the solutions characterizing the optimum filter of order m as specified by relationships (12.23) and (12.24) so as to solve the corresponding equations for the $(m + 1)$-order optimum filter. These next higher-order optimum filter equations are specified by the similar system of equations:

$$\sum_{k=0}^{m+1} r_{\mathbf{xx}}(n - k)h_{m+1}^{\text{o}}(k) = r_{\mathbf{dx}}(n) \qquad \text{for } 0 \leq n \leq m + 1 \tag{12.25}$$

and

$$\sum_{k=0}^{m+1} r_{\mathbf{xx}}(n - k)g_{m+1}(k) = \begin{cases} 0 & \text{for } 0 \leq n \leq m \\ 1 & \text{for } n = m + 1 \end{cases} \tag{12.26}$$

where the subscript $m + 1$ has been appended to reflect the new filter order $m + 1$. It is readily shown by direct substitution that the iterative solution procedure outlined in Table 12.2 will give the required solution to these next-higher-order optimum filter equations. It is to be noted that in the real-valued data case, all the entities appearing in the algorithm are real valued.

The algorithm's initialization is given in step 1, in which the optimum filter of order $m = 0$ is identified. Step 3 constitutes the critical portion of the algorithm, where the solutions to the linear system of equations (12.25) and (12.26) are generated. Upon analysis of step 3, it is seen that on the order of $5m$ multiplication and addition operations are required to complete the mth iteration of the algorithm. Thus on the order of

$$\sum_{m=1}^{p} 5m = \frac{5p(p+1)}{2}$$

multiplication and summation operations are required to synthesize the optimum linear filter of order p. This compares very favorably with the number of operations (on the order of p^3) needed to solve the Wiener–Hopf equations (12.16) if the standard methods were employed. Furthermore, this iterative procedure is typically less sensitive to numerical errors arising from such effects as number truncation. Moreover, the extra dividend of having all the lower-order optimum linear filters (i.e., $m = 0, 1, \ldots, p-1$) and their associated minimum MSE values is obtained as a by-product of this iterative procedure.

EXAMPLE 12.3

Use the algorithm of Table 12.2 to synthesize the optimum filter associated with the problem considered in Example 12.1. It is recalled that the correlation

TABLE 12.2 Iterative Procedure for Solving the Wiener–Hopf Equations

1. $$g_0(0) = \frac{1}{r_{\mathbf{xx}}(0)}$$

$$h_0^{\circ}(0) = \frac{r_{\mathbf{dx}}(0)}{r_{\mathbf{xx}}(0)}$$

$$E\{|\epsilon_0^{\circ}(n)|^2\} = r_{\mathbf{dd}}(0) - \frac{|r_{\mathbf{dx}}(0)|^2}{r_{\mathbf{xx}}(0)}$$

2. $m = 0$

3. $$\alpha_m = \sum_{k=0}^{m} \bar{r}_{\mathbf{xx}}(k+1) g_m(k)$$

$$\beta_m = \sum_{k=0}^{m} r_{\mathbf{xx}}(m+1-k) h_m^{\circ}(k)$$

$$g_{m+1}(k) = \begin{cases} \dfrac{-\alpha_m g_m(m)}{1-|\alpha_m|^2} & \text{for } k = 0 \\[2ex] \dfrac{g_m(k-1) - \alpha_m \bar{g}_m(m-k)}{1-|\alpha_m|^2} & \text{for } 1 \le k \le m \\[2ex] \dfrac{g_m(m)}{1-|\alpha_m|^2} & \text{for } k = m+1 \end{cases}$$

$$h_{m+1}^{\circ}(k) = \begin{cases} h_m^{\circ}(k) + [r_{\mathbf{dx}}(m+1) - \beta_m] g_{m+1}(k) & \text{for } 0 \le k \le m \\ [r_{\mathbf{dx}}(m+1) - \beta_m] g_{m+1}(m+1) & \text{for } k = m+1 \end{cases}$$

$$E\{|\epsilon_{m+1}^{\circ}|^2\} = r_{\mathbf{dd}}(0) - \sum_{k=0}^{m+1} \bar{h}_{m+1}^{\circ}(k) r_{\mathbf{dx}}(k)$$

4. If $m + 1 = p$, stop; otherwise, let $m = m + 1$ and go to step 3.

lag elements characterizing this problem were given by $r_{\mathbf{xx}}(n) = 4(0.5)^{|n|} + 2\delta(n)$ and $r_{\mathbf{dx}}(n) = r_{\mathbf{dd}}(n) = 4(0.5)^{|n|}$. Using these entries, the algorithm as described in Table 12.2 are readily shown to yield the results shown below. These results are in agreement with those found in Example 12.1 obtained using a more direct approach. ■

Iteration Number, m	$g_m(0)$	$g_m(1)$	$g_m(2)$	$h_m^o(0)$	$h_m^o(1)$	$h_m^o(2)$	$E\{\lvert\epsilon_m^o(n)\rvert^2\}$	α_m	β_m
0	$\frac{1}{6}$			$\frac{2}{3}$			$\frac{4}{3}$	$\frac{1}{3}$	$\frac{4}{3}$
1	$-\frac{1}{16}$	$\frac{3}{16}$		$\frac{5}{8}$	$\frac{1}{8}$		$\frac{5}{4}$	$\frac{1}{16}$	$\frac{7}{8}$
2	$-\frac{1}{85}$	$-\frac{1}{17}$	$\frac{16}{85}$	$\frac{53}{85}$	$\frac{2}{17}$	$\frac{2}{85}$	$\frac{106}{85}$		

12.8 LINEAR PREDICTION

Linear predictors are used extensively in signal-processing applications found in such diverse areas as economic, seismology, radar, sonar, and control theory. Although we could directly apply the results of Section 12.5 on general filter theory to the special case of linear prediction, there is much to be gained by taking a more direct approach. This will include an ability to derive the structural properties of the optimum linear predictor as well as providing a lattice implementation.

The fundamental linear prediction problem is concerned with the task of predicting the value assumed by the generally complex-valued signal element $\mathbf{x}(n)$, with this prediction being based on a linear combination of the p most recently observed signal elements. The *linear predictor* of order p therefore takes the form

$$\hat{\mathbf{x}}(n) = -a(1)\mathbf{x}(n-1) - a(2)\mathbf{x}(n-2) - \cdots - a(p)\mathbf{x}(n-p) \quad (12.27)$$

where the prediction coefficients $a(k)$ have purposely been assigned a negative sign in order that they appear with a positive sign in the associated predictor error filter to be introduced next. The degree to which the prediction $\hat{x}(n)$ approximates $\mathbf{x}(n)$ is here measured by the *prediction error*, specified by

$$\begin{aligned} \epsilon(n) &= \mathbf{x}(n) - \hat{\mathbf{x}}(n) \\ &= \sum_{k=0}^{p} a(k)\mathbf{x}(n-k) \end{aligned} \quad (12.28)$$

where the coefficient $a(0) = 1$.

The two operations (12.27) and (12.28) can be interpreted as filters that operate on the signal $\{\mathbf{x}(n)\}$ to produce the prediction $\{\hat{\mathbf{x}}(n)\}$ and the corresponding prediction error $\{\boldsymbol{\epsilon}(n)\}$ responses, respectively. As such, we shall call operator (12.27) the *prediction filter* and operator (12.28) the *prediction error filter*. They are related to one another as shown in Figure 12.3. For reasons that will be made clear shortly, we will be particularly interested in the predictor error filter (12.28) that has the associated transfer function

$$A_p(z) = 1 + a(1)z^{-1} + a(2)z^{-2} + \cdots + a(p)z^{-p} \tag{12.29}$$

The subscript p has been appended to this transfer function so as to explicitly denote its order. This notation is put to use in Section 12.9 when an iterative procedure is developed for obtaining the optimum prediction error filter transfer function of order $p + 1$ given its corresponding pth-order counterpart.

When the generally complex-valued signal to be predicted is wide-sense stationary, the MSE associated with the prediction error (12.28) is readily determined. In particular, we have

$$\begin{aligned} E\{|\epsilon(n)|^2\} &= E\left\{\left[\sum_{k=0}^{p} a(k)\mathbf{x}(n-k)\right]\left[\sum_{m=0}^{p} \bar{a}(m)\bar{\mathbf{x}}(n-m)\right]\right\} \\ &= \sum_{k=0}^{p}\sum_{m=0}^{p} r_{\mathbf{xx}}(m-k)a(k)\bar{a}(m) \end{aligned} \tag{12.30}$$

in which $a(0) = 1$. Thus for each choice of predictor coefficients, this criterion's value provides a measure of the goodness of prediction.

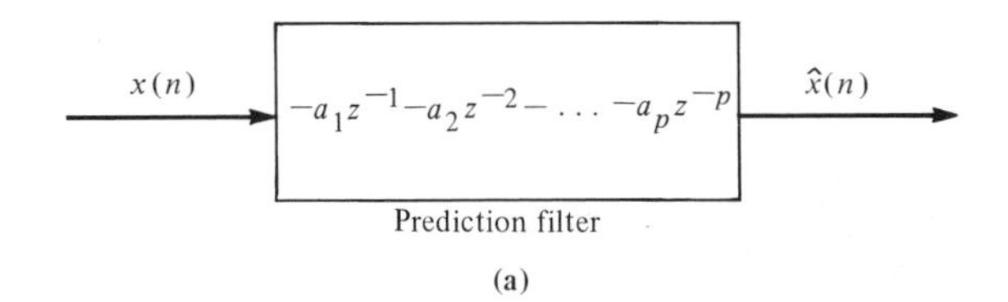

(a)

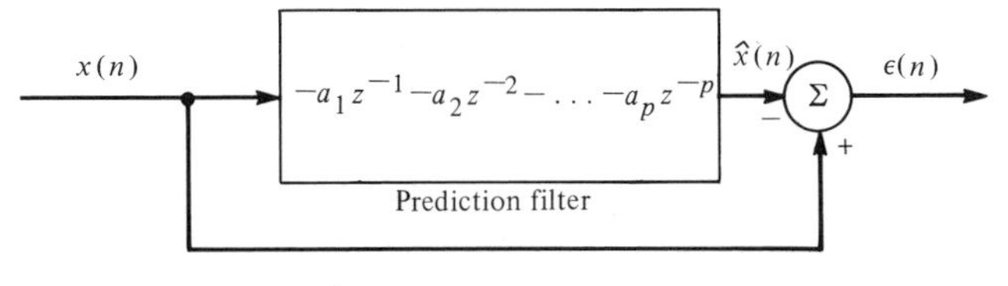

Prediction error filter

(b)

FIGURE 12.3. Prediction operation.

Optimum Predictor Filter

The optimum linear predictor is characterized by that set of predictor coefficients that render the mean-squared prediction criterion (12.30) a minimum. To determine these optimum coefficients, we then simply set to zero the partial derivatives of this criterion with respect to the predictor coefficient's real and imaginary components. Upon setting these derivatives to zero and using the fact that $a(0) = 1$, it is found that the optimal predictor coefficients must satisfy the following consistent linear system of equations:

$$\sum_{k=1}^{p} r_{\mathbf{xx}}(n-k)a^{o}(k) = -r_{\mathbf{xx}}(n) \qquad \text{for } 1 \leq n \leq p \tag{12.31}$$

We may then in principle solve this system of equations for the required optimal predictor coefficients using any convenient solution procedure. When the signal $\{\mathbf{x}(n)\}$ is real valued, these optimum predictor coefficients will also be real valued. Whatever the case, once the optimum coefficients have been determined, the minimum mean-squared prediction error criterion according to expression (12.30) is given by

$$\begin{aligned} E\{|\boldsymbol{\epsilon}^{o}(n)|^2\} &= \sum_{k=0}^{p}\sum_{m=0}^{p} r_{\mathbf{xx}}(m-k)a^{o}(k)\bar{a}^{o}(m) \\ &= \sum_{k=0}^{p} \bar{r}_{\mathbf{xx}}(k)a^{o}(k) \end{aligned} \tag{12.32}$$

Thus the optimal linear predictor of order p is identified by expression (12.31) and its prediction effectiveness is measured by criterion (12.32).

EXAMPLE 12.4

Synthesize the optimum linear predictor of order $p = 2$ that is associated with the wide-sense stationary process $\{\mathbf{x}(n)\}$ whose autocorrelation is specified by $r_{\mathbf{xx}}(n) = 4(0.5)^{|n|}$. The required optimum predictor must satisfy relationship (12.31) for $p = 2$, that is,

$$\begin{aligned} 4a^{o}(1) + 2a^{o}(2) &= -2 \\ 2a^{o}(1) + 4a^{o}(2) &= -1 \end{aligned}$$

It is readily established that the predictor coefficient selection $a_1^{o} = -\frac{1}{2}$ and $a_2^{o} = 0$ uniquely satisfies this system of equations. Moreover, the corresponding minimum mean-squared prediction error criterion is from relationship (12.32) given by

$$E\{\boldsymbol{\epsilon}^{o^2}(n)\} = 3$$

The optimum linear predictor is therefore characterized by

$$\hat{x}(n) = -\tfrac{1}{2}x(n-1)$$

and is seen to be a predictor of order 1 even though we originally selected the order $p = 2$. ■

Whitening Filter Approximation

By selecting the linear predictor's order p large enough, it can be shown that for any wide-sense stationary process $\{\mathbf{x}(n)\}$, the corresponding prediction error signal $\{\boldsymbol{\epsilon}^{o}(n)\}$ can be made to approach a white-noise process [i.e., $r_{\boldsymbol{\epsilon}^{o}\boldsymbol{\epsilon}^{o}}(n) = \sigma_{\boldsymbol{\epsilon}^{o}}^{2}\delta(n)$]. This is readily explained by observing that the ability to predict depends intrinsically on using any correlation existent between elements of a signal. As the prediction performance improves by increasing the predictor's order, the prediction $\hat{\mathbf{x}}(n)$ contains virtually all of the correlated information in the underlying signal. It then follows that the prediction error signal $\boldsymbol{\epsilon}(n) = \mathbf{x}(n) - \hat{\mathbf{x}}(n)$ tends to be a sequence of essentially uncorrelated random variables. The prediction error filter working in this mode can be thought of as being a *whitening filter*. The concept of signal whitening is of great importance in signal processing and will be studied further in later sections. For this reason and others, it is not surprising that the linear predictor plays a prominent role in many signal-processing applications.

12.9 LEVINSON'S ALGORITHM: ITERATIVE SOLUTION OF THE LINEAR PREDICTION EQUATIONS

It was shown in Section 12.8 that the optimal linear predictor of order p was characterized by relationships (12.31) and (12.32). Clearly, these relationships correspond to the Wiener–Hopf equations of Section 12.5 as applied to the specific desirable signal selection $\mathbf{d}(n) = \mathbf{x}(n+1)$. It then follows that the iterative solution procedure described in Section 12.7 may be employed to efficiently compute the optimum predictor coefficients. Because of the inherent simpler algebraic structure possessed by the optimal linear predictor equations, however, this iterative procedure has a more direct implementation as now described.

To describe the new solution procedure, let us first rewrite the optimal linear predictor equations (12.31) and (12.32) of order m as the single system of linear equations

$$\sum_{k=0}^{m} r_{\mathbf{xx}}(n-k)a_m^{o}(k) = \begin{cases} \gamma_m & n = 0 \\ 0 & 1 \le n \le m \end{cases} \tag{12.33}$$

where it is recalled that $a_m^{o}(0) = 1$. In this expression, the subscript m has been appended to the predictor coefficient $a_m^{o}(k)$ to explicitly denote their dependency on the predictor order. Furthermore, the entity γ_m appearing in expression (12.33) is a shorthand notation for the minimum mean-squared prediction error associated with the optimal linear predictor of order m, that is,

$$\begin{aligned}\gamma_m &= E\{|\epsilon^o(n)|^2\} \\ &= \sum_{k=0}^{m} \bar{r}_{\mathbf{xx}}(k)a_m^o(k)\end{aligned} \tag{12.34}$$

The iterative solution procedure now to be described entails the sequential solving of the linear predictor equations (12.33) as the order parameter m takes on the values 1, 2, 3, . . . , p. In particular, the solution to the mth-order linear predictor equations as specified by $a_m^o(1)$, $a_m^o(2)$, . . . , $a_m^o(m)$, γ_m is to be used to efficiently solve the $(m + 1)$st-order linear predictor equations governed by

$$\sum_{k=0}^{m+1} r_{\mathbf{xx}}(n - k)a_{m+1}^o(k) = \begin{cases} \gamma_{m+1} & n = 0 \\ 0 & 1 \le n \le m + 1 \end{cases} \tag{12.35}$$

By using direct substitution, it is readily shown that the iterative solution procedure outlined in Table 12.3 provides the link between the mth-order optimal linear predictor and its $(m + 1)$st-order counterpart. This is readily verified by substituting the predictor coefficients $a_{m+1}^o(k)$ as specified in step 5 into relationship (12.35), where equality is found to hold.

Reflection Coefficients

Central to this iterative solution procedure are the notions of reflection coefficient and the minimum mean-squared predictor error recursion. The *reflection coefficient* is defined by

TABLE 12.3 Levinson's Algorithm: Iterative Optimum Linear Prediction Algorithm for Real-Valued Data

1. $m = 0$
2. $a_0^o(0) = 1$, $\gamma_0 = r_{\mathbf{xx}}(0)$
3. $\rho_{m+1} = -\gamma_m^{-1} \sum_{k=0}^{m} r_{\mathbf{xx}}(m + 1 - k)a_m^o(k)$
4. $\gamma_{m+1} = (1 - |\rho_{m+1}|^2)\gamma_m$
5. $a_{m+1}^o(k) = \begin{cases} 1 & \text{for } k = 0 \\ a_m^o(k) + \rho_{m+1}\bar{a}_m^o(m + 1 - k) & \text{for } 1 \le k \le m \\ \rho_{m+1} & \text{for } k = m + 1 \end{cases}$
6. $m = m + 1$ and go to step 3.

$$\rho_{m+1} = -\gamma_m^{-1} \sum_{k=0}^{m} r_{\mathbf{xx}}(m + 1 - k)a_m^o(k) \tag{12.36}$$

while the minimum mean-squared predictor error recursion is seen to be governed by

$$\gamma_{m+1} = (1 - |\rho_{m+1}|^2)\gamma_m \tag{12.37}$$

Since γ_m can never be negative, we immediately conclude that the reflection coefficients must always have a magnitude less than or equal to 1. This implies further that $\gamma_{m+1} \leq \gamma_m$, which simply indicates that the optimal linear predictor of order $m + 1$ has a prediction performance which is at least as good as its mth-order counterpart. We have therefore established the properties

$$0 \leq |\rho_m| \leq 1 \qquad \text{and} \qquad \gamma_m \leq \gamma_{m-1} \tag{12.38}$$

for all positive integer values of m. Furthermore, it is seen from relationship (12.37) that $\gamma_{m+1} = 0$ whenever $|\rho_{m+1}| = 1$. For such situations, this indicates that the wide-sense stationary time series $\{\mathbf{x}(n)\}$ is perfectly linearly predictable [i.e., $\hat{\mathbf{x}}(n) = \mathbf{x}(n)$] and the required predictor order is $m + 1$. Finally, the reflection coefficients ρ_m are real whenever the signal $\{\mathbf{x}(n)\}$ is real valued. For complex-valued signals, the reflection coefficients are generally complex valued.

Steps 1 and 2 of the iterative solution in Table 12.3 are the initial conditions required to start the algorithm. An analysis of this algorithm indicates that on the order of $2m$ addition and multiplication operations are needed to compute the predictor coefficients, reflection coefficient, and the minimum MSE associated with the optimum mth-order linear predictor. To compute the optimum pth-order linear predictor, p iterations of this algorithm would be required. This would then entail on the order of $\Sigma_{k=1}^{p} 2k = p(p - 1)$ addition and multiplication operations. This compares very favorably with the p^3 computations needed to solve relationship (12.31) directly using standard linear equation solution methods. Furthermore, using the iterative procedure, one obtains the added dividend of knowing all the lowered-ordered (i.e., 1, 2, 3, . . . , $p - 1$) optimal linear predictors as a by-product of the algorithm.

EXAMPLE 12.5

Using the iterative solution procedure specified in Table 12.3, let us design the optimum linear predictor of order $p = 2$ for the signal whose autocorrelation is given by $r_{\mathbf{xx}}(n) = 4(0.5)^{|n|}$. For $m = 0$, we have at step 2

$$a_0^o(0) = 1 \qquad \gamma_0 = 4$$

while steps 3, 4, and 5 yield

$$\rho_1 = -\tfrac{1}{2}, \qquad \gamma_1 = 3, \qquad a_1^o(0) = 1, \qquad a_1^o(1) = -\tfrac{1}{2}$$

For $m = 1$, steps 3, 4, and 5 are carried out and give

$$\rho_2 = 0, \qquad \gamma_1 = 3, \qquad a_2^{\text{o}}(0) = 1, \qquad a_2^{\text{o}}(1) = -\tfrac{1}{2}, \qquad a_2^{\text{o}}(2) = 0$$

These results are in agreement with those obtained in Example 12.4 using a direct but computationally less efficient approach. ■

12.10 REFLECTION COEFFICIENTS AND THE AUTOCORRELATION SEQUENCE

There is a strong interrelationship between the predictor reflection coefficients and the autocorrelation lags of the time series being predicted. More specifically, it is now shown that the autocorrelation elements $r_{\mathbf{xx}}(0), r_{\mathbf{xx}}(1), \ldots, r_{\mathbf{xx}}(k)$ may be systematically recovered from the set of numbers $r_{\mathbf{xx}}(0), \rho_1, \rho_2, \ldots, \rho_k$, and vice versa. As such, the correlation lags and the associated reflection coefficients generated through the algorithm described in Table 12.3 are equivalent. This equivalency is a direct consequence of step 3 of the algorithm, which may equivalently be expressed as

$$r_{\mathbf{xx}}(m+1) = -\rho_{m+1}\gamma_m - \sum_{k=1}^{m} r_{\mathbf{xx}}(m+1-k)a_m^{\text{o}}(k) \qquad \text{for } m \geq 0 \tag{12.39}$$

where the fact that $a_m^{\text{o}}(0) = 1$ has been used. To establish this equivalency, it is noted that at $m = 0$, relationship (12.39) yields

$$r_{\mathbf{xx}}(1) = -\rho_1\gamma_0 = -\rho_1 r_{\mathbf{xx}}(0)$$

Thus the lag element $r_{\mathbf{xx}}(1)$ is recoverable from knowledge of $r_{\mathbf{xx}}(0)$ and ρ_1. Furthermore, the predictor coefficient $a_1^{\text{o}}(1)$ and associated minimum MSE γ_1 may be determined from the algorithm in Table 12.3. We next evaluate relationship (12.39) at $m = 1$, that is,

$$r_{\mathbf{xx}}(2) = -\rho_2\gamma_1 - r_{\mathbf{xx}}(1)a_1^{\text{o}}(1)$$

which is computable due to the preknowledge of ρ_2 and the just-computed values of $r_{\mathbf{xx}}(1)$, $a_1^{\text{o}}(1)$, and γ_1. Furthermore, the algorithm in Table 12.3 may now be used to compute γ_2, $a_2^{\text{o}}(1)$, and $a_2^{\text{o}}(2)$. We may iteratively continue this process to generate the correlation lags $r_{\mathbf{xx}}(3)$, $r_{\mathbf{xx}}(4)$, and so on. Thus the two sets of numbers $r_{\mathbf{xx}}(0), r_{\mathbf{xx}}(1), \ldots, r_{\mathbf{xx}}(k)$ and $r_{\mathbf{xx}}(0), \rho_1, \rho_2, \ldots, \rho_k$ are equivalent in this constructive sense.

The consequences of this observation are of major importance for the class of wide-sense stationary time series that have continuous spectral density functions. Namely, any time series contained within this set may be interpreted as being the response of a recursive filter to a white-noise excitation. We summarize this salient property in the following theorem.

Theorem 12.1. Let $\{\mathbf{x}(n)\}$ be a wide-sense stationary time series that has a *continuous* spectral density function. Then this time series can be interpreted as being the response of a recursive filter of form

$$\mathbf{x}(n) = \boldsymbol{\epsilon}(n) - \sum_{k=1}^{p} a_k \mathbf{x}(n - k)$$

to the white-noise excitation $\{\boldsymbol{\epsilon}(n)\}$. Moreover, the associated set of autocorrelation lags

$$r_{\mathbf{xx}}(n) \qquad \text{for } 0 \leq n < m$$

may be generated from the equivalent number set

$$r_{\mathbf{xx}}(0),\ \rho_1,\ \rho_2,\ \ldots,\ \rho_m$$

through relationship (12.39) for all positive integers m. Finally, the reflection coefficients ρ_m here appearing all have a magnitude of less than 1 due to the continuity of the underlying spectral density function.

It is important to note that the order p of the recursive operator in Theorem 12.1 can be either finite or infinite. Perhaps the most significant aspect of this theorem is the realization that any legitimate autocorrelation sequence $\{r_{\mathbf{xx}}(n)\}$ with a continuous spectral density function may be associated with a reflection coefficient sequence

$$\rho_1,\ \rho_2,\ \rho_3,\ \ldots$$

whose members all have magnitudes of less than 1. Thus there is seen to be a one-to-one mapping between the class of continuous spectral density functions and the number sets $\{r_{\mathbf{xx}}(0), \rho_1, \rho_2, \rho_3, \ldots\}$ in which the constraint $|\rho_m| < 1$ is imposed for all $m \geq 1$.

Finite Length Reflection Coefficient Sequence

An interesting case arises for reflection coefficient sequences that are identically zero beyond some integer: The reflection coefficient set $\{r_{\mathbf{xx}}(0), \rho_1, \rho_2, \ldots, \rho_p, 0, 0, \ldots\}$ may be identified with the all-pole (i.e., autoregressive) recursive model

$$\mathbf{x}(n) = \boldsymbol{\epsilon}(n) - \sum_{k=1}^{p} a_k \mathbf{x}(n - k)$$

whose order p corresponds to the last nonzero element of the reflection coefficient sequence. In the companion volume we shall identify this specific model as being an autoregressive model of order p. Such models play a prominent role in time series analysis.

EXAMPLE 12.6

Construct the autocorrelation lags $r_{\mathbf{xx}}(0)$, $r_{\mathbf{xx}}(1)$, and $r_{\mathbf{xx}}(2)$ that are associated with the entities $r_{\mathbf{xx}}(0) = 4$, $\rho_1 = -\frac{1}{2}$, and $\rho_2 = 0$ as determined in Example 12.5. Evaluating relationship (12.39) at $m = 0$ gives

$$r_{\mathbf{xx}}(1) = -\rho_1 r_{\mathbf{xx}}(0) = 2$$

Given that $r_{xx}(0) = 4$ and $r_{xx}(1) = 2$, we next determine γ_1 and $a_1^o(1)$ from steps 4 and 5 of Table 12.3 at $m = 0$. This yields $\gamma_1 = 3$ and $a_1^o(1) = -\frac{1}{2}$. Evaluating relationship (12.39) at $m = 1$ then gives

$$r_{xx}(2) = -\rho_2\gamma_1 - a_1^o(1)r_{xx}(1) = 1$$

These autocorrelation lags are identical with those used in Example 12.5. ■

12.11 PREDICTOR ERROR FILTER: MINIMUM-PHASE CHARACTERISTIC

There are other benefits to be accrued from the iterative approach taken in Section 12.9. As an example, the transfer function associated with the optimum $(m + 1)$st-order predictor error filter [see equation (12.29)] is obtained by appealing to step 5 of Table 12.3. That is, upon substitution for $a_{m+1}^o(k)$, it is found that

$$\begin{aligned} A_{m+1}^o(z) &= \sum_{n=0}^{m+1} a_{m+1}^o(n)z^{-n} \\ &= \sum_{n=0}^{m} a_m^o(n)z^{-n} + \rho_{m+1}\sum_{n=1}^{m+1} \overline{a}_m^o(m+1-n)z^{-n} \\ &= \sum_{n=0}^{m} a_m^o(n)z^{-n} + \rho_{m+1}z^{-m-1}\sum_{n=0}^{m} \overline{a}_m^o(n)z^{n} \end{aligned}$$

where use of the fact that $a_m^o(0) = 1$ has been made. The first summation on the right is recognized as being the transfer function $A_m^o(z)$ associated with the optimum mth-order predictor error filter, while the second summation is simply $\overline{A}_m^o(\overline{z}^{-1})$. We have therefore established the following iterative optimum predictor error filter transfer function algorithm:

$$A^{o}_{m+1}(z) = A^{o}_{m}(z) + \rho_{m+1}z^{-m-1}\overline{A}^{o}_{m}(\overline{z}^{-1}) \qquad m = 0, 1, 2, \ldots \qquad (12.40)$$

which has the initial condition $A^{o}_{0}(z) = 1$. This transfer function is readily shown to be an $(m + 1)$st-order polynomial in the variable z^{-1}. Given the reflection coefficients $\rho_1, \rho_2, \ldots, \rho_p$, we may then use this iterative relationship to obtain the required optimum pth-order linear predictor error filter transfer function. Clearly, this iterative relationship and the expression in step 5 of Table 12.3 are equivalent. Our main purpose in introducing this transfer function recursion is analytic in basis. In particular, the following theorem is proved by a direct application of this recursion.

Theorem 12.2. Let the random signal $\{\mathbf{x}(n)\}$ be wide-sense stationary. It then follows that the associated optimum mth-order linear predictor is specified by

$$\sum_{k=0}^{m} r_{\mathbf{xx}}(n - k)a^{o}_{m}(k) = \begin{cases} \gamma_m & n = 0 \\ 0 & 1 \le n \le m \end{cases}$$

in which γ_m is the minimum prediction MSE. The following properties hold:

(a) The reflection coefficients ρ_m for $m = 1, 2, 3, \ldots$ generated by expression (12.36) all have a magnitude not exceeding 1, that is,

$$0 \le |\rho_m| \le 1$$

(b) The optimum MSE criterion γ_m is a monotonically nonincreasing sequence of nonnegative real numbers (i.e., $\gamma_{m+1} \le \gamma_m$) governed by

$$\gamma_m = r_{\mathbf{xx}}(0) \prod_{k=1}^{m} (1 - |\rho_k|^2) \qquad m = 1, 2, \ldots, p$$

(c) The associated optimum predictor error filter transfer function $A^{o}_{m}(z)$ has *all* its zeros either on or inside the unit circle.

(d) If the optimum predictor error transfer function $A^{o}_{m}(z)$ has one zero on the unit circle, all its zeros are on the unit circle. Furthermore, the only signal that can give rise to this situation consists of a sum of $m/2$ or fewer pure sinusoids and the associated minimum MSE is zero (i.e., $\gamma_m = 0$).

To prove parts (a) and (b) of this theorem, it is noted that the MSE criterion can never be negative. From relationship (12.37), this in turn implies that $0 \le |\rho_m|^2 \le 1$, which establishes the validity of parts (a) and (b). An inductive method of proof for parts (c) and (d) is now made which presumes that $A^{o}_{m}(z)$ has all its zeros inside the unit circle. We then show that $A^{o}_{m+1}(z)$ as generated by expression (12.40) either has all its zeros inside the unit circle if $|\rho_m| < 1$ or has all its zeros on the unit circle when $|\rho_m| = 1$. To establish this behavior, let us introduce the $(m + 1)$st-order polynomial

$$B_{\rho}(z) = A^{o}_{m}(z) + \rho z^{-m-1}\overline{A}^{o}_{m}(\overline{z}^{-1})$$

in the variable z^{-1}. This polynomial depends on the parameter ρ, where it is seen from expression (12.40) that $B_0(z) = A_m^o(z)$ and $B_{\rho_{m+1}}(z) = A_{m+1}^o(z)$. We now examine the behavior of the zeros of $B_\rho(z)$ as ρ is varied from zero to ρ_{m+1}. Since these zeros are a continuous function of ρ, it is apparent that at $\rho = 0$, $B_\rho(z)$ will have the m original zeros of $A_m^o(z)$, which are hypothesized as being inside the unit circle, and an additional zero at $z = 0$ [due to the $-\rho z^{-m-1}$ term in $B_p(z)$]. As ρ is varied from zero to ρ_{m+1}, these $m + 1$ zeros migrate as a continuous function of ρ. The question to be answered is: Do any of these zeros migrate to or across the unit circle? To answer this question, we evaluate polynomial $B_\rho(z)$ on the unit circle, that is, at $z = e^{j\omega}$. This gives rise to the expression

$$B_\rho(e^{j\omega}) = A_m^o(e^{j\omega}) + \rho e^{-j(m+1)\omega}\overline{A}_m^o(e^{j\omega})$$

where $A_m^o(e^{j\omega})$ is recognized as the Fourier transform of the predictor error filter coefficients. Using the polar representation for $A_m^o(e^{j\omega})$, we have

$$B_\rho(e^{j\omega}) = |A_m^o(e^{j\omega})|[e^{j\phi_m(\omega)} + \rho e^{-j[(m+1)\omega + \phi_m(\omega)]}]$$

Since $A_m^o(z)$ has all its zeros inside the unit circle, it follows that $|A_m^o(e^{j\omega})|$ is positive for all $0 \le \omega \le 2\pi$. Thus the only way for $B_\rho(z)$ to have a zero on the unit circle is for the term multiplying $|A_m^o(e^{j\omega})|$ to be zero. An examination of this term, however, indicates that this can never be achieved as long as $|\rho| < 1$. If ρ_{m+1} has magnitude of less than 1, we therefore conclude that $A_{m+1}^o(z)$ has all its zeros inside the unit circle. On the other hand, if $|\rho_{m+1}| = 1$, the highest-degree term of $A_{m+1}^o(z)$ (i.e., $-\rho_{m+1}z^{-m-1}$) has a unity coefficient. This coefficient, however, is equal to the product of the zeros of $A_{m+1}^o(z)$. Using continuity arguments, we therefore conclude that when $|\rho_{m+1}| = 1$, all the zeros of the optimum predictor error filter transfer function must be on the unit circle. The following lemma is an immediate consequence of this argument.

Lemma 12.1 Let the optimum predictor error filter transfer function $A_m^o(z)$ of order m have all its zeros inside the unit circle. It then follows that the optimum predictor error filter transfer function $A_{m+1}^o(z)$ of order $m + 1$ has its zeros:

(a) All inside the unit circle if the reflection coefficient ρ_{m+1} has a magnitude of less than 1 (i.e., $0 \le |\rho_{m+1}| < 1$). The corresponding minimum MSE is positive, so that $\gamma_{m+1} > 0$.
(b) All on the unit circle if the reflection coefficient ρ_{m+1} has magnitude 1. The corresponding minimum MSE is zero.

12.12 LINEAR PREDICTOR BY MEANS OF LATTICE IMPLEMENTATION

The most natural means for implementing the optimum linear predictor is through the tapped-delay-filter structure in Figure 12.4. The lattice implementation now

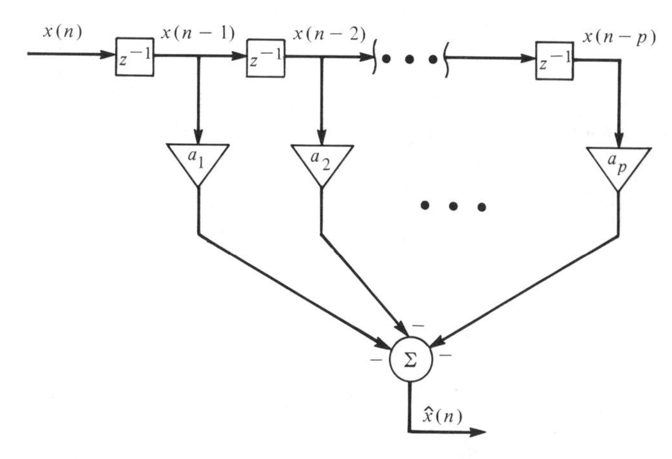

FIGURE 12.4. Tapped delay implementation of linear predictor.

to be described, however, offers a number of important advantages relative to the tapped delay structure. These include the following:

1. The lattice structure is order decoupled in the sense that the optimal $(m + 1)$st-order lattice predictor is obtained simply by cascading an appropriately chosen first-order lattice filter to the optimal mth-order lattice predictor.
2. The lattice structure is of a form that is readily implemented in *very large scale integrated* (VLSI) circuitry.
3. The lattice structure is less sensitive to round-off error than is the tapped-delay equivalent.

These and other reasons provide convincing evidence of the desirability of a lattice implementation. To effect such an implementation, it is necessary to introduce two signals associated with the optimum mth-order predictor error filter. The first of these signals is referred to as the *forward prediction error*, and its z-transform is specifed by

$$F_m(z) = A_m^o(z)X(z) \tag{12.41}$$

while the second signal is called the *backward prediction error* and is described by the z-transform relationship

$$B_m(z) = z^{-m}\overline{A}_m^o(\overline{z}^{-1})X(z) \tag{12.42}$$

The order integer m used in these predictor errors takes on the values 0, 1, 2, . . .

The first of the foregoing two signals, $F_m(z)$, is recognized as being the response of the optimum mth-order predictor error filter to the excitation $X(z)$. We have here referred to $F_m(z)$ as being the *forward* prediction error, since the prediction operation is being made in the positive time (future) direction. The reason for using the descriptive adjective forward is to distinguish $F_m(z)$ from its related backward prediction error signal $B_m(z)$. Specifically, upon expansion of the causal transfer function $z^{-m}\overline{A}_m^{o}(\overline{z}^{-1})$, which gives rise to $B_m(z)$, we have

$$B_m(z) = z^{-m}\left[1 + \sum_{k=1}^{m} \overline{a}_m^{o}(k)z^{k}\right]X(z)$$

After taking the inverse z-transform of this expression, the following time-domain relationship results:

$$b_m(n) = x(n - m) - \sum_{k=1}^{m} [-\overline{a}_m^{o}(k)]x(n - m + k)$$

We have suggestively used the double-negative sign on the summation term to indicate that this summation term can be interpreted as being an estimate of the signal element $x(n - m)$ which is based on a linear combination of the m most immediate future signal elements. Upon using the approach taken in Section 12.5, it is readily shown that this summation represents the optimum linear MSE estimate of $x(n - m)$ based on the m immediate future time series elements. As such, the descriptor *backward predictor error* for the signal $\{b_m(n)\}$ is natural, since the prediction is being made in the negative time direction. From these arguments it is seen that the optimum forward and backward linear predictors have coefficients which are complex conjugate images.

The lattice implementation is obtained straightforwardly by iterating relationships (12.41) and (12.42) by one order. For example, this incrementation for the forward predictor relationship (12.41) gives

$$F_{m+1}(z) = A_{m+1}^{o}(z)X(z)$$

We next substitute into this expression recursive relationship (12.40), which relates the optimum $(m + 1)$st and mth-order predictor error filters. This results in

$$F_{m+1}(z) = [A_m^{o}(z) + \rho_{m+1}z^{-1}z^{-m}\overline{A}_m^{o}(\overline{z}^{-1})]X(z)$$

which upon using relationships (12.41) and (12.42) simplifies to

$$F_{m+1}(z) = F_m(z) + \rho_{m+1}z^{-1}B_m(z)$$

Taking the inverse z-transform of this expression gives the equivalent time-domain relationship

$$f_{m+1}(n) = f_m(n) + \rho_{m+1}b_m(n - 1) \tag{12.43}$$

Thus, given the optimum mth-order forward and backward prediction error signals, this expression provides a simple procedure for computing the optimum $(m + 1)$st-order forward prediction error signal.

To complete the lattice implementation, it is necessary to obtain a similar relationship for the backward prediction error signal $\{b_{m+1}(n)\}$. This is readily achieved by first incrementing by one order relationship (12.42), to give

$$B_{m+1}(z) = z^{-m-1}\overline{A}^{o}_{m+1}(\overline{z}^{-1})X(z)$$

We next substitute in this expression recursive relationship (12.40) with the variable z replaced by z^{-1}. This gives

$$B_{m+1}(z) = z^{-m-1}[\overline{A}^{o}_{m}(\overline{z}^{-1}) + \overline{\rho}_{m+1}z^{m+1}A^{o}_{m}(z)]X(z)$$

where the term in brackets is recognized as being equal to $\overline{A}^{o}_{m+1}(\overline{z}^{-1})$. Upon using identities (12.41) and (12.42), this expression for $B_{m+1}(z)$ simplifies to

$$B_{m+1}(z) = z^{-1}B_m(z) + \overline{\rho}_{m+1}F_m(z)$$

The equivalent time-domain representation is therefore

$$b_{m+1}(n) = b_m(n - 1) + \overline{\rho}_{m+1}f_m(n) \qquad (12.44)$$

which indicates that the optimum $(m + 1)$st-order backward prediction error signal is a simple function of the optimum mth-order forward and backward prediction error signals.

As previously indicated, when the wide-sense stationary signal $\{x(n)\}$ is real valued, the reflection coefficients ρ_m will all be real. For this relevant case, the forward and backward prediction error signals (12.43) and (12.44) are also themselves real valued.

Recursive relationships (12.43) and (12.44) constitute the fundamental building blocks for the required lattice implementation. In particular, it is apparent that these relationships can be implemented by means of the configuration shown in Figure 12.5a. Thus, once the optimum lattice structure for the mth-order prediction error filter has been obtained, the corresponding optimum $(m + 1)$st-order prediction error filter is obtained directly by cascading to it the first-order lattice structure shown in Figure 12.5a. The complete lattice structure is generated by using recursions (12.43) and (12.44) beginning at $m = 0$, where $f_0(n) = b_0(n) = x(n)$, as is evident since $A^{o}_{0}(z) = 1$. Upon letting $m = 1$, recursive expressions (12.43) and (12.44) give the leftmost lattice structure shown in Figure 12.5b. Continuing in this manner, we find that the corresponding optimum mth filter is as shown in Figure 12.5b. It consists of m stages of first-order lattice structures. Each structure is of fixed form in which the reflection coefficient ρ_m is the only variable.

The forward and backward prediction error signals at the various stages of the lattice structure are orthogonal in the sense of the theorem given below. This

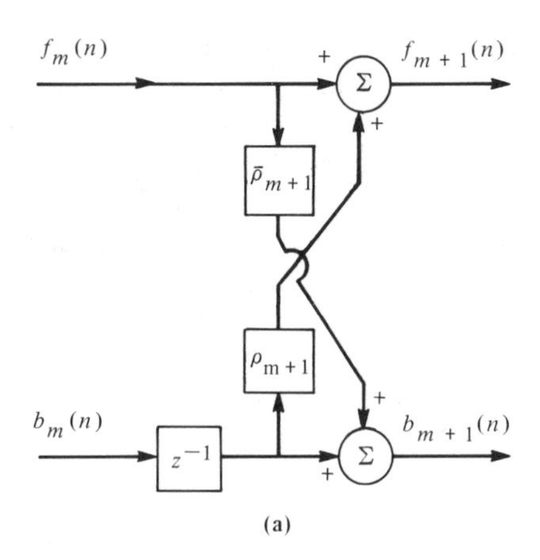

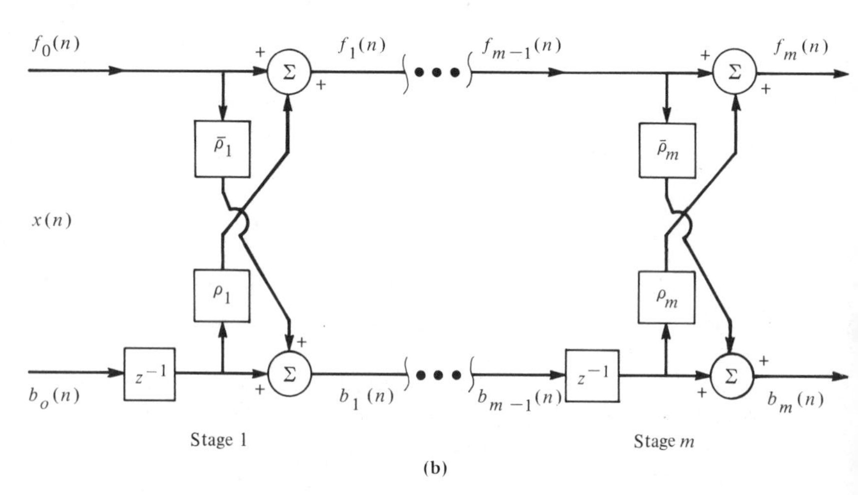

FIGURE 12.5. Lattice implementation: (a) single stage; (b) m^{th}-order multistage structure.

provides yet another reason for preferring a lattice implementation of the optimum linear predictor. Specifically, an objective in many signal-processing applications is the achievement of signal decorrelation. With this in mind, the following theorem is offered.

Theorem 12.3. The forward and backward prediction errors at the different stages of the lattice implementation of the optimum linear predictor are orthogonal in the sense that

$$E\{f_m(n)\bar{x}(n-k)\} = 0 \qquad 0 \le k \le p$$

$$E\{b_k(n)\bar{b}_m(n)\} = \begin{cases} 0 & k \neq m \\ \gamma_m & k = m \end{cases}$$

A proof of this theorem is straightforward and is left as an exercise for the reader.

12.13 SUMMARY

The synthesis of optimal finite memory linear MSE filters, estimators, and predictors was shown to be heavily dependent on linear methods. In particular, the optimum operator was invariably obtained by solving a linear system of equations that had correlation entries as its coefficients. As we shall see in Chapters 13 and 14, these correlation entries play a prominent role in many relevant signal-processing applications.

SUGGESTED READINGS

CHEN, C. T., *One-Dimensional Digital Signal Processing,* New York: Marcel Dekker, Inc., 1979.

FREEMAN, H., *Discrete-Time Systems*. New York: John Wiley & Sons, Inc., 1965.

OPPENHEIM, A. V., and R. W. SCHAFER, *Digital Signal Processing*. Englewood Cliffs, N.J.: Prentice-Hall, Inc., 1975.

TRETTER, S. A. *Discrete-Time Signal Processing*. New York: John Wiley & Sons, Inc., 1976.

WIENER, N., *Extrapolation, Interpolation and Smoothing of Stationary Time Series*. Cambridge, Mass.: The MIT Press, 1949.

PROBLEMS

12.1 Carry out the details that lead to the MSE criterion expression (12.5) and its minimum value as specified by expression (12.7).

12.2 Let the information signal and the uncorrelated noise signal have the correlations

$$r_{\mathbf{ss}}(n) = 3(0.9)^{|n|} \qquad \text{and} \qquad r_{\boldsymbol{\eta\eta}}(n) = 4\delta(n)$$

Design the optimum MSE filters with order $p = 1$ and 2 and determine their associated minimum MSEs. Finally, compute the cross-correlation sequence $\{r_{\boldsymbol{\epsilon}^o\mathbf{x}}(n)\}$.

12.3 Repeat Problem 12.2 for the case in which the information and noise sequences are correlated as $r_{\mathbf{s}\boldsymbol{\eta}}(n) = 2\delta(n)$.

12.4 Let the information signal and the uncorrelated noise signal have autocorrelations

$$r_{\mathbf{ss}}(n) = 2\cos\left(\frac{2\pi n}{8}\right) \qquad \text{and} \qquad r_{\boldsymbol{\eta\eta}}(n) = \delta(n)$$

Synthesize the optimum MSE filter of order $p = 1, 2, 3$ and determine the corresponding minimum MSE values. Evaluate the crosscorrelation sequence $\{r_{\mathbf{e}^o\mathbf{x}}(n)\}$.

12.5 Derive the Wiener–Hopf equations associated with the generally real noncausal finite memory filter specified by

$$y(n) = \sum_{k=-p_1}^{p_2} h(k)x(n - k)$$

in which p_1 and p_2 are nonnegative integers. Determine the corresponding minimum MSE criterion's value, and specify the nature of the orthogonalization between the optimum real error signal $\boldsymbol{\epsilon}^o(n) = \mathbf{d}(n) - \mathbf{y}^o(n)$ and the measured real random signal $\mathbf{x}(n)$.

12.6 Using the signals specified in Problem 12.2, determine the optimum MSE filter of order $p = 2$ when the desired signal is specified by **(a)** $\mathbf{d}(n) = \mathbf{s}(n + 1)$ and **(b)** $\mathbf{d}(n) = \mathbf{s}(n - 1)$. Compute the corresponding minimum MSEs and the cross-correlation sequence $\{r_{\boldsymbol{\epsilon}^o\mathbf{x}}(n)\}$.

12.7 Consider the general optimum MSE linear filtering problem treated in Section 12.5 in which the signals $\{\mathbf{x}(n)\}$ and $\{\mathbf{d}(n)\}$ are wide-sense stationary complex-valued random time series and the filtering weighting coefficients $\{h(n)\}$ can be complex valued. Verify expressions (12.19d)–(12.19g), which characterize the optimum filter.

12.8 Using the iterative algorithm in Table 12.2, synthesize the optimum MSE filters with order $p = 1$ and 2, and determine the corresponding MSE values for the signals described in Problem 12.2.

12.9 Show that the algorithm given in Table 12.2 solves the Wiener–Hopf equations in an order update fashion.

12.10 Repeat Problem 12.8 for the signals described in Problem 12.4.

12.11 Carry out the details of the predictor synthesis outlined in Example 12.5.

12.12 Using the iterative procedure described in Table 12.3, design the optimum linear predictor of order $p = 2$ when the signal being predicted has autocorrelation $r_{\mathbf{xx}}(n) = 2(0.25)^{|n|}$. Find the corresponding optimum prediction error filter transfer function $A_2^o(z)$ and show that all its roots are inside the unit circle.

12.13 Verify the algorithm given in Table 12.3.

12.14 Let $\{\mathbf{x}(n)\}$ be a complex-valued wide-sense stationary signal. Show that the optimal linear predictor is characterized by expression (12.33) while the associated minimum mean-squared prediction error is specified by relationship (12.34).

12.15 Given that $r_{xx}(0) = 1$ and that the first three reflection coefficients are $\rho_1 = 1$, $\rho_2 = \frac{1}{2}$, and $\rho_3 = \frac{1}{4}$, find the associated autocorrelation lags $r_{xx}(1)$, $r_{xx}(2)$, and $r_{xx}(3)$.

12.16 Determine the optimum prediction error filter transfer function $A_3^o(z)$ corresponding to the information given in Problem 12.11. Show that all the roots of $A_3^o(z)$ lie inside the unit circle.

12.17 Sketch a lattice implementation of the optimum predictor error filter transfer function $A_3^o(z)$.

12.18 Give a proof of Theorem 12.3.

CHAPTER 13

Optimum MSE Filtering, Estimation, and Prediction: Infinite Memory

13.1 INTRODUCTION

It was shown in Chapter 12 that algebraic methods play a fundamental role in the synthesis of optimum linear MSE filters, estimators, and predictors that possess finite memory. In some applications, a significant performance improvement can be achieved by relaxing the finite-memory requirement. From a practical viewpoint the user must make a determination as to whether the performance improvement thereby achieved is worth the additional computational requirements that may arise. Fortunately, it is shown that when the relevant spectral density functions are rational, the corresponding optimum infinite memory operators (i.e., Wiener filters) are implementable by finite-order recursive operators. In such situations it is possible to achieve optimum linear performance with a computationally feasible algorithm.

The synthesis of optimum infinite memory filters entails the use of nonalgebraic methods. More specifically, the z-transform is found to be a particularly powerful tool for this purpose. In synthesizing the optimum causal linear operator, the concepts of the *whitening filter* and the *innovations process* arise in a natural manner. To effect the whitening filter, it will be necessary to make a particular z-domain decomposition of the complex spectral density associated with the random signal being whitened.

Our analysis will be directed toward general complex-valued wide-sense stationary time series. By taking this approach, we are at the same time treating the important real-valued time series case. This typically will entail replacing complex conjugated entities by their unconjugated version whenever they appear.

13.2 OPTIMUM MSE FILTERING

We begin our analysis of infinite memory optimum MSE filtering by considering the general linear filter specified by

$$y(n) = \sum_{k=-\infty}^{\infty} h(k)x(n-k) \tag{13.1}$$

in which no restriction as to causality or memory length is imposed. This approach is taken since the resultant optimum filter synthesis is straightforward and provides a useful introduction to the more realistic causal infinite memory problem. Our objective is then to select this filter's weighting coefficients [i.e., the $h(k)$] so that its response $\{\mathbf{y}(n)\}$ to the random signal $\{\mathbf{x}(n)\}$ best matches a random desired signal $\{\mathbf{d}(n)\}$ in the MSE sense. As in our previous investigations, this entails introducing the random error signal

$$\boldsymbol{\epsilon}(n) = \mathbf{d}(n) - \mathbf{y}(n) \tag{13.2}$$

which measures the degree to which the random response signal $\{\mathbf{y}(n)\}$ differs from the random desired signal $\{\mathbf{d}(n)\}$. It is here assumed that the signal to be processed $\{\mathbf{x}(n)\}$ and the desired signal $\{\mathbf{d}(n)\}$ are wide-sense stationary and that the associated correlation sequences $r_{\mathbf{xx}}(n)$ and $r_{\mathbf{dx}}(n)$ are known.

Optimum Filter Synthesis

Under the assumed conditions stated in the preceding paragraph, the MSE associated with the random error signal (13.2) is given by

$$\begin{aligned}
E\{|\boldsymbol{\epsilon}(n)|^2\} &= E\left\{\left[\mathbf{d}(n) - \sum_{k=-\infty}^{\infty} h(k)\mathbf{x}(n-k)\right]\left[\overline{\mathbf{d}}(n) - \sum_{m=-\infty}^{\infty} \overline{h}(m)\overline{\mathbf{x}}(n-m)\right]\right\} \\
&= r_{\mathbf{dd}}(0) - \sum_{k=-\infty}^{\infty} r_{dx}(k)\overline{h}(k) + \overline{r}_{dk}(k)h(k) \\
&\quad + \sum_{k=-\infty}^{\infty}\sum_{m=-\infty}^{\infty} r_{\mathbf{xx}}(m-k)h(k)\overline{h}(m)
\end{aligned} \tag{13.3}$$

where the filter response (13.1) representation for $\mathbf{y}(n)$ has been used. The optimum selection for the filter coefficients is then obtained by setting to zero the derivatives of this MSE criterion with respect to the filter coefficient's real and imaginary components. Upon carrying out this operation, it is found that the optimum linear filter's weighting sequence must satisfy the following consistent set of *infinite* linear equations:

$$\sum_{k=-\infty}^{\infty} r_{\mathbf{xx}}(n - k)h^{o}(k) = r_{\mathbf{dx}}(n) \qquad \text{for } -\infty < n < \infty \tag{13.4}$$

In principle, we can solve this system of *Wiener–Hopf* equations for the optimum $\{h^{o}(n)\}$. The minimum MSE value associated with this optimum selection is then obtained by substituting relationship (13.4) into (13.3) thereby giving

$$E\{|\epsilon^{o}(n)|^2\} = r_{\mathbf{dd}}(0) - \sum_{k=-\infty}^{\infty} r_{\mathbf{dx}}(k)\bar{h}^{o}(k) \tag{13.5}$$

Relationships (13.4) and (13.5) characterize the optimum filter insofar as minimizing the hypothesized MSE criterion.

The task of solving the infinite set of Wiener–Hopf equations (13.4) would appear to be rather imposing were it not for the fact that they appear in the form of a convolution summation. Thus we may appeal to the z-transform-domain representation for convolution relationships (13.4). In particular, in the z-domain these Wiener–Hopf equations may be equivalently expressed as

$$H^{o}(z)R_{\mathbf{xx}}(z) = R_{\mathbf{dx}}(z)$$

where $R_{\mathbf{xx}}(z)$ and $R_{\mathbf{dx}}(z)$ denote the z-transforms of the correlation sequences $r_{\mathbf{xx}}(n)$ and $r_{\mathbf{dx}}(n)$, respectively. Thus the optimum filter transfer function is simply given by the following ratio of complex spectral density functions:

$$H^{o}(z) = \frac{R_{\mathbf{dx}}(z)}{R_{\mathbf{xx}}(z)} \tag{13.6}$$

and the corresponding optimum filter weighting sequence is determined by inverse-z-transforming this result.

Orthogonalization Property

Since the filtering operation (13.1) uses the entire history of the data signal [i.e., $\mathbf{x}(n)$ for $-\infty < n < \infty$] in forming the estimate $\mathbf{y}(n)$, the optimum filter effects a perfect orthogonalization of the filter's error signal and the desired signal. In particular, it is readily shown that

$$r_{\boldsymbol{\epsilon}^{o}\mathbf{x}}(m) = E\{\boldsymbol{\epsilon}^{o}(n)\bar{\mathbf{x}}(n - m)\} = 0 \qquad \text{for all } m \tag{13.7}$$

This is to be contrasted with the *partial* orthogonalization condition that is achieved when employing an optimum finite memory filter, as was the case in Chapter 12. This perfect orthogonalization has arisen because as many filter coefficients are being used [i.e., $h(n)$ for $-\infty < n < \infty$] as there are correlation elements [i.e., $r_{\boldsymbol{\epsilon}^o\mathbf{x}}(n)$ for $-\infty < n < \infty$] to be set to zero. Thus there is seen to be a one-to-one correspondence between the number of filter coefficients and the number of guaranteed zero correlation elements.

EXAMPLE 13.1

Synthesize the optimal infinite memory filter corresponding to the filtering problem considered in Example 12.1. For this case the relevant correlation sequences are given by

$$r_{\mathbf{xx}}(n) = 4(0.5)^{|n|} + 2\delta(n) \qquad \text{and} \qquad r_{\mathbf{dx}}(n) = 4(0.5)^{|n|}$$

where use of the fact that $\mathbf{d}(n) = \mathbf{s}(n)$ has been made. The complex spectral density functions associated with these correlation sequences are

$$\begin{aligned} R_{\mathbf{dx}}(z) &= \sum_{n=-\infty}^{\infty} 4(0.5)^{|n|} z^{-n} \\ &= -4 + 4\sum_{n=0}^{\infty} [(0.5z^{-1})^n + (0.5z)^n] \\ &= \frac{-6z}{(z - 0.5)(z - 2)} \qquad 0.5 < |z| < 2 \end{aligned}$$

in which use of the geometric series summation identity has been made. Similarly,

$$R_{\mathbf{xx}}(z) = \frac{2z^2 - 11z + 2}{(z - 0.5)(z - 2)}$$

In accordance with relationship (13.6), the optimal transfer function is given by

$$\begin{aligned} H^o(z) &= \frac{R_{\mathbf{dx}}(z)}{R_{\mathbf{xx}}(z)} = \frac{-3z}{z^2 - 5.5z + 1} \\ &= \frac{6}{\sqrt{105}}\left[\frac{z}{z - \frac{11}{4} + \frac{1}{4}\sqrt{105}} - \frac{z}{z - \frac{11}{4} - \frac{1}{4}\sqrt{105}}\right] \end{aligned}$$

If this transfer function is to be stable, the first term in this partial-fraction expansion must be causal and the second term must be anticausal. Thus the corresponding optimum unit-impulse response is

$$h^o(n) = \frac{6}{\sqrt{105}}\left[(\tfrac{11}{4} - \tfrac{1}{4}\sqrt{105})^n u(n) + (\tfrac{11}{4} + \tfrac{1}{4}\sqrt{105})^n u(-n - 1)\right]$$

Moreover, the minimum MSE associated with this optimal filter is, by expression (13.5), eventually found to be

$$E\{|\epsilon^o(n)|^2\} = \frac{12}{\sqrt{105}} = 1.17108$$

Although this MSE criterion's value must of necessity be smaller than that achieved with the causal filter of length 3 considered in Example 12.1, it is only marginally better (i.e., 1.171 versus 106/85 = 1.247). This rather negligible improvement has arisen despite the fact that an infinite number of weighting coefficients are incorporated with the filter considered in this example as opposed to the three coefficients used in Example 12.1. Thus for the signal-processing problem being treated here, prudence might suggest utilization of the filter synthesized in Example 12.1. ■

13.3 CAUSAL OPTIMUM MSE FILTERING

We shall now direct our attention to the more difficult but realistic task of synthesizing an optimum causal infinite memory linear filter. Let it be required to find the weighting coefficients $h(k)$ governing the causal filter

$$y(n) = \sum_{k=0}^{\infty} h(k)x(n - k) \tag{13.8}$$

so that its response to the random excitation $\{\mathbf{x}(n)\}$ best matches the desired signal $\{\mathbf{d}(n)\}$ in the sense of minimizing the MSE criterion

$$E\{|\epsilon(n)|^2\} = E\{|\mathbf{d}(n) - \mathbf{y}(n)|^2\}$$

The optimum coefficients $h(k)$ are formally obtained by setting to zero the partial derivatives of this MSE criterion with respect to the coefficient's real and imaginary components. When the signals $\{\mathbf{x}(n)\}$ and $\{\mathbf{d}(n)\}$ are each wide-sense stationary, this is found to lead to the following system of Wiener–Hopf equations:

$$\sum_{k=0}^{\infty} r_{\mathbf{xx}}(n - k)h^o(k) = r_{\mathbf{dx}}(n) \qquad \text{for } 0 \le n \le \infty \tag{13.9}$$

It is to be noted that these equations hold only over the time span $0 \le n \le \infty$. This is a direct consequence of the filter's imposed causality, which requires that the coefficients $h(n)$ for negative n be fixed at zero. For this optimum selection $\{h^o(n)\}$, it is readily shown that the associated minimum MSE value is specified by

$$E\{|\epsilon^o(n)|^2\} = r_{\mathbf{dd}}(0) - \sum_{k=0}^{\infty} r_{\mathbf{dx}}(k)\bar{h}^o(k) \tag{13.10}$$

For real-valued time series, the solution to expression (13.9) will always result in real values for the optimum weighting coefficients $h^o(k)$.

An optimum filter synthesis then entails solving the infinite set of Wiener–Hopf equations (13.9) for the coefficients $h^o(n)$. Although these equations seem like a convolution summation, we cannot directly employ z-transform techniques as we did in the preceding section. This is due to the fact that these equations hold only over the nonnegative time interval $0 \leq n < \infty$, thereby effectively eliminating a simple solution procedure. To obtain the required solution, it is necessary to employ a clever z-domain approach that takes into account the one-sided (i.e., $n \geq 0$) nature of the Wiener–Hopf equations. Furthermore, due to the causal nature of the filter, it follows that the resultant optimum filter error signal is only partially orthogonal with the signal being processed in the sense that

$$r_{\epsilon^o x}(n) = E\{\boldsymbol{\epsilon}^o(k)\bar{\mathbf{x}}(k - n)\} = 0 \qquad \text{for } 0 \leq n < \infty \tag{13.11}$$

This partial orthogonalization is a direct consequence of our restricting the filter to be causal. To achieve complete orthogonalization, it is generally necessary to remove the causal constraint, as was made apparent in the preceding section. Before undertaking the task of solving the system of linear equations (13.9) that constitute the necessary conditions for the optimum filter, it will be beneficial to consider a special case.

Special Case

The Wiener–Hopf equations (13.9) have a particularly simple solution when the excitation signal $\{\mathbf{x}(n)\}$ is white in nature with correlation sequence

$$r_{\mathbf{xx}}(n) = \delta(n) \tag{13.12}$$

Upon substitution of this unit-impulse correlation into expression (13.9), the optimum filter's coefficients are specified directly by

$$h^o(n) = r_{\mathbf{dx}}(n) \qquad \text{for } 0 \leq n < \infty \tag{13.13}$$

For this special situation, the selection of the optimum weighting coefficients is indeed straightforward. Unfortunately, the excitation whiteness assumption that leads to this simple solution is almost never satisfied in practical signal-processing applications. This whiteness concept, however, provides a basis for solving the Wiener–Hopf equation in the more general nonwhite case. We now examine the details of this approach.

Optimal Filter Selection Procedure

Let us consider the system configuration shown in Figure 13.1, in which it is assumed that the causal stable whitening filter's transfer function $W(z)$ can be selected so that the two conditions

$$\text{(a)}\quad r_{\xi\xi}(n) = \delta(n) \quad \text{or equivalently} \quad R_{\xi\xi}(z) = 1 \tag{13.14}$$

$$\text{(b)}\quad \frac{1}{W(z)} \text{ is stable}$$

are satisfied. It is shown in Section 13.4 that such a transfer function can always be straightforwardly determined, provided that $R_{\mathbf{xx}}(z)$ is a rational function. This is readily appreciated by noting from Theorem 11.3 that condition (a) implies that

$$R_{\xi\xi}(z) = W(z)\overline{W}(\overline{z}^{-1})R_{\mathbf{xx}}(z) = 1$$

or equivalently,

$$R_{\mathbf{xx}}(z) = \frac{1}{W(z)\overline{W}(\overline{z}^{-1})} \tag{13.15}$$

Thus the required transfer function $W(z)$ is obtained by factoring the complex spectral density function $R_{\mathbf{xx}}(z)$ in this fashion so that $W(z)$ and its inverse $1/W(z)$ are each causal stable filters.

In the filter synthesis to be now made, the candidate filters are to be expressed in the product form $H(z) = W(z)G(z)$. No restriction is thereby incurred, since *any* causal stable transfer function $H(z)$ may be expressible in the form $W(z)G(z)$ simply by setting $G(z) = H(z)/W(z)$. The transfer function $H(z)/W(z)$ is ensured to be causal and stable, since its constituent component $H(z)$ and $1/W(z)$ are each so characterized [see property (b)]. With this in mind, it follows that the

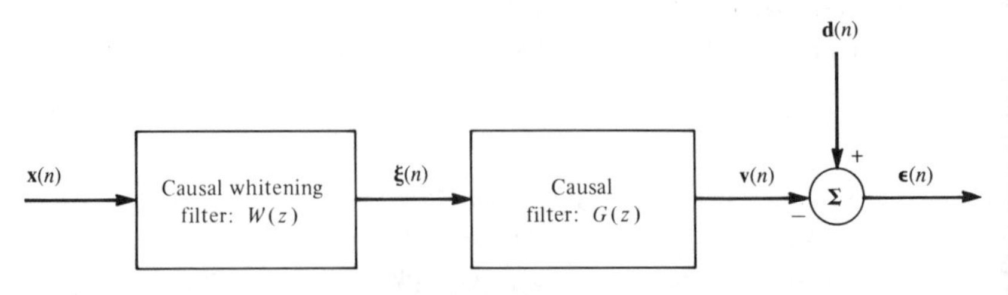

FIGURE 13.1. Optimal infinite memory causal filter.

optimum causal MSE linear filter described at the beginning of this section is obtained by selecting the transfer function $G(z)$ shown in Figure 13.1 so as to minimize $E\{|\boldsymbol{\epsilon}(n)|^2\}$. In accordance with relationship (13.13), the weighting sequence associated with this optimum causal filter is then given by

$$g^{o}(n) = r_{\mathbf{d}\boldsymbol{\xi}}(n)u(n) \tag{13.16}$$

Let us now determine the correlation sequence $r_{\mathbf{d}\boldsymbol{\xi}}(n)$ which characterizes this optimal choice. If we use the fact that $\{\boldsymbol{\xi}(n)\}$ is the response of the whitening filter to the excitation $\{\mathbf{x}(n)\}$, it immediately follows that

$$r_{\mathbf{d}\boldsymbol{\xi}}(n) = \sum_{k=0}^{\infty} \overline{w}(k) r_{\mathbf{dx}}(n + k)$$

where $w(n) = Z^{-1}[W(z)]$. In the z-domain, this relationship becomes

$$R_{\mathbf{d}\boldsymbol{\xi}}(z) = \overline{W}(\overline{z}^{-1}) R_{\mathbf{dx}}(z) \tag{13.17}$$

In accordance with relationship (13.16), the optimum selection of $g(n)$ is then simply equal to the causal component of the inverse z-transform of expression (13.17). It is instructive now to provide a z-domain procedure for effecting this optimum filter synthesis.

As just indicated, the optimal filter is generated by determining the causal time component of the z-transform $R_{\mathbf{d}\boldsymbol{\xi}}(z)$. The general procedure for selecting the causal component of any two-sided z-transform will be represented by the operator notation

$$[X(z)]_+ = \left[\sum_{n=-\infty}^{\infty} x(n) z^{-n}\right]_+ = \sum_{n=0}^{\infty} x(n) z^{-n} \tag{13.18}$$

Very simply, the operator $[\cdot]_+$ corresponds to selecting only those terms of a z-transform associated with nonpositive powers of z. Upon using this operator notation and relationships (13.16) and (13.17), it follows that the optimal selection of $G(z)$ is given formally by

$$\begin{aligned} G^{o}(z) &= [R_{\mathbf{d}\boldsymbol{\xi}}(z)]_+ \\ &= [\overline{W}(\overline{z}^{-1}) R_{\mathbf{dx}}(z)]_+ \end{aligned}$$

Moreover, the optimal filter transfer function is, in accordance with Figure 13.1, given by

$$\begin{aligned} H^{o}(z) &= W(z) G^{o}(z) \\ &= W(z)[\overline{W}(\overline{z}^{-1}) R_{\mathbf{dx}}(z)]_+ \end{aligned} \tag{13.19}$$

In summary, the optimal causal filter transfer function is obtained by implementing the two-step procedure outlined in Table 13.1.

TABLE 13.1 Synthesis of the Optimum Causal Infinite Memory Filter

1. Spectral-factorize the signal's $\{\mathbf{x}(n)\}$ complex spectral density function as

$$R_{\mathbf{xx}}(z) = \frac{1}{W(z)\overline{W}(\overline{z}^{-1})}$$

where $W(z)$ and $1/W(z)$ are each stable causal transfer functions.

2. The optimal filter transfer function is then given by

$$H^{\circ}(z) = W(z)[\overline{W}(\overline{z}^{-1})R_{\mathbf{dx}}(z)]_{+}$$

EXAMPLE 13.2

Synthesize the optimal causal linear filter corresponding to the filtering prob lem considered in Example 12.1. As in Example 13.1, the complex spectra density functions associated with this filter synthesis are given by

$$R_{\mathbf{dx}}(z) = \frac{-6z}{(z - 0.5)(z - 2)} \qquad \text{for } 0.5 < |z| < 2$$

and

$$R_{\mathbf{xx}}(z) = \frac{2z^2 - 11z + 2}{(z - 0.5)(z - 2)}$$
$$= \frac{2(z - \frac{11}{4} + \frac{1}{4}\sqrt{105})(z - \frac{11}{4} - \frac{1}{4}\sqrt{105})}{(z - 0.5)(z - 2)} \qquad \text{for } 0.5 < |z| < 2$$

In accordance with the spectral factorization requirement (13.15) and th conditions (13.14), the resultant causal invertible whitening filter is found t be

$$W(z) = \frac{\sqrt{11 - \sqrt{105}}}{2} \frac{z - 0.5}{z - \frac{11}{4} + \frac{1}{4}\sqrt{105}}$$

as is readily verified by noting that $R_{\mathbf{xx}}(z) = 1/W(z)W(z^{-1})$. How one ma systematically obtain this transfer function will be elaborated on further i Section 13.4. In accordance with relationship (13.19), the optimal filter i therefore specified by

$$H^{\circ}(z) = \frac{\sqrt{11 - \sqrt{105}}}{2} \frac{z - 0.5}{z - \frac{11}{4} + \frac{1}{4}\sqrt{105}}$$
$$\times \left[\frac{\sqrt{11 - \sqrt{105}}}{2} \frac{z^{-1} - 0.5}{z^{-1} - \frac{11}{4} + \frac{1}{4}\sqrt{105}} \frac{-6z}{(z - 0.5)(z - 2)}\right]$$

After canceling the common root $z - 2$ that appears in the numerator and denominator terms within the $[\cdot]_+$ bracket, this expression simplifies to

$$H^o(z) = \frac{\sqrt{11 - \sqrt{105}}}{2} \frac{(z - 0.5)}{(z - \frac{11}{4} + \frac{1}{4}\sqrt{105}}$$

$$\times \left[\frac{-\frac{3}{8}(11 + \sqrt{105})\sqrt{11 - \sqrt{105}}z}{(z - 0.5)(z - \frac{11}{4} - \frac{1}{4}\sqrt{105})} \right]_+$$

To determine the $[\cdot]_+$ term, we make the following partial-fraction expansion of its function argument:

$$\left[\frac{-\frac{3}{8}(11 + \sqrt{105})\sqrt{11 - \sqrt{105}}z}{(z - 0.5)(z - \frac{11}{4} - \frac{1}{4}\sqrt{105})} \right]_+$$

$$= \left[\frac{(3 + \sqrt{105})\sqrt{11 - \sqrt{105}}}{8} \left(\frac{z}{z - 0.5} - \frac{z}{z - \frac{11}{4} - \frac{1}{4}\sqrt{105}} \right) \right]_+$$

The first term on the right-hand side is purely causal in nature; the second term is purely anticausal and therefore is dropped in the $[\cdot]_+$ operation. The optimal transfer function is therefore

$$H^o(z) = \frac{\sqrt{11 - \sqrt{105}}}{2} \frac{z - 0.5}{z - \frac{11}{4} + \frac{1}{4}\sqrt{105}}$$

$$\times \frac{(3 + \sqrt{105})\sqrt{11 - \sqrt{105}}}{8} \frac{z}{z - 0.5}$$

$$= \frac{\sqrt{105} - 9}{2} \frac{z}{z - \frac{11}{4} + \frac{1}{4}\sqrt{105}}$$

The associated optimal unit-impulse response is then

$$h^o(n) = \frac{\sqrt{105} - 9}{2}(\tfrac{11}{4} - \tfrac{1}{4}\sqrt{105})^n u(n)$$

Furthermore, the resultant minimum MSE is obtained by using expression (13.10) and the geometric series summation identity, namely

$$E\{|\epsilon^o(n)|^2\} = \sqrt{105} - 9 = 1.24695$$

Thus the optimal infinite memory causal linear filter provides only a negligible improvement over the optimal causal filter of length 3 considered in Example 12.1 and is not as effective as the optimal noncausal infinite memory filter treated in Example 13.1. ■

13.4 SPECTRAL FACTORIZATION: THE WHITENING FILTER

In synthesizing the optimum causal filter, the main requirement is that of decomposing the observed signal's complex spectral density function into the form

$$R_{\mathbf{xx}}(z) = \frac{1}{W(z)\overline{W}(\overline{z}^{-1})} \tag{13.20}$$

where $W(z)$ represents a stable causal transfer function whose reciprocal $1/W(z)$ is similarly characterized. This decomposition procedure is commonly referred to as *spectral factorization* and is readily carried out when the complex spectral density function is rational, that is,

$$R_{\mathbf{xx}}(z) = \frac{g_0 + \sum_{k=1}^{q} g_k z^{-k} + \overline{g}_k z^k}{f_0 + \sum_{k=1}^{p} f_k z^{-k} + \overline{f}^k z^k} = \frac{G_{2q}(z)}{F_{2p}(z)} \tag{13.21}$$

It is noted that the polynomial terms $G_{2q}(z)$ and $F_{2p}(z)$ constituting this rational expression must of necessity have complex conjugate symmetric coefficients due to the complex spectral density function property $R_{\mathbf{xx}}(z) = \overline{R}_{\mathbf{xx}}(\overline{z}^{-1})$. Moreover, these coefficients are real whenever the time series $\{\mathbf{x}(n)\}$ is itself real.

To effect the required spectral factorization (13.20), it is first necessary to factor the polynomial terms $G_{2q}(z)$ and $F_{2p}(z)$. The roots of this factorization possess a salient structure due to the polynomial coefficient conjugate symmetry indicated above, which gives rise to the properties

$$G_{2q}(z) = \overline{G}_{2q}(\overline{z}^{-1}) \qquad \text{and} \qquad F_{2p}(z) = \overline{F}_{2p}(\overline{z}^{-1})$$

From these properties it follows immediately that if z_k is a root of $G_{2q}(z)$ [i.e., $G_{2q}(z_k) = 0$], then $\overline{z}_k^{-1}$ must also be a root [i.e., $G_{2q}(\overline{z}_k^{-1}) = 0$]. Similarly, if p_k is a root of $F_{2p}(z)$, then $\overline{p}_k^{-1}$ is also a root of $F_{2p}(z)$. With these properties the following spectral factorization theorem has been established.

Theorem 13.1 Let the wide-sense stationary time series $\{\mathbf{x}(n)\}$ have the rational complex spectral density function

$$R_{\mathbf{xx}}(z) = \frac{g_0 + \sum_{k=1}^{q} g_k z^{-k} + \overline{g}_k z^k}{f_0 + \sum_{k=1}^{p} f_k z^{-k} + \overline{f}^k z^k} \tag{13.22}$$

where the numerator and denominator polynomial coefficients appear in the prerequisite symmetrical form. It then follows that this complex density function may always be put into the factored form

$$R_{\mathbf{xx}}(z) = \alpha \frac{\prod_{k=1}^{q} (1 - z_k z^{-1})(1 - \bar{z}_k z)}{\prod_{k=1}^{p} (1 - p_k z^{-1})(1 - \bar{p}_k z)} \tag{13.23}$$

where the normalizing positive scalar α is specified by

$$\alpha = \frac{g_q \prod_{k=1}^{p} (-p_k)}{f_p \prod_{k=1}^{q} (-z_k)} \tag{13.24}$$

Moreover, when the time series $\{\mathbf{x}(n)\}$ is real valued, if z_k is a zero of $R_{\mathbf{xx}}(z)$, then z_k^{-1}, $\bar{z}_k$, and $\bar{z}_k^{-1}$ are also zeros. A similar statement holds for the poles (i.e., the p_k) of $R_{\mathbf{xx}}(z)$.

This theorem indicates that the roots of any rational complex spectral density function will occur in reciprocal–complex conjugate sets for real-valued time series and in reciprocal sets for complex-valued time series. As we now show, this property enables us to achieve the whitening filter factorization required in the synthesis of the causal infinite memory filter examined in Section 13.3.

Whitening Filter Synthesis

An important signal-processing capability entails the whitening of a general wide-sense stationary signal $\{\mathbf{x}(n)\}$. This operation may often be mechanized by applying the signal to an appropriately designed causal linear filter in which the corresponding response $\{\mathbf{y}(n)\}$ is the required whitened signal [i.e., $r_{\mathbf{yy}}(n) = \delta(n)$] shown in Figure 13.2. To see how this may be accomplished, let it be assumed that the signal to be whitened has a complex spectral density function that possess the following properties:

1. $R_{\mathbf{xx}}(z)$ is rational.
2. None of the zeros or poles of $R_{\mathbf{xx}}(z)$ have magnitude 1.

For such spectral density functions, we may apply Theorem 13.1 directly to synthesize the required whitening filter. This is a direct consequence of rela-

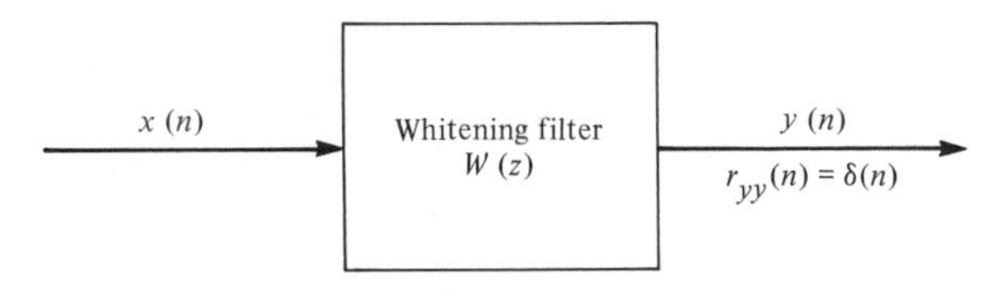

FIGURE 13.2. Process of whitening.

tionship (13.23), which in correspondence with property 2 indicates that $R_{\mathbf{xx}}(z)$ has q zeros and p poles *inside* the unit circle and their reciprocal images lie *outside* the unit circle. Upon using the spectral factorization (13.23), it follows that the whitening filter has transfer function

$$W(z) = \frac{1}{\sqrt{\alpha}} \frac{\prod_{k=1}^{p} (1 - p_k z^{-1})}{\prod_{k=1}^{q} (1 - z_k z^{-1})} \tag{13.25}$$

where the normalizing scalar α is given in expression (13.24). Without loss of generality we have here taken the roots z_k and p_k to be those located *inside* the unit circle. This transfer function and its reciprocal $1/W(z)$ are each seen to have their poles all located inside the unit circle. As such, they each may be implemented by causal, stable recursive relationships. It will be recalled that this capability was required in synthesizing the optimal causal filter considered in the preceding section.

It is a simple matter to establish that filter (13.25) carries out the necessary whitening operation. Specifically, the response of this filter has the associated complex spectral density function

$$R_{\mathbf{yy}}(z) = W(z)\overline{W}(\overline{z}^{-1})R_{\mathbf{xx}}(z)$$

as is apparent from our studies in Chapter 11. The excitation's complex spectral density function, however, is specified by

$$R_{\mathbf{xx}}(z) = \frac{1}{W(z)\overline{W}(\overline{z}^{-1})}$$

We therefore conclude that $R_{\mathbf{yy}}(z) = 1$, which implies that $r_{\mathbf{yy}}(n) = \delta(n)$, as desired. Thus the proposed causal whitening filter (13.25) does in fact realize the whitening operation. The whitening filter synthesis is therefore straightforward when the signal to be whitened has a rational spectral density function and none of its zeros and poles are located on the unit circle. Fortunately, most practical rational wide-sense stationary signals possess this root-location structure.

EXAMPLE 13.3

Make a spectral factorization and determine the causal whitening filter associated with a real-valued wide-sense stationary signal with autocorrelation $r_{\mathbf{xx}}(n) = 4(0.5)^{|n|} + 2\delta(n)$. In Example 13.1 the complex spectral density function associated with this density function was found to be

$$R_{\mathbf{xx}}(z) = 2\frac{z^2 - 5.5z + 1}{(z - 0.5)(z - 2)}$$

Factoring the numerator polynomial and multiplying the numerator and denominator by z^{-1} gives the desired spectral factorization form (13.23):

$$R_{xx}(z) = \frac{11 + \sqrt{105}}{4} \frac{[1 - (\frac{11}{4} - \frac{1}{4}\sqrt{105})z^{-1}][1 - (\frac{11}{4} - \frac{1}{4}\sqrt{105})z]}{(1 - 0.5z^{-1})(1 - 0.5z)}$$

The associated invertible causal whitening filter is therefore, by relationship (13.25),

$$W(z) = \frac{2}{\sqrt{11 + \sqrt{105}}} \frac{1 - 0.5z^{-1}}{1 - (\frac{11}{4} - \frac{1}{4}\sqrt{105})z^{-1}}$$ ■

13.5 INNOVATIONS AND THE WHITENING FILTER

The concept of whitening is of utmost importance in signal-processing theory. We have examined some of the central issues related to the whitening process in previous sections. A new interpretation of this whitening process is now given which is of particular value in the synthesis of the Wiener filter. It is here assumed that the wide-sense stationary time series $\{\mathbf{x}(n)\}$ under examination is regular. A *regular* time series is one whose associated complex spectral density function is expressible in the *spectral decomposition* form

$$R_{xx}(z) = H(z)\overline{H}(\overline{z}^{-1}) \tag{13.26}$$

where the function $H(z)$ and its reciprocal as denoted by

$$W(z) = \frac{1}{H(z)} \tag{13.27}$$

are each analytic outside and on the unit circle (i.e., $|z| \geq 1$). A function $H(z)$ that possesses this analytic property is said to be *minimum phase*. It is well known that a complex spectral density function is regular if it satisfies the Paley–Wiener condition,

$$\int_{-\infty}^{\infty} \ln [R_{xx}(e^{j\omega})]\, d\omega = \int_{-\pi}^{\pi} \ln [S_{xx}(\omega)]\, d\omega > -\infty$$

Fortunately, many time series of interest satisfy this condition.

Due to the minimum-phase nature of the functions $H(z)$ and $W(z)$, it follows that their associated z-transforms must be of the causal form

$$H(z) = \sum_{n=0}^{\infty} h(n)z^{-n} \tag{13.28}$$

and

$$W(z) = \sum_{n=0}^{\infty} w(n)z^{-n} \tag{13.29}$$

in which the characteristic sequences $\{h(n)\}$ and $\{w(n)\}$ are each square summable (i.e., they have finite l_2 norms). This being the case, we may associate these z-transforms with the transfer functions of causal, stable linear systems. In particular, the system with transfer function $H(z)$ is here referred to as the *innovation filter,* while the *whitening filter* corresponds to that system with transfer function $W(z) = 1/H(z)$.

Innovation Time Series

Let us now examine the effects that these two systems have on the random time series under analysis $\{\mathbf{x}(n)\}$ and a related innovation time series $\{\boldsymbol{\epsilon}_x(n)\}$. Specifically, let the time series $\{\mathbf{x}(n)\}$ be applied to the whitening filter (13.29). The corresponding response is then given by the convolution summation

$$\boldsymbol{\epsilon}_x(n) = \sum_{k=0}^{\infty} w(k)\mathbf{x}(n - k) \tag{13.30}$$

This response is commonly referred to as the *innovations* associated with time series $\{\mathbf{x}(n)\}$. It is normalized white-noise time series, as is made evident by appealing to the excitation-response result from Chapter 11 as specified by expression (11.26). This relationship, in conjunction with relationships (13.26) and (13.27), yields the normalized white condition

$$R_{\boldsymbol{\epsilon}_x\boldsymbol{\epsilon}_x}(z) = W(z)\overline{W}(\overline{z}^{-1})R_{\mathbf{xx}}(z) = 1 \tag{13.31}$$

In a similar fashion, let us now apply the innovations time series $\{\boldsymbol{\epsilon}_x(n)\}$ to the innovations filter (13.28). The corresponding response is then found to be given by

$$\mathbf{x}(n) = \sum_{k=0}^{\infty} h(k)\boldsymbol{\epsilon}_x(n - k) \tag{13.32}$$

which arises upon recalling the identity $W(z) = 1/H(z)$ and using the system configuration depicted in Figure 13.3. Thus the two time series $\{\mathbf{x}(n)\}$ and $\{\boldsymbol{\epsilon}_x(n)\}$

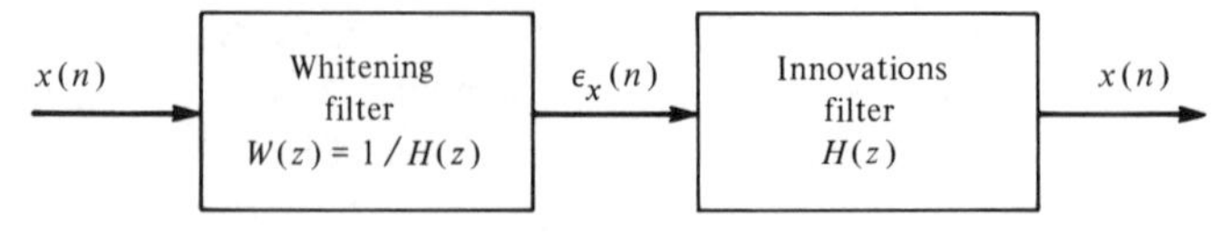

FIGURE 13.3. System relationships between a time series and its innovation equivalent.

are related to one another by causal linear operations. Each of these time series thus contains the same information relative to causal linear operations.

EXAMPLE 13.4

Determine the innovation process that is associated with a real-valued time series $\{\mathbf{x}(n)\}$ whose complex spectral density function is specified by

$$R_{\mathbf{xx}}(z) = 2\left[\frac{z^2 - 5.5z + 1}{(z - 0.5)(z - 2)}\right]$$

The whitening filter associated with this time series was determined in Example 13.3:

$$W(z) = \frac{2}{\sqrt{11 + \sqrt{105}}} \frac{1 - 0.5z^{-1}}{1 - (\frac{11}{4} - \frac{1}{4}\sqrt{105})z^{-1}}$$

If we use the z-transform relationship $\epsilon_x(z) = W(z)X(z)$, it then follows that the required innovations process is generated according to

$$\epsilon_x(n) = \frac{2}{\sqrt{11 + \sqrt{105}}} [x(n) - 0.5x(n - 1)] + (\tfrac{11}{4} - \tfrac{1}{4}\sqrt{105})\epsilon_x(n - 1)$$

This innovation time series is a normalized white-noise process that is equivalent to $\{\mathbf{x}(n)\}$ relative to causal linear operations. ■

13.6 LINEAR PREDICTION AND THE WHITENING FILTER

The most challenging task in Wiener filter synthesis is related to the whitening filter determination. We now examine the whitening-filter operation in the guise of linear prediction. In particular, let us consider the infinite order causal predictor, which takes the form

$$\hat{\mathbf{x}}(n) = \sum_{i=1}^{\infty} a_i \mathbf{x}(n - i) \tag{13.33}$$

where the predictor coefficients a_i are to be chosen so that $\hat{\mathbf{x}}(n)$ best represents the assumed *regular* time series $\{\mathbf{x}(n)\}$ in the mean-squared-error sense. Upon substitution of identity (13.32) into this prediction filter relationship and after manipulation, the following *equivalent* innovations prediction expression is eventually obtained:

$$\hat{\mathbf{x}}(n) = \sum_{k=1}^{\infty} b_k \boldsymbol{\epsilon}_x(n - k) \tag{13.34}$$

in which the z-transforms of the $\{a(n)\}$ and $\{b(n)\}$ coefficients are related by

$$B(z) = A(z)H(z) \tag{13.35}$$

where $H(z)$ is the innovation filter arising from the spectral decomposition (13.26).

The prediction error filter form (13.34) is of particular use when computing the predictor's associated mean-squared error, as is now illustrated. The predictor's mean-squared error is straightforwardly obtained by incorporating relationships (13.32) and (13.34) to obtain

$$\begin{aligned} E\{|\mathbf{x}(n) - \hat{\mathbf{x}}(n)|^2\} &= E\left\{\left|\sum_{k=0}^{\infty} h(k)\boldsymbol{\epsilon}_x(n-k) - \sum_{k=1}^{\infty} b_k\boldsymbol{\epsilon}_x(n-k)\right|^2\right\} \\ &= |h(0)|^2 + \sum_{k=1}^{\infty} |h(k) - b_k|^2 \end{aligned}$$

in which use of the innovation process's normalized whiteness property [i.e., $E\{\boldsymbol{\epsilon}_x(n)\bar{\boldsymbol{\epsilon}}_x(m)\} = \delta(n - m)$] has been employed. Since our objective is to minimize the mean-squared prediction error, it is seen that this is accomplished by setting the coefficients $b_k = h(k)$ for $k \geq 1$. Thus the optimal selection for the transfer function $B(z)$ used in the innovations predictor (13.34) is given by

$$B^{o}(z) = H(z) - h(0) \tag{13.36}$$

while the optimal choice for the predictor's transfer function used in the standard format (13.33) is specified by

$$A^{o}(z) = \frac{H(z) - h(0)}{H(z)} = 1 - \frac{h(0)}{H(z)} \tag{13.37}$$

Upon taking the inverse z-transforms of these two transfer functions, we obtain directly the required optimum a_k^o and b_k^o predictor coefficients. For each of these equivalent predictor filters, the associated minimum mean-squared prediction error is specified by

$$E\{|\mathbf{x}(n) - \hat{\mathbf{x}}^{o}(n)|^2\} = |h(0)|^2 \tag{13.38}$$

In this synthesis process, the need to carry out the spectral factorization operation (13.26) is fundamental. Furthermore, it is important to recognize that this synthesis procedure requires that the time series $\{\mathbf{x}(n)\}$ be *regular*.

EXAMPLE 13.5

Determine the optimum causal infinite memory predictor associated with the time series $\{\mathbf{x}(n)\}$ whose complex spectral density function is specified by

$$R_{\mathbf{xx}}(z) = 2\,\frac{z^2 - 5.5z + 1}{(z - 0.5)(z - 2)}$$

From the spectral density function decomposition found in Example 13.3, we have

$$H(z) = \frac{\sqrt{11 + \sqrt{105}}}{2} \frac{1 - (\frac{11}{4} - \frac{1}{4}\sqrt{105})z^{-1}}{1 - 0.5z^{-1}}$$

To determine the optimum predictor using expression (13.37), the initial-value theorem is first applied to give

$$h(0) = \lim_{z\to\infty} H(z) = \frac{\sqrt{11 + \sqrt{105}}}{2}$$

It then follows from expression (13.37) that the optimum predictor transfer function is

$$\begin{aligned} A^{o}(z) &= 1 - \frac{1 - 0.5z^{-1}}{1 - (\frac{11}{4} - \frac{1}{4}\sqrt{105})z^{-1}} \\ &= \frac{\sqrt{105} - 9}{4} \frac{z^{-1}}{1 - (\frac{11}{4} - \frac{1}{4}\sqrt{105})z^{-1}} \end{aligned}$$

which is seen to correspond to a first-order linear recursive operator. The time-domain representation for this filter is then

$$\hat{x}(n) = \frac{\sqrt{105} - 9}{4} x(n - 1) + \frac{11 - \sqrt{105}}{4} \hat{x}(n - 1)$$

Furthermore, the associated mean-squared prediction error is by expression (13.38) given by

$$E\{[\mathbf{x}(n) - \hat{\mathbf{x}}(n)]^2\} = \frac{11 + \sqrt{105}}{4}$$

■

13.7 WOLD'S DECOMPOSITION THEOREM

In studies related to wide-sense stationary time series, the Wold's decomposition theorem provides an extremely insightful chracterization that is useful in signal-processing applications. This theorem states that any wide-sense stationary time series is always associated with a spectral density function that contains a continuous component, and, a component consisting of Dirac delta impulses (arising from random sinusoidal time series terms).

Theorem 13.2 Wold's Decomposition Theorem. Let $\{\mathbf{x}(n)\}$ be a zero-mean wide-sense stationary time series. Then $\{\mathbf{x}(n)\}$ may be always expressed as the sum of two zero-mean wide-sense stationary processes

$$\mathbf{x}(n) = \mathbf{u}(n) + \mathbf{v}(n) \tag{13.39}$$

in which:

(a) The elements $\mathbf{u}(n)$ and $\mathbf{v}(m)$ are orthogonal, that is,

$$E\{\mathbf{u}(m)\overline{\mathbf{v}}(n)\} = 0 \qquad \text{for all } m \text{ and } n$$

(b) The random time series $\{\mathbf{u}(n)\}$ is *regular* and has the causal representation

$$\mathbf{u}(n) = \sum_{k=0}^{\infty} h(k)\boldsymbol{\epsilon}(n - k)$$

where $\Sigma_{k=0}^{\infty} |h(k)|^2 < \infty$ and $\{\boldsymbol{\epsilon}(n)\}$ is a normalized white-noise process ($\sigma_{\boldsymbol{\epsilon}}^2 = 1$) that is orthogonal with $\{\mathbf{v}(n)\}$, so that $E\{\boldsymbol{\epsilon}(n)\overline{\mathbf{v}}(m)\} = 0$ for all m and n. Moreover, the characteristic sequence $\{h(n)\}$ and the normalized white-noise process $\{\boldsymbol{\epsilon}(n)\}$ are unique.

(c) The random time series $\{\mathbf{v}(n)\}$ is *singular* in the sense that there exists a set of prediction coefficients a_k such that $\mathbf{v}(n)$ is perfectly predictable in the sense that the relationship

$$\mathbf{v}(n) = \sum_{k=1}^{\infty} a_k\mathbf{v}(n - k)$$

is satisfied with a zero-mean squared error.

To prove the Wold decomposition theorem, a number of previously used concepts are employed. For instance, let us first consider the optimum predictor associated with time series $\{\mathbf{x}(n)\}$. This prediction operation produces the related prediction error (or innovations) time series

$$\boldsymbol{\epsilon}(n) = \mathbf{x}(n) - \sum_{k=1}^{\infty} a_k\mathbf{x}(n - k) \tag{13.40}$$

which is white noise in nature, so that $r_{\boldsymbol{\epsilon\epsilon}}(n) = \sigma_{\boldsymbol{\epsilon}}^2\delta(n)$. We next seek that optimum estimate of $\mathbf{x}(n)$ based on a linear combination of $\boldsymbol{\epsilon}(n)$ and its infinite past values. Let this optimum estimate be denoted by

$$\mathbf{u}(n) = \sum_{k=0}^{\infty} h_k\boldsymbol{\epsilon}(n - k) \tag{13.41}$$

where h_k are the optimum estimator coefficients. The associated optimum estimation error is then designated as

$$\mathbf{v}(n) = \mathbf{x}(n) - \mathbf{u}(n) \tag{13.42}$$

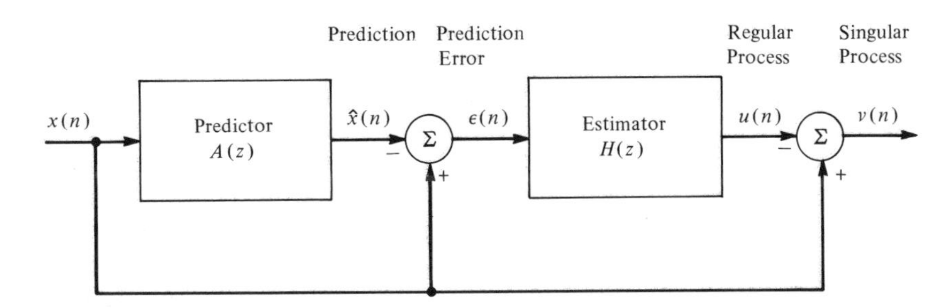

FIGURE 13.4. Generation of regular and singular time series components in the Wold decomposition.

In accordance with the results from Section 13.6, if $\{\mathbf{x}(n)\}$ is a regular time series, the estimation error time series $\{\mathbf{v}(n)\}$ will be identically zero. This regularity assumption is not assumed here. A system depiction of how the time series components $\{\mathbf{u}(n)\}$ and $\{\mathbf{v}(n)\}$ are generated is given in Figure 13.4. We now show that these two time series possess the statistical properties spelled out in Theorem 13.2.

We first establish that the wide-sense stationary time series $\{\mathbf{u}(n)\}$ and $\{\mathbf{v}(n)\}$ are orthogonal. This is made evident by noting that

$$E\{\mathbf{v}(n)\overline{\boldsymbol{\epsilon}}(n - k)\} = 0 \qquad \text{for } k \geq 0 \tag{13.43}$$

since the estimation error $\mathbf{v}(n)$ must be orthogonal to $\boldsymbol{\epsilon}(n - k)$ if $\mathbf{u}(n)$ is an optimum estimate of $\mathbf{x}(n)$ through relationship (13.41). Similarly, we have that

$$E\{\boldsymbol{\epsilon}(n)\overline{\mathbf{u}}(n - k)\} = 0 \qquad \text{for } k \geq 1$$

as is shown by substituting in relationship (13.41) and using the whiteness of the prediction error time series (13.40). The prediction error time series relationship also indicates that

$$E\{\boldsymbol{\epsilon}(n)\overline{\mathbf{x}}(n - k)\} = 0 \qquad \text{for } k \geq 1$$

Subtracting these last two expressions and making use of identity (13.42), we conclude that

$$E\{\boldsymbol{\epsilon}(n)\overline{\mathbf{v}}(n - k)\} = 0 \qquad \text{for } k \geq 1$$

This expectation expression in conjunction with equation (13.43) indicates that the two time series $\{\boldsymbol{\epsilon}(n)\}$ and $\{\mathbf{v}(n)\}$ are orthogonal, that is,

$$E\{\boldsymbol{\epsilon}(n)\overline{\mathbf{v}}(m)\} = 0 \qquad \text{for all } k \text{ and } m \tag{13.44}$$

In turn, this orthogonality property, in conjunction with relationship (13.41), indicates that $\{\mathbf{u}(n)\}$ and $\{\mathbf{v}(n)\}$ are also orthogonal time series, and part (a) of the theorem is proven.

To establish part (b), it is seen from relationship (13.41) that

$$r_{\mathbf{xx}}(0) = E\{|\mathbf{x}(n)|^2\} \geq E\{|\mathbf{x}(n) - \mathbf{u}(n)|^2\} = E\{|\mathbf{v}(n)|^2\} = \sigma_{\boldsymbol{\epsilon}}^2 \sum_{k=0}^{\infty} |w(k)|^2$$

which indicates that the estimation filter coefficients $w(n)$ are square summable and therefore $\{\mathbf{u}(n)\}$ must be a regular time series that has the representation (13.41), as postulated.

Finally, part (c) is proved by introducing the auxiliary time series $\{\mathbf{s}(n)\}$ as specified by

$$\mathbf{s}(n) = \mathbf{v}(n) - \sum_{k=1}^{\infty} a_k \mathbf{v}(n - k) \tag{13.45}$$

and showing that the variance of $\mathbf{s}(n)$ is zero. Upon taking the z-transform of this expression and those of (13.40) to (13.42), it is found that

$$\begin{aligned} S(z) &= [1 - A(z)]V(z) \\ &= [1 - A(z)][X(z) - U(z)] \\ &= [1 - [1 - A(z)]W(z)]\in(z) \end{aligned}$$

where $A(z)$ and $W(z)$ are the transfer functions associated with the optimal predictor (13.40) and the optimal estimator (13.41). Taking the inverse z-transform of this latter expression gives the equivalent causal time-domain relationship

$$\mathbf{s}(n) = \sum_{k=0}^{\infty} g(k)\boldsymbol{\epsilon}(n - k) \tag{13.46}$$

in which $G(z) = 1 - [1 - A(z)]W(z)$. If we use the orthogonal condition (13.44) in conjunction with relationship (13.45), it follows that $E\{\mathbf{s}(n)\boldsymbol{\epsilon}(m)\} = 0$ for all m and n. Using this orthogonality property and relationship (13.46), we obtain $g(k) = 0$ for all k, which establishes that $\mathbf{s}(n) = 0$, thereby establishing part (c).

Identification of Singular Time Series

Wold's decomposition theorem is of practical value in applications entailing the detection of sinusoidal signal components (e.g., Doppler processing). From the developments just made, it is clear that any zero-mean wide-sense-stationary time series may be decomposed as

$$\mathbf{x}(n) = \mathbf{u}(n) + \sum_{k=1}^{m} A_k \sin(\omega_k n + \boldsymbol{\theta}_k) \tag{13.47}$$

where $\{\mathbf{u}(n)\}$ is a regular time series and the sum of sinusoids represents the singular time series component $\{\mathbf{v}(n)\}$. The random phase angles $\boldsymbol{\theta}_k$ here appearing are independent and uniformly distributed on the interval $[-\pi, \pi)$.

To detect the presence of sinusoidal signal terms, we first determine the optimum causal predictor associated with the time series $\{\mathbf{x}(n)\}$. If any zeros of the optimum predictive error filter's transfer function $1 - A(z)$ should lie on the unit circle (i.e., at $e^{\pm j\omega_k}$ for $1 \leq k \leq m$) the existence of sinusoidal component (at frequencies ω_k for $1 \leq k \leq m$) is thereby indicated. Thus predictor error filter zeros on the unit circle serve the purpose of sinusoidal detection. The predictor zeros associated with the regular time series component $\{\mathbf{u}(n)\}$ are always located inside the unit circle.

13.8 SUMMARY

We have considered the important topic of synthesizing optimal linear filters that minimize a MSE criterion. The different filtering problems addressed share a commonality which will now be briefly summarized. In particular, the linear filter to be synthesized is governed by a convolution summation of the form

$$y(n) = \sum_{k \in I} h(k)x(n - k) \tag{13.48}$$

where I is a set of integers specifying the domain of the weighting coefficients. For example, the selections $0 \leq k \leq p$ and $0 \leq k < \infty$ for I correspond to the filtering problems considered in Chapter 12 and in this chapter, respectively. We have sought values for the $h(k)$ weighting coefficients that minimized the MSE criterion

$$E\{|\boldsymbol{\epsilon}(n)|^2\} = E\{|\mathbf{d}(n) - \mathbf{y}(n)|^2\} \tag{13.49}$$

Under the assumption that the random signals $\{\mathbf{x}(n)\}$ and $\{\mathbf{d}(n)\}$ are each wide-sense stationary, the optimum weighting coefficients were found to satisfy the Wiener–Hopf equations

$$\sum_{k \in I} h^o(k) r_{\mathbf{xx}}(n - k) = r_{\mathbf{dx}}(n) \qquad \text{for } n \in I \tag{13.50}$$

Moreover, the optimum filtering error associated with this optimum filter possessed the orthogonal property

$$E\{\boldsymbol{\epsilon}^o(n)\overline{\mathbf{x}}(n - m)\} = 0 \qquad \text{for } m \in I \tag{13.51}$$

Clearly, the degree to which $\{\boldsymbol{\epsilon}^o(n)\}$ and $\{\mathbf{x}(n)\}$ are orthogonal depends on the indexing set I. Complete orthoganization is realized in general only for the choice $I = (-\infty, \infty)$.

Various procedures were proposed for solving the Wiener–Hopf equations (13.50) for the optimal weighting coefficients. When the indexing set $I = [1, p]$, a computationally efficient iterative method was outlined in Chapter 12. On the other hand, a choice for I of $(-\infty, \infty)$ and $[0, \infty]$ leads quite naturally to a z-transform type solution, as shown in this chapter.

SUGGESTED READINGS

See the Suggested Readings in Chapter 12.

PROBLEMS

13.1 Let the measurement signal be given by $\mathbf{x}(n) = \mathbf{s}(n) + \boldsymbol{\eta}(n)$, in which $r_{\boldsymbol{\eta\eta}}(n) = \delta(n)$ and $r_{\mathbf{ss}}(n) = (0.5)^{|n|}$. Design the optimum infinite memory filter that minimizes the MSE where $\boldsymbol{\epsilon}(n) = \mathbf{y}(n) - \mathbf{s}(n)$.

13.2 Carry out the details of Example 13.1.

13.3 Given the real wide-sense stationary signal $\mathbf{x}(n) = \mathbf{u}(n) + \mathbf{w}(n)$, in which the relevant complex spectral density functions are specified by

$$R_{\mathbf{uu}}(z) = \frac{1}{(1 - 0.5z^{-1})(1 - 0.5z)}, \qquad R_{\mathbf{ww}}(z) = 1, \qquad R_{\mathbf{uw}}(z) = 0$$

Find the optimal linear filter that minimizes $E\{[\mathbf{u}(n - 1) - \mathbf{y}(n)]^2\}$, where $\{\mathbf{y}(n)\}$ represents the filter's response to the excitation $\{\mathbf{x}(n)\}$.

13.4 Let the real wide-sense stationary signal $\{\mathbf{x}(n)\}$ have the complex spectral density function

$$R_{\mathbf{xx}}(z) = 4\left[\frac{(1 - 0.2z^{-1})(1 - 0.2z)}{(1 - 0.8z^{-1})(1 - 0.8z)}\right]$$

Find the optimum linear two-step predictor [i.e., $\mathbf{d}(n) = \mathbf{x}(n + 2)$].

13.5 If the wide-sense stationary signal $\{\mathbf{x}(n)\}$ has an all-pole complex spectral density function, that is,

$$R_{xx}(z) = \frac{\sigma^2}{A_p(z)A_p(z^{-1})}$$

where $A_p(z) = 1 + a_1z^{-1} + \cdots + a_pz^{-p}$ has all its roots inside the unit circle, show that the optimum linear one-step predictor [i.e., $\mathbf{d}(n) = \mathbf{x}(n+1)$] has transfer function

$$H^{\circ}(z) = -a_1 z^{-1} - a_2 z^{-2} - \cdots - a_p z^{-p}$$

Moreover, the associated mean-squared prediction error is given by σ^2.

13.6 Let the real wide-sense stationary signal $\mathbf{x}(n) = \mathbf{u}(n) + \mathbf{w}(n)$ have the underlying complex spectral density functions $R_{\mathbf{ww}}(z) = 1$, $R_{\mathbf{uw}}(z) = 0$, and

$$R_{\mathbf{xx}}(z) = \frac{\sigma^2}{A_p(z)A_p(z^{-1})}$$

where $A_p(z) = 1 + a_1 z^{-1} + \cdots + a_p z^{-p}$ has all its roots inside the unit circle. Synthesize the optimum linear one-step predictor in which $\mathbf{d}(n) = \mathbf{x}(n+1)$.

13.7 Consider the wide-sense stationary signal whose complex spectral density function is given by

$$R_{\mathbf{xx}}(z) = 2\left[\frac{z^2 - 6.5z + 1}{z^2 - 2.5z + 1}\right]$$

Determine the whitening filter associated with this signal.

13.8 Carry out the details of Example 13.2.

13.9 Show that if $\{\mathbf{x}(n)\}$ is a complex-valued wide-sense stationary time series, then the causal whitening filter is obtained by performing the spectral decomposition.

$$R_{\mathbf{xx}}(z) = \frac{1}{W(z)\overline{W}(\overline{z}^{-1})}$$

where $W(z)$ has all its zeros and poles located inside the unit circle.

13.10 Determine the innovations filter associated with the wide-sense stationary signal described in Problem 13.7.

13.11 Using the approach taken in Section 13.6, synthesize the optimum one-step predictor associated with the wide-sense stationary signal whose complex spectral density function is given by

$$R_{\mathbf{xx}}(z) = 2\left[\frac{z^2 - 6.5z + 1}{z^2 - 2.5z + 1}\right]$$

CHAPTER 14

Matched Filters

14.1 INTRODUCTION

A signal-processing problem of interdisciplinary interest is concerned with the task of determining whether a prescribed signal is present in noise-contaminated data. For example, the prescribed signal might be a spread-spectrum signal in a communication application, a weak fetal heartbeat in a pregnant woman's EKG recording, or a reflected target signal in a radar return. To describe this general situation, the data to be characterized are hypothesized to take on one of the two forms

$$\begin{aligned} H_0: \quad \mathbf{x}(n) &= \mathbf{s}(n) + \boldsymbol{\eta}(n) \\ H_1: \quad \mathbf{x}(n) &= \boldsymbol{\eta}(n) \end{aligned}$$

where $\{\mathbf{s}(n)\}$ and $\{\boldsymbol{\eta}(n)\}$ represent the prescribed signal and a wide-sense stationary noise background, respectively. The task is then to determine which of these two hypotheses (i.e., the prescribed signal is or is not present) holds by carefully processing the data provided. In this chapter various means of achieving this objective are investigated. We shall here restrict the data to be real valued, although the results to be given now are readily extended to the complex-valued data case.

14.2

FINITE MEMORY-MATCHED FILTER: FINITE-LENGTH KNOWN SIGNAL

When the background noise is relatively weak, the problem of deciding whether the prescribed signal is or is not present is usually straightforward. When this is not the case, however, it is generally necessary to employ sophisticated signal-processing techniques such as matched filtering. We now examine an important class of problems that fall into the category described in Section 14.1. In particular, it is now assumed that the prescribed real-valued signal $\{s(n)\}$ is deterministic in nature and has length $p + 1$, that is,

$$s(0), s(1), \ldots, s(p) \tag{14.1}$$

where it is tacitly assumed that $s(n) = 0$ for n not in the integer set $0 \le n \le p$. In the problem being examined, it is presumed that a time-shifted version (with unknown shift) of this signal is present under hypothesis H_0. That is, the data to be processed under this hypothesis are presumed to take the form

$$\mathbf{x}(n) = s(n - \mathbf{n}_0) + \boldsymbol{\eta}(n) \tag{14.2}$$

where the unknown shift integer variable $\mathbf{n}_0$ denotes the time instant at which the prescribed signal first appears. The signal is therefore present for the time span $\mathbf{n}_0 \le n \le \mathbf{n}_0 + p$ that is unknown to the data analyst.

To reflect most typical situations, the *onset time* instant $\mathbf{n}_0$ will be taken to be an integer-valued *random variable*. Thus $s(n - \mathbf{n}_0)$ represents a random-shifted replication of the deterministic signal $s(n)$. Since the onset time instant is not known a priori, it will be necessary to employ a signal-processing algorithm that searches for the prescribed signal on a continuous basis. A logical means for accomplishing this objective is to apply the given real-valued data $\{\mathbf{x}(n)\}$ to a linear filter and then examine the corresponding response for the presence or absence of the prescribed signal. Because of the finite length of the shifted signal $\{s(n - \mathbf{n}_0)\}$, it is natural to employ a nonrecursive filter as specified by

$$y(n) = \sum_{k=0}^{q} h(k)x(n - k) \tag{14.3}$$

where the filter order q is selected to satisfy $p \le q$. That is, the filter's memory should at least be equal to the prescribed signal's length.

Our objective is to select this filter's weighting coefficients $h(k)$ so as to enhance the prescribed signal's detection. To characterize this enhancement feature analytically, let us substitute the hypothesized data (14.2) into the filtering operation (14.3). The corresponding response may be therefore decomposed as

$$\mathbf{y}(n) = \mathbf{y}_s(n) + \mathbf{y}_\eta(n)$$

in which $\mathbf{y}_s(n)$ and $\mathbf{y}_\eta(n)$ denote the signal and corruptive noise-generated response components as specified by

$$\mathbf{y}_s(n) = \sum_{k=0}^{q} h(k)\mathbf{s}(n - \mathbf{n}_0 - k) \tag{14.4}$$

and

$$\mathbf{y}_\eta(n) = \sum_{k=0}^{q} h(k)\boldsymbol{\eta}(n - k) \tag{14.5}$$

respectively. Unlike the standard filtering problem, it is not required that the response component $\{\mathbf{y}_s(n)\}$ look like a replication of the prescribed signal $\{\mathbf{s}(n)\}$. In particular, since our primary objective is that of *detecting* the presence of $\{\mathbf{s}(n)\}$, we shall instead consider the task of making the signal response component $\mathbf{y}_s(n)$ peak at a given time instant relative to the background noise response component. If such a filter operation can be implemented, the prescribed signal present decision will be made whenever the filter response signal contains a sharp peak that exceeds a given threshold level. This behavior is depicted in Figure 14.1, where a large peak is seen to occur in a neighborhood of the onset time instant $\mathbf{n}_0$.

Examination of expression (14.4) suggests that the required peak behavior should occur when the total set of signal elements $\mathbf{s}(k)$ for $0 \le k \le p$ appear

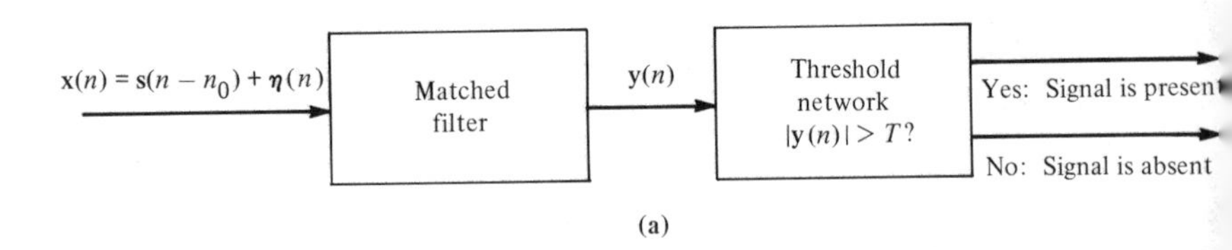

(a)

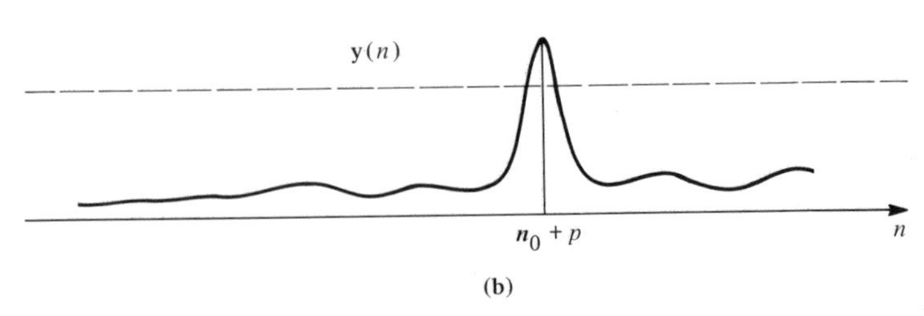

(b)

FIGURE 14.1. Matched filtering operation.

as elements in the convolution summation. Upon reflection, this is seen to be accomplished for any choice of time $\mathbf{n}_0 + p \leq n \leq \mathbf{n}_0 + q$. For our purposes, the peak time instant is arbitrarily selected to be the first possible time instant $\mathbf{n}_0 + p$. With these thoughts in mind, let us now synthesize a linear filter of form (14.3), which maximizes the *signal-to-noise ratio* (S/N) at time $n = \mathbf{n}_0 + p$. This ratio is formally defined by

$$\rho(\mathbf{h}) = \frac{[y_s(\mathbf{n}_0 + p)]^2}{E\{[\mathbf{y}_\eta(\mathbf{n}_0 + p)]^2\}} \tag{14.6}$$

where the numerator term provides a measure of instantaneous signal response power, while the denominator term measures the mean-squared noise response power. The dependency of this S/N-ratio criterion on the filter's weighting coefficients has been explicitly recognized by the appearance of the filter weighting coefficient vector $\mathbf{h} = [h(0), h(1), \ldots, h(q)]'$ as an argument in $\rho(\mathbf{h})$. Upon inserting relationships (14.4) and (14.5) into expression (14.6), the signal-to-noise-ratio criterion becomes

$$\rho(\mathbf{h}) = \frac{[\Sigma_{k=0}^{q} h(k)s(p - k)]^2}{\Sigma_{k=0}^{q} \Sigma_{m=0}^{q} r_{\eta\eta}(m - k)h(k)h(m)} \tag{14.7}$$

where $r_{\eta\eta}(k) = E\{\eta(k + m)\eta(m)\}$ designates the autocorrelation of the additive-noise signal.

EXAMPLE 14.1

Determine the instantaneous signal-to-noise ratio associated with the case in which the prescribed signal is specified by $s(n) = 3\delta(n) - \delta(n - 1)$ and the additive noise has autocorrelation $r_{\eta\eta}(n) = (0.5)^{|n|}$. In accordance with relationship (14.7), the instantaneous signal-to-noise ratio is therefore given by

$$\rho(\mathbf{h}) = \frac{[-h(0) + 3h(1)]^2}{\Sigma_{k=0}^{q} \Sigma_{m=0}^{q} (0.5)^{|m-k|}h(k)h(m)}$$

This S/N ratio is seen to be a ratio of quadratic functions of the filter's weighting coefficients. ■

14.3

FINITE MEMORY-MATCHED FILTER SYNTHESIS: FINITE-LENGTH KNOWN SIGNAL

A *matched filter* corresponds to a filter of form (14.3) whose weighting coefficients are selected to render the signal-to-noise ratio (14.7) a maximum. To effect this maximization in a computationally efficient manner, it is first noted that the signal-to-noise ratio is weighting-coefficient-size independent. That is, this ratio takes on the same value for the parameter selection $h(0), h(1), \ldots, h(q)$ as it does for the scalar-multiplied version $\alpha h(0), \alpha h(1), \ldots, \alpha h(q)$, where α is an arbitrary nonzero scalar. This property suggests that the maximization of criterion (14.7) can be obtained equivalently by solving the related constrained maximization problem

$$\max_{\sum_{k=0}^{q} \sum_{m=0}^{q} r_{\eta\eta}(m - k)h(k)h(m) = 1} \left[\sum_{k=0}^{q} h(k)s(p - k) \right]^2 \tag{14.8}$$

That is, a choice of filter weighting coefficients that solves this constrained problem also maximizes the signal-to-noise criterion.

To obtain a solution to the constrained maximization problem above, let us employ the Lagrange multiplier approach. This first entails introducing the auxiliary function

$$g(\lambda, \mathbf{h}) = \left[\sum_{k=0}^{q} h(k)s(p - k) \right]^2 + \lambda \left[1 - \sum_{k=0}^{q} \sum_{m=0}^{q} r_{\eta\eta}(m - k)h(k)h(m) \right]$$

where λ is a Lagrange multiplier scalar used to reflect the maximization problem's constraint. Upon taking the derivative of this auxiliary function with respect to the filter weighting coefficients and setting these derivatives to zero, we obtain the following set of necessary conditions for the matched filter's weighting coefficients:

$$\sum_{k=0}^{q} s(p - k)s(p - m)h^{o}(k) = \lambda \sum_{k=0}^{q} r_{\eta\eta}(m - k)h^{o}(k) \qquad \text{for } 0 \le m \le q \tag{14.9}$$

Thus the optimum filter's weighting coefficients must satisfy this system of linear equations. Furthermore, after multiplying each side of this expression by $h^{o}(m)$ and then summing these equalities for $0 \le m \le q$, it is found that the Lagrange multiplier is specified by

$$\lambda = \frac{\left[\sum_{k=0}^{q} s(p - k)h^{o}(k)\right]^2}{\sum_{k=0}^{q} \sum_{m=0}^{q} r_{\eta\eta}(m - k)h^{o}(k)h^{o}(m)} \tag{14.10}$$

This expression, however, corresponds to the signal-to-noise ratio (14.7). We therefore conclude that the maximum signal-to-noise ratio is associated with the largest choice of the Lagrange multiplier for which the set of necessary conditions (14.9) has a nontrivial solution.

To obtain the solution of the necessary conditions (14.9), it is beneficial to employ standard matrix-vector concepts. The reader unfamiliar with these notions is referred to the companion volume, which puts a heavy emphasis on the vector-space approach to signal processing. To effect the desired vector representation, let us introduce the $(q + 1) \times 1$ *reversed signal* and *filter weighting coefficient* vectors, respectively:

$$\mathbf{s} = [s(p), s(p-1), \ldots, s(0), 0, 0, \ldots, 0]' \tag{14.11}$$

$$\mathbf{h} = [h(0), h(1), \ldots, h(q)]' \tag{14.12}$$

and the $(q + 1) \times (q + 1)$ additive-noise correlation matrix $R_{\eta\eta}$, whose elements are specified by

$$R_{\eta\eta}(m, k) = r_{\eta\eta}(m - k) \qquad \text{for } 1 \leq k, m \leq q + 1 \tag{14.13}$$

The prime used in the vector representations above denotes the operation of vector transposition. Upon using these entities, it follows that the necessary conditions (14.9) for the optimum filter weighting coefficients can be equivalently expressed as

$$\mathbf{s}\mathbf{s}'\mathbf{h}^o = \lambda R_{\eta\eta}\mathbf{h}^o \tag{14.14}$$

A direct evaluation of this expression will verify this equivalency.

To obtain the optimum filter's weighting coefficients, it will be assumed initially that the noise correlation matrix is invertible. Upon left-multiplying each side of relationship (14.14) by the inverse of $R_{\eta\eta}$, the following equivalent system of equations is found to arise:

$$R_{\eta\eta}^{-1}\mathbf{s}\mathbf{s}'\mathbf{h}^o = \lambda\mathbf{h}^o \tag{14.15}$$

This relationship is seen to be in a standard eigenvector-eigenvalue format relative to the $(q + 1) \times (q + 1)$ matrix $R_{\eta\eta}^{-1}\mathbf{s}\mathbf{s}'$, in which λ is the largest eigenvalue and $\mathbf{h}^o$ is its associated eigenvector. To obtain the optimum solution to expression (11.14), it is then required to compute the eigenvalue-eigenvector characterization:

$$R_{\eta\eta}^{-1}\mathbf{s}\mathbf{s}'\mathbf{h}_k = \lambda_k\mathbf{h}_k \qquad 1 \leq k \leq q + 1 \tag{14.16}$$

Due to the structure of the matrix $R_{\eta\eta}^{-1}\mathbf{s}\mathbf{s}'$, its constituent eigenvalues and eigenvectors are readily identified as is now described.

Because of the appearance of the rank 1 *outer product* matrix $\mathbf{s}\mathbf{s}'$ in eigenrelationship (14.16), it follows that q of the $q + 1$ eigenvalues must be zero. This is a direct consequence of the fact that it is always possible to find q linearly independent $(q + 1) \times 1$ vectors which are each orthogonal to the prescribed reversed signal vector $\mathbf{s}$. Let $\mathbf{h}_1, \mathbf{h}_2, \ldots, \mathbf{h}_q$ denote such a linear independent set (i.e., $\mathbf{s}'\mathbf{h}_k = 0$ for $1 \leq k \leq q$). Upon substitution of these $\mathbf{h}_k$ into relationship (14.16), it is found that

$$R_{\eta\eta}^{-1}\mathbf{s}\mathbf{s}'\mathbf{h}_k = \mathbf{0} \qquad 1 \leq k \leq q$$

We have therefore established that $\mathbf{h}_1, \mathbf{h}_2, \ldots, \mathbf{h}_q$ are linearly independent eigenvectors which each have the same eigenvalue $\lambda = 0$. The remaining eigenvector is specified by

$$\mathbf{h}_{q+1} = R_{\eta\eta}^{-1}\mathbf{s} \tag{14.17a}$$

and its associated positive eigenvalue is given by

$$\lambda_{q+1} = \mathbf{s}' R_{\eta\eta}^{-1}\mathbf{s} \tag{14.17b}$$

This is readily shown upon substitution of the proposed eigenvector (14.17a) into eigenrelationship (14.16).

Since expressions (14.14) and (14.15) are equivalent, it is clear from the arguments above that there exist precisely two values for the Lagrange multiplier (i.e., 0 and λ_{q+1}) for which expression (14.14) has nontrivial solutions. Using this observation and the comments preceding equation (14.16) relative to the largest eigenvalue, we have proved the following theorem. In this theorem the matched filter associated with the case in which the noise matrix $R_{\eta\eta}$ is singular is also given. The proof of this result is left to the reader as an exercise.

Theorem 14.1. An optimum, matched filter, weighting coefficient vector that maximizes the instantaneous signal-to-noise ratio (14.7) is specified by

$$\mathbf{h}^{\mathrm{o}} = R_{\eta\eta}^{-1}\mathbf{s} \tag{14.18}$$

provided the $R_{\eta\eta}$ is invertible. Furthermore, the associated maximum S/N ratio is given by

$$\rho(\mathbf{h}^{\mathrm{o}}) = \mathbf{s}' R_{\eta\eta}^{-1}\mathbf{s} \tag{14.19}$$

When $R_{\eta\eta}$ is not invertible, it is always possible to select $\mathbf{h}^{\mathrm{o}}$ so that $\rho(\mathbf{h}^{\mathrm{o}}) = \infty$. A selection that achieves this infinite signal-to-noise ratio is characterized by:

$$\text{(a)} \quad \text{if } R_{\eta\eta}\mathbf{s} = \mathbf{0}, \text{ select } \mathbf{h}^{\mathrm{o}} = \mathbf{s}. \tag{14.20}$$

and

$$\text{(b)} \quad \text{if } R_{\eta\eta}\mathbf{s} \neq \mathbf{0}, \text{ select } \mathbf{h}^{\mathrm{o}} \text{ to satisfy the two conditions } R_{\eta\eta}\mathbf{h}^{\mathrm{o}} = \mathbf{0} \text{ and } \mathbf{h}^{\mathrm{o}\prime}\mathbf{s} \neq 0. \tag{14.21}$$

Although the matched filter problem has been directed toward the task of detecting the presence of a known prescribed signal in additive noise, the approach taken here can also be adopted for synthesizing an optimal prescribed signal. In particular, we are often required to generate a prescribed signal of fixed power that in a given noise environment causes the corresponding maximum signal-to-noise ratio to be the largest possible. A straightforward procedure for effecting this signal synthesis is detailed in Problem 14.5.

EXAMPLE 14.2

Design a matched filter of length 2 (i.e., $q = 1$) in which the prescribed signal and additive noise are as specified in Example 14.1. For $q = 1$ the noise correlation matrix is invertible, and by relationship (14.18), the matched filter's weighting coefficient vector is specified by

$$\mathbf{h}^o = \begin{bmatrix} 1 & 0.5 \\ 0.5 & 1 \end{bmatrix}^{-1} \begin{bmatrix} -1 \\ 3 \end{bmatrix} = \begin{bmatrix} -\frac{10}{3} \\ \frac{14}{3} \end{bmatrix}$$

and

$$\rho(\mathbf{h}^o) = \tfrac{52}{3}$$

From these results, a matched filter is therefore implemented by the nonrecursive expression

$$y(h) = -\tfrac{10}{3}\,x(n) + \tfrac{14}{3}\,x(n - 1)$$ ■

14.4 CORRELATOR AND THE FINITE MEMORY-MATCHED FILTER

One of the more frequently occurring matched filtering applications arises when the additive noise is white in nature. For such situations the additive noise has its autocorrelation specified by $r_{\eta\eta}(n) = \sigma^2\delta(n)$, where σ^2 designates the noise power. It then follows directly that the noise correlation matrix is given by

$$R_{\eta\eta} = \sigma^2 I \tag{14.22}$$

in which I denotes the $(q + 1) \times (q + 1)$ identity matrix. In accordance with relationship (14.18), the associated matched filter has its weighting coefficient vector specified by

$$\mathbf{h}^o = R_{\eta\eta}^{-1}\mathbf{s} = \sigma^{-2}\mathbf{s} \tag{14.23a}$$

while the corresponding maximum instantaneous signal-to-noise ratio for this matched filter is

$$\rho(\mathbf{h}^o) = \mathbf{s}'R_{\eta\eta}^{-1}\mathbf{s} = \sigma^{-2}\mathbf{s}'\mathbf{s} \tag{14.23b}$$

As suggested in Section 14.3, any nonzero scalar multiple of $\mathbf{h}^o$ will provide the same signal-to-noise ratio. With this in mind, let us take the optimum weighting coefficient vector to be specified by

$$\mathbf{h}^{o} = \mathbf{s} \tag{14.24}$$

The algorithm that implements this particular choice of weighting coefficients is then characterized by the nonrecursive relationship

$$\begin{aligned} y(n) &= \sum_{k=0}^{q} h^{o}(k)x(n-k) \\ &= \sum_{k=0}^{p} s(p-k)x(n-k) \end{aligned} \tag{14.25}$$

where the prescribed reversed signal vector form (14.11) has been used in arriving at this result. This nonrecursive algorithm is commonly referred to as a *correlator* and is implemented as shown in Figure 14.2. The correlator is seen to consist of a sequence of p delay units (or a shift register), a set of p multipliers, and a summation unit. This network is called a correlator because the constituent multiplying weights correspond to the signal element values being detected. The prescribed signal is deemed to be present whenever the correlator's output signal $\{y(n)\}$ has a magnitude that exceeds some specified threshold level.

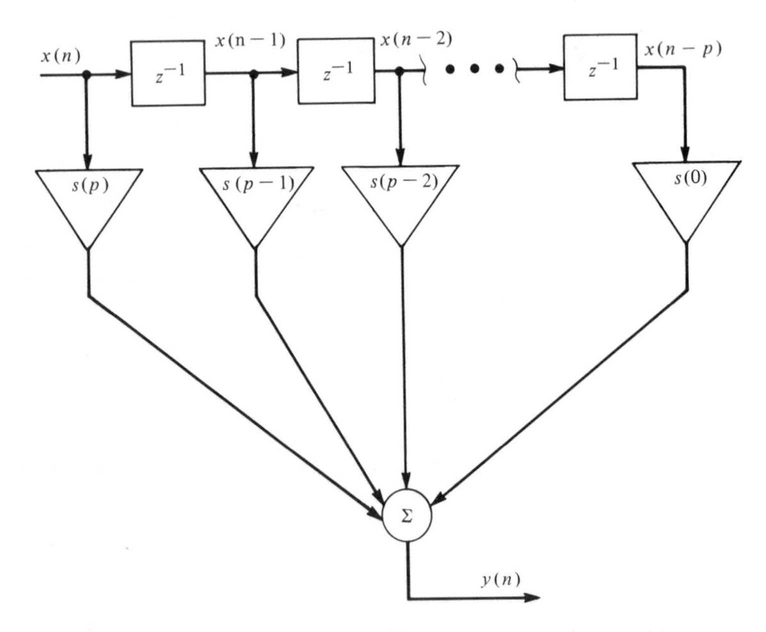

FIGURE 14.2. Correlator interpretation of matched filter.

14.5 FINITE MEMORY-MATCHED FILTER: FINITE-LENGTH RANDOM SIGNAL

A more general form of matched filtering arises when the elements of the prescribed signal are random in nature. In particular, it will be now assumed that each of the prescribed signal components

$$\mathbf{s}(k) \qquad \text{for } 0 \le k \le p \tag{14.26}$$

is a real-valued random variable in which the expected values

$$r_{ss}(i, j) = E\{\mathbf{s}(i)\mathbf{s}(j)\} \qquad 0 \le i, j \le p + 1 \tag{14.27}$$

are assumed to be known. By formulating the prescribed signal in this manner, we are able to treat a more relevant class of problems. It is rare indeed that we know the exact form of the prescribed signal a priori. More typically, the prescribed signal is known to be contained in a class of signals that share common statistical features (e.g., the set of fetal EKG waveforms). We now seek to use these common features to detect the prescribed signal's presence or absence.

Using the approach taken in Sections 14.2 and 14.3, we now seek to determine the $h(k)$ weighting coefficients of the qth-order nonrecursive filter (14.3) so as to maximize the signal-to-noise ratio. Since $\{\mathbf{s}(n)\}$ is random in nature for the problem being considered here, the response term $\mathbf{y}_s(n_0 + p)$ is itself random. It is therefore necessary to reformulate the signal-to-noise ratio as

$$\rho(\mathbf{h}) = \frac{E\{\mathbf{y}_s^2(n_0 + p)\}}{E\{\mathbf{y}_\eta^2(n_0 + p)\}} \tag{14.28}$$

Upon substituting the response components (14.4) and (14.5) into this criterion and using the expected values (14.27) for the now presumed random prescribed signal, we find that

$$\rho(\mathbf{h}) = \frac{\mathbf{h}' R_{ss} \mathbf{h}}{\mathbf{h}' R_{\eta\eta} \mathbf{h}} \tag{14.29}$$

In this expression, $\mathbf{h}$ denotes the nonrecursive filter's $(q + 1) \times 1$ weighting coefficient vector as specified by

$$\mathbf{h} = [h(0), h(1), \ldots, h(q)]' \tag{14.30}$$

while R_{ss} and $R_{\eta\eta}$ designate the $(q + 1) \times (q + 1)$ prescribed signal and noise correlation matrices, respectively, whose elements are given by

$$\begin{aligned} R_{ss}(i, j) &= r_{ss}(i - 1, j - 1) \\ R_{\eta\eta}(i, j) &= r_{\eta\eta}(i - j) \end{aligned} \qquad 1 \le i, j \le q + 1 \tag{14.31}$$

It is important to note that when $p < q$, the elements of the last $q - p$ rows and columns of R_{ss} are all zero.

The matched filter is defined to be one whose weighting coefficient vector maximizes the modified signal-to-noise ratio (14.29). Using the same reasoning adopted in Section 14.3, we obtain the solution by first making the eigenvalue-eigenvector characterization

$$R_{\eta\eta}^{-1} R_{ss} \mathbf{h}_k = \lambda_k \mathbf{h}_k \qquad 1 \le k \le q + 1 \tag{14.32a}$$

where the nonnegative real eigenvalues are ordered in the monotonically increasing fashion

$$0 \le \lambda_1 \le \lambda_2 \le \cdots \le \lambda_{q+1} \tag{14.32b}$$

The following theorem characterizes the matched filter for the random signal problem being considered here.

Theorem 14.2. An optimum matched filter weighting coefficient vector that maximizes the signal-to-noise ratio

$$\rho(\mathbf{h}) = \frac{\mathbf{h}' R_{ss} \mathbf{h}}{\mathbf{h}' R_{\eta\eta} \mathbf{h}}$$

for the case in which $R_{\eta\eta}$ is invertible is specified by

$$\mathbf{h}^{\circ} = \mathbf{h}_{q+1} \qquad \text{and} \qquad \rho(\mathbf{h}^{\circ}) = \lambda_{q+1} \tag{14.33a}$$

where λ_{q+1} is the largest eigenvalue of the matrix $R_{\eta\eta}^{-1} R_{ss}$ and $\mathbf{h}_{q+1}$ is the associated eigenvector. When $R_{\eta\eta}$ is not invertible, two separate cases must be considered.

Case (a): The null space of $R_{\eta\eta}$ is a proper subset of the null space of R_{ss}. In this situation, an optimum, matched filter's weighting coefficient vector corresponds to an eigenvector of the $(q + 1) \times (q + 1)$ matrix

$$\left[\sum_{k=1}^{r} \frac{1}{\sigma_k} \mathbf{v}_k \mathbf{v}_k' \right] R_{ss} \tag{14.33b}$$

associated with its largest eigenvalue. The corresponding maximum signal-to-noise ratio is equal to this largest eigenvalue. In expression (14.33b), σ_k and $\mathbf{v}_k$ are the r ($\le q$) nonzero eigenvalues and associated normalized eigenvectors (i.e., $\mathbf{v}_k' \mathbf{v}_k = 1$) of matrix $R_{\eta\eta}$.

Case (b): In all other situations where $R_{\eta\eta}$ is not invertible, there exists a nonzero vector $\mathbf{h}^o$ such that $R_{\eta\eta}\mathbf{h}^o = \mathbf{0}$ and $R_{ss}\mathbf{h}^o \neq \mathbf{0}$. The maximum signal-to-noise ratio for this selection is $\rho(\mathbf{h}^o) = \infty$.

14.6 INFINITE MEMORY-MATCHED FILTER: KNOWN SIGNAL

For certain types of matched filtering problems, it is feasible to use filters that possess an infinite memory. To determine what performance benefits thereby accrue, let us examine the case in which the real-valued data to be analyzed are of the additive form

$$\mathbf{x}(n) = s(n) + \boldsymbol{\eta}(n) \tag{14.34}$$

where $\{s(n)\}$ is a known deterministic signal which may or may not have finite length, and $\{\boldsymbol{\eta}(n)\}$ is a wide-sense stationary noise process. To detect the presence of the prescribed signal, let us apply these data to the infinite memory filter

$$y(n) = \sum_{k=-\infty}^{\infty} h(k)x(n-k) \tag{14.35}$$

It is to be noted that this filter is not constrained to be causal or to have a finite-length unit-impulse response. The task at hand is to select the $h(k)$ weighting coefficients of this filter so as to enhance the detection of the prescribed signal. As in previous sections, this capability will be measured by the signal-to-noise ratio

$$\rho(\mathbf{h}) = \frac{|y_s(n_0)|^2}{E\{\mathbf{y}_\eta^2(n_0)\}} \tag{14.36a}$$

where $\mathbf{y}_s(n_0)$ and $\mathbf{y}_\eta(n_0)$ denote the signal and noise components of the response at the (arbitrary) time instant n_0. Upon inserting the data form (14.34) into expression (14.35), this signal-to-noise ratio relationship is given by

$$\rho(\mathbf{h}) = \frac{|\sum_{k=-\infty}^{\infty} h(k)s(n_0 - k)|^2}{\sum_{k=-\infty}^{\infty}\sum_{m=-\infty}^{\infty} r_{\eta\eta}(m-k)h(k)h(m)} \tag{14.36b}$$

in which $r_{\eta\eta}(n)$ denotes the noise process's autocorrelation sequence.

To maximize criterion (14.36) by a proper selection of the $h(k)$, we shall utilize the philsophy taken in Section 14.3. This is predicted on first noting that for any weighting coefficient selection, the equality $\rho(\mathbf{h}) = \rho(\alpha\mathbf{h})$ holds for all real $\alpha \neq 0$. Thus the maximization of $\rho(\mathbf{h})$ is equivalent to solving the constrained minimization problem

$$\min_{\sum_{k=-\infty}^{\infty} h(k)s(n_0-k)=1} \sum_{k=-\infty}^{\infty}\sum_{m=-\infty}^{\infty} r_{\eta\eta}(m-k)h(k)h(m) \tag{14.37}$$

This constrained minimization problem is readily solved by using the Lagrange multiplier method, where the auxiliary functional

$$g(\mathbf{h}, \lambda) = \sum_{k=-\infty}^{\infty} \sum_{k=-\infty}^{\infty} r_{\eta\eta}(m - k)h(k)h(m) + \lambda\left[1 - \sum_{k=-\infty}^{\infty} h(k)s(n_0 - k)\right]$$

is introduced. Setting the derivative of this expression with respect to the $h(n)$ coefficients equal to zero, we eventually find that a necessary condition for minimizing criterion (14.37) is specified by

$$\sum_{k=-\infty}^{\infty} r_{\eta\eta}(n - k)h^{o}(k) = \lambda s(n_0 - n) \qquad \text{for all } n \tag{14.38}$$

To obtain the optimum filter coefficients, it is then necessary to solve these infinite sets of linear equations. This solution is obtained straightforwardly by taking the z-transform of expression (14.38) to yield

$$H^{o}(z)R_{\eta\eta}(z) = \lambda z^{-n_0}S(z^{-1})$$

where $S(z) = Z\{s(n)\}$ and $R_{\eta\eta}(z) = Z\{r_{\eta\eta}(n)\}$. Using the argument that if $\{h^{o}(n)\}$ is an optimum selection, then so will be $\{(1/\lambda)h^{o}(n)\}$, we conclude that the optimum transfer function is given by

$$H^{o}(z) = z^{-n_0}\frac{S(z^{-1})}{R_{\eta\eta}(z)} \tag{14.39}$$

In situations where the z-transforms $S(z^{-1})$ and $R_{\eta\eta}(z)$ are each rational, we may conceptually take the inverse z-transform of expression (14.39) to determine $\{h^{o}(n)\}$. The maximum signal-to-noise ratio is then obtained by substituting this $\{h^{o}(n)\}$ into relationship (14.36).

EXAMPLE 14.3

Design an optimum general matched filter associated with the signal $s(n) = 3\delta(n) - \delta(n - 1)$, in which the additive noise has the autocorrelation $r_{\eta\eta}(n) = (0.5)^{|n|}$ as treated in Example 14.2. For this problem, let us select $n_0 = 0$. Using expression (14.39), we obtain the required z-transforms:

$$S(z) = 3 - z^{-1}$$

$$R_{\eta\eta}(z) = \sum_{n=-\infty}^{\infty} (0.5)^{|n|}z^{-n}$$

$$= \frac{0.75}{(1 - 0.5z^{-1})(1 - 0.5z)}$$

Thus the matched filter transfer function (14.39) is given by

$$\begin{aligned} H^o(z) &= \tfrac{4}{3}(3 - z)(1 - 0.5z^{-1})(1 - 0.5z) \\ &= 0.5z^2 - 2.75z + 4.25 - 1.5z^{-1} \end{aligned}$$

The matched filter's optimum weighting coefficients are then specified by $h(-2) = 0.5$, $h(-1) = -2.75$, $h(0) = 4.25$, $h(1) = -1.5$, and are otherwise zero. This matched filter is governed by the following nonrecursive algorithm

$$y(n) = 0.5x(n + 2) - 2.75x(n + 1) + 4.25x(n) - 1.5x(n - 1)$$

Moreover, after using these optimal $h^o(k)$ elements in expression (14.36), the maximum signal-to-noise ratio is found to be

$$\rho(h^o) = \tfrac{62}{3}$$

which is superior to that achieved by the optimum finite memory filter of Example 14.2. In summary, from the class of infinite memory filters, the finite memory filter above is found to provide the maximum signal-to-noise ratio for the problem being considered. It should be noted, however, that the optimum matched filter will generally have an infinite memory. ■

14.7 INFINITE MEMORY-MATCHED FILTER: RANDOM SIGNAL

Let us now examine the case in which the real-valued signal being detected is a member of a wide-sense stationary process with known autocorrelation behavior. To detect the presence or absence of the prescribed signal, we shall use an infinite memory linear filter whose unit-impulse response is denoted by $\{h(n)\}$. When this filter's response to the provided data exceeds a given threshold level, the prescribed signal is deemed to be present. Due to the wide-sense stationarity assumption on the signal, however, it is necessary that this threshold level be exceeded on a continuous-time basis. This is to be contrasted with the matched filtering problems considered previously, where only an instantaneous exceeding of the threshold level was required.

To synthesize a filter that will achieve this detection capability, we select the filter's weighting coefficents to maximize the steady-state signal-to-noise ratio as defined by

$$\rho(\mathbf{h}) = \frac{E\{\mathbf{y}_s^2(n)\}}{E\{\mathbf{y}_\eta^2(n)\}}$$

In this expression $\mathbf{y}_s(n)$ and $\mathbf{y}_\eta(n)$ denote the signal and noise components of the filter response, respectively. It is readily shown that this criterion may be expressed as

$$\rho(\mathbf{h}) = \frac{\sum_{k=-\infty}^{\infty} \sum_{m=-\infty}^{\infty} r_{ss}(m - k)h(k)h(m)}{\sum_{k=-\infty}^{\infty} \sum_{m=-\infty}^{\infty} r_{\eta\eta}(m - k)h(k)h(m)} \tag{14.40}$$

in which $r_{ss}(n)$ and $r_{\eta\eta}(n)$ designate the autocorrelation sequences associated with the desired signal and additive-noise components, respectively.

The task of maximizing criterion (14.40) is very imposing and a closed-form solution for the optimal weighting sequence $\{h^o(n)\}$ is, in general, difficult to obtain. A useful insight into the optimal selection is made possible, however, by employing Parseval's relationship developed in Section 3.8. In particular, the signal-to-noise ratio criterion (14.40) may be equivalently expressed in the frequency domain as

$$\rho(\mathbf{h}) = \frac{\int_{-\pi}^{\pi} |H(e^{j\omega})|^2 S_{ss}(\omega)\, d\omega}{\int_{-\pi}^{\pi} |H(e^{j\omega})|^2 S_{\eta\eta}(\omega)\, d\omega} \tag{14.41}$$

where $H(e^{j\omega})$, $S_{ss}(\omega)$, and $S_{\eta\eta}(\omega)$ are the Fourier transforms of the sequences $\{h(n)\}$, $\{r_{ss}(n)\}$, and $\{r_{\eta\eta}(n)\}$, respectively. As we now show, this relationship readily leads to good-quality approximations of the ideal matched filter.

From the frequency-domain representation for $\rho(\mathbf{h})$ as specified by expression (14.41), it is intuitively apparent that a desired filter's frequency response magnitude $|H(e^{j\omega})|$ should be relatively large over those frequencies for which the real nonnegative symmetric *signal-to-noise spectral density ratio* as defined by

$$\gamma(\omega) = \frac{S_{ss}(\omega)}{S_{\eta\eta}(\omega)} \tag{14.42}$$

takes on its largest values. This is indeed the case, as is made evident by the following upper-bounding lemma.

Lemma 14.1. The signal-to-noise ratio (14.41) is bounded above by

$$\rho(\mathbf{h}) \le \gamma_{\max} \tag{14.43}$$

where

$$\gamma_{\max} = \max_{-\pi<\omega\le\pi} \gamma(\omega)$$

when $\gamma(\omega)$ has a maximum on $(\pi, \pi]$. Otherwise, the max operator is replaced by the smallest upper-bound (or "sup") operator.

To prove this lemma, we simply apply basic integration rules to yield the inequality

$$\int_{-\pi}^{\pi} |H(e^{j\omega})|^2 S_{ss}(\omega)\, d\omega = \int_{-\pi}^{\pi} |H(e^{j\omega})|^2 \gamma(\omega) S_{\eta\eta}(\omega)\, d\omega a$$

$$\leq \gamma_{\max} \int_{-\pi}^{\pi} |H(e^{j\omega})|^2 S_{\eta\eta}(\omega)\, d\omega$$

from which the upper bound (14.43) is established. This inequality arose due to the replacement of $\gamma(\omega)$ by $\gamma_{\max}$ in the right integral in the first line and then noting that $0 \leq \gamma(\omega) \leq \gamma_{\max}$ for all $-\pi < \omega \leq \pi$.

Although Lemma 14.1 does provide an upper bounding on the signal-to-noise ratio, the synthesis of a filter that achieves this performance is not feasible in most applications. To understand why this is so, let us consider a typical situation in which $\gamma(\omega)$ is a continuous function whose maximum occurs at the two frequencies $\pm\omega_m$ shown in Figure 14.3. It then follows that the bandpass filter as specified by

$$\tilde{H}(e^{j\omega}) = \begin{cases} 1 & \text{for } |\omega - \omega_m| \leq \Delta \\ 0 & \text{otherwise} \end{cases} \tag{14.44}$$

will have a signal-to-noise ratio which closely approximates the upper-bound $\gamma_{\max}$ provided that the bandwith parameter Δ is selected suitably small. This is

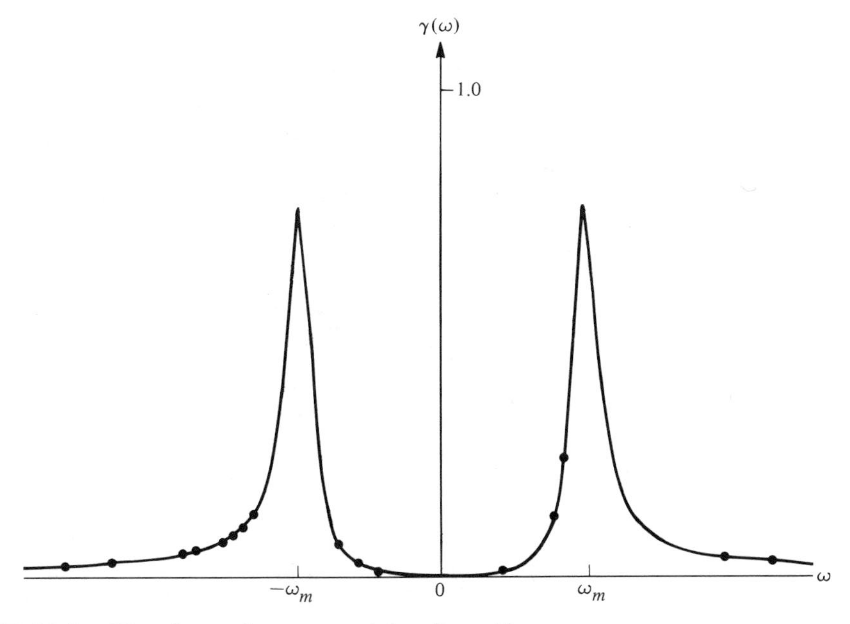

FIGURE 14.3. Signal-to-noise spectral density ratio.

made evident upon substitution of this magnitude behavior into criterion (14.41), which yields

$$\rho(\tilde{\mathbf{h}}) = \frac{2\int_{\omega_m-\Delta}^{\omega_m+\Delta} \gamma(\omega)S_{\eta\eta}(\omega)\,d\omega}{2\int_{\omega_m-\Delta}^{\omega_m+\Delta} S_{\eta\eta}(\omega)\,d\omega} \approx \gamma_{\max}$$

where the symmetry of $\gamma(\omega)$ and $S_{\eta\eta}(\omega)$ has been used in equating the behavior at $-\omega_m$ with that at ω_m. The ratio of integrals above is approximately equal to $\gamma_{\max}$, since for Δ sufficiently small, the function $\gamma(\omega)$ in the numerator integral is itself approximately $\gamma_{\max}$ for $|\omega - \omega_m| \le \Delta$.

Thus a bandpass filter with center frequency ω_m and a sufficiently small bandwith provides a good performance approximation to that of the theoretical optimum matched filter. From a practical viewpoint it is beneficial to select Δ large enough to pass sufficient signal power but small enough so that $\gamma(\omega) \approx \gamma_{\max}$ in the passband of the bandpass filter. This result is intuitively appealing in that it indicates that a pseudo-matched filter is one whose passband corresponds to those frequencies where the prescribed signal's power is largest relative to that of the additive noise.

Although the development above was made for the case in which $\gamma(\omega)$ is continuous and has a unique maximum, its philosophy can be applied to other situations as well. For instance. if $\gamma(\omega)$ should have multiple maximums, a filter with multiple passbands centered at the maximum frequencies would suffice. In another situation, suppose that $\gamma(\omega) = \gamma_{\max}$ over a frequency interval(s) denoted by Ω. It then follows that the corresponding matched filter (not an approximation) has a frequency response equal to 1 for all $\omega \in \Omega$ and is zero otherwise. Finally, if the signal's spectral density function has impulses (e.g., a random sinusoid), a filter with passbands centered at the impulses frequency locations yields a reasonable choice.

EXAMPLE 14.4

Analyze the matched filter associated with the case in which the wide-sense stationary signal and additive noise have autocorrelations given by

$$r_{ss}(n) = \sigma_s^2 (r_s)^{|n|} \cos(\omega_s n)$$

$$r_{\eta\eta}(n) = \sigma_\eta^2 (r_\eta)^{|n|}$$

respectively, where the positive parameters r_s and r_η each have values less than 1. Taking the Fourier transforms of these autocorrelations gives the correspond spectral density functions

$$S_{ss}(\omega) = \frac{2\sigma_s^2(1 - r_s^2)[1 + r_s^2 - 2r_s\cos(\omega - \omega_c) - 2r_s\cos(\omega + \omega_s)]}{[1 + r_s^2 - 2r_s\cos(\omega - \omega_s)][1 + r_s^2 - 2r_s\cos(\omega + \omega_c)]}$$

$$S_{\eta\eta}(\omega) = \frac{\sigma_\eta^2(1 - r_\eta^2)}{1 + r_\eta^2 - 2r_\eta\cos\omega}$$

A plot of the signal-to-noise spectral density ratio $\gamma(\omega) = S_{ss}(\omega)/S_{\eta\eta}(\omega)$ is shown in Figure 14.3 for the case $\sigma_s^2 = 1$, $\sigma_\eta^2 = 10$, $r_s = 0.95$, $r_\eta = 0.9$, and $\omega_s = 0.3\pi$. It is seen to have maximums at $\omega_m = \pm 0.3\pi$, where $\gamma_{max} = 3.01$. Thus a bandpass filter with center frequencies at 0.3π and a narrow bandwidth will yield a good approximation to this upper bound on the signal-to-noise ratio. It is to be noted that the signal-to-noise ratio at the filter's input $\sigma_s^2/\sigma_\eta^2 = 0.1$ has been increased to approximately $\gamma_{max} = 3.01$ at the bandpass filter's output. ■

14.8 SUMMARY

A number of matched filtering problems have been considered in which the common assumption of wide-sense stationary additive noise is made. These different versions involve choices as to the filter memory and the determinancy or randomness of the signal prescribed. The nature of the resultant matched filter was greatly influenced by these choices. For example, in the case of a wide-sense stationary signal, it was found that normally there does not exist a filter that achieves the optimum performance. Fortunately, in such situations, prudent utilization of bandpass filtering operations can often yield nearly optimal performance.

SUGGESTED READINGS

HELSTROM, C. W., *Statistical Theory of Signal Detection*. Elmsford, N.Y.: Pergamon Press, Inc., 1960.

TURIN, G. L., "An Introduction to Matched Filters" (a review), *IRE Trans., Information Theory*, Vol. IT-6, June 1960, pp. 311–329.

TURIN, G. L. "An Introduction to Digital Matched Filters," *Proceedings of the IEEE*, Vol. 64, No. 7, July 1976, pp. 1092–1112.

VANTREES, H. L., *Detection, Estimation, and Modulation Theory*, Part III. New York: John Wiley & Sons, Inc., 1971.

WOODWARD, P. M., *Probability and Information Theory with Applications to Radar*, 2nd ed. Elmsford, N.Y.: Pergamon Press, Inc., 1964.

PROBLEMS

14.1 Prove the validity of the signal-to-noise-ratio expressions (14.6) and (14.7).

14.2 Let the prescribed signal be specified by $s(n) = 3\delta(n) + 2\delta(n-1)$ and the additive noise have autocorrelation $r_{\eta\eta}(n) = (0.5)^{|n|}$. Evaluate the signal-to-noise ratio (14.7) for the following filter weighting coefficient selections.

(a) $h_1(0) = 1$, $h_1(1) = -0.5$

(b) $h_2(0) = 2$, $h_2(1) = 1$
(c) $h_3(0) = 1$, $h_3(1) = 3$
Which of these filters has the largest signal-to-noise ratio?

14.3 Design a finite memory-matched filter of order $q = 2$ for the case in which the prescribed signal is specified by $s(n) = 3\delta(n) - \delta(n-1)$ and the additive noise has autocorrelation $r_{\eta\eta}(n) = (0.5)^{|n|}$. Compute the associated maximum signal-to-noise ratio and show that it is larger than that obtained in Example 14.2. Comment as to why this is so.

14.4 Consider a digital communication system in which the signal (or code) to be detected consists of a known sequence of plus and minus A's of length $q + 1$ as denoted by

$$s(0), s(1), \ldots, s(q)$$

in which $s(k) = \pm A$ for $0 \leq k \leq q$. Furthermore, let the additive noise be a zero-mean white-noise process with variance σ^2. For the case $p = q$, show that a matched filter is specified by

$$h^o(k) = s(q - k) \qquad \text{for } 0 \leq k \leq q$$

and the corresponding maximum signal-to-noise ratio is given by

$$\rho(\mathbf{h}^o) = \frac{(q + 1)A^2}{\sigma^2}$$

Thus by increasing the length of the signal, we can cause $\rho(h^o)$ to be as large as desired.

14.5 Consider the matched filter problem described in Section 14.3, in which $p = q$ and the $(q + 1) \times (q + 1)$ autocorrelation matrix $R_{\eta\eta}$ is known. It is desired to synthesize a prescribed reversed signal vector $\mathbf{s}^o$ of length $q + 1$ for which the associated maximum signal-to-noise ratio (14.7) is the largest possible for all possible choices of $\mathbf{s}$ in which the signal power constraint $\mathbf{s}'\mathbf{s} = 1$ is imposed. Show that this optimal signal corresponds to an eigenvector associated with the smallest eigenvalue of matrix $R_{\eta\eta}$. Moreover, show that corresponding largest maximum signal-to-noise ratio is equal to the inverse of the smallest eigenvalue.

In solving this problem, use is made of the fact (developed in the companion volume) that the noise autocorrelation matrix may be decomposed as

$$R_{\eta\eta} = \sum_{k=1}^{q+1} \lambda_k \mathbf{v}_k \mathbf{v}_k'$$

where the eigenrelationships $R_{\eta\eta}\mathbf{v}_k = \lambda_k \mathbf{v}_k$ for $1 \leq k \leq q + 1$ hold, in which the eigenvalue ordering $0 \leq \lambda_1 \leq \lambda_2 \leq \cdots \leq \lambda_{q+1}$ is adopted and the eigenvecor normalization $\mathbf{v}_k'\mathbf{v}_k = 1$ for $1 \leq k \leq q + 1$ is imposed. Using this notation, show that an optimum prescribed signal, the associate matched filter weighting coefficient vector, and the largest maximum signal-to-noise ratio are specified by, respectively

$$\mathbf{s}^o = \mathbf{v}_1, \qquad \mathbf{h}^o = \frac{\mathbf{v}_1}{\lambda_1}, \qquad \rho(\mathbf{h}^o) = \frac{1}{\lambda_1}$$

14.6 For the case in which the autocorrelation matrix is singular, show that the optimal matched filter relationships (14.20) and (14.21) are valid.

14.7 Show that the signal-to-noise ratio specified in expression (14.28) has the equivalent representation (14.29).

14.8 Consider a random signal of length 2 (i.e., $p = 1$), in which the required expectations are specified by

$$R_{ss}(1, 1) = R_{ss}(2, 2) = 1$$

$$R_{ss}(1, 2) = R_{ss}(2, 1) = 0.9$$

Synthesize a matched filter of length 2 (i.e., $q = 1$) for the case in which the additive noise has autocorrelation
(a) $r_{\eta\eta}(n) = 0.5^{|n|}$
(b) $r_{\eta\eta}(n) = \sigma^2\delta(n)$

and evaluate the corresponding values for $\rho(\mathbf{h}^o)$.

14.9 Prove the noninvertible $R_{\eta\eta}$ cases (a) and (b) of Theorem 14.2. In the proof of case (a), justify and use the fact that all candidate $\mathbf{h}$ vectors must be linear combinations of the nonzero eigenvectors of matrix $R_{\eta\eta}$.

14.10 Synthesize an infinite memory filter for the case in which the prescribed signal is specified by

$$s(n) = \cos(\omega_1 n)u(n)$$

where ω_1 is a known radian frequency. The additive noise has its autocorrelation specified by
(a) $r_{\eta\eta}(n) = \sigma^2\delta(n)$
(b) $r_{\eta\eta}(n) = (0.5)^{|n|}$

Discuss the significance of your results.

Computer Program Problems

1 Write a FORTRAN program that implements the convolution of

$$h(n) = w_8(n) \qquad \text{and} \qquad x(n) = \cos\left(\frac{\pi}{4}\right) n w_{17}(n)$$

using the FFT and IFFT. The rectangular window $w_N(n) = 1$ for $0 \leq n \leq N - 1$ and is zero otherwise. Sketch the output.

2 Using the FFT, find the causal FIR filter coefficients (128 points) that approximate the high-pass filter with linear phase:

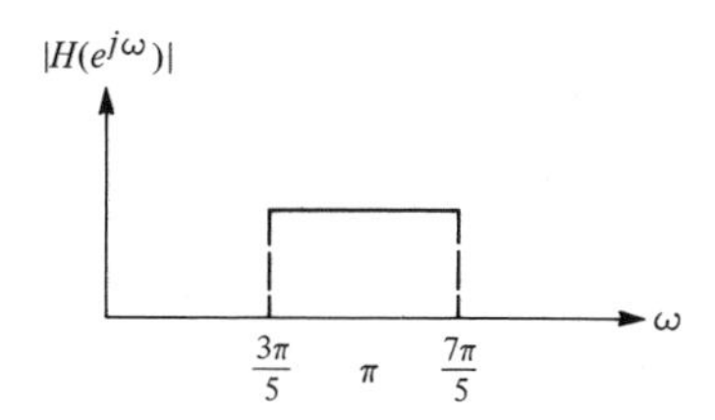

Sketch the impulse response.

3 Design a digital low-pass Butterworth filter with cutoff frequency at 500 Hz (sampling frequency = 200 Hz). The stopband starts at 550 Hz, with at least 20-dB attenuation, while the passband ends at 450 Hz with maximum attenuation of 1 dB. Show that the required filter order is $p = 15$. Write a FORTRAN program to compute the filter poles and sketch the cascade implementation.

4 Compute the unbiased estimate of the autocorrelation function of $y(n)$, given by

$$\mathbf{y}(n) = 0.5\mathbf{y}(n - 1) + \mathbf{w}(n)$$

where $\mathbf{w}(n)$ is white noise with zero mean and unit variance. Use the FFT, find the Fourier transform of $\hat{r}_{xx}(n)$.

5 Consider the following signal:

$$\mathbf{x}(n) = [1 + \mathbf{w}(n)]u(n)$$

where $\mathbf{w}(n)$ is white noise with zero mean and $\sigma_{\mathbf{w}}^2 = 1$. Find the optimal nonrecursive filter estimating the constant $(= 1)$ such that for the output:

$$P\{|\mathbf{y}(n) - 1| < \epsilon\} \geq 0.9 \qquad \epsilon = 0.1$$

Write a program that finds the filter order and computes the mean and variance of $y(n)$. Show correspondence with the theoretical predictions.

6 Carry out the following steps in the described matched filtering problem:

(a) Generate the signal

$$\mathbf{y}(n) = \tfrac{1}{2}\mathbf{y}(n - 1) + \mathbf{w}(n)$$

where $\mathbf{w}(n)$ is white noise with zero mean and variance $\sigma_{\mathbf{w}}^2 = 0.1$.

(b) Consider $\mathbf{y}(n)$ to be the noise in a detection problem, where the signal to be detected is

$$s(n) = u(n) - u(n - 3)$$

(c) Compute the optimal second- and ninth-order matched filters after estimating the noise correlation matrix.

(d) Pass $\mathbf{y}(n) + s(n)$ through the matched filter; plot the result and compare it with the input.

(e) Estimate the output SNR.

7 Carry out the following steps in the described prediction problem:

(a) Generate the signal

$$\mathbf{x}(n) = -\tfrac{1}{2}\mathbf{x}(n - 1) + \mathbf{w}(n)$$

with $\mathbf{w}(n)$ white noise with mean zero and variance $\sigma_{\mathbf{w}}^2 = 1$.

(b) Estimate the autocorrelation sequence using a standard unbiased estimate, and compute the second-order prediction coefficients and the optimal least-squared prediction error. Compare with the theoretical results.

Answers to Selected Problems

Chapter 1

1.1 **(a)** Nondeterministic **(b)** Deterministic **(d)** Nondeterministic

1.2 **(b)** $\tilde{x}(n) = 3 + nT$ for $n \geq 0$, $1 - n^2T^2$ for $n < 0$

1.6 **(a)** $y(n) = \delta(n) - \delta(n-1)$ **(b)** $y(n) = u(n-1) + u(-n-1)$

1.9 **(a)** $\mathcal{M}_x = 3, \mathcal{E}_x = 12, \mathcal{P}_x = 0$ **(e)** $\mathcal{M}_x = 1, \mathcal{E}_x = \infty, \mathcal{P}_x = \frac{1}{2}$

1.12 **(a)** Anticausal **(d)** Mixed; causal component: $-5\delta(n-1)$, anticausal component: $2\delta(n+3)$ **(g)** Mixed; causal component: $-u(n-5)$, anticausal component: $u(-n)$

1.17 **(a)** $y(n) = \delta(n) - 2\delta(n-1) + \delta(n-2)$ **(c)** $y(n) = (\frac{1}{2})^n u(n)$

1.21 **(a)** (1) $y(n) = \delta(n) + \delta(n+1) - 2\delta(n-1)$;
(2) $y(n) = u(n) + u(n+1) - 2u(n-1) = 2\delta(n) + \delta(n+1)$

1.22 **(b)** $h(n) = -(-4)^{-n}u(-n-1)$

1.25 **(a)** $y(n) = 3\delta(n) - 5(\frac{1}{4})^n u(n-1)$ **(b)** $y(n) = 8\delta(n) + 5(4)^{-n}u(-n-1)$

1.26 **(a)** $y(n) = 3\delta(n) + \delta(n-1) - (\frac{1}{3})^{n-2}u(n-2)$
(b) $y(n) = 4\delta(n-1) + 12\delta(n) + 9(3)^{-n}u(-n-1)$

1.28 (a) $h_c(n) = \delta(n) + 4\delta(n-1) + 9(2)^{n-2}u(n-2)$
(b) $h_a(n) = -\frac{1}{2}\delta(n-1) - \frac{5}{4}\delta(n) - \frac{9}{8}(\frac{1}{2})^{-n-1}u(-n-1)$

Chapter 2

2.1 (a) $\dfrac{1}{3z-1}, |z| > \frac{1}{3}$ (d) $\dfrac{az}{(z-a)^2}, |z| < |a|$ (i) $\dfrac{z}{e^{j\omega_0} - z}, |z| < 1$

2.3 (c) $\dfrac{z(13z - 53.2)}{(7-z)(z-0.7)}$, for $0.7 < |z| < 7$

2.9 The poles (p_k) and zeros (z_k) are changed to $-p_k$ and $-z_k$, respectively.

2.13 (a) (1) $-\frac{2}{3}\delta(n) + \frac{6}{7}(\frac{1}{2})^n u(n) - \frac{4}{21}(-3)^n u(n)$
(2) $-\frac{2}{3}\delta(n) - \frac{6}{7}(\frac{1}{2})^n u(-n-1) + \frac{4}{21}(-3)^n u(-n-1)$
(c) (1) $57\delta(n) + 12\delta(n-1) - \frac{297}{4}(\frac{1}{3})^n u(n) + \frac{87}{4}u(n) - \frac{9}{2}(n+1)u(n)$
(2) $57\delta(n) + 12\delta(n-1) + \frac{297}{4}(\frac{1}{3})^n u(n-1) - \frac{87}{4}u(n) + \frac{9}{2}(n+1)u(-n-1)$

2.14 (a) $-\frac{1}{2}[1 + (-1)^n]u(-n-1)$ (c) $\frac{2}{3}u(n) - \frac{1}{3}(4)^n u(-n-1)$

2.16 (a) $H(z) = \dfrac{z^4 - 2z^2 - 1}{z^2 - z + \frac{1}{4}}$, $h(n) = \delta(n+2) + \delta(n+1) - 4\delta(n) + \frac{17}{2}(\frac{1}{2})^n u(n) - \frac{23}{4}(n+1)(\frac{1}{2})^n u(n)$

2.19 (a) $H(z) = (1 - \frac{1}{4}z^{-1})(1 + \frac{1}{2}z^{-1})$

Chapter 3

3.1 (a) $\dfrac{1}{3e^{j\omega} - 1}$ (c) $\dfrac{ae^{j\omega}}{(e^{j\omega} - a)^2}$ (e) $\pi\delta(\omega - \omega_0) + \pi\delta(\omega + \omega_0)$

3.3 (b) $2\pi\delta(\omega - \omega_1)$

3.5 (a) $\dfrac{a\omega e^{-j\omega}}{(1 - ae^{-j\omega})^2}$

3.6 (b) $-3\delta(n+2) + 7\delta(n-4)$ (c) $u(n)$

3.7 $\text{BPF} = \dfrac{1}{\pi n}[\sin(\omega_{c2}n) - \sin(\omega_{c1}n)]$

3.15 (a) $X(\omega_a) = \dfrac{T}{1 - (\frac{1}{2})^T e^{-j\omega_a T}}$

Chapter 4

4.1 (a) $X(k) = 1 - 3e^{-jk\pi}$ (c) $X(k) = -4 - e^{-j3\pi k/2}$

4.3 $X_N(k) = e^{-j\pi km/N}\sin[\pi k(m+1)/N]/\sin(\pi k/N)$

4.6 $X_8(5) = 3 - j,\ X_8(6) = 0,\ X_8(7) = 1 + j2$

4.14 $$h(n) = \frac{\sin [5\pi(n - \frac{15}{2})/16]}{16 \sin [(n - \frac{15}{2})\pi/16]}$$

4.16 $y(0) = 2,\ y(1) = -5,\ y(2) = 1,\ y(3) = 2,\ Y_4(k) = 2 - 5e^{-jk\pi/2} + e^{-jk\pi} + 2e^{-j3k\pi/2}$

Chapter 5

5.1 $$H(z) = \frac{z(z - 0.5)}{(z - 0.9)(z + 0.9)}$$

5.2 $H(z) = 1 - \frac{1}{2}z^{-1} + z^{-2} - \frac{1}{2}z^{-3} = (1 - \frac{1}{2}z^{-1})(1 + z^{-2})$

5.14 $H_2(z) = 4 - 9z^{-1} + 2z^{-2},\ H_3(z) = 2 - 9z^{-1} + 4z^{-2},\ H_4(z) = 1 - \frac{9}{4}z^{-1} + \frac{1}{2}z^{-2}$

5.17 $$H_1(z) = H(z)H_{ap}(z) = \frac{z^3 + 2z + 5}{3(z - 0.5)(z - \frac{1}{3})}$$

5.18 **(a)** $$H(z) = H_1(z)H_{ap}(z) = \frac{(2z - 1)(z - 2)}{4(z - \frac{1}{2})(z - \frac{1}{2})}$$

(b) $$H(z) = H_2(z)H_{ap}(z) = \frac{z}{5(z - \frac{1}{5})(z - \frac{1}{3})}$$

Chapter 6

6.1 $$h_d(n) = \frac{1}{n\pi}[\sin(\omega_1 n) - \sin(\omega_2 n) + \sin(\omega_3 n)]$$

6.2 $$h_d(n) = \frac{\omega_1}{n\pi}\cos(\omega_1 n) - \frac{1}{n^2\pi}\sin(\omega_1 n)$$

6.5 $$q = 80,\ h_\Delta(n) = \frac{\sin(0.3\pi n)}{\pi n}$$

6.10 $p = 6$

6.17 $y(n) - \frac{1}{4}y(n - 1) = x(n) + \frac{1}{2}x(n - 1)$

Chapter 7

7.1 **(a)** $AB^c = \{-2, 1\}$ **(c)** $(A \cup BC)^c = \{-5, -4, -3, -1, 0, 2, 4, 5\}$

7.2 **(a)** $AB^c = \{x\colon -3 \le x \le -2\}$
(c) $(A \cup BC)^c = \{x\colon -20 \le x < -3,\ 3 < x \le 10\}$

7.4 **(a)** $A = \{(-1, 2), (0, 2), (1, 2)\}$

7.6 $A = \{h, th, tth, ttth\}$

7.12 $S = \{123, 132, 213, 231, 312, 321\}$, $P = \frac{2}{3}$

7.13 $P = \frac{1}{15}$

7.14 $P(A) = \dfrac{2}{n}$

7.16 **(a)** 120 **(b)** 625

7.20 **(a)** $\frac{8}{13}$ **(b)** $\frac{5}{13}$

7.21 $P(A) = \frac{1}{6}$, $P(B) = \frac{1}{3}$, $P(C) = \frac{1}{2}$

7.22 **(a)** $P(A/C) = \frac{1}{3}$ **(b)** $P(B/C) = \frac{2}{15}$

7.24 $P(B) = \frac{1}{3}$

7.25 **(a)** $p = 0.3$ **(b)** $p = 0.5$

Chapter 8

8.1 $P[1.5 \le X \le 2.2] = \frac{27}{128}$

8.2 **(a)** $\alpha = \frac{10}{3}$ **(b)** $F_x(x) = \frac{1}{3}x^4(5 - 2x)$, $0 \le x \le 1$, **(c)** 0.96416

8.4 **(a)** $F_{\mathbf{X}}(x) = \dfrac{2x}{1 + x}$, $f_X(x) = \dfrac{2}{(1 + x)^2}$, $0 \le x \le 1$ **(b)** $\frac{2}{3}$

8.6 $f_{\mathbf{X}}(x) = 6x[h_1 + x(h_2 - h_1)/r\}/r^2(h_1 + 2h_2)$, $0 \le x \le r$

8.8 **(a)** $1 - e^{-7x}$, $x \ge 0$ **(b)** $\frac{1}{3}u(x + 4) + \frac{1}{5}u(x + 1) + \frac{1}{3}u(x - 1) + \frac{2}{15}u(x - 7)$

8.11 **(a)** $P[|\mathbf{X} - \mu| \ge k\sigma] = 1 - \dfrac{k}{\sqrt{3}}$

8.12 **(a)** e^{-2a2} **(b)** $f_{\mathbf{Y}}(y) = ae^{-ay}u(y)$

8.13 **(b)** $f_{\mathbf{Z}}(z) = \frac{1}{4}\delta(z + 1) + \frac{3}{4}\delta(z - 1)$

8.16 **(a)** $P[\mathbf{X}_1 > \frac{1}{2}] = \frac{5}{6}$
(d) $f_{\mathbf{X}_1}(x_1) = 2x_1^2 + \frac{2}{3}x_1$, $0 < x_1 < 1$; $f_{X_2}(x_2) = \frac{1}{3} + \frac{1}{6}x_2$, $0 < x_2 < 2$

8.18 **(b)** $f_{\mathbf{X}_1}(x_1) = 2x_1$, $0 < x_1 < 1$; $f_{X_2}(x_2) = 6x_2(1 - x_2)$, $0 < x_2 < 1$

8.19 **(a)** $f_{\mathbf{X}_1}(x_1) = e^{-x_1}$, $0 < x_1 < \infty$; $f_{\mathbf{X}_2}(x_2) = x_2e^{-x_2}$, $0 < x_2 < \infty$

8.21 (a) $f_{\mathbf{x}_2}(x_2) = \frac{1}{24} \ln \dfrac{24}{24 - x_2}$, $0 < x_2 < 24$;

$f_{\mathbf{x}_1\mathbf{x}_2}(x_1, x_2) = \dfrac{1}{24(24 - x_1)}$, $0 < x_1 < x_2 < 24$

Chapter 9

9.1 $\mu_{\mathbf{x}} = \frac{8}{5}$, $\mu_{\mathbf{y}} = -\frac{8}{15}$, $\sigma_{\mathbf{x}}^2 = \frac{8}{75}$, $\sigma_{\mathbf{y}}^2 = \frac{136}{225}$, $r_{\mathbf{xy}} = -\frac{8}{9}$, $c_{\mathbf{xy}} = -\frac{8}{225}$, $\rho_{\mathbf{xy}} = -\dfrac{1}{\sqrt{51}}$

9.3 $\mu_{\mathbf{x}} = \frac{13}{18}$, $\mu_{\mathbf{y}} = \frac{10}{9}$, $\sigma_{\mathbf{x}}^2 = \frac{73}{1620}$, $\sigma_{\mathbf{y}}^2 = \frac{26}{81}$, $r_{\mathbf{xy}} = \frac{43}{54}$, $c_{\mathbf{xy}} = -\frac{1}{162}$, $\rho_{\mathbf{xy}} = -\dfrac{\sqrt{10}}{26}$

9.6 (a) $\hat{\mathbf{Y}} = -3\mathbf{X} + \frac{64}{15}$ (c) $\hat{\mathbf{Y}} = -\frac{10}{73}\mathbf{X} + \frac{265}{219}$

9.11 $\rho_{\mathbf{uv}} = \sigma_{\mathbf{y}}^2 / \sqrt{(\sigma_{\mathbf{x}}^2 + \sigma_{\mathbf{y}}^2)(\sigma_{\mathbf{y}}^2 + \sigma_{\mathbf{z}}^2)}$

9.12 $E\{\mathbf{X}\} = \dfrac{1}{1 - \alpha\gamma}[\alpha\delta + \beta]$, $E\{\mathbf{Y}\} = \dfrac{1}{1 - \alpha\gamma}(\gamma\beta + \delta)$

9.23 $\mu_{\mathbf{Y}} = A\mu_{\mathbf{x}} + b$, $C_{\mathbf{yy}} = AC_{\mathbf{xx}}A'$

Chapter 10

10.1 $\mu_{\mathbf{x}} = 2\mu_{\mathbf{v}} - 3$, $r_{\mathbf{xx}}(n_1, n_2) = 4[\mu_{\mathbf{v}}^2 + \sigma_{\mathbf{v}}^2\delta(n_1 - n_2)] - 12\mu_{\mathbf{v}} + 9$

10.3 $\mu_{\mathbf{x}} = -7n^2\mu_{\mathbf{v}} + 3n - 5$
$r_{\mathbf{xx}}(n_1, n_2) = 49n_1^2n_2^2[\mu_{\mathbf{v}}^2 + \sigma_{\mathbf{v}}^2\delta(n_1 - n_2)] - 7n_1^2(3n_2 - 5)\mu_{\mathbf{v}} - 7n_2^2(3n_1 - 5)\mu_{\mathbf{v}} + (3n_1 - 5)(3n_2 - 5)$

10.5 $E\{\mathbf{z}(n)\} = \alpha E\{\mathbf{x}(n)\} + \beta E\{\mathbf{y}(n)\}$
$r_{\mathbf{zz}}(m, n) = \alpha^2 r_{\mathbf{xx}}(m, n) + \beta^2 r_{\mathbf{yy}}(m, n) + 2\alpha\beta E\{\mathbf{x}(n)\}E\{\mathbf{y}(n)\}$

10.7 (a) $4(\sigma_{\mathbf{v}}^2 + \mu_{\mathbf{v}}^2) - 12\mu_{\mathbf{v}} + 9$

10.12 (a) $3 + 2\pi\left[\delta\left(\omega - \dfrac{\pi}{2}\right) + \delta\left(\omega + \dfrac{\pi}{2}\right)\right]$

Chapter 11

11.3 (a) $\mu_{\mathbf{y}}(n) = 0$ (b) $r_{\mathbf{yy}}(n_1, n_2) = \begin{cases} \frac{4}{3}\sigma_{\mathbf{w}}^2(\frac{1}{2})^{n_2 - n_1}[1 - (\frac{1}{4})^{n_1 + 1}], & n_2 \geq n_1 \\ \frac{4}{3}\sigma_{\mathbf{w}}^2(\frac{1}{2})^{n_1 - n_2}[1 - (\frac{1}{4})^{n_2 + 1}], & n_1 \geq n_2 \end{cases}$

11.5 (a) $\mu_{\mathbf{x}}(n) = (\frac{1}{2})^n u(n) + 2(\frac{1}{2})^{n-1}u(n - 1) = 5(\frac{1}{2})^n u(n) - 4\delta(n)$

11.7 (a) $\frac{52}{3}$ (c) $\frac{128}{21}$

11.9 (a) $\mu_{\mathbf{y}}(n) = 0$

(b) $S_{\mathbf{yy}}(\omega) = \dfrac{1}{(1.16 - 0.8 \cos \omega)(1.81 - 1.8 \cos \omega)}$,

$S_{\mathbf{yx}}(\omega) = \dfrac{1}{(1 - 0.4e^{-j\omega})(1 - 0.9e^{-j\omega})}$

(c) $\rho = 13.315$

11.10 (a) $S_{\mathbf{yy}}(\omega) = \sigma_{\mathbf{w}}^2 \dfrac{9(1 - \cos \omega)}{5 - 3 \cos \omega}$, $S_{\mathbf{yx}}(\omega) = \sigma_{\mathbf{w}}^2 \dfrac{1 - e^{-j\omega}}{1 - \frac{1}{3}e^{-j\omega}}$

11.11 $b_1 = b_2 = b_3 = \frac{1}{3}$

11.13 $a = -\frac{3}{5}$, $b = \frac{2}{5}$

11.18 (a) $S_{\mathbf{yy}}(\omega) = \dfrac{30(1 + \cos \omega)}{5 - 4 \cos \omega} \dfrac{1}{(4 - e^{j\omega})(1 - \frac{1}{4}e^{-j\omega})}$

$S_{\mathbf{yx}}(\omega) = \dfrac{15}{4} \dfrac{1 + e^{-j\omega}}{1 - \frac{1}{2}e^{-j\omega}} \dfrac{1}{(4 - e^{j\omega})(1 - \frac{1}{4}e^{-j\omega})}$

(b) $r_{\mathbf{yy}}(n) = 12.84(\frac{1}{2})^{|n|} - 7.14(\frac{1}{4})^{|n|}$,

$r_{\mathbf{yx}}(n) = 1.43(4)^n u(-n - 1) - 5(\frac{1}{4})^n u(n) + 6.43(\frac{1}{2})^n u(n)$

(c) $\sigma_{\mathbf{y}}^2 = 5.7$

Chapter 12

12.2 $p = 1$, $y(n) = 0.3287x(n) + 0.25893x(n - 1)$, $J(h^o) = 1.31479$

$p = 2$, $y(n) = 0.29015x(n) + 0.20999x(n - 1) + 0.16542x(n - 2)$, $J(h^o) = 1.16061$

12.3 $p = 1$, $y(n) = 0.41958x(n) + 0.14247x(n - 1)$, $J(h^o) = 0.51743$

$p = 2$, $y(n) = 0.4021x(n) + 0.12167x(n - 1) + 0.10222x(n - 2)$, $J(h^o) = 0.4126$

12.4 $p = 2$, $y(n) = \dfrac{8}{15} x(n) + \dfrac{\sqrt{2}}{5} x(n - 1) - \dfrac{2}{15} x(n - 2)$

12.6 (a) $y(n) = 0.26113x(n) + 0.18899x(n - 1) + 0.1488x(n - 2)$, $J(h^o) = 1.5101$

12.12 $\hat{x}(n) = -\frac{1}{4}x(n - 1)$, $A_2^o(z) = 1 - \frac{1}{4}z^{-1}$

12.15 $r_{\mathbf{xx}}(1) = -1$, $r_{\mathbf{xx}}(2) = 1$, $r_{\mathbf{xx}}(3) = -1$

12.16 $A_3^o(z) = 1 + \frac{13}{8} z^{-1} + \frac{7}{8}z^{-2} + \frac{1}{4}z^{-3} = (1 + z^{-1})(1 + \frac{5}{8}z^{-1} + \frac{1}{4}z^{-2})$

Chapter 13

13.1 $h^o(n) = \dfrac{\sqrt{3}}{4} [(2 - \sqrt{3})^n u(n) + (2 + \sqrt{3})^n u(-n - 1)]$

13.3 $h^{o}(n) = -2\delta(n) + \frac{4}{\sqrt{65}}\left[\frac{1}{9+\sqrt{65}}\left(\frac{9+\sqrt{65}}{4}\right)^{n} u(-n-1) + \frac{1}{9-\sqrt{65}}\left(\frac{9-\sqrt{65}}{4}\right)^{n} u(n)\right]$

13.4 $h^{o}(n) = 0.48(0.2)^{n}u(n)$

13.7 $W(z) = \frac{2}{\sqrt{13+\sqrt{153}}} \cdot \frac{1-0.5z^{-1}}{1-\frac{13-\sqrt{153}}{4}z^{-1}}$

13.10 $\hat{x}(n) = \frac{\sqrt{153}-11}{4}x(n-1) + \frac{13-\sqrt{153}}{4}\hat{x}(n-1)$

Chapter 14

14.2 **(a)** $\rho(h) = \frac{1}{3}$ **(b)** $\rho(h) = 7$ **(c)** $\rho(h) = \frac{121}{13}$

14.3 $h^{o} = [-\frac{10}{3} \quad \frac{17}{3} \quad -2]'$, $\rho(h^{o}) = \frac{61}{3}$

14.8 **(a)** $h^{o} = [2.46 \quad 1]'$, $\rho(h^{o}) = 0.95$

INDEX